Agro's Dictionary of MICROBIOLOGY

A. Krishnan

2006

SECOND ENLARGED EDITION

Agrobios (India)

Published by:
Agrobios (India)
Agro House, Behind Nasrani Cinema
Chopasani Road, Jodhpur 342 002
Phone: 91-0291-2642319, Fax: 2643993
E-Mail: agrobios@sify.com
Web Site: www.agrobiosindia.com

AGROBIOS (INDIA)

First Published: 1997
Reprinted: 1999
Second Edition: 2000
Reprinted 2003
Reprinted 2006

9 788177 540185
Rs. 2995.00 / US$ .00
Published by:
AGROBIOS (INDIA)

Published by: Updesh Purohit for Agrobios (India), Jodhpur
Lasertypeset at: Yashee Computers, Jodhpur
Cover Design by: Shyam
Printed at: Bharat Printers, Jodhpur

Preface

Microbes are the part of our lives in more ways than most understand. They have shaped the present environment and their activities will greatly influence our future. Microbiology is one of the most applied of all the biological sciences.

The science of microbiology is subdivided into a number of diverse fields: *Bacteriology* is the study of the bacteria, *Virology* the study of viruses, *Protozoology* the study of protozoans, *Phycology* the study of algae, and *Mycology* is the study of fungi. These fields center around the kinds of organisms under study. Other fields of microbiology, center around where the microbes grow and have the most significant effects like medical microbiology, food microbiology, agricultural microbiology, industrial microbiology, environmental microbiology, waste treatment technology, microbial-biotechnology, etc. The terms related to the fields, which were coined/emerged in recent past years, have been included.

A. Krishnan

A

A. Abbreviation for an Angstrom, 10^{-4} of a micron; there are 10,000,000 Angstroms in a millimeter. After A. J. Angstrom, a Swiss physicist; equals 10^{-7} millimeter.

A23187. An Ionophore which transports divalent cations, particularly Ca^{2+}; it can effect the transmembrane exchange of $1Ca^{2+}$ (or $1Mg^{2+}$) for $2H^{+}$ without causing perturbation in the gradients of other monovalent cations.

Abbe condenser. A simple two or three-lens sub stage Condenser which is uncorrected for spherical or chromatic aberrations.

ABC immunoperoxidase method. An immuno-peroxidase method involving the use of a preformed avidin-biotin-peroxidase complex (ABC) which has surplus biotin-binding capacity. Initially, a (`primary') antiserum is raised against the required antigen; if the primary antiserum is derived from e.g. a rat, a `secondary' anti-rat antiserum is prepared, and the anti-rat Ig antibodies are Biotinylated. To locate a specific antigen, the section is treated with primary antiserum, washed, and then treated with secondary antiserum; the subsequent addition of ABC localizes peroxidase at the site of specific antigen (since the ABC adheres non-specifically to biotin). Peroxidase (and hence antigen) is detected by incubating the section with e.g. H_2O_2 and diaminobenzidine (which results in the antigenic site being stained brown) or H_2O_2 and 4-chloro-1-naphthol (resulting in a blue stain).

The ABC method can be used for paraffin embedded sections, frozen sections, and smears. Endogenous (tissue or cell) peroxidase may be quenched e.g. with H_2O_2 in methanol.

Abdomen. (L.: *belly*). In arthropods, the posterior portion of the body; in mammals that part of the body containing the intestines and most other internal organs, posterior to the thorax.

ABE process. An industrial process in which acetone, butanol and ethanol are produced by the fermentation of e.g. molasses by Clostridium acetobutylicum.

Abelson murine leukaemia virus (AbMuLV) A replication-defective, v-onc+ Murine leukaemia virus isolated from a prednisolone-treated BALB/c mouse inoculated with *Maloney murine leukaemia virus* (MoMuLV). Ab-MuLV apparently arose by recombination between Mo-MuLV and mouse c-abl sequences; the v-abl product has tyrosine kinase activity. Ab-MuLV induces B-cell lymphoid leukaemia with a short latent period (3-4 weeks).

Abequose (3,6-dideoxy-D-galactose). A sugar, first isolated from Salmonella abortusequi, which occurs in the O-specific chains of the Lipopolysaccharide in certain *Salmonella* serotypes and which contributes to the specificity of O antigen 4 in group B salmonellae.

Aberation. Disturbance of the rays of light so they are unable to be brought to a sharp focus.

Abhymenial. Of or pertaining to a region opposite or away from the Hymenum.

Abiogenesis (Spontaneous generation). The spontaneous formation of living organisms from non-living material; apart from its application to the evolutionary origin or life, this doctrine has long been abandoned.

Abiotic. Non-living; of non-biological origin.

Abl. An Oncogene originally identified as the transforming determinant of *Abelson Murine Leukaemia Virus* (Ab-MuLV). The v-abl product has tyrosine kinase activity. In humans. c-abl normally occurs on chromosome 9, but is translocated to chromosome 22q-(the Philadelphia chromosome) in cells from patients with chronic myelogenous leukaemia (CML); c-abl may be involved in the pathogenesis of CML.

Ablastin. Antibody, which specifically inhibits reproduction of epimastigote forms of Try-panosoma lewisi in the vertebrate host.

Abomasitis. Inflammation of the abomasum.

Aboral. Away from, or opposite to, the mouth.

Abortifacient. Able to cause abortion.

Abortive infection (virol.). A viral infection of (non-permissive) cells which does not result in the formation of infectious progeny virions, even though some viral genes (e.g. early genes) may be expressed.

Abscess. A localized accumulation of pus in tissues.

Abscisic acid (L. *abscissio*: breaking off). A plant growth substance having growth-inhibiting action.

Absolute containment. Capacity for keeping or retaining completely any specified substance.

Absolute temperature scale. A temperature scale in which the degree is the same size as in the Celsius (centigrade) scale, and zero is the state of no molecular motion. Absolute zero is –273° on the Celsius scale.

Absorption spectrum. A graph of light absorption versus wavelength of light; shows how much light is absorbed at each wavelength.

Absorption. (1) *Of light:* complete retention, without reflection or transmission. (2) *Of liquids:* soaking up (taking in through pores or cracks).

Abyssal zone. That portion of the deep ocean where no light penetrates.

Acanthamoeba. A genus of amoebae (order Amoebida) in which the pseudopodia each have a broad hyaline zone from which arise several to many slender, tapering, flexible, and sometimes forked projections (acanthopodia). Polyhedral or roughly circular cysts with cellulose-containing walls are formed. Species are widespread and common in soil and fresh water, where they prey on e.g. bacteria, yeasts etc. Some strains can be opportunist pathogens, causing e.g. eye infections and Meningoencephalitis.

Acantharea. A class of marine, mostly planktonic protozoa (Super class-Actinopoda) which have elaborate `skeletons' composed of strontium sulphate; typically, the skeleton consists of 10 spines arranged diametrally in the (more or less spherical) cell, or 20 spines which radiate from the cell centre (where they may or may not be joined at their bases, according to species). In many species the cell contains a central capsule (cf. Radiolaria); many species contain zooxanthellae. Five orders are recognized; genera include e.g. *Acanthochiasma, Acanthometra, Astrolophus, Gigartacon.*

Acarospora. A genus of lichens (order Lecanorales). *Thallus*: crustose, areolate, with prominent areolae. Apothecia are embedded in the areolae; ascospores: very small, many per ascus. All species are saxicolous, some are endolithic; *A. smaragdula* occurs on rocks and slag rich in heavy metals.

Acarpomyxea. A class of protozoa (Super class Rhizopoda) with characteristics intermediate between those of the naked amoebae and the plasmodial slime moulds: they form small plasmodia (or large uninucleate plasmodium-like forms) which are usually branched and which sometimes anastomose to form a coarse reticulum. Spores, fruiting bodies and tests are absent; cysts are produced by some species. Orders: Leptomyxida (soil and freshwater organisms, e.g. *Leptomyxa* and *Stereomyxida.*

Accessory cells (Immunol.). Those cells which, together with B lymphocytes and/or T lymphocytes, are involved in the expression of humoral and/or cell-mediated immune responses; they include e.g. macrophages, dendritic cells, and langerhans' cells.

Accessory pigments. In photosynthesis: those pigments contained in light-harvesting complexes.

Acclimation. A change in an organism which improves its ability to tolerate a changed environmental situation.

AcCoA. Acetyl-Conzyme A.

Acellular (non-cellular). (1) Refers to an organism, usually a protozoon, which consists essentially of a single cell but in which occur functionally specialized regions sometimes regarded as analogous to the organs and tissues of a differentiated multicellular organism. (2) Refers to an organism (e.g. a virus) or structure (e.g. the stalk of *Acytostelium*) which is not cellular in any sense. (3) Not divided into cells (as e.g. in a *Plasmodium*).

Acerococcus. A genus of Gram-positive bacteria of the family Streptococcaceae. Cells: non-motile cocci, commonly in pairs and tetrads. Growth occurs optimally under micro aerobic conditions; glucose is fermented homofermentatively to L (+)-lactic acid. Some strains form a pseudocatalase. *A. viridans* (formerly *Gaffkya homari* and *Pediococcus homari*) is the causal agent of Gaffkaemia.

Acervulus. A flat or saucer-shaped fungal STROMA supporting a mass of typically short and densely-packed conidiophores; acervuli commonly develop subcuticularly or subepidermally in a plant host, becoming erumpent at maturity, i.e., rupturing the overlying plant tissue to allow dispersal of the conidia. Some acervuli bear setae.

Acetabularia. A genus of Dasycladalean algae. The vegetative thallus consists of a single cell in which the cell wall contains mannan as a major component and is generally more or less heavily calcified; the cell is differentiated into an erect stalk or axis (up to several centimeters tall) anchored to the substratum by a branching rhizoid. The single nucleus is located in one branch of the rhizoid. As the stalk grows, whorls of sterile `hairs' develop around the tip; these hairs are eventually shed, leaving rings of scars around the stalk. When the thallus is mature, gametangia develop as an apical whorl of elongated sac-like structures, which, depending on species, may, or may not be joined to form a characteristic cap (giving rise to the popular name `mermaid's wine-glass'). Once the gametangial sacs have developed, the primary nucleus in the rhizoid grows to ca. 20 times its original size; it then undergoes meiosis, and numerous small secondary nuclei are formed. These migrate from the rhizoid to the gametangia by cytoplasmic streaming. Within a gametangial sac, each nucleus becomes surrounded by a resistant wall, resulting in the formation of many resistant cysts; the cyst walls contain cellulose rather than mannan, and are often heavily calcified. The cysts are liberated into the sea and then undergo a period of dormancy before liberating numerous biflagellate isogametes; pairs of gametes fuse to form zygotes which then develop into new vegetative thalli.

Acetic acid bacteria. (1) *Acetobacter* spp. (2) Any bacteria capable of acetification, including *Acetobacter* spp and *Gluconobacter* sp.

Aceticlastic. Able to catabolize acetate.

Acetification. The aerobic conversion of ethanol to acetic acid by bacteria (usually *Acetobacter* spp). Ethanol is converted to hydrated acetaldehyde ($CH_3CH(OH)_2$) which is then dehydrogenated to give acetic acid. Acetification is an exothermic process.

Acetivibrio. A genus of bacteria (family Bacteroidaceae) whose natural habitat is unknown. Cells: straight to slightly curved rods, 0.5-0.9 × 1.5-10.0 μm; in motile species the concave side of the cell has either a single flagellum or a number of flagella which arise in a line along the longitudinal axis of the cell wall of the type species resembles those of Gram-positive bacteria. The major products of carbohydrate fermentation typically include acetic acid, ethanol, CO_2 and H_2;

butyric, lactic, propionic and succinic acids are not formed. GC%: ca. 37-40. Type species: *A. cellulolyticus*.

Acetobacter. A genus of Gram type-negative bacteria of the family acetobacteraceae, the organisms occur e.g. on certain fruits and flowers, are responsible for some types of beer spoilage and wine spoilage, and are used e.g. in the manufacture of VINEGAR. Cells: typically ovoid or rod-shaped, 0.6-0.8 x 1.0-4.0 æm, non-motile or with peritrichous or lateral flagella. Most strains are catalase-positive. Typically, ethanol is oxidized to acetic acid, and acetic acid is oxidized ('overoxidation') to CO_2 (cf. gluconobacter). Principal substrates include e.g. ethanol, glycerol and lactate; most strains grow well on glucose-yeast extract-CaCO3 agar (GYC agar), forming round pale colonies. Some strains form cellulose. Sugars appear to be metabolized primarily via the Hexose monophosphate pathway and the TCA cycle; phosphofructokinase seems to be absent. The Entner-doudoroff pathway appears to occur only in cellulose-synthesizing strains. Growth on Hoyer's medium appears to involve enzymes of the glyoxylate shunt. Optimum growth temperature: 25-30°C. GC%: ca. 51-65. Type species:

aceti. Ketogenic with glycerol or sorbitol substrates; 5-ketogluconic acid (but not 2,5-diketogluconic acid) formed from D-glucose. No diffusible brown pigments are formed on GYC agar. Grows on sodium acetate.

hansenii. Ketogenic with glycerol or sorbitol substrates; 5-ketogluconic acid (but not 2,5-diketogluconic acid) is formed by some strains from D-glucose. No growth on sodium acetate. No diffusible brown pigments are formed on GYC agar. (cf. *A. xylinum*).

liquefacens. Brown diffusible pigments are formed on GYC agar. 2,5-Diketogluconic acid is formed from D-glucose. Ketogenic with glycerol as substrate.

pasteurianus. Ketogluconic acids are not formed from D-glucose. No brown diffusible pigments are formed on GYC agar. Some strains (formerly called *A. peroxydans*) are catalase-negative. (cf. *A. xylinum.*)

xylinum. Cellulose-producing strains formery classified as a subspecies of A. aceti, then distributed between the two species *A. hansenii* and *A. pasteurianus*; *A. xylinum* has now been accepted as a revived name for cellulose-forming and cellulose-less, acetate-oxidizing strains.

Acetobacteraceae. A family of aerobic, oxidase-negative, chemoorganotrophic, Gram type-negative bacteria which typically oxidize ethanol to acetic acid. Metabolism: strictly respiratory (oxidative), with O_2 as terminal electron acceptor. Growth occurs optimally at ca. pH 5-6. The organisms occur e.g. in acidic, ethanol-containing habitats. GC%. ca. 51-65. Two genera: *Acetobacter* (type genus), *Gluconobacter*.

Acetobacterium. A genus of Gram-negative, obligately anaerobic bacteria which occur in marine and freshwater sediments. *Cells*: polarly flagellated ovoid rods, ca. 1.0 × 2.0 μm, often in pairs. The type species, *A. woodii*, can carry out a homoacetate fermentation of e.g. fructose, glucose or lactate, or can grow chemolithoautotrophically; it contains group B peptidoglycan. Optimum growth temperature: 30° C. GC%: ca. 39.

Acetogen. (1) Any bacterium - e.g. *Acetobacterium woodii, Clostridium aceticum* - which can produce acetate, as the main product, from CO_2 and H_2 and/or from certain sugars. (2) (hydrogenogen; proton-reducing acetogen) Any bacterium which can use protons as electron acceptors for the oxidation of certain substrates (e.g. ethanol, lactate, fatty acids) to acetate with concomitant formation of H_2. Obligate hydrogenogens include e.g. *Syntrophomonas.* Some sulphate-reducing bacteria appear to be facultative hydrogenogens is thermodynamically favourable only when the partial pressure of hydrogen is very low - e.g. in the presence of a hydrogen-utilizing methanogen.

Acetogenesis. Acetate formation. Acetate can be formed, as a major or minor product, by a variety of microorganisms in fermentations such as e.g. the mixed acid fermentation and the propionic acid fermentation. (cf. Acetification.)

Acetogens. Such as *Acetobacterium woodii, Clostridium aceticum* and *C. thermoaceticum* can form acetate, as the main product, from e.g. certain hexoses. In this process (homoacetate fermentation) a hexose is metabolized to pyruvate (via the Embden-Meyerhof Parnas pathway) and thence to acetate and CO_2. Additional acetate is synthesized as follows. Some of the CO_2 is reduced to formate which is bound to tetrahydrofolate (THF) and further reduced (in an ATP-requiring reaction) to yield 5-methyl-THF; the methyl group is then transferred to coenzyme B_{12}. The remainder of the CO_2 is reduced to carbon monoxide (by CO dehydrogenase) which, with methyl-coenzyme B_{12}. CO dehydrogenase disulphide reductase and coenzyme A, yields acetyl-CoA; acetyl-CoA is converted to acetate and CoASH with concomitant substrate-level phosphorylation to yield ATP.

Some acetogens (e.g. *A. woodii, C. aceticum,* some strains of *C. thermoaceticum*) can form acetate from CO_2 and H_2; this process resembles the latter part of the pathway described above: CO is derived from CO_2, $2H^+$ and $2e^-$, while 5-methyl-THF is generated from THF, CO_2 and hydrogen.

Acetone-butanol fermentation (solvent fermentation). A fermentation carried out by certain saccharolytic species of *Clostridium* (e.g. *C. acetobutylicum*), in which the products include acetone (or isopropanol) and n-butanol (collectively referred to as solvent). Glucose is initially metabolized via the butyric acid fermentation, but subsequently the pH drops to ca. 4.5-5.0 and acetone and n-butanol are formed as major end products. This fermentation is carried out on an industrial scale to a limited extent.

Acetylcholine. A neurotransmitter substance that carries information across vertebrate neuromuscular junctions and some other synapses.

Achlya. A genus of aquatic fungi (order Saprolegniales) in which the thallus is characteristically a branched, coenocytic mycelium; the width of the hyphae varies with species. Although *Achlya* species are typically saprotrophic some have been reported to parasitize rice plants.

Acholeplasma. A genus of facultativey anaerobic, urease-negative bacteria (family Acholeplasmataceae) which are associated with various vertebrates (and possibly with invertebrates and plants), and which also occur e.g. in soil and sewage and as contaminants in tissue culture. Cells: non-motile cocci (minimum diam. ca. 300 nm) or filaments (typically ca. 2-5 μm in length); carotenoid pigments occur in some species. The organisms resemble *Mycoplasma* spp in their general properties, but differ e.g. in that their growth is sterol-independent, and in that NADH oxidase occurs in the cytoplasmic membrane rather than in the cytoplasm. *Acholeplasma* spp are susceptible to various Acholeplasmaviruses. GC%: ca. 26-36. Type species: *A laidlawii*: other species: *A axanthum, A. equifetale, A. granularum, A. hippikon, A. modicum, A. morum, A. oculi*.

Acholeplasmataceae. A family of bacteria of the order Mycoplasmatales; species of the sole genus, *Acholeplasma*, differ from the other members of the order e.g. in that their growth is not sterol-dependent.

Achromabacter. An obsolete bacterial genus. For *A. anitratum*.

Achromat. (achromatic objective) An objective lens in which chromatic aberration has been corrected for two colours (usually red and blue), and spherical aberration has been corrected for one colour (usually yellow-green). (cf. Apochromat.) a flatfield objective lens of this type is called a planachromat.

Achromogenic. Refers to an organism (or e.g. reagent) which does not produce pigment (or colour), used e.g. of non-pigmented strains of normally chromogenic organisms.

Aciculoconidium. A genus of fungi (class Hyphomycetes) which form budding ovoid or ellipsoidal cells (occurring singly or in short chains or clusters) as well as branched septate hyphae. Conidia are formed terminally and are acicular, rounded at one end and pointed at the other. NO_3^- is not assimilated. One species: *A. aculeatum* (formerly *Trichosporon aculeatum*), isolated from *Drosophila* spp.

Acid curd. Coagulation of casein in milk by an acid pH of 4.7 or lower.

Acid. (L. *acidus*: sharp, sourl). A substance that can release a proton. (contrast with base.)

Acidaminococcus. A genus of Gram-negative bacteria (family Veillonellaceae) which occur e.g. in the intestine in humans and pigs. Cells: typically kidney-shaped cocci, 0.6-1.0 °μm diam, occurring in pairs. Amino acids are the main sources of carbon and energy; all strains need e.g. arginine, glutamate, tryptophan and valine, and most need e.g. cysteine and histidine. In general, the organisms metabolize carbohydrates weakly or not at all. Optimum growth temperature: 30-37° C. Optimum pH: 7.0. GC%: ca. 57. Type (only) species: *A. fermentans*.

Acid-fast. Not decolorized with alcohol which contains 3% HCl after staining with carbol fuchsin.

Acidic. Having a pH of less than 7 (a hydrogen ion concentration greater than 10-7 molar).

Acidophile. Acid-loving microorganism.

Acidophilus milk. A sour, medicinal beverage made by fermenting heat-treated, partially skimmed milk with *Lactobacillus acidophilus*. (Viable *L. acidophilus* appears to have a therapeutic effect on some intestinal disorders.) The main fermentation product is lactic acid, which reaches a level of ca. 1.0%. A more palatable preparation, *sweet acidophilus milk*, is made by adding *L. acidophilus* to milk at ca. 5 °C; under these conditions the cells remain viable but lactic acid is not produced.

Acidosis. (1) (lactic acidosis) (vet.) A (sometimes fatal) condition which may occur in ruminants fed excessive amounts of readily fermentable carbohydrates (e.g. starch, sugars - found e.g. in grain and beet, respectively) or when the transfer from a roughage to a `concentrate' diet is made too quickly. Under these conditions the rate of acid production in the RUMEN is very high; the resulting fall in pH in the rumen (due mainly to the accumulation of lactic acid) inhibits cellulolytic bacteria and protozoa, and favours the growth of certain Lactic acid bacteria -so that the pH falls still further. A gradual transition from roughage to concentrate may permit the somewhat more acid-tolerant bacterium *Megasphaera elsdenii* to metabolize the lactic acid and maintain a normal pH in the rumen. (2) (med., vet.) A pathological condition characterized by an abnormally low pH in the blood and tissues.

Acidothermus. A genus of oxidase-negative, strictly aerobic, Gram type-negative bacteria (family Neisseriaceae) which occur e.g. in soil and water, and which may act as opportunist pathogens in man. Cells: rods, 0.9-1.6 × 1.5-2.5 μm (coccoid in stationary phase cultures), which may exhibit twitching motility. Metabolism is respiratory (oxidative), with O_2 as terminal electron acceptor; no growth occurs anaerobically, with or without nitrate. Most strains can grow on a minethanol or lactate as the sole source of carbon and energy, and nitrate as the sole source of nitrogen; some strains can use amino acids (e.g. L-leucine, ornithine) and/or pentoses (e.g. L-arabinose, D-xylose), and some can degrade e.g. benzoate, n-hexadecane, and alicyclic compounds. *Acinetobacters* appear to contain all the enzymes of the TCA cycle and the glyoxylate cycle. Few strains can use glucose as a source of carbon-one that does uses the Entner-Doudoroff pathway. (While unable to use glucose as a growth substrate, some strains-including those formerly referred to as *Achro-mobacter anitratum*, *Bacterium anitratum* and *Herellea vaginicola*- can acidify glucose-containing media by oxidizing glucose to gluconic acid; strains unable to do this include those formerly referred to as *Acinetobacter lwoffi* and *Mina polymorpha*.) Colonies are typically non-pigmented. Optimum growth temperature: typically 33-35° C. GC%: ca. 38-47. Type species: *A. calcoaceticus*.

Acidulent. A substance used to increase acidity.

Acne. A chronic skin disorder characterized by increased sebum production and the formation of comedones (`blackheads' and `whiteheads') which plug the hair follicles. Propionibacterium acnes, present in the pilosebaceous canal, may play a causal role; it produces a lipase that hydrolyses sebum triglycerides to free fatty acids, and these can cause inflammation and comedones. Treatment: e.g. topical Salicylic acid or benzoyl peroxide; the latter has keratinolytic activity and exerts bactericidal action on P. acnes by releasing free-radical oxygen.

Acquired immunity. Resistance resulting from infection with a specific microorganism. It may be active or passive. It is acquired during life, in contrast to natural or innate immunity.

Acrasin. In cellular slime moulds: a generic term for a chemotactic substance which is produced by cells and which serves as a chemoattractant for cell aggregation. Acrasins are a diverse group of substances; they include cAMP in *Dictyostelium discoideum* (q.v.), a pterin in *Dictyostelium lacteum*, and a dipeptide, *glorin*, in *Polysphondylium violaceum* (q.v.).

Acrasiomycetes. (acrasid cellular slime moulds; acrasids) A class of cellular slime moulds (division Myxomycota) in which the vegetative phase consists of amoeboid cells that form lobose pseudopodia; the amoebae aggregate (without streaming) to form a pseudoplasmodium which is not slug-like and does not migrate (cf. Dictyosteliomycetes). The pseudoplasmodium gives rise to multispored fruiting bodies which may have long or short stalks (but no cellulosic stalk tube) bearing e.g. simple globular sori or branched or unbranched chains or spores. Flagellated cells have been observed in only one species (*Pocheina rosea*). Sexual processes are unknown. *Acrasids* occur in various habitats: e.g. dung, tree-bark, dead plant materials, etc. Genera include *Acrasis, Copromyxa, Copromyxella, Fonticula, Guttulinopsis, Pocheina* (formerly *Guttulina*). (Zoological taxonomic equivalents of the Acrasiomycetes include the class Acrasea of the Mycetozoa, and the class Acrasea of the Rhizopoda.)

Acremonium. A genus of fungi of the class Hyphomycetes; teleomorphs occur in e.g. *Emericellopsis* and *Nectria*. The genus includes organisms formerly classified as species of *Cephalosporium*. *Acremonium* spp form septate mycelium; conidia, often in gelatinous masses, are produced from phialides which develop from simple, single branches of the vegetative hyphae. A. kiliense (=*Cephalosporium acremonium*) produces cephalosporin C.

Acridine orange (basic orange; euchrysine; 3,6-bis (dimethylamino)-acridinium chloride). A basic dye and Fluorochrome used e.g. in fluorescence microscopy to distinguish between dsDNA (which fluoresces green) and ss nucleic acids (which fluoresces orange-red). Sublethal concentrations of the dye are used for curing plasmids.

Acridines. Heterocyclic compounds which include acridine and its derivatives. At low concentrations, aminoacridines (e.g. proflavine (3,6-diaminoacridine), quinacrine) appear to bind to dsDNA (or to double-stranded regions of ssDNA) primarily as Intercalating agents. At higher concentrations there is also a weaker, secondary type of binding in which the acridine binds to the outside of dsDNA or to ssDNA or ssRNA; the two types of binding may account for the differential staining of DNA and RNA by Acridine orange. Acridines inhibit DNA and RNA synthesis and cause e.g. Frameshift mutations. They are used e.g. as antimicrobial agents, as mutagens, and as fluorescent stains for nucleic acids; they also have potential antitumour activity. As **antimicrobial agents**, acridines are active against a wide range of bacteria, but they are not sporicidal; some are active against certain parasitic protozoa and inhibit the replication of certain viruses. Activity is not significantly affected by proteinaceous matter. As **mutagens**, acridines may be effective in replicating bacteriophages but are generally not effective in bacteria. However, compounds in which an acridine nucleus is linked to an alkylating side-chain-ICR compounds (ICR = Institute for Cancer Research).

Acriflavine (acriflavin; euflavin) 3,6-Diamino-10-methylacridinium chloride. A mixture of this compound and 3,6-diaminoacridine (proflavine). Acriflavine is soluble in water and in ethanol, and has been used as an antiseptic.

Acrolein (CH_2=CH-CHO). An aldehyde used e.g. for pre-fixation; it penetrates tissues more rapidly than glutaraldehyde.

Acronematic. Refers to a eukaryotic flagellum which is smooth and tapers to a fine point.

Acropetal development. Development from the base, or point of attachment, towards the tip; e.g., in a chain of acropetally developing spores the first-formed spores occupy positions in the chain nearest the base of the spore-bearing structure, while spores formed later occupy positions in the distal parts of the chain. (cf. Basipetal development.)

Acropleurogenous. Located both at the tip and on the sides of an elongated structure.

Acrosiphonia. A genus of branched, filamentous, siphonocladous green algae (division Chlorophyta).

Acrosome. (Gr. *acros*: highest or outermost + *soma*: body). The structure at the forward tip of an animal sperm which is the first to fuse with the egg membrane and enter the egg cell.

ACTH (Adrenocorticotrophic hormone). A pituitary hormone that stimulates the adrenal cortex.

Actin. A protein, found in most types of eukaryotic cell, which can polymerize (reversibly) to form non-contractile filaments (microfilaments) that are involved e.g. in maintaining cell shape and structure and in capping, amoeboid movement, cytoplasmic streaming, phagocytosis, and (in higher animals) muscle contraction.

Actinobacilli. It occurs as commensals in the alimentary, respiratory and/or genital tracts of animals: *A. lignieresii* in cattle and sheep, *A. equuli* in horses, *A. suis* in pigs (?) and horses, *A. capsulatus* in rabbits (?), *A. actinomycetemcomitans* in man. All can be opportunist pathogens. (*A. muris* = *Streptobacillus moniliformis*; *A. mallei* = *Pseudomonas mallei*; *A. ureae*.

Actinobacillosis. Any animal (or human) disease caused by a species of *Actinobacillus*. *A. lignieresii* causes granulomatous lesions in and around the mouth - particularly the tongue (`wooden tongue') - in cattle; in sheep *A. lignieresii* is associated with suppurative lesions in the skin and internal organs. *A. equuli* is pathogenic for horses and pigs; in pigs symptoms may include fever, haemorrhagic or necrotic skin lesions, arthritis and endocarditis. *A. suis* causes septicaemia and localized lesions in pigs.

Actinobacillus. A genus of Gram-negative bacteria of the pasteurellaceae. Cells; mostly rod-shaped (ca. 0.3-0.5 × 0.6-1.4 μm), but a coccal form often occurs at the end of a rod, giving a characteristic `Morse code' form; filaments may occur in media containing glucose or maltose. Extracellular slime is often produced. Cells stain irregularly. Glucose, fructose, xylose, and (most strains) lactose are fermented (no gas). Growth occurs only on complex media; all species (except *A. actino-mycetemcomitans*) can grow on MacConkey's agar. Most species are non-haemolytic, but *A. suis* and some strains of *A. equuli* exhibit clear haemolysis on sheep blood agar; *A. suis* causes partial haemolysis on horse blood agar. GC%: 40-43. Type species: *A. lignieresii.*

Actinobifida. An obsolete genus of actino-mycetes which included species with dichot-omously-branching sporophores; at least some strains were transferred to thermomonospora.

Actinobolina. A genus of carnivorous ciliates (subclass Gymnostomatia). *Cells*: roughly ovoid, with uniform somatic ciliature, an apical cytostome, toxicysts, and retractable tentacles distributed evenly over the body.

Actinomadura. A genus of bacteria (order Actinomycetales, wall type III; group: Maduromycetes) which occur e.g. in soil; some species (*A. madurae, A. pelletieri*) can be pathogenic in man.

Actinomyces. A genus of asporogenous bacteria (order Actinomycetales; wall type varies with species); species occur in warm-blooded animals e.g. as part of the microflora of the mucous membranes (particularly in the mouth) and can act as opportunist pathogens. The organisms occur as rods, branched rods or filaments, or as a rudimentary mycelium. All species can grow anaerobically, or under reduced partial pressure of oxygen.

Actinomycetales. An order of Gram positive, typically aerobic bacteria; species range from those which occur as cocci and/or rods to those which form a well-developed, branching Substrate mycelium and/or Aerial mycelium, and which may form sophisticated structures such as sclerotia, sporangia and synnemata. (cf. Actinomycete). Most members of the order have a GC%> 55, thus distinguishing them from species of the other major sub-branch of Gram-positive bacteria: the Clostridium-Bacillus-Thermoactinomyces line (but cf. *Corynebacterium*, *Renibacterium* and Thermoactinomyces). Phylogenetic relationships between actinomycetes are indicated by 16S rRNA oligonucleotide cataloguing and nucleic acid hybridization; within the order, groups of genera can be distinguished on the basis of e.g. the chemical nature of the cell wall and the lipid profiles of the organisms.

Actinomycetes are widespread in nature, occurring typically in soil, composts and aquatic habitats; most species are free-living and saprotrophic, but some form symbiotic associations and others are pathogenic in man other animals, and plants. The organisms are chemoorganotrophs; collectively they can degrade a wide range of substances which include e.g. agar, cellulose, chitin, keratin, paraffins and rubber, Some species produce important antibiotics.

Ultrastructure and staining. The cell structure is that of a Gram-positive prokaryote; most species give an unequivocally positive reaction in the Gram stain, and some species are acid-fast. Cytoplasmic inclusions observed in at least some species include e.g. granules of poly-á-hydroxybutyrate, polyphosphate, and polysaccharide, and globules of lipid. The cell wall commonly appears to be either uniformly electron-dense or three-layered, the electron-density of the middle layer being somewhat less than that of the layer on either side of it. The wall contains Peptidoglycan and other polymers, e.g. Teichoic acids - although the latter appear not to occur in the Nocardioform Actinomycetes; the cell wall is commonly surrounded by a layer of diffuse or (in sporoactinomycetes) fibrous material. Depending on the presence of certain amino acids in the peptidoglycan, and the identity of the cell wall sugars, eight wall types (chemotypes I-VIII) of actinomycetes can be distinguished.

In most species which form non-fragmenting mycelium (e.g. Streptomyces spp) the vegetative hyphae are largely aseptate, although septa (cross-walls) can be present - particularly in the older parts of the mycelium. The septa in non-fragmenting mycelium have been designated type 1 septa; each septum consists of a single layer which develops centripetally from the cell wall. Such septa may contain microplasmodesmata, each 4-10 nm in diameter.

In fragmenting mycelium each septum consists of two distinct layers, each layer eventually forming a terminal wall of one of the two neighbouring cells; such septa are designated type 2 septa.

Spore formation. Spores are formed by the septation and fragmentation of hyphae, the spore wall being formed, at least in part, from all the wall layers of the sporogenous hypha. Spore-delimiting septa are of various types, and different types may occur even within a given genus; such septa have been designated type I (two layers developing centripetally), type II (two layers which develop centripetally on a single, initially-formed annulus), and type III (a single, thick layer which develops centripetally). Spore chains are reported to develop acropetally (in e.g. Pseudonocardia), basipetally (in e.g. Micropolyspora), randomly (in e.g. Nocardiopsis), or more or less simultaneously (in e.g. Streptomyces).

In some actinomycetes the spores are formed within sporangia: Actinoplanes, Amorphosporangium, Ampullariella, Dac-Tylosporangium, Frankia and Pilimelia.

Genetic aspects. Genetic exchange has been studied in various actinomycetes, particularly Streptomyces spp. Actinomycetes are hosts to a number of Actinophages, and generalized transduction with phage SV1 has been recorded in strains of Streptomyces. Actinomycetes can contain various transmissible or non-transmissible plasmids, some of which are involved in antibiotic production. Genetic analyses have been carried out by methods involving e.g. conjugation and protoplast fusion.

***Genera include*:** *Actinomadura, Actinomyces, Actinoplanes, Actinopolyspora, Actinosynnema, Agromyces, Amorphosporangium, Ampullariella, Arachnia, Arcanobacterium, Arthrobacter, Brevibacterium, Caseobacter, Cellulomonas, Corynebacterium, Curtobacterium, Dactylosporangiu, Dermatophilus, Excellospora, Frankia, Geodermatophilus, Intrasporangium, Kineosporia, Microbacterium, Microbispora, Micromonospora, Micropolyspora, Microtetraspora, Mycobacterium, Nocardia, Nocardioides, Nocardiopsis, Oerskovia, Pilimelia, Planobispora, Planomonospora, Promicromonospora, Pseudonocardia, Renibacterium, Rhodococcus, Rothia, Saccharomonospora, Saccharopolyspora, Spirillospora, Sporichthya, Streptoalloteichus, Streptomyces, Streptosporangium, Streptoverticillium, Thermomonospora.*

Actinomycin D (Actinomycin C1). An Antibiotic from *Streptomyces* sp; it contains a (red) substituted phenoxazone chromophore linked to two identical pentapeptide lactone rings. All cell types are potentially lactone rings. All cell types are potentially susceptible, any resistance being due to low permeability of cells to the drug. Actinomycin D specifically inhibits DNA-directed RNA synthesis. It binds specifically to B-DNA as an Intercalating Agent (ca. 26° C).

Actinomycosis. (1) Any human or animal disease caused by a species of Actinomyces: *A. israelii* in man, *A. bovis* in cattle. Infection is probably endogenous. Dense nodular lesions are formed, mainly around the jaw (`lumpy jaw'), developing into pus-discharging abscesses. Abscesses may also occur in the lungs, brain or intestine. Lab. diagnosis: the pathogen may be isolated from small yellow granules (`sulphur granules') present in the pus. Chemotherapy: e.g. penicillins. (2) Any human or animal disease caused by an Actinomycete: e.g. actinomycosis.

Actinophage. Any Bacteriophage whose host(s) are member(s) of the Actinomycetales. Actinophages, which include both temperate and virulent types, can be isolated from e.g. soils and composts; most have a wide host range, but some (e.g. Bacteriophage ec, Bacteriophage Vp5) can infect only one or a few species.

Actinophryida. An order of protozoa (class Heliozoea) in which the cells have no skeleton and no centroplast (cf. Centrohelida). Some members have flagellated stages. Sexual processes have been observed in some species. Genera include e.g. *Actinophrys, Actinosphaerium, Ciliophrys.*

Actinophrys. A genus of heliozoa (order Actinophryida). *A. sol* is common among vegetation in freshwater ponds and lakes.

Actinoplanes. A genus of aerobic, sporogenous bacteria (order Actinomycetales, wall type II) which occur e.g. in soil, plant litter and aquatic habitats.

Actinopoda. A superclass of protozoa (subphylum Sarcodina) which are typically more or less spherical, typically have axopodia (filopodia in some members), and are usually planktonic. Classes: Acantharea, Heliozoea, Phaeodarea and Polycystinea.

Actinopolyspora. A genus of bacteria (order Actinomycetales, wall type IV); the sole species, *A. halophila*, was isolated from a salt rich bacteriological medium.

Actinorrhiza. A bacterium-plant root association in which nitrogen-fixing root nodules are formed in certain non-leguminous angiosperms infected (through root hairs) by *Frankia* strains; the plants involved are typically woody pioneers of nutrient-poor soils in cold or temperate regions in the northern hemisphere. There are at least two morphological types of actinorrhizal root nodules are coralloid (i.e., thickened and dichotomously branched). In the Myrica type, formed e.g. in species of Myrica, Casuarina and Rubus, the nodule is clothed with upward-growing (negatively geotropic) rootlets which may aid aeration in boggy habitats.

Actinosphaerium. A genus of heliozoa (order Actinopheryida) in which the cells are multinucleate.

Actinosporea. A class of protozoa (phylum Myxozoa) which are parasitic in invertebrates (particularly annelid worms).

Actinosynnema. A genus of bacteria (order Actinomycetales, wall type III) which occur e.g. on vegetable matter in aquatic habitats.

Action potential. An impulse in a neuron, propagated without change in magnitude, taking the form kof a wave of depolarization or reverse polarization imposed on a polarized cell surface.

Action spectrum. A graph of biological activity versus wavelength of light. It compares the effectiveness of light of different wavelengths.

Action. (Gr. *aktis*: a ray). One of the two major proteins of muscle; it makes up the thin filaments. Found in other cell types as well. (Contrast with myosin.)

Activated sludge. Method of sewage treatment where flocs absorb organic material.

Activating enzyme. An enzyme that couples a low-energy compound with ATP to yield a high-energy derivative. As used in this book, an enzyme that condenses ATP with a specific amino acid to give enzyme-bound AMP-amino acid for used in protein synthesis.

Activation (Immunol.) (1) (of lymphocytes) A process which begins with Blast Transformation and continues with proliferation (cell division) and differentiation; some authors use `activation' to refer specifically to stage (s) preceding proliferation. (2) (of complement)

Activation energy. The energy barrier that blocks the tendency for a set of chemical substances to react. A reaction is speeded up if this energy barrier is lowered by providing a different reaction pathway with the aid of a catalyst. Designated by the symbol Ea.

Active immunity. Acquired immunity resulting from the presence of antibody or immune lymphoid cells produced in response to an antigenic stimulus.

Active site. The region on the surface of an enzyme where the substrate binds, and where catalysis occurs.

Active transport. The transport of a substance across a biological membrane against a concentration gradient-that is, from a region of low concentration (of that substance) to a region of high concentration. Active transport requires the expenditure of energy and is a saturable process. (Contrast with facilitated diffusion, free diffusion.)

Acute. Most severe stage of a disease; usually of short duration.

Acyclovir. (ACV; acycloguanosine; Zovirax) An Antiviral Agent, 9-(2-hydroxyethoxymethyl) guanine, which is active against alphaherpesviruses.

Acytostelium. A genus of cellular slime moulds (class Dictyosteliomycetes) in which the sorocarp stalk is acellular, cellulosic, slender, and apparently tubular; no myxamoebae are sacrificed in stalk formation (cf. Dictyostelium).

Adansonian analysis. Numerical type of analysis of the characteristics of an organism; each property of an organism (e.g., motility, sugar fermentations) is enumerated. Strains yielding the same or a similar total number are considered to be taxonomically closely related.

Adansonian taxonomy. A method of biological classification, proposed in the 18th century by Michel Adanson, in which relationships between organisms are defined by the number of characteristics, which the organisms have in common; the same degree of importance (`weighting') is attached to each characteristic. (cf. *Numerical Taxonomy*.)

Adaptation. In evolutionary biology, a particular structure, physiological process, or behaviour that makes an organism more fit to survive and reproduce. Also, the evolutionary process that leads to the formation of such a trait.

Adaptive radiation. The division of a single species into many species specialized to diverse ways of life.

ADCC (antibody-dependent cell-mediated cytotoxicity). A cytotoxic reaction in which Fc receptor-bearing killer cells recognize target cells via specific antibodies.

Adenine. A purine constituent of nucleic acids.

Adenosine triphosphate (ATP). Compound in which energy is stored in high-energy phosphate bonds; its components are the purine, adenine, D-ribose, and three phosphoric acid groups.

Adenoviridae (Adenovirus family). A family of non-enveloped, icosahedral, linear dsDNA-containing viruses which infect mammals (genus *Mastadenovirus*) or birds (genus *Aviadenovirus*).

Adenylate cyclase. An enzyme which catalyses the synthesis of Cyclic Amp (cAMP) from ATP. In general, intracellular levels of cAMP are determined largely by regulation of adenylate cyclase activity and/or synthesis.

Adenylate kinase. An enzyme (EC 2.7. 4.3) which catalyses the reversible conversion of two molecules of adenosine 5'-diphosphate (ADP) to one molecule each of ATP and AMP.

ADEPT (Antibody Directed Enzyme Prodrug Therapy). This is a way to target a drug to a specific tissue. The targeting mechanism and the drug are administered separately. The drug is administered as an inactive prodrug that does not itself have any effect. The prodrug can be converted into an active drug by an enzyme. In ADEPT, the converting enzyme can be, and indeed preferably is, one that does not occur normally in humans. Instead, it is ;administered with a second injection, coupled to an antibody that concentrates it in the target tissue. When the enzyme has arrived at the target tissue, the prodrug is activated there to form the active drug, while elsewhere it remains inactive.

Adhesin. Adherence polypmers found on the surface of microorganisms.

Adiabatic. (Gr. *adiabatos*: impassable (to heat)). Occurring without the addition or subtraction of heat.

Adiaspiromycosis (adiaspirosis; haplomycosis). A non-infectious Mycosis which primarily affects animals, rarely affecting man. It is caused by *Chrysosporium parvum* var. *parvum* (formerly *Emmonsia parva* or *Haplosporangium parvum*) or by *C. parvum* var. *crescens* (formerly *E. crescens*).

Adjuvant (1) (Immunol.) Any substance which, when administered with or before an antigen, heightens and/or affects qualitatively the immune response in terms of antibody formation and/or the cell-mediated response. (2) Any substance which is added to a drug or other chemical (e.g. a disinfectant) to enhance its activity.

Adjuvant 65 is a water-in-oil emulsion Adjuvant made by emulsifying peanut oil with mannide monooleate and stabilizing with aluminium monostearate.

Adjuvants. Insoluble materials which act to keep antigens in tissues for longer periods, and thus cause a longer stimulation of antibody production.

Adk gene (DNA W gene). In *Escherichia coli*: the gene for adenylate kinase.

Adler test A test used for the identification of *Leishmania* spp. The organisms are cultured in an immune serum; in the presence of homologous antibodies, promastigotes develop in clusters or syncytia.

Adnate (1) Of the region of lamella-stipe attachment in an agaric: extending for a length equal to much or most of the depth of the lamella. (2) Of flagellar spines: flattened against the flagellum from which they arise.

Adnexed (Mycol.) Of the region of lamellastipe attachment in an agaric: extending for a length equal to only a small fraction of the depth of the lamella.

ADP-ribosylation. The transfer of an ADP-ribosyl group from NAD^+ to a protein, catalysed by an ADP-ribosyl transferase. In eukaryotic cells, various proteins may be ADP-ribosylated-apparently as a normal regulatory mechanism; certain bacterial toxins act by exerting an ADP-ribosyl transferase function: ADP-ribosylation also occurs e.g. in cells of *Escherichia coli* infected with Bacteriophage T_4, the host RNA polymerase undergoing phage-induced ADP-ribosylation.

ADRY reagents. Certain substituted thiophenes which can e.g. stimulate the photooxidation of cytochrome b_{559} in photosystem II

Adsorption. Binding of a gas or a solute to the surface of a solid.

Adult T-cell leukaemia (ATL; adult T-cell leukaemia/lymphoma, ATLL). An acute or subacute T-cell leukaemia which affects human adults; it is endemic in S.W. Japan and also occurs in the Caribbean and in parts of the USA, S. America and Africa.

Aegyptianella. A genus of Gram-negative bacteria of the family Anaplasmataceae.

Aeration. The process of introducing oxygen so that organic material can be degraded by aerobic bacteria. The procedure is used in sewage treatment.

Aerial Mycelium. In many actinomycetes: mycelium which projects above the level of the medium; no aerial mycelium is formed e.g. by *Arachnia* or *Intrasporangium*, while *Sporichthya* forms only aerial mycelium.

Aerobacter. An obsolete bacterial genus. Most strains formerly regarded as *Aerobacter aerogenes* are referable (e.g. on the basis of motility) to *Klebsiella pneumoniae* or to *Enterobacter aerogenes*.

Aerobe. An organism requiring oxygen for growth.

Aerobic. (Gr. *aer*: air + *bios*: life). In the presence of oxygen, or requiring oxygen.

Aerobiosis (1) The state or condition in which oxygen is present. (2) Life in the presence of air.

Aeromonas. A genus of Gram-negative bacteria of the Vibrionaceae.

Aerosols. Vapors or particles suspended in air; chemical aerosols may be used to kill microorganisms. A colloid of liquid or solid particles suspended in a gas, usually air.

Aerotaxis. A Taxis in which a (motile) cell migrates along an oxygen concentration gradient to a location where the concentration is optimal for that cell. In bacteria, aerotaxis appears to be axhibited by all aerobic and facultatively aerobic motile species; a swimming cell senses a change in concentration by means of oxygen receptors, presumed to be cytochromes, and taxis is believed to be mediated by changes in proton motive force resulting from changes in electron flow in the electron transport chain.

Aerotolerant. Refers to an Anaerobe which, in the presence of air, can either survive (but not grow) or grow at sub-optimal rates.

Aesculin (esculin). The 6-α-D-glucosyl derivative of 6,7-dihydroxycoumarin.

Aetiology (etiology). The study of causation. The aetiological agent of an infectious disease is the organism which initiates/causes the disease.

AFC (antibody-forming cells). Functionally equivalent to plasma cells.

Afferents. (L. *ad*: to + *ferre*: to bear). Neurons that carry impulses to the central nervous system. (Contrast with efferent.)

Affinity chromatography. This is a method of separating molecules by using their ability to bind specifically to other molecules. This is of particular use in the separation of biological molecules because so many biomolecules bind very strongly and specifically to other molecules - their substrates, inhibitors, regulators, etc. Collectively, the molecule that binds to a larger molecule is called a ligand.

Affinity labeling. A technique used to identify the Combining Site of an antibody. In one method, the antibody is allowed to combine with a homologous Hapten (the `affinity label') that carries an azide side-chain; on subsequent irradiation with ultraviolet light the azide forms a highly reactive nitrene group that combines with any of a range of organic groups in the combining site. The nature of the combining site may subsequently be determined e.g. by chemical analysis.

Affinity maturation. The increase in average antibody affinity frequently seen during a secondary immune response.

Affinity tag. This is also sometimes called a purification tag, an affinity tail, or an affinity handle. An affinity tag is a bit of the amino acid sequence of a protein that has been engineered into the protein to make its purification easier. These can work in a number of ways.

Affinity. Measure of the binding strength between an antigenic determinant (epitope) and an antibody combining site (paratope).

Aflatoxicosis. A Mycotoxicosis caused by ingestion of Aflatoxins.

Aflatoxins. A group of Mycotoxins, produced by strains of *Aspergillus flavus* and *A. parasiticus*, which contain a bifuran moiety fused at the 4,5 ring positions to a substituted coumarin. (cf. Sterigmatocystin.) Aflatoxins are soluble in organic solvents, and some exhibit fluorescence on UV irradiation; they are extremely heat-stable. Aflatoxins are toxic to a wide range of eukaryotes; they can retard germination and growth in plants, inhibit the germination of several moulds (e.g. *Mucor, Neurospora, Penicillium*), and are hepatotoxic and hepatocarcinogenic in animals. Severe outbreaks of disease have occurred in domestic animals and poultry fed with aflatoxin-contaminated feedstuffs-particularly groundnut and cereal feeds. Aflatoxins can affect the immune system, reducing resistance to infection and possibly increasing the risk of hepatocellular carcinoma due to Hepatitis B Virus infection. Aflatoxins inhibit mitosis in tissue culture systems and appear to act at many different sites in both nucleus and cytoplasm, inhibiting e.g. DNA, RNA and protein synthesis.

African horse sickness. An infectious Horse Disease caused by an Orbivirus and transmited by insects (e.g. *Culicoides* spp); it occurs in Africa, parts of the Middle East, and the Mediterranean region. The disease may be acute, with an incubation period of ca. 5-7 days, followed by fever, laboured breathing, severe paroxysms of coughing, and a profuse nasal discharge of yellowish, frothy serous fluid; death usually occurs within 4-5 days of onset.

Agammaglobulinemia. Defect in antibody synthesis due to failure to develop peripheral lymphoid tissue.

Agar diffusion test. A Diffusion Test which differs from the Disc Diffusion Test in that, instead of employing antibiotic-impregnated discs, a solution of each antibiotic is allowed to diffuse from a separate `well' cut into the agar.

Agar. (Malay, agar-agar): A gelatinous substance obtained from certain species of red algae, e.g., *Gelidium*. It is a polysaccharide of seaweed; used as a solidifying agent in bacteriological media.

Agaricales. An order of terrestrial (typically humicolous or lignicolous), mainly saprotrophic fungi (subclass Holobasidiomycetidae) most of which form mushroom-shaped, gymnocarpic or semiangiocarpic, fleshy fruiting bodies.

Agaricoid. Refers to the type of fruiting body which is characteristic of fungi of the Agaricales: gymnocarpic, with the hymenium forming a layer on lamellae and giving rise to ballistospores. (cf. Gasteroid.)

Agaricus (formerly Psalliota) A genus of fungi (Agaricales, Agaricaceae), most (not all) species of which are edible.

Agent. Any power, principle, or substance capable of acting upon the organism, whether curative, morbific, or other.

Agglutination. Clumping of cells, as occurs with a particulate antigen and its antibody, agglutinin.

Agglutinin. Antibody that causes agglutination of its particulate antigen, as bacteria or other cells.

Aggressin. Any product or component of a pathogenic microorganism which promotes the invasiveness of that organism Hyaluronate Lyase. (cf. Streptokinase).

Aging cask. A barrel-shaped vessel of staves, headings, and hoops usually of oak, used to store wine for a period during which taste and aroma develop before it is bottled.

Agmenellum. A phycological genus of `blue-green algae' currently included in the *Synechococcus* complex.

Agnoprotein. A 61-amino acid, highly basic polypeptide encoded by a gene (the agnogene) in the late leader region of Simian Virus 40; it appears to play a role in viral assembly.

Agonomycetales (Mycelia Sterilia) An order of fungi (class Hyphomycetes) which include species that form neither conidia nor sexual structures.

Agranulocyte. Any white blood cell which has non-granular cytoplasm, e.g., a Lymphocyte.

Agrobacterium. A genus of Gram-negative bacteria of the Rhizobiaceae. *Agrobacterium* spp occur in soil, mainly in the rhizosphere. All (except *A. radiobacter*) can infect a wide range of dicotyledonous plants (and some gymnosperms) and induce the formation of self-proliferating galls or adventitious roots. The species are defined primarily or solely on the basis of their pathogenic characteristics: *A. radiobacter* is non-pathogenic, *A. rhizogenes* causes Hairy Root, *A. rubi* causes Cane Gall, and *A. tumefaciens* causes Crown Gall. However, pathogenicity depends on the presence of a plasmid(s) and can readily be altered or lost; hence the currently recognized species do not reflect true taxonomic relationships among the agrobacteria.

Agrobacterium tumefaciens. *Agrobacterium tumefaciens* is a bacterium that causes crown gall disease in some plants. The bacterium infects a wound, and injects a short stretch of DNA into some of the cells around the wound. The DNA comes from a large plasmid - the Ti (tumour induction) plasmid. A short region of the Ti plasmid (called T-DNA) is transferred to the plant cell, where it causes the cell to grow into a tumour-like structure. The T-DNA contains genes that, among other things, allow the plants cells to make two unusual compounds, nopaline and octopine, which are characteristic of transformed cells. The transformed cells form the gall, a hard lumpy growth on the plant, which is a benign home for the bacterium.

This DNA transfer mechanism has been harnessed as a way of genetically engineering plants. The Ti plasmid is engineered so that a foreign gene is transferred into the plant cell along with the nopaline synthesis genes. When the bacterium is cultured with isolated plant cells or with wounded plant tissue, the 'new' gene is injected into the cells and ends up integrated into the chromosomes of the plant.

Agrobacterium tumefaciens usually only infects some dicotyledons, because their response to wounding is compatible with *A. tumefaciens* DNA transfer mechanism. When they are wounded, dicotyledons make specific phenolic chemicals that are part of their wound-protection mechanism.

Agrobacterium tumefaciens has been used particularly to get DNA into trees. Trees are difficult plants to breed because of their size and long life cycles, and so genetic engineering techniques offer unusual advantages of speed and the ability to engineer millions of clones. Walnut, poplar, apple, and plum trees have all had DNA transferred to them using *A. tumefaciens.*

Agrocins. Antibiotics which are produced by certain strains of *Agrobacterium* and which are active against other strains of the same genus; being non-protein in structure, agrocins are not strictly Bacteriocins.

Agroinfection. A method for introducing into a plant infectious DNA (or a cDNA copy of RNA) from a plant virus: the DNA (more than one genome's length) is incorporated into the T-DNA portion of a Ti plasmid, the plasmid is introduced into *Agrobacterium tumefaciens*, and the plant is then infected with this organism; the viral DNA is transferred to the plant cells with the T-DNA.

Agromyces. A genus of microaerophilic to anaerobic, catalase-negative, asporogenous bacteria (order Actinomycetales, wall type VII-).

AIDS (acquired immune deficiency syndrome) A transmissible, fatal human disease which affects the immune system; it was first recognized in 1981 and is becoming increasingly prevalent among certain `high-risk' groups of people. AIDS is characterized by a severe defect of Cell-Mediated Immunity: there is a profound lymphocytopenia.

AIDS is transmitted via blood or blood products, sexually (particularly by male homosexual practices), and via the placenta; thus, groups at particular risk of infection are promiscuous homosexual males, intravenous drug abusers, haemophiliacs dependent on blood products, patients receiving transfusions of infected blood, and infants and sexual partners of members of these groups. Initial infection may, in some cases, result in an acute, often mild, transient illness of infection involving lymphadenopathy with malaise, fever, gastrointestinal symptoms, etc (lymphadenopathy syndrome); however, the full AIDS syndrome generally does not develop until months or years after infection (in adults), and only a small proportion of infected individuals will develop AIDS within the first few years.

The causal agent of AIDS is an exogenous retrovirus which has been isolated from blood, saliva and semen of AIDS patients.

Air incinerator. An electric or fuel-fired furnace for the sterilization of microbiologically contaminated air (or other gases).

Air lock. An unventilated space isolated by doors, used to separate areas with different levels of contamination and at different air pressures, which permits passage of personnel and/or equipment without air flow.

Airlift fermenter. A Loop Fermenter in which the circulation of the culture is typically achieved by pumping air in at the bottom of a Draft Tube (or at the base of the annulus) - the air being voided via an opening in the top of the column; the air bubbles lower the hydrostatic pressure of culture in the draft tube so that culture continually flows down the annulus and up into the draft tube.

AIV process. A process for preparing Silage by direct acidification of vegetable matter (to ca. pH 3.5) with a mixture of dilute HCl and H_2SO_4. In the Penthesta process HCl and H_3PO_4 are used.

Ajellomyces. A genus of fungi of the Gymnoascales. *A. dermatitidis = teleomorph* of *Blastomyces dermatitidis* (q.v.); it is heterothallic.

Akabane virus disease. A cattle Disease (which can also affect sheep and goats) caused by the *Akabane* virus. It transmitted by midges (*Culicoides* spp) and mosquitoes; it occurs e.g. in Africa, Australia and Japan.

Akinete. Dormant thick-walled reproductive cell of some cyanobacteria.

AKV. A replication-competent Murine Leukaemia Virus which occurs endogenously in various strains of mice (e.g. AKR, BALB/c).

Alafosfalin (alaphosphin; L-alanyl-L-1-amino-ethylphosphonic acid). A synthetic Antibiotic which is taken up by the LL-dipeptide transport system of a sensitive cell; it is subsequently hydrolysed intracellularly to release the inhibitory component, 1-amino-ethylphosphonic acid (ala-P), which itself cannot cross the cytoplasmic membrane. (cf. Warhead Delivery.).

Alamethicin. A water-soluble peptide antibiotic (MWt ca. 2100) which is produced-often together with a related antibiotic, suzukacillin-by strains of *Trichoderma viride*. It can act as an Ionophore

Alarmone. A low-MWt molecule, synthesis of which serves as a trigger or signal for the redirection of cellular metabolism in response to a particular type of stress; an example is ppGpp in Stringent Control.

Albert's stain. A stain used to demonstrate Metachromatic Granules.

Albicidin. An antibiotic, produced by *Xanthomonas albilineans*, which inhibits DNA synthesis in *Escherichia coli*.

Albugo. A genus of obligately plant-parasitic fungi (order Peronosporales) which are distinguished by their production of basipetally formed chains of zoosporangia. *A. candida* (economically the most important species) is the causal agent of `white rust' of crucifers (=`crucifer white blister' or `white blister disease').

Albumen. The white of an egg. (cf. Albumins.)

Albumins. A class of low-MWt proteins which are soluble in dilute salt solutions and (unlike globulins) readily soluble in water.

Alcaligenes. A genus (incertae sedis) of catalase-positive, oxidase-positive, Gram-negative bacteria which occur e.g. in soil, water, the alimentary tract in vertebrates, and in clinical specimens.

Alcian blue. A basic dye used e.g. for staining glycoproteins and polysaccharides.

Alcohol oxidase. (alcohol:oxygen oxidoreductase; EC 1.1.3.13) An enzyme which oxidizes alcohols, giving the corresponding aldehydes and H_2O_2. It is obtained e.g. from *Pichia pastoris*.

Alcoholic fermentation (ethanol fermentation). A type of fermentation, carried out by various yeasts and other fungi (e.g. species of *Saccharomyces, Pichia, Aspergillus, Fusarium, Mucor*) and by certain bacteria (e.g. *Zymomonas*), in which ethanol is formed by D-glucose (or certain other sugars, depending e.g. on organism).

Ale. Beer rewed with top-fermenting yeast.

Alectoria. A genus of Lichens (order Lecanorales); photobiont.

Aleuria. A genus of fungi (order Pezizales) which form sessile or stipitate, discoid or cup shaped, minute or conspicuous apothecia in which the hymenium may be red, orange (as in *A. aurantia*, the `orange peel fungus'), or yellow.

Aleuriospore (Mycol.) A term which has been used to refer to various types of spore-including e.g. thick-walled and thin-walled, pigmented and non-pigmented blastoconidia-and which has become meaningless owing to indiscriminate use.

Aleutian disease of mink. A progressive disease of mink caused by an autonomous Parvovirus; it is a virus-induced immune complex-mediated disease and is characterized by glomerulonephritis, artertis, plasmacytosis and hypergamma-globulinaemia

Alfalfa mosaic virus (AMV). A multicomponent ssRNA-containing plant virus which has a wide host range and is transmitted via seeds (in some plants), by aphids (non-persistently), and mechanically (under experimental conditions).

Alga. (L.: *seaweed*). A heterogeneous group of unicellular and multicellular eukaryotic photosynthetic organisms. Any one of a wide diversity of plants belonging to the phyla Pyrrophyta, Chrysophyta, Phaeophyta, Rhodophyta, and Chlorophyta (and, formerly, Cyanophyta- bluegreen algae).

Algae. *(Commercial uses).* Algae have been used for centuries is Asia to produce food ingredients and other materials, but their use elsewhere to produce biotechnology products is relatively new. Algae comprise the seaweeds and a host of smaller plants, including single-celled organisms. Among their uses are:

Production of gums. Many of the gums used in food, medicine (e.g. in wound dressings), and for a variety of other things. Such as research reagents and components of printing inks, come from seaweed. Among the more common compounds prepared commercially from seaweed are: alginate (an acetylated polymer of mannuronic acid and glucuronic acid), carageenan, agar, and agarose.

Production of chemicals. Some chemicals are produced commercially from algae, such as the food dyes beta-carotene and astaxanthin: the latter is added to salmon food to give farmed fish pinker flesh (which wild fish get from eating plankton). A variety of human food ingredients are made from cultured algae, such as Omega-3 fatty acids DHA (docosahexaenoic acid) and ARA (arachidonic acid) from dinoflagellates by Merck. Single-celled algae have also been used as food - the algae Chlorella, Scenedesmus (in Japan), *Spirulina* (in Israel, USA, and Mexico) are used to make algal 'biomass' for single-cell protein (SCP).

Algicides. Chemical agents which kill algae. Algicides include e.g. copper sulphate and TBTO. Some herbicides (e.g. Diquat, Paraquat, Terbutryne) are also effective against at least some algae, as are various general disinfectants (e.g. chlorine in Water Supplies).

Alginate. A salt of alginic acid: a linear polymer.

Alimentary toxic aleukia. A severe, usually lethal Mycotoxicosis caused by ingestion of mouldy grain contaminated with certain Trichothecenes-usually T-2 toxin produced by *Fusarium tricinctum.* Symptoms include extreme leucopenia and multiple haemorrhages.

Alkalescens-Dispar group. Non-motile strains of *Escherichia coli* in which glucose is fermented anaerogenically and lactose fermentation is delayed or absent.

Alkalophile. Microorganism having optimum growth at alkaline pH.

Alkylating agents Agents which react with nucleophilic groups (e.g. amino, carboxyl, hydroxyl, phosphate, and/or sulphhydryl groups) in e.g. proteins and nucleic acids, substituting them with alkyl groups. (As commonly used, the term `alkylating agent' is also applied to agents which substitute nucleophilic groups with derivatives of alkyl groups: e.g. hydroxyethyl groups in the case of Ethylene Oxide.) Bifunctional alkylating agents have two reactive groups and can cause cross-linking between nucleophilic groups in proteins and/or nucleic acids.

Allele (allelomorph). Intraspecies variance at a particular gene locus. Any of one or more alternative forms of a given Gene; both (or all) alleles of a given gene are concerned with the same trait or characteristic, but the product or function coded for by a particular allele differs, qualitatively and/or quantitatively, from that coded for by other alleles of that gene.

Allergen A more specific term for the antigens that are involved in allegic or hypersensitivity reactions An antigen (or autocoupling Hapten - e.g., certain drugs) which can initiate a state of Hypersensitivity (commonly Immediate Hypersensitivity) or which can provoke a hypersensitivity reaction in individuals already sensitized with the allergen.

Allergy. Originally defined as altered reactivity on second contact with antigen, now usually refers to a Type 1 hypersensitivity reaction. (atopy)

Allerton disease. An African Cattle Disease which involves a mild febrile condition followed by the appearance of skin nodules; it is caused by the bovine mammillitis virus and is apparently similar or identical to pseudo-Lumpy skin disease.

Alloantigen. Antigen from a genetically different individual of the same species.

Allochromasy. Gradual, spontaneous chemical modification which occurs in the solutions of certain dyes-a single dye becoming a mixture of dyes.

Allochthonous. Referring to organisms not indigenous to a given environment.

Allogeneic Derived from a genetically different individual of the same species. (cf. Syngeneic; Xenogeneic.)

Allograft (homograft). A tissue or organ graft between members of the same species who are not genetically identical.

Allomonas. A genus of bacteria (family Vibrionaceae) which have been isolated from fresh water, sewage and faeces.

Allomyces. A genus of fungi (order Blastocladiales) which occur in moist soils, muds, and water.

Allopatric. (Gr. *allos*: other + *patria*: father-land). Existing in different environments or geographical regions (cf. Sympatric).

Allopolyploid. (Gr. *allos*: other + *poly*: many + *ploos*: fold). A polyploid constructed of sets of chromosomes that originated from two or more species.

Allosteric enzyme. An enzyme whose activity is changed due to reaction with a compound at a nonactive site. Allosteric enzymes are subject to feedback inhibition, in which the end product of a series or reactions will react directly with the enzyme of the first reaction and inactivate it.

Allostery. (Gr. *allos*: other + *stereos*: structure). Regulation of the activity of an enzyme by binding, at a site other than the catalytic active site, of an effector molecule that does not have the same structure as any of the enzyme's substrates.

Allotype. The protein product of an allele which may be detectable as an antigen by another member of the same species.

Allylamines. A group of synthetic antifungal agents.

Alphaherpesvirinae. (herpes simplex virus group) A subfamily of viruses of the Herpes-Viridae. Alphaherpesviruses have a short replication cycle (<24 h).

Alphanochaete. A genus of filamentous, prostrate, branched green algae (division Chlorophyta) which grow as epiphytes on other freshwater algae.

Alpha-richness. The number of species present in a small, local homogeneous area. (Contrast with betarichness, gamma-richness.)

Alphavirus (arbovirus group A) A genus of viruses of the family Togaviridae (q.v. for replication cycle etc); nearly all members are transmitted by mosquitoes, and many can cause disease (commonly encephalitis or fever with rash and arthralgia) in man and/or animals.

Alsever's solution. A solution containing D-glucose (20.5 g), sodium citrate dihydrate (8.0 g) and NaCl (4.2 g) dissolved in distilled water (1.0 litre); the pH is adjusted to 6.1 with citric acid. The solution is sterilized by filtration and used for the preservation of sheep blood; blood is added to Alsever's solution (1:1 by volume) and stored at 4 °C.

Alternaria, A genus of fungi (class Hyphomycetes) which include many plant-pathogenic species- e.g. *A.solani* and *A. radicina* (causal agent of black rot of carrot seedlings and stored carrots).

Alternaria rot. A firm, dark rot produced in various plant hosts by species of Alternaria.

Alternaric acid. A complex compound, produced by *Alternaria solani*, which contains a diketotetrahydropyran group linked to a long-chain fatty acid; it inhibits germination of the spores of certain fungi, and causes wilting and necrosis in the tissues of higher plants.

Alternate host. (of heteroxenous rust fungi) (1) The secondary host (2) Sometimes, loosely: either host of a heteroxenous rust.

Alternation of generations. The succession of haploid and diploid phases in a sexually reproducing organism. In the life cycles of some organisms: the alternating formation of one or more generations of mature haploid individuals and one or more generations of mature diploid individuals; in this context `mature' refers to the ability of the organisms to produce reproductive cells (gametes or spores). An alternation of generations occurs e.g. in some fungi (e.g. *Allomyces* spp), in certain protozoa of the *Foraminiferida*, and in many algae (e.g. *Laminaria*).

Alternative (alternate) complement pathway. Pathway in which the complement cascade is activated at the C_3 level; immune system products are not required for the activation of this pathway.

Alteromonas. A genus (incertae sedis) of aerobic, chemoorganotrophic, Gram-negative bacteria which occur in coastal and marine waters.

Alu sequences. In the human genome: a family of closely related, dispersed sequences, each ca. 300 nt long, many of which contain a common cleavage site for the restriction enzyme AluI; Alu-like sequences occur in the genomes of other mammals and of certain lower eukaryotes.

Alveolus. (L. *alveus*: cavity). A small, bag-like cavity, especially the blind sacs of the lung.

Amanita. A genus of fungi (Agaricales, Amanitaceae) which occur in deciduous and/or coniferous woodlands; some species form mycorrhizal associations.

Amanitin. An amatoxin which, at low concentrations, specifically inhibits eukaryotic RNA Polymerase II; RNA polymerase III is inhibited at high concentrations.

Amantadine. (1-adamantanamine hydrochloride; 1-aminoadamantane hydrochloride) A polycyclic antiviral agent (adamantane = tricyclodecane, $C_{10}H_{16}$) which inhibits the replication of certain viruses in tissue culture.

Amastigote. A form assumed by the cells of many species of the Trypanosomatidae (q.v.) during certain stages of their life cycles.

Ambisense RNA. A viral ssRNA genome or genome segment which is positive-sense with respect to some genes but negative-sense with respect to others. Ambisense genome segments have been found in the Arenaviridae and in the genus Phlebovirus. In these cases the –ve sense sequences are transcribed directly into (viral-complementary) subgenomic mRNA; however, +ve sense sequences are expressed only after genome replication, subgenomic mRNA being transcribed from the RNA strand complementary to the genomic strand.

Ambrosia fungi. Fungi which grow in the tunnels made by wood-boring ambrosia beetles, e.g. *Xyleborus* spp (Scolytidae).

Ambrosiozyma. A genus of fungi (family Saccharomycetaceae) which form budding yeast cells.

Ambruticin. An antifungal antibiotic (a cyclo-propylpolyene-pyran acid) obtained from a strain of *Polyangium cellulosum*.

Ameboid (movement). A type of movement in which the protoplasm of the cell changes shape to produce feet (pseudopodia) that help the cell move.

Amensalism. Interaction in which one animal is harmed and the other is unaffected. (Contrast with commensalism, mutualism.)

American foulbrood. A Bee Disease which affects the larvae of *Apis mellifera* - usually after they have spun their cocoons; the causal agent is Bacillus larvae.

Ames test. A procedure using bacteria to determine the potential carcinogenicity of a substance.

Amikacin. A semi-synthetic derivative of Kanamycin.

Amino acid. An organic compound of the general formula H_2N - CHR - COOH, where R can be one of twenty or more different side groups. An amino acid is to named because it has both a basic amine group-NH_2 and an acidic carboxyl group, -COOH, Proteins are polymers of amino acids.

Aminoadipic acid pathway (AAA pathway) A pathway for lysine biosynthesis which occurs only in certain lower fungi (*Blastocladiales, Chytridiales, Mucorales*), ascomycetes (including e.g. *Saccharomyces*), basidiomycetes, and euglenoid flagellates.

Aminobenzoic acid (PABA; PAB). A component of Folic Acid. PABA is synthesized from chorismate. Certain microorganisms -e.g. *Clostridium* and *Lactobacillus* spp - require PABA as a growth factor.

Aminoglycoside antibiotics. A class of broad-spectrum Antibiotics; a typical aminoglycoside antibiotic contains an aminosugar and either Streptidine or (more usually) 2-Deoxystereptamine. The class is generally taken to include e.g. Amikacin, Apramycin, butirosin, Framycetin, Gentamicin, hygromycin B, Kanamycin, Kasugamycin, Lividomycin, Neamine, Neomycin, netilmicin, Paromomycin, Ribostamycin, Sisomycin, Spectinomycin, Streptomycin and Tobramycin.

Aminoquinolines. A group of antimalarial Agents which includes 4-aminoquinolines (e.g. chloroquine, amodiaquine) and 8-aminoquinolines (e.g. primaquine)-(chemotherapy).

Amitosis. In certain eukaryotes: an atypical form of Mitosis (q.v.) in which the chromosomes do not condense, a spindle is not formed, the nuclear membrane persists throughout division, and the nucleus divides by constriction.

Amixis. In certain haploid organisms: a type of reproduction, considered to be a deviant form of the sexual process, in which karyogamy and meiosis do not occur even though morphologically differentiated structures (e.g. spore-containing asci) are formed; thus, e.g., the eight haploid ascospores formed (per ascus) by *Podospora arizonensis* are said to be derived by mitotic divisions of each nucleus of the dikaryon in the ascus mother cell. (cf. Apomixis.)

Ammonia assimilation. Ammonia, which can be used as the sole source of nitrogen by many microorganisms, may be obtained from an external source or produced intracellularly- e.g. by deamination reactions or by Assimilatory Nitrate Reduction; NH_3 can be taken up readily through the cytoplasmic membrane, but NH_4^+ uptake requires a transport system. Within the cell NH_3 may be assimilated in various ways.

Ammonification. Breakdown of organic nitrogen compounds with the release of ammonia.

Ammonium salt sugars. A range of media used for determining which (if any) of a series of carbohydrates are metabolized by a given organism; these media are used e.g. for those organisms which media are used e.g. for those organisms which form excess alkaline products from peptone media and which, therefore, cannot be tested in Peptone-Water Sugars.

Ammonotelic. (Gr. *telos*: end). Describes an organism in which the final product of breakdown of nitrogen-containing compounds (primarily proteins) is ammonia. (Contrast with ureotelic, uricotelic.)

Amnion. (Gr.: *sac*). A membranous sac containing a watery fluid in which the embryo is suspended (in reptiles, birds, and mammals).

Amoeba. (Gr. *amoibe*: change). Any one of a large number of different kinds of unicellular animals belonging to the phylum Sarcodina characterized among other features by its ability to change shape frequently through the protrusion and retraction of cytoplasmic extensions called pseudopodia.

Amoebiasis. (1) Any disease of man or animals caused by a Protozoon of the Amoebida (b) and Meningoencephalitis. (2) Syn. amoebic Dysentery.

Amoebicide Strictly. Any chemical agent which kills amoebae; in practice, the term is commonly used for any drug which is active against (particularly) *Entamoeba histolytica*, even when the drug may inhibit, rather than kill, the parasite (i.e., it may act as an amoebistat).

Amoebida, An order of naked, typically uninucleate (cf. *Chaos*) amoebae (class Lobosea) in which mitochondria are typically present (cf. Pelobiontida) and flagellate stages do not occur (cf. Schizopyrenida). Most species occur in fresh water and/or moist soil (e.g. *Acanthamoeba, Amoeba, Echinamoeba, Mayorellia, Rosculus, Saccamoeba, Thecamoeba, Trichamoeba, Vannella, Vexillifera*); some are marine (e.g. *Flabellula, Paramoeba*, most *Platyamoeba* spp), and a few are parasitic.

Amoeboflagellate. An essentially amoeboid protozoon which can-at least under certain circumstances-form flagella as well as pseudopodia.

Amoeboid. Like an amoeba; constantly changing shape by the protrusion and retraction of pseudopodia.

Amoebostome. A sucker-like structure present on free-living cells of *Naegleria fowleri*; amoebostomes are apparently used to engulf food particles.

Amorphosporangium. A genus of bacteria (order Actinomycetales, wall type II) which occur e.g. in leaf litter.

Amorphous. Without forms; characteristics of plasmodium stage of slime molds.

Amphi- (Gr. *both*). A prefix used to denote a character or kind of organism that occupies two or more states. For example, amphibian (an animal that lives both on the land and in water).

Amphibiotic. (1) Amphibious. (2) Able to behave either as a commensal or as a parasite in a given host.

Amphibolic pathway. A metabolic pathway which can have both catabolic and anabolic functions: e.g. the TCA Cycle.

Amphiesma. The wall-including the vesicles, thecal plates (if present), etc. of a Dinoflagellate.

Amphigenous. Situated or growing on both sides of, or all over, a given structure.

Amphipathic. Having both hydrophilic and hydrophobic regions.

Amphispore. A type of dark, thick-walled uredospore formed e.g. by some species of Hyalospora and Puccinia.

Amphitrichous. Refers to the presence of a single flagellum (or e.g. fimbria) at each pole of a cell.

Amphitrichous. Polar type of flagellation, with a single at each end of the cell.

Amphitrophic. Refers to an organism which can grow photosynthetically in the light and chemotrophically in the dark.

Ampholyte. An electrolyte whose molecule has both acidic and basic groups; the pH of a solution of a given ampholyte depends on the relative strengths of its acidic and basic groups.

Amphora. A genus of pennate Diatoms in which the cingulum is much wider on one side (the *dorsal* side) than on the other, resulting in the two valves facing in more or less the same direction; the raphes in both valves can thus both make contact with the substratum. Species are aquatic, benthic organisms capable of Gliding Motility.

Amphotericin B. A Mycosamine-containing heptaene Polyene Antibiotic produced by *Streptomyces nodosus*.

Amphotropic. Refers to a retrovirus which can replicate both in its host of origin and in foreign host cells (e.g. a murine retrovirus which can replicate in both murine and non-murine mammalian cells).

Amplification (gene). Cellular mechanism for increasing the dosage of specific genes by making numerous DNA copies of the gene (as in case of genes for producing ribosomal RNA).

Ampulla (1) (Protozool.) In certain ciliates (e.g. some hymenostomes): a channel, or the wider part of a channel, leading into a contractile vacuole. (2) (Protozool.) In certain hypostome ciliates: an organelle which has an adhesive function. (3) (mycol.) The expanded, terminal region of some types of conidiogenous cell. (4) (mycol.) The expanded, terminal region of some types of conidiophore.

Ampullariella. A genus of bacteria (order Actinomycetales, wall type II) which occur e.g. in soil and freshwater habitats.

AMU (atomic mass unit). The basic unit of mass on an atomic scale, defined as 1/12 the mass of a carbon-12 atom. There are 6.023 × 1023 AMU in one gram. This number is known as Avagadro's number.

Amygdalin. A α-glycoside, present e.g. in bitter almonds, consisting of Gentiobiose linked to mandelonitrile. On α-glucosidase action (or acid hydrolysis), amygdalin yields hydrogen cyanide, glucose and benzaldehyde.

Amylase. An enyme that hydrolyses the -1,4 glucosidic bonds of starch, yielding dextrins and maltose.

Amyloid (1) Starch-like; staining blue or blue-black with iodine (e.g. Meltzer's Reagent). (2) (med.) An abnormal protein deposited, intercellularly, in various tissues in certain pathological conditions.

Amylovorin. An extracellular polysaccharide which is produced by strains of *Erwinia amylovora* and which may be involved in the production of wilt symptoms in Fireblight.

Amylum stars. Star-shaped aggregates of starch-filled cells which develop from the lower nodes in Charophytes and function in vegetative propagation.

Amytal. A barbiturate which acts as a mitochondrial Respiratory Inhibitor: its binding site(s) and mechanism of action appear to be similar to those of Rotenone. Many bacterial systems are partially or completely insensitive to amytal.

Anabaena A genus of filamentous Cyanobacteria (section IV) in which the trichomes are made up of spherical, ovoid or cylindrical vegetative cells; heterocysts may be intercalary or terminal.

Anabolism (Gr. *ana*: up, throughout + ballein: to throw). Synthetic reactions of metabolism, in which complex molecules are formed simpler ones. (contrast with catabolism.).

Anacystis. A genus of cyanobacteria; strains designated *A. nidulans* are apparently strains of Synechococcus (restricted sense).

Anaerobe. Organism that lives in the absence of air (Oxygen).

Anaerobic digestion. The anaerobic breakdown of complex animal and/or plant material to simple substances which include a high proportion of gaseous and soluble products; it involves a diverse range of bacteria whose metabolic pathways are closely interrelated, and it occurs in benthic muds, in the Rumen, in certain types of Sewage Treatment plant, etc.

Anaerobic jar. A container used for the incubation of materials (e.g. inoculated media) in the absence of oxygen or, in general, under gaseous conditions other than atmospheric.

Anaerobic respiration. Respiration under anaerobic conditions, the terminal electron acceptor being (according to species and/or environmental conditions): fumarate, nitrate, nitrite, nitrous oxide, sulphur, sulphate etc.

Anaerobic. (Gr. *an*: not + *aer*: air + bios: life). Occurring without the use of molecular oxygen, O_2.

Anaerobiosis. (1) The complete absence, or (loosely) the paucity, of gaseous or dissolved elemental oxygen in a given place or environment. In nature, such conditions occur e.g. in the bottom muds of ponds and rivers and in the intestinal tracts of animals. In aerobic habitats, anaerobic microenvironments may be created by the consumption of locally available oxygen e.g. by facultatively anaerobic organisms. (2) Life in the absence of air.

Anaerobiospirillum. A genus of Gram-negative bacteria (family Bacteroidaceae).

Anaeroplasma. A genus of obligately anaerobic. cell wall-less, sterol-requiring bacteria (class Mollicutes) which occur e.g. in the Rumen.

Anaerovibrio. A genus of Gram-negative bacteria (family Bacteroidaceae) which occur e.g. in the Rumen.

Ana-holomorph (mycol.) A fungus for which no sexual stage or state has been identified.

Analogy. (Gr. *analogia*: resembling). A resemblance in function, and often appearance as well, between two structures which is due to convergence in evolution rather than to common ancestry. (Contrast with homology.)

Anamnestic response. In persons or animals: a heightened response to the second or subsequent administration of a particular antigen given some time after the initial administration; thus, e.g., the concentration of specific antibody increases at a rate much greater than that in response to the first dose of the antigen.

Anamnestic response. Recall phenomenon, or secondary response to an antigen with very rapid production of antibodies. The recall or the *remembering* by the immune system of a prior response to a given antigen.

Anamorph. (mycol.) The asexual (=imperfect) stage or state of a given fungus. (cf. Ana-holomorph; Holomorph; Synanamorph; Teleomorph.)

Anaphylactic shock. A manifestation of immediate hypersensitivity involving a Type I Reaction. In man, the symptoms include bronchospasm (contraction of bronchial smooth muscle) and cyanosis; in severe cases death may occur within minutes or hours.

Anaphylatoxin. Complement peptides (C3a and C5a) which cause mast cell degranulation and smooth muscle contraction. The fragments C3a and C5a formed during Complement Fixation; they induce the release of Histamine from Mast Cells, and C5a is chemotactic for neutrophils and monocytes.

Anaphylaxis. An antigen specific immune reaction mediated primarily by IgE which results in vasodilation and constriction of smooth muscles, including those of the bronchus, and which may result in death of the animal.

Anaplasma. A genus of Gram-negative bacteria (family Anaplasmataceae)-once thought to be protozoa of the Piroplasmea.

Anaplasmataceae. A family of bacteria of the order Ricketisiales. Species occur as parasites in or on erythrocytes, or free in the plasma, in a range of vertebrate hosts-in which they may or may not cause disease; arthropod vectors include ticks, fleas and lice.

Anaplasmosis. Any disease caused by an *Anaplasma* sp; transmission occurs mainly via ticks. In cattle, A. marginale can cause a severe anaemia with debility and wasting; in sheep and goats infection with *A. marginale* or *A. ovis* is usually subclinical.

Anaplerotic sequences. Sequences of reactions which serve to replenish `pools' of intermediates which have been depleted as a consequence of biosynthesis.

Anaplerotic. A reaction or series of reactions that replenishes intermediates used in biosynthetic reactions.

Anaptychia A genus of foliose or fruticose Lichens (order Lecanorales). Photobiont: a green alga.

Anastomosis. The fusion of fungal hyphae, which results in cytoplasmic fusion (plasmogamy). It does not inevitably lead to immediate fusion of nuclei (karyogany).

Ancalomicrobium. A genus of chemoorganotrophic, facultatively anaerobic Prosthecate Bacteria found in aquatic habitats.

Andersen sampler. An instrument used e.g. for sampling the airborne microflora.

Andersen sieve sampler. A sampler consisting of six impaction stages positioned in series, each containing a plate perforated with a pattern of 400 holes. A Petri dish containing agar is placed beneath each stage, but it is smaller in each succeeding stage. Therefore, as air is drawn through the sampler, the jet velocity is uniform for each stage, but increase with each succeeding stage. The result is particles are collected on the agar plates in a districution according to size.

Andrade's indicator. A pH indicator used e.g. in Peptone-Water Sugars. To 100 ml 0.5% aqueous acid Fuchsin is added 15 ml 4% NaOH; the solution is allowed to stand at room temperature for a day, or until its colour changes to a very pale yellow. (A further 1-2 ml NaOH may be added if this has not occurred after 24h.) This solution (before or after autoclaving) is added to the medium (1 ml in 90 ml); the indicator becomes red at ca. pH 5.5.

Androphage (male-specific phage). Any Bacteriophage which infects only host cells that contain a conjugative plasmid. Androphages include members of the genus Inovirus and of the family Leviviridae. (cf. Female-Specific Phage.)

Anemia. A reduction below an established range in the erythrocytes, hemoglobin, or hematocrit.

Anergy. The absence of a hypersensitivity reaction (particularly a Delayed Hypersensitivity reaction) in a Primed subject exposed to the relevant antigen; for example, an anergic response to a tuberculin test is given by some tuberculous individuals-particularly those with extensive pulmonary tuberculosis.

Aneuploid (anorthoploid). Having one or more complete chromosomes in excess of, or less than, the normal haploid, diploid or polyploid number characteristic of the species.

Angiocarpic development. (angiocarpous development) (mycol.) The mode of development of a fruiting body in which the spores reach maturity within a closed chamber or cavity in the fruiting body.

Ang-kak (ang-quac, angkak; red rice). A Chinese product (used mainly as a colouring agent for foods) made by the fermentation of rice by *Monascus purpureus* strains; the fungus produces two pigments: monascorubrin (red) and monascoflavin (yellow).

Animal cage rack. Stack of steel shelves, generally movable, used to hold animal cages; sometimes equipped for UV irradiation and sometimes provided with exhaust manifold to accommodate ventilated cages.

Animal viruses. Viruses which can infect and replicate in the cells of animals-including mammals, birds, poikilothermic vertebrates and/or invertebrates. The host range for a given animal virus varies from very narrow (e.g. in nature, rubella and variola viruses apparently infect only man) to very wide (e.g. rabies virus can infect most mammals, while Arboviruses can replicate in both invertebrates and mammals). A virus may be transmitted from host to host directly (by body contact), via aerosols (Droplet Infection), via food or water via a Vector, etc: some viruses can be transmitted transovarially or can cross the placenta and infect the foetus. Probably, many virus infections in animals are asymptomatic; but some can result in diseases ranging from mild and self-limiting (e.g. most cases of the Common Cold) to severe or fatal (e.g. Aids, Rabies, Yellow Fever). The virus may remain more or less localized at the site of infection (e.g. in some forms of Herpes Simplex, and in Papillomas), or it may spread systemically via the bloosdstream, lymphatic system or nervous system-often replicating preferentially or only in a specific tissue or organ. Disease symptoms may be due e.g. to disturbances in host cell structure and/or metabolism, host cell lysis, immune responses in the host animal, etc. Animal viruses have been implicated in certain cancers and in other diseases of unknown causation-e.g. atherosclerosis, Multiple Sclerosis, and even psychiatric disorders.

Animal. (L. *animus*: breath, soul). A member of the kingdom Animalia. In general, a multicellular eukaryote that obtains its food by ingestion.

Animalcules (Animaliculae) An archaic term for microscopic organisms, particularly protozoa.

Anion. A negatively charged ion.

Anisogamy. The union of gametes which differ in form and/or physiology. When the gametes differ in size, the larger (macrogamete) is regarded as female, the smaller (microgamete) as male.

Anisogamy. (Gr. *aniso*: unequal + *gamos*: marriage). The existence of two dissimilar gametes (egg and sperm).

Anisokinetic sampling. The taking of air samples under conditions in which the velocity and direction of the air ambient air. When sampling velocity is greater than that of the ambient air, smaller particles are selectively collected over larger particle's, and vice versa, resulting in counting errors proportional to the particle size and distribution.

Anisomycin [2-(p-methoxybenzyl)-3,4-pyrrolidinediol 3-acetate] An antibiotic which inhibits protein synthesis in eukaryotes, binding to 80S ribosomes and inhibiting the peptidyl transferase reaction. Many archaebacteria (e.g. species of *Halobacterium, Methanobacterium, Methanospirillium* and *Thermoplasma*) are also sensitive to anisomycin, although *Methanococcus* and *Methanosarcina* spp are relatively resistant.

Anisotropic inhibitor. A type of compound (e.g. monoazide ethidium) which inhibits energy transduction in Oxidative Phosphorylation by binding to certain negatively charged sites on the outer surface of the mitochondrial inner membrane.

Ankistrodesmus. A genus of unicellular, nonmotile green algae (division Chlorophyta); the cells are generally long and thin and may be straight, curved, helical etc, sometimes occurring in bundles.

Annelate ascus. An Ascus whose wall has a small, annular, apical thickening that is often amyloid or light-refractive; at maturity, the ascospores are released via a pore in the centre of the thickening.

Annelid. A member of the phylum Annelida; one of the segmented worms, such as an earthworm or leech.

Annulus (microbiol.). Any of various structures which are ring-shaped (either actually or in cross-section). Examples include e.g. the outermost layer (= sheath) of the contractile stalk in some peritrich ciliates, the periseptal annulus, and the annular remnant of the partial veil.

Anoxygenic. Refers primarily to photosynthetic processes in which oxygen is not present or produced.

Anoxyphotobacteria. A proposed class of anoxygenic photosynthetic prokaryotes (division Gracilicutes).

Ansamycins. A group of structurally related Antibiotics consisting of a macrocyclic ring which comprises an aromatic moiety spanned by an aliphatic bridge.

Antagonism. The inhibition of one entitiy by another of similar kind.

Anterior station (parasitol.) The anterior part of the alimentary canal and/or the salivary glands of an arthropod vector; infective forms of parasites in the anterior station can be transmitted to the vertebrate host by the bite of the vector. (cf. Inoculative Infection.)

Anterior. Toward the front.

Antheridium. (Gr. *antheros*: blooming). The multicellular structure that produces the sperm in bryophytes and ferns.

Anthracnose. Any of various plant diseases-particularly those caused by fungi of the Melanconiales-in which discrete, dark-coloured, necrotic lesions develop on the leaves and/or fruits. For example, anthracnose of beans (*Phaseolus* spp) is caused by *Colletotrichum lindemuthianum* and is characterized by brown or black lesions (sometimes bearing masses of pink conidia) on the leaves, pods and seeds.

Anthracycline antibiotics. Antibiotics which have a tetrahydrotetracenequinone chromophore (containing three flat, coplanar, 6-membered rings). They function as Intercalating Agents, inhibiting DNA and RNA synthesis. Daunomycin (=rubidomycin, daunorubin), produced by Streptomyces sp, is an anticancer agent which binds to dsDNA (ca. 12°) with no apparent base preference; it contains an aminosugar (daunosamine) which projects into the major groove and stabilizes the drug-DNA complex by electrostatic interaction with phosphate in the DNA. Adriamycin is very similar to daunomycin in structure and function. Other anthracyclines: nogalamycin (antitumour agent, ca. 18°), cinerubin, rhodomycin.

Anthramycin. A member of the pyrrolo-(1,4)- benzodiazepine group of antibiotics; other members include sibiromycin and tomaymycin. Anthramycin (produced by *Streptomyces refuineus*) has antibacterial and antitumour activity; it binds covalently to DNA, causing e.g. inhibition of DNA and RNA synthesis. The modes of action of sibiromycin and tomaymycin are apparently similar.

Anthrax. A disease-primarily of animals but also of man - caused by strains of *Bacillus anthracis* which are both capsulated and toxinogenic. In man the disease is usually localized (cutameous anthrax, anthrax boil, malignant pustule), infection occurring via skin lesions; septicaemia may occur in untreated cases.

Anthrax toxin. A Plasmid-encoded Toxin, produced by *Bacillus anthracis*, comprising 3 proteins: oedema factor (EF, factor I), protective antigen (PA, factor II), and lethal factor (LF, factor III); each component alone has no biological activity, but, in the presence of PA, EF causes oedema in guinea pigs, and LF is rapidly lethal on injection into rats.

Anthroponosis. Any infectious disease which can be contracted by man and in which the pathogen is normally maintained in a reservoir consisting of human population(s). (cf. Zoonosis.)

Antibacterial agents. Agents which kill, or inhibit the growth of, bacteria-e.g. certain Antibiotics, Antiseptics and Disinfectants.

Antibiogram. The pattern of sensitivities of a given microorganism towards a range of antibiotics.

Antibiogram. Test for determining antibiotic susceptibility pattern of microorganisms.

Antibiosis. The antagonism of one organism towards another or others. The production by one species of an organic compound that is toxic at low concentrations to another species.

Antibiotic. (1) Originally, any microbial product which, in low concentrations (of the order of μg/ml), can inhibit or kill (susceptible) microorganisms. The term was used by Waksman in the 1940s to distinguish the (then) newly discovered drug penicillin (a product of the fungus *Penicillium* sp) from the synthetic (i.e., man-made) sulphonamides which had been developed in the 1930s. Difficulties in terminology arose when the penicillins and other natural (i.e., microbial) antimicrobial products were modified-or synthesized de novo-in the laboratory. Currently, antibiotic is generally used to refer to natural, semi-synthetic and wholly synthetic antimicrobial compounds which are effective at low concentrations.

Antibiotic-sensitivity tests. Laboratory tests are often carried out to determine the susceptibility of a pathogen to a range of antibiotics; the results of such tests may enable the clinician to select optimally active chemotherapeutic agents, and may be useful in identifying the pathogen.

Antibodies. Antibodies are proteins made by the immune system to combat infection. Each antibody is made to recognize one target antigen molecule. If the antigen is a small molecule, then the antibody will recognize only a part of the antigen, which will be called that antibody's 'epitope'. The binding site of the antibody latches on to this antigen very specifically and very strongly. This latching-on allows the body to recognize the antigen as part of something that should not be there - a virus, a bacterium, a toxin - and so start the process of removing it.

Mammals make antibodies to almost anything that is a 'non-self' molecule, i.e. which is not normally part of the body. Thus, you can get a mammal to make an antibody against almost any molecule by injecting it into the bloodstream. The immune system recognizes it as foreign material and makes a suitable antibody. In fact, it makes a whole range of slightly different antibodies: the blood of most people usually contains a vast host of different antibody molecules targeted at the various disease agents and other foreign molecules that have got into their blood in the past. For this reason the antibodies that are prepared from mammalian blood are called polyclonal antibodies, because they come from a large number of 'clones' (i.e. identical sets) of cells. This contrasts with the synthetic monoclonal antibodies.

Antibodies have been enormously useful to biotechnology because of their ability to latch tightly on to only one antigen, ignoring all others. For example, they would accurately distinguish sucrose from glucose, right-hand from left-hand amino acids, human blood proteins from ape proteins, etc. Thus they are the basis of many processes where great discrimination is needed.

Technically, the antibody proteins are called immunoglobulins. There are four types usually mentioned:

1. **IgM** - the first type made by the body when it encounters a foreign material.
2. **IgG** - the most common type, made after a prolonged encounter (as during a disease).
3. **IgE** - the type responsible for allergic reactions.
4. **IgA** - a rarer type which is present in saliva and some other non-blood fluids. Antibodies are made by lymphocytes - they are actually made by B lymphocytes (B cells) in a process helped by T cells.

Antibody structure. Antibodies have a well-defined structure. Antibodies are made of equal numbers of 'light' chains and 'heavy' chains: IgG antibodies (the sort most commonly used as reagents in biology and biotechnology) contain two of each. The antigen-binding region or binding site ('complementarity determining region') lies at the end of the light and heavy chains; it is therefore formed from both chains. The chains fold up into discrete blobs called domains: a 'single domain antibody' (Dab) is just one domain of an antibody.

The amino-terminal domains of both heavy and light chains are called the variable domains because they vary between antibodies. The other domains are constant domains, i.e. they are the same between antibodies of the same class and subclass.

The antibody can be cut by proteases into several fragments known as Fab, Fab', and Fc (for historical reasons). These also feature in some biotechnological literature.

Anti-idiotype antibodies. Anti-idiotype antibodies are antibodies that recognize the binding sites of other antibodies. Their binding sites are complementary to the binding site of another immunoglobulin. They are important to biotechnology in three ways.

First, they occur in normal blood. When we become immune to something, we not only acquire antibodies against that something, we also acquire antibodies against those antibodies (and antibodies against those antibodies, etc.). This forms a network of antibodies that can all bind to each other to various degrees, a network that helps to regulate the immune response. It is possible that allergic responses are in part a result of the breakdown of this sort of regulation. Thus anti-idiotype antibodies are important to the regulation of the immune system, and understanding how and why they are produced is an important part of understanding how the immune system works.

If an antibody is a 'key' exactly selected to fit the 'lock' of a virus or a bacterium, then an anti-idiotype antibody is a 'lock' exactly selected to fit that 'key" in other words, it must have some similarity to the original antigen, the material that the original antibody reacted to. This means that making anti-idiotype antibodies could be a way of replicating the functional properties of such proteins as hormones or hormone receptor molecules. By raising an antibody against that molecule, and then raising an anti-idiotype antibody against the antibody, you will create an immunoglobulin with some of the functional characteristics of the original hormone or hormone receptor molecule, but which can be produced more easily and is chemically quite distinct.

In practice, antibodies only recognize a small region of the surface of a protein. Thus, an anti-idiotype antibody can only mimic the properties or functions of that region of the protein, and these are likely to be rather limited. Thus, for example, an anti-idiotype antibody that binds to an antibody against insulin (and hence will have a binding site looking like some of the insulin molecule) will sometimes bind to the insulin receptor molecule. However, it does not necessarily make a cell respond in the same way as it would to insulin, because the receptor is not bound in the same way as when the insulin molecule itself binds. These relatively subtle differences have limited the use of anti-idiotype antibodies.

Anti-idiotype antibodies also have potential as vaccines. Here again, they are used to mimic a protein, in this case part of the surface of a virus or a bacterium. However, it is not so critical in this case to mimic the whole viral coat protein. Provided the anti-idiotype antibody mimics a part of the virus' surface that is readily accessible to the immune system (so that it is easily recognized in the final virus), then it could be used to stimulate the immune system to make a suitable antibody. This is a good idea because it allows a vaccine to be developed without over using the live virus to make it, so there is no risk of contamination the vaccine with live virus. However, the link of the virus being used to make antibody, which is used to make anti-idiotype antibody, which is injected to get the body to make antibody, seems to be too tenuous in the experiments done so far for the resulting antibody to recognize the virus properly.

Antibody excess. The presence of a high antibody-to-antigen ratio. In an agglutination test gross antibody excess may give rise to a Prozone-possibly due to the saturation of antigenic determinants by antibodies in the absence of free antigen with which to form links.

Antibody formation. One form of Immune Response. When stimulated by a given Antigen, and by TH cells (and/or their products), certain B Lymphocytes undergo Blast Transformation and proliferate; cells of the enlarged clone may differentiate to become Antibody-secreting Plasma Cells or may become memory cells, i.e., cells which have been `primed' to react rapidly (in terms of antibody production) on a second or subsequent stimulation by the specific antigen. This constitutes the primary response. On subsequent stimulation by the same antigen the enlarged clone can respond, promptly, with an increased amount of specific antibody (the secondary or anamnestic response)

The mechanism of antibody formation is not completely understood. The process can vary in detail according to e.g. the nature of the antigen and the particular type of B cell which responds to it; Histocompatibility Restriction appears to be involved, at least at some stage. The following two models give an outline (only) of some of the important features of antibody formation which may be observed in particular instances.

Lyb 5- B Lymphocytes produce antibody in response to Thymus-Dependent Antigens, but only with Cognate Help from helper T cells (TH cells). The given antigen, carried at the surface of an Antigen-Presenting Cell, is recognized and bound by a specifically-reactive TH cell-the TH cell concurrently recognizing and binding to a class II MHC antigen on the APC itself; under such conditions the APC appears to release a lymphokine which activates the TH cell. The TH cell then binds to a B cell which has bound the given antigen and which bears the same class II MHC antigen as that displayed by the APC; the B cell then proliferates, and some cells of the clone differentiate to form plasma cells. The nature of cognate help is unknown at the molecular level. Lyb 5- cells also respond to Type 1 Antigens.

Lyb 5+ cells respond to Thymus-Independent Antigens (both type 1 and type 2) and to some TD antigens. In the response to TI antigens, antigen-binding occurs (separately) to the B cell and to a macrophage, and a TH cell binds to the macrophage via (i) the given antigen, and (ii) class II MHC antigen on the macrophage cell surface; once linked to the TH cell, the macrophage releases Interleukin 1 which induces the TH cell to synthesize and release Lymphokines: e.g. B cell differentiation factors, BCDFs (which promote the differentiation of proliferating B cells to antibody-secreting plasma cells). In this process (factor-dependent activation) the TH cell helps the B cell by supplying soluble factors-i.e. there is no direct B cell-T cell contact. (A given TH cell may be able to provide either cognate or factor-dependent help.)

The immune response to TD antigens usually involves an initial production of IgM antibodies followed by the formation of IgG (or IgA) antibodies-although subclass IgG3 antibodies are commonly elicited poorly. It has been customary to regard TI antigens as elicitors of mainly or solely IgM antibodies, with little ability to induce memory; however, recent evidence suggests that these antigens can induce memory, and that initial TI antigen-induced IgM responses can be switched to IgG (or IgA) responses-the switch apparently depending on the type of T help involved. When produced in response to invasion by pathogens (or their products-e.g. toxins) antibodies commonly have a protective function-cf. Immediate Hypersensitivity and Delayed Hypersensitivity.

Antibody-antigen. Combination occurs by contact between specific antibody Combining Sites and antigenic Determinants, and may promote Complement

Fixation. In vivo, antibody-antigen combination may be beneficial if the antigen combination may be beneficial if the antigenic stimulus derives from an invading pathogen but can be harmful

Anticoagulant. Any agent which inhibits the coagulation (clotting) of blood- e.g. Edta, Heparin, sodium citrate, sodium oxalate. SPS is used in Blood Culture.

Anticodon. A "triplet" of three nucleotides in transfer RNA that is able to pair with a complementary triplet (a codon) in messenger RNA, thus aligning the transfer RNA on the proper place on the messenger. The codon (and, reciprocally, the anticodon) codes for a specific amino acid.

Anticomplementary (immunol.) Refers to any substance (e.g. Zymosan), or to any system (e.g. pre-formed antigen-antibody complexes) other than the specific combination of antibody with antigen, which brings about Complement Fixation, If present in a Complement-Fixation Test, anticomplementary factors can-by fixing complement-simulate the effects of antigen-antibody combination, giving a false-positive reaction.

Antifungal agents. Chemical agents (including certain Antibiotics) which either kill or inhibit the growth of fungi (fungicidal and fungistatic action, respectively). (Antifungal agents are often called `fungicides' regardless of whether their action is fungicidal or fungistatic.) Antifungal agents are used e.g. for the treatment or prevention of fungal disease of man, animals and plants, and as Preservatives for a wide range of materials.

Medical and veterinary antifungal agents (`antimycotics') are used for the treatment of local and systemic mycoses such agents include e.g. Allylamines, Azole Antifungal Agents, Cycloheximide, Flucytosine, griseofulvin, Nitrofurans, Polyene Antibiotics, Salicylin Acid, Tolnaftate. Many general Disinfectants and Antiseptics also have antifungal activity.

Agricultural antifungal agents (used against fungal diseases of crop plants) may function as Eradicants and/or as Protectants; they fall into two broad categories: contact (surface or non-systemic) antifungal agents, which adhere to plant surfaces but do not penetrate the plant tissues, and systemic antifungal agents, which are taken up by the plant and translocated within its tissues-usually in the transpiration stream (from roots to leaves). In general, a contact agent typically acts as a general enzyme inhibitor, while a systemic agent usually acts at a specific target site (process or structure) in the fungal cell. Fungi may rapidly acquire resistance to an agent which has a single target site, and continued use of such agents has led in many cases to the emergence and spread of resistant strains of fungal pathogens; the risk of emergence of resistant strains can be reduced by using mixtures containing two or more agents which have different modes of action. Agricultural antifungal agents include e.g. Benzimidazoles (a), Bisdithiocarbamates, Carboxin, Chloroneb, Chloronitrobenzenes, copper sulphate Dicarboximides, Dichlofluanid, Dichlone, Dinitrophenols, Dithiocarbamates, Dodine, Ethazole, Glyodin, Hydroxypyrimidine Antifungal agents, Imazalil, Morpholine Antifungal Agents, certain Organophosphorus Compounds, Phenylamide Antifungal Agents,

Phithalimide Antifungal Agents, Prochlraz, Pyracarbolid, Quinone Antifungal Agents, Sulphur, Thiram, Triforine.

Antigen (ag). Any agent which can elicit an Immune Response (cf. Hapten); `antigen' may refer to an individual macromolecule or to a homogeneous or heterogeneous population of antigenic macromolecules. A given antigen usually contains more than one Determinant. An antigen may be soluble (e.g. microbial toxins, extracts) or particulate. The most effective antigens are proteins and polysaccharides in general, large molecules are more antigenic than are smaller ones, and antigenicity is enhanced by the presence of an Adjuvant. Minute quantities (e.g. a few ag) of antigen may be sufficient to elicit amount required depending e.g. on the route of administration.

The surface of a microorganism typically consists of repeating patterns of antigens, and the classification of some groups of microorganisms is based on differences between the antigens of different strains.

Antigen-binding cell (ABC) (1) Any cell which can bind only a specific antigen. (2) Any cell (including e.g. macrophages) which can bind various antigens.

Antigen deletion. In a microorganism: the loss of particular antigenic determinant(s) due e.g. to a mutation or to the loss of a plasmid. (cf. Antigen Gain.)

Antigen excess. The presence of a high antigen-to-antibody ratio. Under conditions of gross antigen excess precipitation does not occur since, according to the Lattice Hypothesis, all the combining sites of antibodies are bound to otherwise unattached antigens-leaving no combining sites free to form cross-links and hence larger (insoluble) complexes.

Antigen gain. In a microorganism: the formation of new antigenic determinant(s). Antigen gain may be due e.g. to a Mutation, to the acquisition of a Plasmid, or to Bacteriophage Conversion; thus, e.g., the sex factor in a conjugative plasmid specifies certain donor-specific antigens, while certain cell-surface antigens in Salmonella strains are present only when the cells are lysogenized by a particular phage. (cf. Antigen Deletion.)

Antigen-binding site. The variable Fab portion of the antibody molecule that binds a specific or closely related antigenic determinant.

Antigenic competition. Inhibition of the immune response to a given antigen or Determinant due to exposure to another (different) antigen or determinant; antigenic competition has been regarded as a form of natural immunosuppression.

Antigenic determinant. A specific region of an antigen, which is recognized by and binds to a specific antibody.

Antigenic drift. (immunological drift) In a given species, strain or type of microorganism: minor changes in antigenic specificity which occur over an extended period of time (e.g. years)

Antigenic shift. In a given species, strain or type of microorganism: major change(s) in antigenic specificity which apparently occur abruptly, For example, antigenic shifts in strains of Influenzavirus have been reported to coincide with pandemics of influenza. (cf. Antigenic Drift; Antigenic Variation.)

Antigenic variation. Successive changes which occur in the surface antigens in certain microorganisms typically giving rise to an extensive range of antigenically distinct forms.

Antigenic. Capable of stimulating the formation of antibodies.

Antigen-presenting cell (APC) Any of a range of cells which bind antigen and then present it to particular types of T Lymphocyte-thereby activating the T cells. APCs, which have MHC-specified class II cell-surface antigens include macrophages, dendritic cells, Langerhans' cells and Kupffer cells. APC function is not fully understood. Macrophages, at least, produce Interleukin 1 which causes T cell proliferation.

Antigen-specific T helper cell. A type of T helper cell which is activated by specific antigen. Most antigen-specific T helper cells appear to be Carrier-Specific T Helper Cells; others, when stimulated by specific antigen borne on Antigen-Presenting Cells, release factors which act polyclonally on B cells-activating specific antigen-binding B cells without direct B cell-T cell contact.

Antiglobulin. (Coombs' reagent) Antibody homologous to the antigenic determinants of serum globulins, particularly Immunoglobulins; antiglobulin-Ig combination can occur without involvement of the specific Combining Sites of the Ig molecule.

Antiglobulin consumption test (serol.). Any Consumption Test in which the presence of cell-or particle-bound Ig is detected by the consumption of added Antiglobulin. For example, the in vitro combination of autoantibodies with homologous cell-surface antigens can be detected by washing the cells free of uncombined antibody and incubating them with antiglobulin; if autoantibody-antigen combination has occurred, antiglobulin will be consumed and may cause agglutination by the formation of cell-autoantibody-antiglobulin complexes. The amount of antiglobulin consumed can be determined e.g. by titrating the unconsumed antiglobulin against Ig-coated erythrocytes.

Anti-idiotype. Antiserum to the variable-region domains of an Ig molecule (cf. Idiotype); it contains antibodies homologous to each of the individual determinants (Idiotopes), though the different types of antibody may differ quantitatively and in their Affinity for their respective determinants.

Antilymphocyte serum (ALS) Antiserum to the cell-surface antigens of Lymphocytes; it is used e.g. for Immunosuppression. Antilym-phocyte globulin (ALG) is the globulin fraction of ALS.

Antimalarial agents. (`antimalarials') Drugs effective in the chemotherapy and/or chemo-prophylaxis of Malaria. Most drugs are effective against only some stages of the malaria parasite; for this reason, and to inhibit the emergence of resistant strains, combinations of drugs are commonly used.

Antimetabolite. A substance which competitively inhibits the utilization, by an organism, of an exogenous substrate or endogenous metabolite.

Antimicrobial. An agent (chemical and physical) that inhibits growth of or kills microorganisms.

Antimony. (as an antimicrobial agent) Antimony is a Heavy Metal whose compounds have long been used as therapeutic agents-though their toxic effects were often greater than the benefits they bestowed. **Antimonials** have found use in the treatment of certain diseases of protozoal aetiology, e.g. leishmaniasis, and other diseases, e.g. schistosomiasis.

Antimycin. A An antibiotic which acts as a Respiratory Inhibitor in Complex III of the mitochondrial Electron Transport Chain, apparently blocking electron flow between b-type cytochromes and cyt c1; antimycin A and HOQNO appear to act at a common site which may be associated with the b-type cytochromes.

Antioxidant (food sci.). A substance added to certain foods to inhibit oxidation of the lipids in those foods-thus enhancing stability and prolonging shelf-life. Antioxidants currently used include e.g. ascorbic acid, butylated hydroxyanisole, butylated hydroxytoluene, carotene, propyl-gallate and tertiary butylhydroquinone. Some antioxidants have antimicrobial properties.

Antiport (counter transport, exchange diffusion) The transmembrane transport of a solute (e.g. a given ion) coupled, directly, to the transmembrane transport of a different solute in the opposite direction. (cf. Symport, Uniport;)

Antiprotozoal agents. Chemical agents which kill or inhibit protozoa; they include e.g. Aminoquinolines; antimonials arsenicals certain Diamidines; Diloxanide; Emetine; 8-Hydroxy-Quinoline; Metronidazole; Monensin; Nitrofurans; Pyrimethamine; Sulphonamides; Suramin.

Antisense RNA. In the original (narrow) sense: any ssRNA strand which, on hybridizing with complementary bases in an RNA or DNA molecule,·inhibits the latter's role if the `target' is a transcript, the antisense RNA is countertranscript RNA (ctRNA), and if it is mRNA the antisense RNA is mRNA-interfering complementary RNA (micRNA).

Antisense. Antisense RNA or DNA is a single-stranded nucleic acid that is complementary to the coding, or 'sense', strand of a gene, and hence is also complementary to the mRNA, it hybridizes to it, forming a double helix. This double helical RNA cannot then be translated by ribosomes to make protein. Thus, antisense RNA can be used to block the expression of genes that make proteins.

Antisepsis (degerming). Prophylactic or therapeutic treatment of human or animal tissues with Antiseptics. (cf. Sepsis.)

Antiseptic paint. Antibody-containing secretions which bathe the mucous surfaces of the body.

Antiseptic. A chemical agent that prevents growth, by either inhibiting the growth of or destroying microorganisms; used for topical application to living tissue, primarily skin.

Antiseptics. Chemical agents used for treating human and animal tissues (particularly skin) with the object of killing or inhibiting pathogens. Antiseptics are typically non-injurious to living tissues (cf. Disinfectants). For examples of antiseptics

Antiserum (immune serum). Serum containing antibodies to one or more particular antigens.

Antiserum. Blood serum containing specific antibodies against an antigen, such as a bacterial cell, a virus, or an exotoxin.

Antistreptolysin O test (ASLT; ASO test; ASOT). A test used for detecting/quantifying serum antibodies to Streptolysin O; it is of use e.g. in the differential diagnosis of Rheumatic Fever and rheumatoid arthritis.

Antitermination. A process in which, during Transcription, the RNA polymerase fails to recognize a transcription termination signal and therefore continues transcription beyond the terminator. Antitermination may involve various mechanisms and can serve as a means for the regulation of gene expression and Operon (attenuator control) and for the prevention of erroneous transcription termination.

Antitoxin. Antibodies that are formed in response to bacterial toxins or toxoids and neutralize or inactivate those toxins (1) An antibody to a toxin. (2) An antiserum which contains antibodies to one or more particular toxins.

Antiviral agents. Compounds which more or less specifically inhibit the replication of viruses in cells, tissues or organisms-either directly (e.g. by inhibiting a particular viral enzyme) or by inducing Interferon synthesis (e.g. analogues of 2-5A, Tilorone).

Antiviral compounds. One of the areas in which biotechnology is playing a substantial role in developing new drugs is in producing antiviral compounds. The work has focused on a range of technical approaches.

Anus. Opening through which digestive wastes are expelled, located at the posterior end of the gut.

APC (antigen-presenting cell). A variety of cell types which carry antigen in a from which can stimulate lymphocytes.

Apex. The tip or highest point of a structure, as the apex a growing stem or root.

Aphanizomenon. A genus of filamentous Cyanobacteria; the cells are cylindrical with closely abutting ends, and the filaments aggregate to form raft-like floating colonies. Gas Vacuoles are present, and heterocysts occur but are often sparse. Trichomes commonly terminate in tapering, colourless `hair cells'. *A. flos-aquae* is an important freshwater Bloom-forming species in temperate regions; it may also occur in brackish waters. The filaments of this species typically form wispy, flake-like colonies, the formation of which, in culture at least, appears to depend on the presence of iron. Strains of A. flos-aquae produce toxins (*aphantoxin*) which may apparently include e.g. Saxitoxin and neosaxitoxin; `aphantoxin' acts as a neuromusucular blocking agent, and is toxic for amphibia, fish, etc.

Aphanomyces. A genus of predominantly aquatic fungi of the order Saprolegniales (q.v.). A species known as the `MG fungus' has been reported to cause a mycotic granuloma of fish while *A. astaci* causes Crayfish Plague Plant-pathogenic species include *A. brassicae*, *A. camptostylus, A. cochlioides* (on sugar-beet), *A. euteiches* (on peas) and *A. raphani* (on radish).

Aphanothece. A phycological genus of unicellular `blue-green algae' (Chroococcales) apparently refrable to Cyanothece.

Aphidicolin (Apc). A tetracyclic diterpenoid, isolated from *Cephalosporium aphidicola* which-in eukaryotic cells-is a potent and specific inhibitor of DNA Polymerase.

Aphthovirus. A genus of viruses (family Picornaviridae) which infect mainly cloven-hooved animals, causing Foot and Mouth Disease.

Aphyllophorales (Polyporales) An order of terrestrial (typically humicolous or lignicolous), mainly saprotrophic fungi (subclass Holobasidiomycetidae) which characteristically form tough (rather than fleshy).

Apical complex. A collective term for the Conoid, Micronemes, Polar Ring and Rhoptreis-some or all of which are found in members of the Apicomplexa, particularly in motile stages. (Some authors include the Micropores and/or the subpellicular tubules as part of the apical complex.)

Apical. Pertaining to the apex, as the apical meristem, which is the actively growing tissue at the tip of a stem or root.

Apicomplexa. A phylum of the Protozoa. The organisms are parasitic. They lack both cilia and flagella (althogh microgametes are flagellated in some species).

Apochlorotic. Having lost chlorophyll-either during evolutionary development or during growth under certain conditions. Apochlorotic organisms may have lost either chloroplasts or chlorophyll; in the latter case a colourless plastid (leucoplast) may be formed.

Apochromat. (apochromatic objective) An objective lens in which chromatic aberration has been corrected for three colours (red, green, and blue), and spherical aberration has been corrected for two colours (red and blue). (cf. Achromat.) A Flat-Field Objective Lens of this type is called a planapochromat.

Apoenzyme. The protein portion of enzyme and requiring coenzyme before becomming catalytically active.

Apomixis (microbiol.). In certain diploid organisms, a type of reproduction, considered to be a deviant form of the sexual process, in which meiosis, gamete formation and fertilization do not occur even though morphologically differentiated structures (e.g. spore-containing asci) are formed; progeny produced apomictically from a given cell (the apomict) are genetically identical to that cell.

Apoptosis. Apoptosis, or *programmed cell death*, is the process by which cells deliberately destroy themselves.

Apothecioid. Of an Ascocarp: having, at maturity, a hymenium more or less open to the environment-as in a typical Apothecium; apothecioid ascocarps may be e.g. cup-shaped, discoid (circular and flat), or lirelliform.

Appertization. A method of Food Preservation in which certain types of food are exposed to a temperature/time regime which renders them safe for the consumer and microbiologically stable for extended, or indefinite, periods of time on subsequent storage in hermetically sealed containers at temperatures below 40° C.

Appetitive behaviour. Searching behaviour, often of a specific nature, which terminates when the animal contacts the appropriate signal from the environment (sign stimulus) and performs the corresponding instinctive act (consummatory act).

Apple canker. A canker of apple trees caused by *Nectria galligena* f. sp. *mali*; other trees e.g. pear, poplar, hawthorn-may also be affected.

Apple scab. An important disease of apple trees (*Malus*) caused by *Venturia inaequalis.*

Appressorium A specialized, flattened region of a hypha or germ tube formed by some plant-pathogenic fungi (e.g. *Puccinia graminis, Venturia inaequalis*).

Apramycin. An Aminoglycoside Antibiotic which is resistant to many of the enzymes which inactivate antibiotics of this group; it is inactivated by e.g. chromosome-and plasmid-specified acetyltransferases.

A-protein, In tobacco mosaic virus: a mixture of small oligomers and monomers of coat protein subunits which occur in equilibrium with the larger `disc' aggregates under conditions of physiological pH and ionic strength; coat protein occurs mainly as A-protein under conditions of high pH and low ionic strength.

Aprotinin. A basic polypeptide which inhibits many serine Proteases, including e.g. trypsin, chymotrypsin, and some bacterial proteases.

APS. Adenosine-5'-phosphosulphate (or adenosine-5' - sulphatophosphate or adenylylsulphate).

APT agar. An agar-based medium containing tryptone. yeast extract, glucose, Tween 80, citrate, and various inorganic salts. It is used for isolating organisms such as lactobacilli and Brochothrix from e.g. meat and meat products.

Aptamer. Aptamers are nucleic acid molecules, usually short lengths of DNA, that have been selected to bind to a target other than another nucleic acid. Just as a chain of amino acids can fold to form a shape that exactly binds to some target molecule, so a nucleic acids can fold to form a specific binding shape. Nucleic acids that bind to specific proteins (such as thrombin, an experiment performed by Gilead Sciences) or small molecules such as NADPH or ATP have been developed by Darwinian cloning.

APW, Alkaline peptone water: 1% peptone and 1% NaCl in distilled water, pH adjusted to 8.6-9.0, used e.g. for isolating *Vibrio cholerae*.

Aquaspirilla. occur in a wide range of freshwater habitats- e.g., stagnant waters rich in organic matter; none appears to be pathogenic.

Aquaspirillum. A phylogenetically heterogeneous genus of Gram-negative, asporogenous bacteria formerly included in the genus *Spirillum.*

Aquatic. (L. *aqua*: water). Living in or on water, or taking place in or on water.

Aqueous. Containing water, or dissolved in water.

Ara operon (arabinose operon). An Operon concerned with the metabolism of Arabinose; when used without qualification, the term `ara operon' usually refers to the AraBAD operon.

Arabinose. An aldopentose which occurs e.g. as a component of hemicelluloses and Pectic Polysacharides, and in the cell walls of certain bacteria (e.g. *Nocardia* spp).

Arabinosyl nucleosides. A group of compounds which inhibit various enzymes-including DNA polymerases and reverse transcriptase-by acting as analogues of biological nucleosides and nucleotides.

Arabinosyl nucleosides. Inhibit DNA synthesis in a variety of eukaryotic cells. and have been used as antitumour and Antiviral Agents; in general, viral replication is more sensitive to these agents than are the host cells. Arabinosyl nucleosides probably must be converted to the corresponding nucleoside 5'-triphosphate for activity.

Arabinoxylans. Heteroglycans containing both xylose and arabinose (cf. Xylans).

Arabitol. A Pentitol formed e.g. by the reduction of Arabinose. It occurs, often with Mannitol, in various basidiomycetes and ascomycetes.

Arachidonic acid. In mammals: an essential C_{20} straight-chain fatty acid-$CH_3(CH_2)_3(CH_2CH{=}CH)_4(CH_2)_3COOH$; it occurs e.g. as a component of cell membrane phospholipids, and (bound to protein) in blood.

Arachinia. A genus of anaerobic (but aerotolerant), catalase-negative, asporogenous bacteria (order Actinomycetales, wall type I). The organisms grow as a substrate mycelium which fragments into branched filaments and irregularly-shaped rods;. aerial hyphae are not formed.

Ara-T (1-α-D-arabinofuranosylthymine) An Arabinosyl Nucleoside isolated from a marine sponge (*Cryptoethya crypta*). It has antiviral activity.

Arboviruses. Viruses transmitted by arthropod vectors. Arthropod-borne viruses which has a non-taxonomic category.

Arcanobacterium. A genus of asporogenous bacteria (order Actinomycetales, wall type VI).

Arcella. A genus of testate amoebae (order Arcellinida) in which the test is predominantly or entirely organic.

Arcellinida An order of freshwater and soil amoebae (class Lobosea).

Archaebacteria A phylogenetic category (kingdom) of prokaryotic organisms. (cf. Eubacteria; Urkingdom.) *Archaebacteria* differ from both eubacteria and eukaryotes e.g. in the base sequences of their 16S rRNA and in that the archaebacterial cytoplasmic membrane contains branched, isoprenoid, either-linked lipids rather than fatty acid ester-linked lipids. Archaebacteria include the Methanogens, halophiles (e.g. *Halobacterium*), the Sulphus-dependent *Archaebacteria*, and the genus *Thermoplasma.*

Archicarp. In ascomycetes: the cell(s) which give rise to a fruiting body or to a part of it.

Arcyria. A genus of slime moulds (class Myxomycetes) which form stalked, globose to cylindrical sporangia.

ARD. Acute respiratory disease: a general term for any such disease affecting closed populations of people (e.g. military recruits, school children). Major causal agents of ARDs are adenoviruses (usually Ad3, Ad4, Ad7, Ad14, or Ad21-).

Arenaceous. Resembling or (referring e.g. to the test of Difflugia) incorporating or bearing grains of sand.

Arenaviridae (arenavirus group). A family of enveloped ssRNA viruses.

Arginine biosynthesis. In *Escherichia coli* the genes encoding enzymes for arginine biosynthesis constitute a Regulon (the arg regulon); four of the genes occur in a cluster (argECBH), the remainder occur singly at loci scattered around the chromosome.

Arginine dihydrolase (ADH). An enzyme system which catalyses the catabolism of arginine, with a concomitant substrate-level phosphorylation; the system occurs in a range of bacteria.

Arildone. {4-[6-(2-chloro-4-methoxyphenoxy)-hexyl]-3-5-heptanedione} An Antiviral Agent which is active against various DNA and RNA viruses in vitro; it blocks uncoating in polioviruses and apparently also in herpes simplex viruses. Arildone may be useful clinically e.g. in the topical treatment of Herpes Simplex infections.

Arizonosis. A Poultry Disease, caused by *Salmonella arizonae* (= *Arizona hinshawii*), characterized by malaise, diarrhoea, and often symptoms of CNS involvement.

Armillaria. (Armillariella) A genus of mainly lignicolous fungi (Agaricales, Tricholomataceae).

ARS Autonomously replicating sequence: a DNA sequence (first described from Saccharomyces cerevisiae) which, when linked to

Arsenic (as an antimicrobial agent) Arsenic is a Heavy Metal whose (aromatic) compounds include some effective antimicrobial agents which have found use as therapeutic agents; however, owing to their toxicity, some of these `arsenicals' have been largely superceded by antibiotics.

ART. Automated reagin test: a qualitative or quantitative Standard Test For Syphilis similar in principle to the RPR TEST.

Artefact. (arfifact) Any feature which does not occur in a specimen under natural conditions, but which may be seen in that specimen during experimentation. Artefacts are due to the disturbance introduced by the process of experimentation or observation; they may occur e.g. as a result of Fixation.

Arterivirus. A genus of viruses (family Togaviridae) which currently includes only equine arteritis virus. This virus infects equines, causing necrosis of the small arteries with various clinical manifestations (e.g. rhinitis, oedema, enteritis, bronchopneumonia, etc). Infection of pregnant mares commonly results in abortion. Transmission occurs horizontally and vertically; vectors are unknown. The virus can infect a range of vertebrate cells in vitro.

Arthoniales. An order of fungi of the Ascomycotina; members include crustose Lichens (photobiont green, often trentepohlioid) and lichenicolous and saprotrophic fungi. Ascocarps are Apothecioid and may be lirelliform, irregular, etc; they contain paraphysoids. Asci are bitunicate, usually clavate. Genera: e.g. Arthonia.

Arthopyrenia. A genus of crustose Lichens of the order Dothideales; ascospores have one to several septa. *A. halodytes* (photobiont Hyella) is a marine species which grows endolithically in calcareous rocks and on the shells of limpets, barnacles etc in the intertidal zone; it occurs in Europe and N. America.

Arthroascus. A genus of fungi (family Saccharomycetaceae) which form yeast cells (which bud, often bipolarly, on a wide base) and true mycelium (which tends to break up into arthrospores).

Arthrobacter. A genus of obligately aerobic, catalase-positive, asporogenous bacteria (order Actinomycetales, wall type VI) which occur in the soil.

Arthropod. An invertebrate animal with jointed legs, such as an insect or a crustancean.

Arthrospores. Asexual spores formed by fragmentation of a filament in certain fungi.

Arthus reaction. Reaction resulting from injection of an antigen into the skin; inflammation results from activation of complement by IgG-antigen complexes.

Artificial active immunity. Resistance to disease which develops after administration of a vaccine or inactivated toxin.

Artificial key. A key not extensively based on evolutionary relationships, but based on convenient characters.

Artificial medium. A medium for microbial growth that contains complex substances such as beef extract, yeast extract, tryptones, and blood.

Artificial sweeteners. Many substances are used to make food taste sweeter without increasing its calorie value. Among those of interest to biotechnology are:

Thaumatin. A protein produced by Thaumatococcus danielli in its fruit. Thaumatins are 3000 times as sweet as sugar, and at low concentrations enhance other flavours as well. Because they are proteins they can and have been produced by genetic engineering in bacteria, so avoiding having to go to the tropics to harvest the fruit. Thaumatins have been produced in *Escherichia coli*, *Bacillus subtilis*, *Streptomyces lividans*, and *Saccharomyces cerevisiae*, and the genes have been put into higher plants too.

Aspartame. Known as nutrasweet, this is one of the most commonly used commercial artificial sweeteners. It is a dipeptide (aspartate-phenylalainine-methyl). Because it is made from two amino acids, there are two parts of its manufacture that may be of interest to a biotechnologist. One amino acid, phenylalanine more efficiently is a worthwhile goal as part of the production of aspartame. The other aspect is the synthesis of the dipeptide by enzymes, and particularly by using a protease to join the two amino acids together (rather than its more normal reaction of separating them). Both areas are undergoing continuing commercial development.

Ascocarp. Sexual fruiting structure of higher ascomycetes.

Ascogonium. Female reproductive structure of an ascomycete.

Ascospores. Haploid spores produced inside a diploid cell called the ascus. Since meiosis in the ascus is often followed by mitosis, there are commonly eight ascospores per ascus. Found exclusively in the class ascomycetes.

Ascothecium. A container formed from the sterile hyphae which surround and protect developing asci. Various forms are perithecia (flask shaped), Cleisto-thecia (spherical) and apothecia (disk shaped) (Synonym, ascocarp.

Ascus. The cell in which nuclear fusion occurs in ascomycetes and in which ascospores subsequently develops. Asci arise from distinct ascogenous hyphae, in which (usually) genetically dissimilar nuclei share a common cytoplasm.

Asepsis. A condition in which there are no infectious or contaminating microorganisms; the absence of sepsis.

Aseptic meningitis. Non-bacterial meningitis.

Aseptic processing. Aseptic processing is processing materials so that no microorganisms can get into them. This either means that the production process machinery must keep the material sterile, or the whole manufacturing environment must be aseptic, i.e. a clean-room environment. In many pharmaceutical production processes, both approaches are adopted. UHT (ultra heat treated) milk and pasteurized fruit juice is usually packaged aseptically.

Aseptic technique. The performance of a procedure or operation in a manner that prevents the introduction of septic material.

Aseptic. Free of microorganisms that can cause contamination of infection.

Asexual spores. Spores generated by one cell, without fusion of two cells.

Asexual. Without sex.

Assay. An assay is a specific chemical test for something. In biotechnology this usually means a specific biomolecule in a mix of other ingredients. This means that an assay must be able to select the analyte molecule (the thing to be measured) from among a host of others, and measure its concentration. This in turn means that at least one of the reagents involved must be very specific for the analyte. Lots of assays are developed by and used in biotechnology. Key parameters in an assay are its sensitivity and specificity. There are two meanings for each of these words. Chemical sensitivity is a simple measure of how little of the analyte the assay can respond to. Some PCR assays, for example, can detect a single molecule of their target DNA. Their sensitivity is one molecule. Usually sensitivities are given as concentrations (micrograms per millilitre) or as molar concentrations (moles/litre). There is no standard way of measuring the chemical specificity of an assay, so usually it is tested against all the chemicals that might give a false result, and thus show that those chemicals have no effect. Diagnostic sensitivity and specificity are slightly different. These ask how good the assay is at telling me whether or not something has happened. Assays are usually quantitative, giving a quantitative answer. They can be converted to a qualitative, yes/no test by setting a threshold above which a response is taken to be a 'yes' (positive), and below which it is a 'no' (negative). But how reliable is this dichotomy ? Any test will give true positives (i.e. tests that show up positive when the sample really is positive - P), false positives (showing up positive when the sample is negative - F), true negatives (not showing positive when there is nothing there - N), and false negatives (not showing up positive when there is something there - G).

Assembly (L. *assimulare*, to bring + together): The phase during replication of viruses in which the process of integration of viral components into virus particles takes place.

Assimilation. The absorption of food and its conversion into cells or tissue.

Atmosphere. The gaseous mass surrounding our planet. Also; a unit of pressure, equal to the normal pressure of air at sea level.

Atom. (Gr. atmos: indivisible). The smallest unit of a chemical element. Consists of a nucleus and one or more electrons.

Atopic hypersensitivity. Immediate-type, IgE-mediated common clinical allergies for which there is a genetically determined predisposition for formation.

Atopy. The chemical manifestation of type 1 hypersensitivity reactions including eozema, asthma and rhinitis. Where there is genetic predisposition.

Atoxyl (p-aminophenylarsonic acid, NH_2. C_6H_4.$As(OH)_2O$) was the first arsenical to be used against trypanosomiasis; arsenicals still used in the later stages of certain forms of trypanosomiasis include tryparsamide (p-N-phenylglycineamidoarsonic acid) and melarsoprol (a derivative of benzenearsenous acid).

Attenuation. Change; used in reference to certain vaccines in which the organism or its products are reduced in virulence.

Attenuator. A transcription termination site located in the interior of an operon.

Attic. An important utility service area for the laboratories containing much service equipment.

Autoallergy. Immune response against an individual's own antigens.

Autochthonous. Referring to the indigenous members of an ecological site.

Autoclave (Steam sterilizer). A chamber used for heat sterilization of materials and equipment by direct exposure to steam under pressure. Standard conditions for laboratory glassware are 15 labs/in.2 (250 °F; 121 °C) for 15 to 20 minutes.

Autogamy. Union of gametes produced by one individual.

Autograft. Tissue graft from one area to another area of the body.

Autoimmune disease. Due to antibodies produced against the individual's own tissues; failure to differentiate between self and non-self.

Autologous. Part of the same individual.

Autolysis. Self-lysis, or disintegration of cells due to the action of their enzymes.

Autonomic nervous system. The system (which in vertebrates comprises sympathetic and parasympathetic sub-systems) that controls such involuntary functions as those of guts and glands.

Autonomously replicating sequence. DNA sequence (first described from Saccharomyces cerevisiae) which, when linked to a non-replicative DNA fragment, can confer on the fragment the ability to replicate autonomously in a yeast cell. An ARS-containing fragment can transform yeast cells at much higher frequencies than can a fragment without an ARS, since the former can replicate autonomously in the recipient cell while the latter must undergo (rare) homologous recombination with the chromosome before its genes can be expressed. However, transformants formed by recombination are highly stable, whereas those with ARS-containing DNA are extremely unstable and readily lose the transforming DNA even under selective conditions.

Autopolyploid. (Gr. *auto*: self + *poly*: many + *ploos*: fold). A polyploid constructed entirely of sets of chromosomes that originated from the same species.

Autoradiograph. (1) Radioactive labels are often detected by autoradiography (autorad for short), a technique where as X-ray film is laid against something in the dark, and the film darkens where the radiation struck it, making an image of the radioactive regions of the object. (This was how Antoine Becquerel discovered radioactivity to start with, in 1896). Autorads were a standard way of detecting DNA or protein samples in a wide variety of gels from electrophoresis, and are still common, although they have been supplanted to a degree by non-radioactive methods. Typically, autorads produce dark patches of exposure on transparent film, and so are presented as black spots or bands on a white background. In some experiments, the film emulsion is actually laid on to the experiment directly, so that, for example, the radiation trapped in specific sections of cells can be detected.

Autosomes. Chromosomes other than the X or Y sex chromosomes.

Autotrophic. Pertains to an organism that is able to manufacture its own food. They have the ability to use carbon dioxide or bicarbonate as the sole source of carbon.

Auxospore. The reproductive cell of diatoms.

Auxostat. An auxostat is a chemostat in which the dilution rate can vary. Usually a chemostat, a closed culture vessel into which new medium is continuously fed in and old medium and organisms continuously removed, has a fixed rate of dilution, i.e. a fixed rate at which the new material is added and the old removed. This determines how fast the organism in the chemostat will grow. In an auxostat the rate at which new material is added and old material removed is determined by some feature of the culture. For example, the amount of bacteria may be measured by measuring the cloudiness (turbidity) of the culture, and the amount of material added adjusted to keep the turbidity constant. Alternatively, if the bacteria reduce the pH of the culture as they grow (as bacteria often do), the pH may be used to control the rate of dilution. The former is called a turbidostat, the latter a pH auxostat.

Auxotroph. (Gr. *auxamein*: to grow + *trophe*: food). A mutant form of an organism that requires a nutrient or nutrients not required by the wild-type, or reference, form of the organism. (Contrast with prototroph.)

Auxotrophic. Refers to mutants, generally of microorganisms, which grow in culture only when specific requirements are added to the culture medium.

Aviarty. The functional combining strength of an antibody with its antigen, which is related to both the affinity of the reaction between the epitopes and paratopes and the valencies of the antibody and antigen.

Avirulent. Lacking disease-producing ability.

Axenic. Pure or uncontaminated culture.

Axial filament. Fibrillar structure that spirals around the organism, is attached at each end of a spirochete.

B

B cell. (B lymphocytes) a major population of lymphocytes; the B cell stem cells are found in bone marrow (in the bursa of Fabricius in avian species); mature B lymphocytes have cell-surface Ig receptors for specific antigenic determinants; antigen-stimulated B cells proliferate and differentiate into antibody-producing plasma cells.

B1H. A cofactor of the complement system recently renamed Factor H.

B2-microglobulin. A polypeptide which constitutes part of some membrane proteins including the Class 1 MHC molecules.

Babes-ernst bodies. Metachromiatic granules, or volutin; granules of long chains of inorganic phosphate; characteristic of Corynebacterium diphtheriae.

Bacillus. A cylindrical or rod-shaped bacterium; Bacillus is genus name for aerobic spore-forming rods.

Bacillus anthracis. Aerobic spore-forming rod; the causative agent of anthrax.

Backflow preventer. A manufactured piping device of the type that has two spring-loaded vertical check valves and one spring-loaded, diaphragm-activated differential pressure relief valve. It is installed in a water supply line to prevent reversal of water flow in case the supply pressure falls below the downstream pressure.

Bacteremia. The presence of bacteria in the blood.

Bactericidal. Bacteria-killing.

Bactericide. An agent which kills vegetative bacteria and some spores; synonymous with germicide.

Bacteriocin. Antimicrobial substance produced by bacteria that kills sensitive members of related strains.

Bacteriology. The science that deals with bacteria.

Bacteriophage. Literally eater of bacteria; a virus that infects specific bacteria, multiplies there-in and ultimately destroys the host bacterial cell. A bacteriophage is a virus that attacks bacteria. They have been used extensively in DNA cloning work, where they are the basis of convenient vector molecules. The bacteriophage (or 'phage', for short) used most are derived from two 'wild' phages called m13 and lambda: Pl is another, less commonly used, phage.

Lambda phages are used to clone large pieces of DNA or RNA. They are a 'lytic' phage, i.e. they replicate by breaking open (lysing) their host cell. If a few phage are spread on a huge number of bacterial cells, they lyse open the cells they hit, releasing more phage, and so on. On a bacteriological plate, this results in a small clear zone, a plaque, where each original phage landed. In a bulk liquid culture it results in an enormous number of phage particles, 1014 per litre in some cases. Both plaques and bulk culture are useful sources of large amounts of the bacteriophage DNA, for analysis. Some lambda vectors have also been developed that are expression vectors, producing foreign protein from a cloned gene.

The other principal bacteriophage vector is the m13 system. This phage can grow inside a bacterium as a plasmid, so that it does not destroy the cell it infects but causes it to make new phage continuously. It is a single-stranded DNA phage, and is used for the Sanger dideoxy DNA sequencing method (which requires single-stranded DNA as a starting material). Messing has developed a well-known series of m13 vectors for cloning pieces of DNA into m13 for sequencing.

Both these phage grow on *Escherichia coli* as a host bacterium. Many other phages, both of *E. coli* and of other bacteria, are used in more specialist research applications.

Bacteriostat. An agent which stops the growth and multiplication of bacteria, but does not necessarily kill them. Usually growth resumes when the bacteriostat is removed.

Bacteriostatic. Inhibiting growth or multiplication of bacteria.

Bacterium. (Gr. *bakterion*: little rod). A prokaryote. An organism with chromosomes not contained in nuclear envelopes.

Bacteroid. Irregular shape of bacterium (Rhizobium) associated with symbiotic nitrogen fixation in roots of plant.

Baculovirus. Baculovirus is a class of insect virus that has been used to make DNA cloning vectors for gene expression in eukaryotic cells. The vector system was derived from Autographa californica nuclear polyhedrosis virus to enable biotechnologists to make large quantities of proteins from cloned genes in insect cells (the cells usually used are a cell line derived from the fall armyworm). Baculoviruses have a gene that is expressed late on during their infection cycle at very high levels, filling the nucleus of the cell with many-sided bodies full of a protein that is not necessary to produce more viruses, but is necessary for the virus' spread in the wild. In the vector cloning system this gene is replaced by the one that we want expressed.

Production of the protein can be up to 50% of the cells' protein content, and several proteins can be made at once so that multi-subunit enzymes could (in principle) be made by this system. However, the main strength of the baculovirus system is that, as a genuine animal expression system, it produces proteins that are glycosylated like the proteins in animals. This, in combination with the relatively high levels of expression can make this an attractive option for proteins that are to be used as biopharmaceuticals, outweighing the problems of growing a virus in mammalian cell culture. In addition, baculovirus is non-infective and non-pathogenic to vertebrates.

The baculovirus DNA is very large (100-150 kb), and so conventional recombinant DNA methods cannot be used to engineer it. Instead, plasmids containing the desired gene are recombined with the virus in vivo through homologous recombination.

A novel use for baculovirus systems is as viral insecticides. A gene is inserted into the virus that is lethal to an insect (e.g. the endotoxin gene from *Bacillus thuringiensis*) but does not affect isolated cells. This is then used to produce infective virus, which can, in principle, infect insects and kill them. There are technical problems with this, however (such as whether the virus is still infective in a real organism), as well as regulatory ones.

Baeocyte. Reproductive cells produced by species of cyanobacteria.

Balance polymorphism. (Gr. *polymorphos*: having many forms). Retention of both extreme phenotypes in a population, usually resulting when the heterozygote is superior to both homozygotes.

Balantidium coli. Protozoan ciliate; causative agent of balantidiasis, a type of dysentery.

Basal body. Structure at the base of flagella and cilia.

Base plate. The hexagonal plate at the end of the helical tail to which the tail fibres are attached.

Base. A substance which can accept a proton (H^+). (Contrast with acid.)

Basic. Having a pH greater than 7.0 (having a hydrogen ion concentration lower than 10^{-7} molar).

Basidiospores. Spores produced externally by meiosis from a diploid cell called the basidium. Found exclusively in the class Basidiomycetes.

Basidium. In fungi of the class Basidiomycetes, the characteristic sporangium in which four spores are formed by meiosis and then borne externally before being shed.

Basophil. A blood leukocyte that is rich in vasoactive substances; the blood counterpart of mast cells. Stains with basic dyes.

BCG (bacille calmette guerin). An attenuated strain of Mycobacterium tuberculosis used as a vaccine, an adjuvant or a biological response modifier in different circumstances.

Bdellovibrio. Very small microscopic procaryotic cells which parasitize other bacteria.

Behavioral biology. The scientific study of all aspects behaviour, including neurophysiology, ehtology, and socio-biology.

Bence jones protein. Free immunoglobulin light chain dimers found in the serum and urine of patients and animals with multiple myeloma.

Benefit. An improvement in survival and reproductive. success resulting from a behaviour. (Contrast with cost.)

Benign. Mild; not malignant.

Benthic zone. (Gr. *benthos*: bottom of the sea). The bottom of the ocean. (Contrast with pelagic zone.)

Benthos. Sediment, or pertaining to life within the sediment.

Beta-lacuam. The ring structure associated with compounds such as the antibiotics penicillin and cephalosporin.

Beta-oxidation. Catabolic process in which two carbon units of a long chain fatty acid are removed to form acetyl CoA.

Beta-richness. The rate at which new species are added to a biota. (Contrast with alpha-richness, gamma richness.)

Bilateral symmetry. The condition in which only the right and left sides, divided exactly down the back, are mirror images of each other. (Contrast with radial symmetry.)

Binal. Refers to bacteriophages with a combination of two different symmetryes; for example, bacteriophages T4 with a cuboidal head and a helical tail.

Binary fission. An asexual process in which a cell splits into two cells of equivalent size.

Binding. Much of biochemistry and molecular biology is about molecules binding to each other. This means that they stick to each other because the exact shape and chemical nature of parts of their surfaces mean that they are 'complementary': a common model is 'lock-and-key', much used to describe how enzymes fit around their substrate. Biology only works because many biological molecules bind extremely tightly and specifically to other molecules - enzymes to their substrates, antibodies to their antigens, DNA strands to their complementary strands, and so on. This binding is entirely spontaneous, and depends on the chemical nature of the molecules concerned.

Binding can be characterized by a binding constant or association constant (Ka), or its inverse the dissociation constant (Kd).

Binomial. Consisting of two names; for example, the binomal nomenclature of biology which gives the name of the genus followed by the name of the species.

Bioaccumulation. This is the accumulation of materials that are not critical components of an organism by that organism. Usually it refers to the accumulation of metal. Many organisms (e.g. plants, fungi, protists, bacteria) accumulate metals when grown in a solution of them. Sometimes this is part of their defence mechanism against the poisonous effect of those metals, sometimes it is a side-effect of the chemistry of their cell walls.

In a few cases, bioaccumulation is economically important as part of the microbial mining cycle. Using this biosorption process, metals present in very low concentrations in water can be accumulated in the cell walls of living organisms and subsequently harvested. Bioaccumulation and the use of bacteria to remove toxic metals from waste water as a purification ('bioremediation') process are clearly closely related.

Bioassay. A bioassay is a way of measuring something that has a living system as a key component. Usually it means a way of measuring the concentration of chemical, although bioassays for magnetic fields (using homing pigeons or magnetic bacteria), ionizing radiation (measuring mutation), or other physical effects are possible.

Many bioassays have been in traditional use - the proverbial canary in the coal mine was bioassay for poisonous gas, the canary being the biological element. Animals have been used extensively in drug research in bioassays for the pharmacological activity of drugs. However, new bioassays are usually developed using bacteria or animal or plant cells, as these are normally much easier to handle than whole animals or plants and are cheaper to make and keep. Thus, bacterial bioassays for biological oxygen, demand (BOD) and poisons in general are in use in the water industry. Here, bacteria are mixed in with a sample of water, and how well they can metabolize (and hence use up oxygen, produce carbon dioxide, or, in one case, give out light) is measured by an instrument.

Many of the cytokines and growth factors that biotechnologists are now producing using recombinant DNA methods were originally identified using bioassays in which mammalian cells were used to detect minuscule amounts of the compounds concerned through their potent effects on the cell's behaviour. They are often still measured by bioassay, because this is the only reliable way to distinguish the functional, active protein from similar proteins or degraded material. Bioassays also reflect the mode of action of the material, i.e. what it meant to do, rather than an arbitrary definition of chemical purity. Many very active biological substances are still measured in 'international units', which are defined as the amount required to have a specific effect in a bioassay.

On the borderline between bioassays and chemical assays are immunoassays and enzyme assays. These use proteins, made from a biological system, in an otherwise entirely standard chemical-type measurement. Bioassays can be turned into sensor systems, although as bioassays are no more convenient to use than any other chemical reaction, this is quite a difficult task.

Bioconversion. Bioconversion is the conversion of one chemical into another by living organisms, as opposed to their conversion by enzymes (which is biotransformation) or chemical processes. Synonyms are biological transformation or microbial transformation. Bioconversion has been used for a long time to make a few chemicals such as alcohol (from sugar) and, more recently, the drug ephedrine. However, it is only since the Second World War that bioconversion methods have become common place. The usefulness of bioconversion is much the same as that of biotransformation, especially its extreme specificity and ability to work in moderate conditions. However, bioconversion has several different properties. The positive points are that bioconversions can involve several chemical steps, and the enzymes involved can be quite unstable, because the cell continuously remakes them as they break down. The problem with bioconversion is that most bacteria either convert chemicals very inefficiently, in which case they are not much use to the biotechnologist, or they convert them very efficiently into more bacterium, which is also not much use. Thus, to make an effective bioconversion process the bacterial strain has to be optimized so that it converts substrate to product efficiently but does not convert the product into something else.

Bioconversions of a number of chemicals have been reported, and some are used commercially. A major commercial application is in the manufacture of steroids. The 'basic' steroid molecule, often isolated from plants, is itself a very complicated molecule, and not one that it is easy to modify by normal chemical means in order to produce the very specific molecules needed for drug use. However, a variety of bioconversions that attack only specific bits of the molecule can be used. Bioconversion is particularly useful for introducing chemical changes at specific points in large, complex molecules, such as steroids. In many cases bioconversion is used together with more traditional organic chemistry to complete a complex synthesis.

Among the applications of bioconversion are bioremediation and bacterial mining (or biohydrometallurgy), which are discussed elsewhere. Other applications centre around the degradation of compounds that are difficult to attack chemically. A major class of such compounds are the hydrocarbons in oil, which bioconversion seeks to transform into more reactive alcohols and aldehydes. This can be done chemically, but requires extreme conditions and metallic catalysts, and usually results in a complex mixture of products. Bioconversion takes place under much milder conditions, and produces primarily a single product. Bacterial oxidation systems that can convert hydrocarbons to alcohols, aldehydes, or acids are known in many bacteria, such as Pseudomonas oleovorans, which soil bacterium has been the subject of a lot of work to make it more industrially efficient. Pseudomonas species contain a wide variety of plasmids that allow them to break down many organic chemicals, and so they could be of use in bioconversion processes.

Bioconversion in organic solvents. Many chemical reactions that are to be targeted for bioconversion or biotransformation are conventionally carried out in organic solvents, not in water, because the reagents are insoluble in water or because water causes undesirable side reactions. Enzymes can also be used in organic solvents, but interest is growing in the use of bacteria in solvents other than water.

Some bacterial bioconversions can be carried out in mixed phases, where the bacterium is tough enough to survive living next to droplets of the solvent. This has the advantage that a large number of enzymes, or very unstable enzymes that would not survive on their own in a bioreactor, can be used for a bioconversion. The disadvantage is that the bacterium must be kept alive, and that the bacteria produce all sorts of other metabolites in addition to the one you are after.

Biocosmetics. A term for cosmetics with a biotechnological ingredient, or an activity or action based on biological knowledge (rather than cosmetic industry experience or belief, or marketing ploys).

Biocosmetics can be divided into three areas: biomaterials, biologically based ingredients, and medically rationalized products. The last class includes hypoallergenic products and UV-blocking agents, whose mode of action is supported by medical research but that are not 'biotechnological' as such. It also includes products such as liposome based preparations, which may or may not have the effects claimed for them, but whose use of biotechnology products or buzzwords undoubtedly gives them a marketing edge.

Biomaterials include the use in cosmetics of collagen and collagen hydrolysate, a wide range of lipids as 'moisturizers' (including liposomes, which are claimed to carry active ingredients to the skin), fibronectin, and hyaluronic acid. These, and especially the last, are water-retention agents, which are meant to prevent the skin drying and wrinkling. Lipids such as gamma-linolenic acid are also claimed to have anti-inflammatory effects.

Biological ingredients include biotin, cyclodextrins, sphingosine, and a range of pigments. These are all 'natural' products, i.e. are made in a living organism rather than by chemical synthesis, and so can be produced biotechnologically: however, their actual effect is debated by medical professionals.

Cosmetics cannot be too effective since any cosmetic that had a substantial effect on skin physiology would be classed as a drug, and so have to undergo all the rigours proof of efficacy and safety that drugs must undergo. Cosmetics that are deliberately being developed to have drug-like effects are called cosmeceutics. An example is suntan lotion which blocks UV light (rather than just replacing skin lipids). More conventional drug-like activity includes the use of retinoids for reducing skin wrinkling, and hydroxy-acid compounds, called exfoliation agents, which increase the rate at which the skin cells are replaced and so give a more youthful appearance.

Biodegradable materials. Biotechnology has preceded the 'green' bandwagon by some years in developing biodegradable materials. These endeavours generally fall into three areas. 1. Developing organisms that can break down normal materials, especially **plastics**. 2. Developing composite materials. Most 'biodegradable' plastics are composites of a conventional plastic laced with a biodegradable material such as starch. They break down when soil bacteria digest the starch, leaving small granules of the plastic behind. There is controversy over whether this is an improvement, especially since these materials are much weaker than unadulterated plastic, so you need to use much more of them to make bottles and containers of the required strength. 3. **Biopolymers.** Most living things produce polymers to make cell walls or other structural materials. Some of these can be used to make things: however, most are easily wetted, and tend to fall apart if left in the rain for a long time. A few exceptions have been found, the most well-developed of which are polyhydroxybutyrate (PHB), developed by ICI, and polycaprolactone (PCL).

One strong, flexible, water-resistant, biodegradable polymeric material not usually talked about is wood. Quite a lot of plant biotechnology is aimed at trees, and biotechnology is indeed working on genetic engineering of trees. However, growing wood deliberately as an 'industrial' production process is itself controversial.

Biodegradation. The biological destruction of organic compounds.

Biodiversity. This means the diversity of life. Usually, the measure of biodiversity is genetic diversity, a measure of how many distinct genetic species there are in an area or population. These are sometimes called genetic resources, because they can be used as the starting point for new products or processes. Biodiversity is used in two contexts: the diversity of natural organisms, and the diversity of farmed crops or animals.

In the larger world, biodiversity concerns the vast range of plants (and animals, although these are considered less important from a biotechnological point of view) currently alive. Many of them may produce something useful to humans: a new drug, a new foodstuff, a new material. If they are allowed to be wiped out (and most species of plants live in the tropics, which are now under substantial threat) that potential will be lost for ever. There is substantial concern that the biodiversity of 'Third World' (i.e. not rich) countries is being exploited by rich ones.

On the farm, biodiversity is considered to be a 'good thing'. If a country plants only one type of crop (for example), then a pathogen can sweep the entire country's crop from the fields. This happened to US wheat farmers in a wave of epidemics in the 1960s. Thus, not having a country plant just a single type (or cultivar) of a crop is protection against pandemics.

The role of biotechnology in this is double-edged. If biotechnology produces a new, wonder-wheat, then that will be planted in place of all other cultivars, and the wheat-growing world will end up being a monoculture, biodiversity will have been reduced. On the other hand, biotechnological methods are such that if you can transform one cereal with a gene you can transform a lot more, so biotechnology could actually increase biodiversity by increasing the number of crops with desirable genes in them. And it has been argued that the 'green revolution', including biotechnology, has been so successful that farmers are no longer under pressure to grow monocultures of the most productive crop: many farmers in Europe are paid to let fields lie fallow to reduce production, and so would be under pressure to diversify.

On the rain forest question biotechnology is less vociferous, but one of the key technologies in plant biotechnology is plant cloning, storage, and micropropagation, which could be used to store and propagate rare of endangered species.

Bioethics. Bioethics is the branch of ethics, philosophy, and social commentary that deals with the life sciences and their potential impact on society. At one extreme it can be enormously useful in focusing attention on problems that need to be solved. At the other extreme it can become a name-calling exercise between the 'pro-biotechnology' and 'anti-biotechnology' schools of thought which, as it reduces complex arguments to cliches and abuse, makes better sound bites. The Human Genome Project has set aside around 3% of its budget solely to consider ethical issues. The medical genetics and pharmaceutical communities have employed ethical experts for years. Thus the industry and regulators of biotechnology have a substantial concern in bioethics. The more serious side of bioethics is sometimes called ELSI-ethical, legal, and social issues.

Bioethics is not strictly confined to classical ethics, but rather spills over into social policy and even politics. Issues of current concern are almost all to do with how far biotechnology should be encouraged or (as is usually believed to be the case) prevented from doing something. These issues include: The validity of making animal models for human diseases; for example, transgenic models of cancer. (Animal rights as a whole is also a major issue for some biotechnology companies, but is not really a biotechnology-specific issue.) (1) The use and abuse of information about people's genetic make-up. (2) The problem of trading off a new drug's potential side-effects with a need to have patients benefits from it as quickly as possible.

How people should be informed for trials of experimental procedures and products. (This goes beyond drug trials into food safety and environmental issues).

Bioethicists have identified a number of common themes among the arguments that are brought to bear on biotechnological issues. The most controversial is probably the 'yuk factor'. Others include the need for individuals to determine their own fate, the need to protect the vulnerable from the unscrupulous, and so on, which are common to many wider ethical arguments.

There is also a strong component of public opinion in bioethics, although often the reason that people feel some particular way about the technology s not examined.

Biofilms. A biofilm is a layer of microorganisms growing on a surface, in a bed of polymer material which they themselves have made. Biofilms tend to form wherever a surface on which bacterial can grow is exposed to some suitable medium and a supply of bacteria. Thus biofilms form on sites as diverse as domestic plumbing, hyperbolic cooling towers in power stations, sewage plants, and teeth.

The bacteria stick to the surface by a combination of corrosion and glue. The bacterial film is rarely a single type of organism-rather they are communities (or 'consortia') of different organisms. Some corrode the surface. This process, called biocorrosion, can go on, on its own. It leaves the surface rougher and more chemically 'sticky'. Other bacteria synthesize extensive networks of sticky mucopolysaccharide polymers to stick themselves, and any other bacteria nearby, to the surface. The resulting films can be amazingly hard to get off. They also increase the surface's roughness (and so the amount of pumping needed in a pipe, for example), and block the diffusion of oxygen through membranes.

The process of covering surfaces with biofilm in this way is called biofouling. It is a particular problem where liquid is recirculated round a closed loop of pipework (so any bacteria washed off the film can get the chance to stick back on next time round0, or when filter membranes are exposed to bacteria. Unlike normal fouling of membranes by solids or large molecules, biofouling is an active process which, once under way, cannot easily be reversed by cross-filtering or reversing the flow through the membrane. The biocorrosion can also start to break down the membrane, making it leaky. Thus there is a lot of interest in the use of biocides (both in the solution and impregnated into the surface) to stop biofilm production.

Biofouling and **biocorrosion** can affect nearly all materials known. Bob Tatnell of DuPont estimates that about 50% of all metal corrosion world-wide is caused, at least in part, by biocorrosion. However, biofilms can also be used; some biosensors use a biofilm of cells to detect when the water flowing past them contains a poison, and biofilms growing on permeable membranes have been used to break down organic wastes.

Biofilms form rapidly where non-sterile water containing nutrients flows over a surface (the slime forming on stones on the bottom of streams is an example and also shows that, if the water is flowing fast enough, film cannot form). However, biofilm has been seen even when no obvious nutrient was available in ultra-pure water.

Biofilter. A biofilter is a large-scale filtering system where the surface area of the filter provides a support for organisms. Waste fluid flowing through the filter is broken down into harmless products by the organisms. There are three variants.

Biofilters for treatment of waste gases. Gases from food and chemical factories, especially for processes that produce a lot of organic volatiles such as acetone or hydrocarbons, can be adsorbed efficiently by a large filter mest of microorganisms. These have to be kept moist and supplied with nitrogen and trace elements, but otherwise the filter is self-sustaining, with the organisms getting carbon from the gases and oxygen from air. This is often cheaper than conventional chemical scrubbers.

Trickle bed sewage treatment. This uses the same principle, but for treating liquid waste.

Reed beds. A variant on the biofilter idea, reed beds are becoming increasingly used for bioremediation of contaminated water. The reeds, planted in gravel or loose soil, grow from hollow, air-containing roots (rhizomes) and the hollow-stemmed shoots grow from these. Oxygen passes down the plant and out through the root hairs, so the water immediately around the root is highly oxygenated. This supports the growth of lots of bacteria, which can in turn use up the carbon pollutants in high BOD (biological oxygen demand) water. The bacteria can also remove nitrogen and phosphorus, pathogenic bacteria, and some metals and other toxins. The region around the roots is called the rhizosphere (in reed beds and other plant ecosystems too).

Two types of system are used. In horizontal flow systems water flows through a dense reed bed as in a stream, and usually flows just below the surface (because the surface is full of reed stems). This is used to treat relatively lightly contaminated water. More heavily contaminated water is passed through a vertical bed system. Here, water is pumped on to the top of a reed bed based in sand or gravel and air is pumped into the base of the bed. Water is taken out of the base. After a few days such a bed becomes clogged, so it is used in 'shifts' of a few days each to give it time to recover. This is similar to the trickle bed system for sewage treatment, but with reeds as well as bacteria in the bed.

Reed beds are usually used for secondary treatment of farm and municipal waste; they cannot handle raw sewage. They have the advantage of being relatively cheap to build and very 'green' in the literal and political senses.

Biogas. Biogas is the name given to methane ('natural gas') produced by fermenting waste, and particularly sewage. This is an alternative method of waste disposal to landfill or conventional sewage plants. Waste is incubated with suitable bacteria in a digestor in the complete absence of air (an anaerobic fermentor). The organic matter in the waste is converted mainly into methane and carbon dioxide, and the methane can be burned to provide power, heat, etc. In sewage treatment plants using anerobic fermentation, methane is often used as a power source for the plant itself. The process is also called anaerobic digestion. **Anaerohbic sewage** disposal has some advantages over conventional systems (such as the activated sludge system). It produces less microbial mass to get rid of, and does not require aeration (which is expensive because it requires power). However, it only works well on concentrated wastes: solid food waste or sewage sludge. Anaerobic fermentation is rarely a practical option for treatment of raw sewage, which is rather dilute.

The bacteria responsible for generating methane from waste are the methanogenic bacteria, an unusual group which can turn a limited number of carbon substrates into carbon dioxide and methane. To break the waste down into things that the methanogens can' eat requires other bacteria. Thus an anaerobic digestor needs a specialized population of bacteria to work well. In practice, waste digestion processes tend to use whatever organisms are on the waste, with consequently lower efficiency.

Biogenesis. (Gr. *bios*: life + *genesis*: source). The origin of living things from other living things.

Biogeochemical cycles. Movement of elements through living organisms and the physical environment.

Biogeochemistry. The science that deals with the role of living organisms in affecting the chemistry of the earth and its atmosphere.

Biogeography. The scientific study of the geographic distribution of organisms. Ecological biogeography is concerned with the habitats in which organisms live, historical biogeography with the complete geographic ranges of organisms and the historical circumstances that determine the ranges.

Biohlm. A layer of microorganisms attached to a substrate.

Biohydrometallurgy. This is the use of bacteria to perform processes involving metals. It encompasses a wide range of industrial processes, including microbial mining, oil recovery, desulfurization, and a range of physiological processes, including biosorption and the redox metabolism of bacteria. It is also the study of how bacteria corrode metal and metal containing surfaces, a process known as biocorrosion. Biohydrometallurgy includes two broad areas of bacterial activity.

Biosorption. This is the selective absorption of metal ions by bacteria and bacterial materials (such as their isolated cell walls).

Redox reactions. These are reactions where the bacterium uses the metal ion, or a mineral in which the metal is immobilized, for its metabolism. A major use is the oxidation of sulfide to sulfate, which reaction some bacteria use as a source of energy (the reaction releases substantial chemical energy when carried out in air). As sulfides are frequently insoluble, and sulfates are often soluble, this is a fairly general method of releasing the metals in sulfide ores. The same reactions can be used to oxidize sulfides in one compound, releasing sulfuric acid, which then dissolves another compound, or to pre-oxidize a metal ore to make it more amenable to further processing. These topics are covered in separate entires on microbial mining and its subcomponents.

Bacteria can also oxidize or reduce metals themselves. The manganese nodules on the sea floor and the Banded Iron Formation rock strata (laid down 1000 million years ago) are probably the result of bacterial reduction of manganese and oxidation of iron, respectively.

Bioinformatics. This is the use and organization of information of biological, almost always molecular biological, interest. In particular, it is concerned with organizing biomolecular databases, in getting useful information out of such databases, and in integrating information from disparate sources. *Bioinformatics* covers four distinct types of work.

Data acquisition and analysis. This is the realm of LIMS systems.

Database development. The rapid growth of DNA sequencing drove the initial requirement for databases that could store lots of biological data. There is a lot of computer science expertise out there concerning databases, but it needs adaptation to work on the analogue, 'woolly' concepts of biology. In the 1990s DNA sequence databases have been equaled in size by databases of 3D molecular structure, and surpassed by databases from high throughput screening work (HTS) in drug discovery.

There is a debate about the 'curation' of public databases, the extent to which a curator manages and prunes the data. Some databases, such as dbEST (which collects expressed sequence tag sequences), have very little curation-providing the data is entered in the right format, the database managers will include it. Some, such as Swiss Prot (protein sequence information), have substantial curation, with the database 'owner' performing many checks on the data, sending it to experts for additional commentary, and adding their own comments and links to other data. This makes the data entries much more valuable to the user, but slows down the process of releasing the data and means that some data just does not appear. Several databases have got into public trouble with the scientific community for too much or too little curation (often the same databases, sometimes at the same time).

By mid-1996 there were well over 100 biotechnology-related databases available free over the Internet, ranging from DNA sequence, to biomedical literature, to product safety data, to fish farming tips. Many pharmaceutical and biotechnological pharmaceutical companies also have very substantial in-house, private databases of genomic, HTS, or other data. There is also much debase about the 'correct' structure for databases for specific data, which means the format best suited to: (a) efficient storage, (b) efficient search, or (c) efficient analysis. There are dozens of different formats for many types of data. This makes searching the data automatically very tiresome: in practice, scientists use just a small number of databases that they have become familiar with. Most databases are relational (i.e. are based on tables of data and associated tables of links that relate pieces of data to each other) or object-oriented (i.e. treat data as separate 'objects' that have a set of predefined properties). However these are approaches to constructing a database, not prescriptions for them; saying that something is a relational database does not tell you very much about what is stored in it or how it could be used.

Integration and links. Scientists need links between database to point to where related information is stored. This is difficult to do, especially if the database you are pointing to changes its numbering scheme (or even disappears completely). Building stable links and 'pointers' (pieces of information that tell you unambiguously that this entry in my database refers to the same gene or protein as that entry in your database) is a major task of the database curators. A related term is interoperability, which means that data from one database can make sense in the context of another (they both define 'gene' the same way, for example).

Analysis. Once you have got all that data, you want to analyse it. This usually means: **searching** for similar sequence or structural data and comparing the items with each other. The programs BLAST, FASTA, and BLITZ are fast methods for searching DNA or protein sequence databases for things that are similar to a 'query' sequence that you type in predicting structure from sequence. This is a major area of research, and is edging closer to being realistic. Generally, a combination of many methods will give quite good results unless your sequence is like nothing seen before. One expert system-based predictor is even accessible over the World-Wide Web. Nearly all bioinformatics resources were developed as public domain 'freeware' until the early 1990s, and much is still available free over the Internet. Some companies have developed proprietary databases or analytical software. In some cases, they have done so on the back of public domain work,, which has raised scientific hackles: for example, in 1996, the database company MDL started to sell a commercial molecular modelling programme based around RasMol, a popular 'freeware' modelling tool that had been sponsored by drug giant Glaxo. MDL claimed that they had added a lot of functionality to the program. Scientists were not convinced.

Biolistics. This is a method, developed at Cornell University and commercialized by DuPont, to get DNA into cells. The DNA is mixed with small metal particles, usually tungsten, a fraction of a micrometer across. These are then fired into a cell at very high speed. They puncture it and carry the DNA into the cell. In the original system a .22 cartridge was used to drive the particles, hence this is also called a 'particle gun' system.

Biolistics has an advantage over transfection, transduction, etc. because it can be applied to any cell, or indeed to parts of a cell. Thus, biolistics has put DNA into animal, plant, and fungal cells, and into the mitochondria inside cells.

The force to drive the particles into the cells can also be electrical. A spark is used to vaporize a water droplet, which explodes like a small cartridge. This has the advantage that the current, and hence the energy of the explosion, can be varied at will. However, it is more complicated to set up.

As well as getting DNA into isolated cells, biolistic have been used to transfect DNA into animal tissues. Mouse skin and ears have been transfected by a suitably modified biolistic gun in whole, live mice, suggesting that this could be a route to somatic cell gene therapy in humans. The key to getting this to work is to limit the damage the gun like propulsion causes: curiously, this is not caused by the particles themselves, but by the blast of air or gas that accompanies them. The DNA was only active for a few days, however, before the cells broke it up.

Biologic oxidation. Any chemical reaction occurring within a cell that results in a release of energy, respiration. Oxidation results from the addition of oxygen, or the loss of hydrogen and electrons.

Biological containment. This re restricting how far an organism can travel by arranging for their biology to prevent them from growing outside the laboratory. Biological containment can take two forms: making the organism unable to survive in the outside environment, or making the outside environment inhospitable to the organism. The latter is rarely suitable for bacteria, which, in principle, could survive almost anywhere. Thus, for bacteria and yeasts the favoured approach is to mutate the genes in the bug so that they need to have a supply of a nutrient that is usually only found in the laboratory. If they get out, they then cannot grow. Other mutants may have weakened cell walls so that they fall apart outside the laboratory, or may even have 'destructor genes' in them which destroy the cells if the temperature becomes lower or higher than the laboratory optimum. Making the environment unfriendly to the organism is partly biological control, partly a physical one. For example, some of the first genetically engineered rice strains were developed in England (which is too cold to grow rice) and tried in the field in Arizona (where it is too dry). Thus there was no rice growing nearby to cross-pollinate with the genetically engineered rice, and if any rice 'escaped' it would die. This is containment based on the biology of the plant, but without altering the plant specially.

Biological control. Also called biocontrol, this is the control of one species by another which has been deliberately introduced for that purpose. The most famous example is the introduction of myxamatosis into Australia to control rabbits, although biological control is much more ancient, dating back at least to the ancient Chinese, who used pharaohs ants to combat destructive insects in their grain stores. Biotechnology has looked at many potential biological control agents: sometimes these overlap with biopesticides. For example, *Bacillus thuringiensis* produces an anti-lepidopetric (caterpillar-killer) protein. *B. thuringiensis* has been used as a biocontrol agent for many years, but recently biotechnology has isolated the protein responsible to make it into a biopesticide. Half-way house versions have been tested, such as the baculovirus (*Autographica californica*) which contained a scorpion toxin gene-tested on cabbages in the UK in 1993. The virus is still a living entity, but has been engineered to be a pathogen to the caterpillars infesting the cabbages. (This particular experiment raised substantial public opposition, as it was a field trial of a genetically engineered virus, i.e. a deliberate release experiment.). Biotechnologists approach biological control in several ways. Fungi, viruses, or bacteria that are known to attack a pest can be cultured in large amounts and sprayed on to a crop, to kill that particular pest. Entamophagous fungi (fungi that infect insects) are favourites here, as they can infect insects through the cuticle, and so do not have to be eaten to be effective. Such fungi are termed mycoinsecticides, and about a dozen are under scale-up development. Some mycoinsecticide fungi produce short epidemics (called epizooics) among the target pest population without creating a stable presence in the ecology: they can only continue to spread while there is a high density of the insect pathogen around after this they die out. Bacteria, such as *Bacillus popillae* (which attacks the Japanese beetle *Popilla japonicum*) have been used. Insects can also be controlled with nematodes, as is done against sciarid fly and phorid fly, which attack commercial mushroom farms. In essence culturing fungal pathogens is the same as culturing any other fungus, with the additional constraint that the fungi usually require very specific and unusual culture media. Fungi, bacteria, and insects are also considered as weed-control agents: microorganisms to attack northern joint vetch and milkweed vine (weeds of rice and citrus trees, respectively) are in use, and others are under development. Such agents are called mycoherbicides. Biocontrol can also be aimed at pathogenic fungi: Gary Strobel gained some notoriety in 1987 when he inoculated elm trees with a genetically engineered bacterium designed to protect them from Dutch elm disease without obtaining proper Federal approval. Monsanto performed field trials of a bacterial biocontrol agent against the fungus that causes the wheat disease 'take-all' in 1988. Biotechnology has been more visible is producing viral biocontrol agents. Here, genetic engineering and advances in culturing viruses in insect cells have enabled biotechnologists to manipulate insect viruses to be, potentially, more effective biocontrol agents. The aim is to increase or alter the host range of a virus by altering the specificity of the viral proteins that bind to the cell surface, or to increase the virulence of a normally benign but very infectious virus by engineering in toxin gene or the 'pathogenesis' genes from another virus. In practice these aims are rather hard to achieve, since viral infection is a very complex process. Some trials have marked viruses with marker genes so that their spread can be monitored: this gives a measure of how well a simpler form of viral control - growing large amounts of the virus and then spraying it on to the crop - is working. Such field trials have been carried out, most famously is Scotland where pine trees were sprayed with an anti-moth viral biocontrol agent without (as it turned out subsequently) permission being given for release of this genetically modified organism (GMO). The key to any biocontrol programme is to isolate an effective organism, one that can spread rapidly and effectively through the target pest population but that will not spread to other species (and hence become a pest in its own right). As pests are often foreign organisms introduced into an area where they have no natural enemies (e.g. water hyacinths in much of Africa, tumble-weed in the US, Dutch elm disease in many temperate climates), the best source of a potential biocontrol agent is often the original home of the pest. Biotechnology helps in this by providing tools for cultivating strange organisms in the laboratory, and for following them in the environment, even if the organisms themselves are not altered in any way.

Biochemical Oxygen Demand (BOD). Biochemical oxygen demand (BOD) is the most commonly used parameter to define the strength of a municipal or organic industrial wastewater. Widest application is in measuring waste loading to treatment plants and in evaluating the efficiency of such treatment systems. In addition, the BOD test is used to determine the relative oxygen requirements of treated effluents and polluted waters. However, it is of limited value in measuring the actual oxygen demand of surface waters, and extrapolation of test results to actual stream oxygen demands is highly questionable, since the laboratory environment cannot reproduce the physical, chemical, and biological stream conditions.

BOD is by definition the *quantity of oxygen utilized by a mixed population of microorganisms in the aerobic oxidation* (of the organic matter in a sample of wastewater) at a temperature of 20°C. Measured amounts of a wastewater, diluted with prepared water, are placed in 300-ml BOD bottles. The dilution water, containing phosphate buffer (pH 7.2), magnesium sulfate, calcium chloride, and ferric chloride, is saturated with dissolved oxygen. Seed microorganisms are supplied to oxidize the waste organics if sufficient microorganisms are not already present in the wastewater sample. The general biological reaction that takes place in E. 3-12. The wastewater supplies the organic matter (biological food), and dilution water furnishes the dissolved oxygen. The primary reaction is metabolism of the organic matter and uptake of dissolved oxygen by bacteria, releasing carbon dioxide and producing a substantial increase in bacterial population. The secondary reaction results from the oxygen used by the protozoa-consuming bacteria, a predator-prey reaction. Depletion of dissolved oxygen in the test bottle is directly related to the amounts of degradable organic matter. The BOD of a wastewater microorganisms are already present in the sample, recurring are already present in the sample, requiring no outside seed, is calculated using (The standard test has an incubation period of five days at 20°C).

$$\text{Organic matter} \xrightarrow{\text{dissolved oxygen}} CO_2 + \text{Bacterial cells} \xrightarrow{\text{dissolved oxygen}} CO_2 + \text{Protozoal cells} \quad (25.1)$$

$$BOD = \frac{D1 - D2}{P} \quad (25\text{-}2)$$

Where BOD = biochemical oxygen demand, milligrams per litre

D_1 = initial DO of the diluted wastewater sample about 15 min after preparation, milligrams per litre

D_2 = final DO of the diluted wastewater sample after incubation for five days, milligrams per litre

P = decimal fraction of the wastewater sample used

$$BOD = \frac{\text{milliliters of wastewater sample}}{\text{mililiter volume of the BOD bottle}}$$

Comparison of BOD and COD

Biological Oxygen Demand (BOD)	Chemical Oxygen Demand (COD)
It is the amount of oxygen used for biochemical oxidation by microorganism of water. This test has been developed for five days at 20° C. is expressed as ppm of oxygen taken.	It is the amount of oxygen required by organic matter in a sample of water for its oxidation by a strong chemical oxidant, and solution of potassium dichromate in two hours.
BOD value approximates the amount of oxidizable organic matter, and therefore, used as a measure of degree of water pollution and waste strength	This value is a poor measure of strength of organic matter as oxygen is also consumed in the oxidation of inorganic matter such as nitrates, sulphates, reduced metal ions, and also that some organic materials such

Biological Oxygen Demand (BOD)	Chemical Oxygen Demand (COD)
	as benzene, pyridine and few other cyclic organic compounds are not oxidized by this test.
BOD values are useful in process design and loading calculations, measure of treatment efficiency and operation, stream pollution control and in evaluating self purification capacity of the stream	It is very important parameter in management and design of the treatment plants because of its rapidity in determination. Values are taken as basis for calculation of in the standards for discharging efficiency of treatment plants and also figure industrial/domestic effluents in various kinds of waters.
Types of microorganisms, pH, presence of toxins some and nitrification process are the important factors that influence the BOS test.	Presence of toxins and other such unfavourable conditions for the organisms donot affect COD values.

On an average basis, the demand for oxygen, i.e. BOD value, is proportional to the amount of organic waste present in water. The BOD value can be used as a measure of waste strength and also an indicator of degree of pollution. The BOD values are thus very useful in process design and loading calculation as well as the measure of efficiency and operation of treatment plants.

The BOD value is also useful in stream pollution control managements and in evaluating the self-purification capacity of a water body. Along with the BOD, the quantity of oxygen dissolved in a body of water (DO) is indicated by the kind of biotic life, which lives there. When dissolved oxygen is reduced below 4 to 5 ppm of water, fish are scarce. Further reduction of the amount of oxygen results in an increase in anaerobic bacteria. However, BOD values should not be used equivalent to the organic load regardless of the presence of nondegradable organic matter, presence of toxins and local changes in populations of microorganisms. BOD test should be restricted to only suitable wastes in management of treatment plants. However, for other kinds of wastes chemical oxygen demand (COD) values would be more appropriate.

***Stimulation of algal growth*:** Stimulation of algal growth and also a shift in the algal flora to the blue-green algae, leading to production of obnoxious blooms, floating scums or blankets of algae, results into phenomenon known as eutrophication. Most of the bloom forming blue-green algae does not seem to be utilized as food by invertebrates or zooplanktons, thereby, minimizing the predatory control. Biological decomposition of such algal masses in turn leads to further oxygen depletion. In a poorly oxygenated condition, with increased carbon dioxide, fishes and other animals die, and clean river is turned into a stinking drain.

The BOD value is useful in stream pollution control management and in evaluating the self-purification capacity of a water body. Along with the BOD, the quality of oxygen in a body of water (DO) is indicated by the kind of biotic life, which lives there. When dissolved oxygen is reduced below 4 to 5 ppm of water, fishes are scare. Further reduction of the amount of oxygen results in an increase in anaerobic bacteria. However, BOD values should not be used equivalent to the organic load regardless of the presence of non-degradable organic matter, presence of toxins and local changes in populations of microorganisms. BOD test should be restricted to only suitable wastes in management of treatment plants.

***Chemical Oxygen Demand* (COD).** Chemical oxygen demand (COD) is the measure of oxygen required in oxidizing the organic compounds present in water by means of chemical reactions involving oxidizing substances such as potassium dichromate and potassium permanganate. The estimation of COD is of great importance for water having unfavourable conditions for the growth of microorganisms, such as presence of toxic chemicals. In such water BOD can not be determined accurately. However, COD too is not a perfect index of organic compounds present in water because, in this reaction, many inorganic compounds are also oxidized, and at the same time some organic compounds remain unaffected.

There are both physical and chemical conditions that define a healthy environment for aquatic organisms. Among the physical conditions are temperature, pressure, osmotic pressure, light, and turbidity. Although a few organisms can live in hot water, most prefer a moderate temperature. Because there are so many varieties of aquatic life, it is impossible to generalize about an optimum temperature for a healthy aquatic environment. Among bacteria, some prefer warm water, some cold, and others grow best at intermediate temperatures. Biological response modifiers. This is a very general term, usually meaning proteins that affect how the immune system works. In this meaning it is almost synonymous with 'cytokine'. The term is widely used because of the existence of the FDA Biological Response Modifiers Advisory Committee, which oversees biopharmaceuticals that modify biological response mechanisms (i.e. all of them to date). Normally, biological response modifiers act in concert, not as isolated chemical entities. Thus, there has been much agonizing about how the components of biological response modifier drugs, cloned as pure proteins but used in combinations, can be regulated by the drug regulatory agencies, and particularly by the FDA (Federal Drug Administration). Cetus had well-publicized problems trying to get its IL-2 (interleukin-2) approved as an anticancer drug, because IL-2 was not effective on its own. Cetus wanted to use it in combination with other biopharmaceuticals, but as it was not effective enough as an isolated entity it was refused approval.

Biological safety cabinet. A ventilated enclosure that provides a physical barrier between a worker and a hazardous operation. It may be used with an open front (or open glove ports) and a high rate of ventilation away from the operator, like a fume hood, or with a closed front and attached rubber gloves. In the latter use, protection depends upon a negative pressure maintained within the cabinet. The ventilation air exhausts through a high-efficiency filter. A gastight ventilated safety cabinet maintained at negative pressure, in which work is done using attached rubber gloves should be used for highly infectious material. Ultrahigh-efficiency filters on the inlet and exhaust, and provision for exhaust air incincration is required.

Biological spill alarm. A system provided in large infectious disease buildings to warn building occupants that of hazardous material has occurred. Alarm switches are conveniently located throughout the building and give a coded audible signal and actuate a warning light.

Biology. (Gr. *bios*: life + *logos*: discourse). The scientific study of life in all its forms.

Bioluminescence. The production of light by biochemical processes in an organism.

Biomass. The total weight of all the living organisms, or some designated group of living organisms, in a given area. Biomass means the mass of organic material in any bulk biological material, and, by extension, any large mass of biological matter. Fermentations generate biomass as well as the fermentation product, and it is important to control the amount of biomass to maintain the rate of fermentation. The most important thing to measure is the active biomass, which is the amount of living cells in fermentation. However, this can be hard to measure, as a dead cell looks very like a living cell. Biomass products are ones made from the biomass of your process. Usually, in a fermentation system, the biomass is a side product, or even a waste. In brewing, for example, we want the alcohol, not the yeast. Biomass products are made from the mass of cells themselves, and some processes grow plants, fungi, or bacteria specifically for their biomass. Such biomass production is split into several areas of interest.

Single-cell protein (SCP). Algal biomass. Single-celled plants such as *Chlorella* and *Spirulina* are grown commercially in ponds to make food materials. *Spirulina* enjoyed a vogue as a health food a few years ago, because of an unfounded belief that it was extraordinarily nutritious. Like most algae (including some seaweeds) it is quite a good food, but *Spirulina* is not outstanding. *Chorella* is grown commercially to make into fish food: it is fed to zooplankton (microscopic animals), and these in turn are harvested to feed the fish in fish farms. This is a way of converting sunlight into food in a more convenient and controllable way than normal farming.

Biomaterials. Biomaterial is a general term for any biologically derived material that is used for the sake of its material property, rather than because it is a catalyst or a pharmaceutical. Thus, DNA could be a biomaterial if you used it to make paper-clips or cranes, rather than using it to store information. 'Biomaterial' is also sometimes used to mean the material that is compatible with biology, and particularly with medical devices such as inhalers or catheters. This is not the usual use in a biotechnological context. The most common biomaterials are some proteins, many carbohydrates, and some specialized polymers. The proteins considered for biomaterials applications are usually proteins used as structural elements in animals or, occasionally, plants. Collagen, the protein in bone and connective tissues in a wide variety of animals, is a common candidate, and has been used (controversially) as a cosmetic biomaterial, being used as a 'natural filler' for plastic surgery operations. Fibroin, the protein in silk, has been suggested as a protein with sufficient strength to rival nylon, and spider silk could challenge even Kevlar as a structural material. Most of these structural materials have fairly simple amino acid sequences, as they are made of short blocks of amino acids repeated many times. Thus the rigid, central sections of the collagen molecule, which give it its elastic strength, are made mostly of repeats of the three-amino acid unit glycine-X-proline (where X can be one of several amino acids). Biotechnologists have therefore made synthetic proteins with simple repeating patterns in the search for new biomaterials. All these proteins can be produced as amorphous materials by conventional cloning methods, and then spun into ordered fibres. Carbohydrates have been used as structural materials for millennia: the strength of paper and papyrus derives from the properties of the carbohydrate, principally cellulose, components. Biotechnology has produced a wide range of carbohydrates with modified properties, which act as lubricants for biomedical applications or as texture modifying or bulking agents in food manufacture. These include rare but natural materials made from bacteria, such as polydextrose, carbohydrates modified by enzymes to have improved properties, and entirely artificial polymers. Other polymers include 'natural' plastics, such as polyhydroxybutyrate, or rubbers produced by bacteria or fungi. The company Enzymol has developed the enzyme soybean peroxidase to make polyphenol resins, a common industrial plastic. The properties of a polymer that are crucial in determining whether it will make a 'good' biomaterial for a specific application include: **tensile strength** (both elasticity and breaking strength), **hydration** (How much water does it bind? How much does it need to bind to keep its properties?), **visco-elastic properties**, and viscosity.

Biome type. One of a broad category of biomes, such as all of the grasslands of the world taken together.

Biome. A major portion of the living environment of a particular region, such as a fir forest or grassland, characterized by its distinctive vegetation and maintained by local conditions of the climate.

Biomimetic. Literally meaning 'imitating life', this is the area of chemistry that seeks to develop reagents that perform some of the functions of biological molecules. The reason for doing this is that many biological molecules are chemically inconvenient to produce, handle, or apply in large amounts and using cheap processes. By using chemical mimics of such molecules the biotechnologist hopes to achieve a more flexible and more commercially useful way of achieving the same ends.

Biomineralization. This is the deposition of minerals by living organisms. In some applications it is related to microbial mining (the breakup of minerals by microorganisms) and hence is part of biohydrometallurgy. However, biomineralization extends beyond this. There are two general areas of interest to the biotechnologist. Microbial biomineralization. This is the laying down of minerals by microorganisms. If the minerals are deposited inside the bacterial cell, then they are, of necessity, laid down as extremely small crystals or granules. The magnetite made by magnetic bacteria is of this sort; this magnetic mineral is made as tiny inclusion bodies inside some bacteria, which are thereby enabled to swim preferentially along lines of magnetic field. (This enables them to swim towards the bottom of ponds in temperate zones.) Many larger mineral forms are also made partly by bacteria, and this has been suggested as a way in which minerals could be isolated or purified using biotechnology. **Multicellular biomineralization**. In many plants and animals minerals are used to give strength. Thus, vertebrate bone contains calcium phosphate, and many grasses have silica in their leaves to give them a hard cutting edge to dissuade animals from eating them. The control of biomineralization is of substantial interest in several human diseases, notably osteoporosis, a disease in which too much calcium and phosphorus is lost from the bones. Biomineralization is of interest to the materials scientist as well. Biological systems manage to deposit minerals in unusually useful forms. Thus, bone and teeth are much stronger than 'raw' calcium phosphate. Additional strength and specific crystal forms are potentially useful as ways of extending the range of mineral materials available for construction, electronics, and chemical industries. Living things achieve these feats by incorporating specific proteins into the growing mineral, to force crystal growth into the form they require or to reduce the propagate

Binomial System. Taxonomists name biological organisms by using the binomial system of the Swedish botanist Carl von Linne, or Carolus Linnaeus as he is often called. The *Latinized, italicized* name consists of two parts.

The first part, which is capitalized, is the generic name, and the second is the uncapitalized specific epithet or name (e.g., *Escherichia coli*). The specific name is stable; the oldest epithet for a particular organism takes precedence and must be used. In contrast, a generic name can change if the organism is assigned to another genus because of new information.

For example, *Streptococcus pneumoniae* was originally called *Diplococcus pnemoniae*. Often the name is shortened by abbreviating the genus name with a single capital letter, for example *E. coli*. A list of approved bacterial names are published periodically. *Bergey's Manual of Systematic Bacteriology*, has been published in four volumes and contains the currently accepted system of bacterial taxonomy. A new edition of *Bergey's Manual of Determinative Bacteriology*, a single volume indended for us in identifying bacteria, has recently been published in 1994

Biopesticide. A biopesticide is a pesticide, i.e. a compound that kills animal pests, that is based on specific biological effects, and not on broader chemical poisons. Specific types are also called bioinsecticides and biofungicides. Biopesticides are different from biocontrol agents in that biopesticides are not living, and so cannot reproduce themselves in a target population, whereas biocontrol agents are active, living things that seek out the pest to be destroyed. There are a wide range of materials that plants produce to foil pest attack - the caffeine in coffee beans is probably such a material. However, the most attractive for biotechnologists are protein anti-pest materials, such as the much-hyped Bacillus thuringiensis (B.t. sometimes B.t.k.) toxin, which specifically interferes with the absorption of food from the guts of some insects but is harmless to mammals. This protein (which has been used as a pesticide for some time as a bacterial suspension) has been 'cloned' into more amenable bacteria. The gene for the protein has also been inserted into petunia by Calgene, making a plant that is more resistant to pest attack. The rationale behind developing biopesticides, in contrast with conventional pesticides, is threefold. First, they are more likely to be biodegradable than chemical entities that are not normally found in nature. Secondly, they are intended to be more specific (and sometimes, as a consequence, more potent), since they are targeted at specific elements of the pest's metabolism. This is an important aspect in Third World applications, where toxicity to people is a major problem. (It also means that it is not worth stealing them for use as broad-spectrum pesticides for crop protection, which is the fate of some of the more conventional pesticides used to control malaria and sleeping sickness in poor countries.) Thirdly, their targets are less likely to evolve resistance to them. This is speculative, and some resistance of Culex mosquitoes to B.k.t. toxin has been recorded. Biological control agents are sometimes also known as biopesticides. By the end of 1991 there were 45 biopesticides or biocontrol agents aimed at insects (mostly bacteria, bacterially derived proteins, or viruses), 10 at organisms that cause plant diseases, and two at weeds. Biopesticides are sometimes grouped with probiotics, biocontrol agents, and other biologically derived products in being called agrobiologicals.

Biopolymers. This can mean two things. Biopolymers can be made synthetically, by polymerizing a simple biological monomer. A popular example is the hyaluronate polymers (hyaluronic acid is a sugar derivative), which are used in a range of artificial tissues as the base material on which to grow cells, and as a material in advanced wound dressings. Another is the range of dextrans, which have uses ranging from chromatography columns to the fillings in low calorie chocolate bars. The other, related, meaning is a polymer made by a living organism. Most living things produce polymers to make cell walls or other structural materials, or to store energy or food. Cellulose is a biopolymer, but the more common meaning of the word is the plastic or rubber like materials produced by bacteria and fungi. The most well-developed examples are polyhydroxybutyrate (PHB), developed by ICI, and polycaprolactone (PCL). Both materials can be moulded like other plastics, and are water-resistant and water-tight. However, their structures can be attacked slowly by bacteria, and so after a period of months to years will be completely broken down. PHB is one of a class of polymers called polyhydroxalkanoates (PHAs), which also includes polyhydrovaleric acid (PHV). A copolymer of PHB and PHV is produced by fermentation on *Alcaligenes eutrophus*, and sold by Zeneca as Biopol. (The Biopol product line was sold to Monsanto at the end of 1996.) The only problem remaining is what to make out of it. For demonstration purposes, Zeneca has made entirely biodegradable coffin handles; however, this will not alter the waste budget of the Western world significantly. Hundreds of tonnes of PBH are produced annually. It is adapted to a range of applications by mixing it with small amounts of other biodegradable polymers.

Biopreservation. This is usually the preservation of food using biological materials. The leading example is nicin, a bacterial protein that acts as a broad-spectrum antibiotic and which has been approved for food use in Japan. There are a range of other bacteriocins that could be developed, but the main barrier to their use is that they have to undergo severe regulatory tests before manufacturers are allowed to use them in food. Other approaches to biopreservation are:

Probiotics. The growth of bacteria on food can be suppressed by other bacteria. Some experimental products have tried spraying 'good' bacteria on to food to block the growth of 'bad' bacteria. This has been marginally successful, but has not been accepted by regulators as a safe treatment method.

Enzymatic suppression of fermentation effects. Often, some bacterial growth is not harmful in itself, but generates undesirable side products. The smelly amines produced by rotting fish are an example. A biopreservation approach is to use enzymes to break down these products, so that the bacterial growth does not matter.

Bioprospecting. This is searching for new biological resources, usually new plant or microbial strains, that could be sources of natural products and phytopharmaceuticals. There is a substantial argument about who actually owns the resources, between the countries in which these genetic resource reside and the companies who turn them into valuable products. Shaman Pharmaceuticals and Phytogenetics are among the specialist biotechnology companies prospecting in this way. Countries resent having their resources taken and essentially getting nothing in return, and call such collection 'biopiracy'. This is illustrated by the 1995 row over the Neem tree. The US company W.R.Grace was granted a patent for a new process for manufacturing a pesticide extract from this tree, which has been used in India for pest control and traditional medicine for centuries. The Indian government and a range of pressure groups objected furiously, although, in fact, it was the manufacturing process, and not the tree, that was the subject of the patent. Regulation of the use of native biodiversity (a fancy term for things that grow where you live) is the aim of the Convention on Biodiversity (CBD), signed, with various delays, after the Earth Summit conference in Rio de Janeiro in 1992 (as of April 1996, the USA had not ratified the treaty). It provides rules concerning access to biological material, the sustainable use of those resources, and fair distribution of the resulting benefit. It does not seek to say how that benefit should be protected, and so does not address patent or ownership issues directly. The preferred approach is 'bilateralism' (which will no doubt be called 'biolateralism' in time), where the company taking the genetic resources has a long-term commitment to give back some of the value resulting to the originating country. Most companies in the field have agreements of this sort.

Bioreactor control. Very small and simple bioreactors, such as garden composts or 'home brewing', can run successfully without any control on what happens. Any larger process requires control to make sure that the conditions (chemical additions, pH, gas content, temperature control, and so on) remain appropriate for the reaction. Large fermentation systems can have thousands of sensors linked to hundreds of control valves through a substantial computer. Since the exact efficiency of a bioreactor can be critical to whether it is economic to run or not, such control systems are a very important topic in a range of biotechnological areas. Bioreactor control is complex for two reasons. Firstly, most biological systems are themselves complicated. A fermentor growing a microorganism must be monitored to keep the concentration of substrate chemicals, pH, gas levels, temperature, cell mass, and amount of product material within tight boundaries, otherwise, at best, the yield of product will decline, and, at worst, the organisms will die completely. In particular, in continuous bioreactors these conditions must be monitored in real time (i.e. as they happen, also called 'online'), and cannot be monitored offline (i.e. put on a laboratory bench and analysed tomorrow). Secondly, many of the key parameters are hard to measure. The method used to measure something must be sterilizable, because it is going to measure it inside the bioreactor. This means, usually, that it must stand up to autoclaving, live steam, corrosive chemicals, or all three. Thermometers and pH meters can be built to do this quite easily. Chemical and biomass measurements are very much harder to do. A great deal of ingenuity has gone into devising 'non-contact' or 'non-invasive' sensors for bioreactor control. Among the success stories are:

Capacitative and inductive sensors for biomass. Alternating electric fields are affected in a characteristic way by biomass in the liquid between two electrodes. This can be used to find out the amount of biomass there. This does not work with only small amounts of biomass.

Real-time offline sensors. These take a sample out of the bioreactor and measure its properties. Providing the test is fast, this works OK, and the test does not have to be sterilized. However, it is expensive and complicated to engineer, since the sample has to be taken out of the reactor in a way that guarantees that nothing gets in. Many things that a scientist wants to measure about a fermentation, such as the concentration of a particular chemical, simply cannot be measured quickly using any existing technology. They can fall back on chemometrics, which is the use of sophisticated statistical techniques to identify the concentration of one chemical species from the spectral 'signature' of a complicated mixture, or they can use 'surrogate measures', things that behave in the same way as what you are trying to measure, but are not that thing. An example would be to measure cloudiness (turbidity) in a microorganism culture as a surrogate for biomass; this is not actually measuring biomass, but usually comes out with a pretty close approximation. Bioreactor control also needs complex algorithms to make it work. These are among the few applications of artificial intelligence techniques to biotechnology. Expert systems (computer programs that capture human knowledge as a set of formal rules) and neural nets (programs that learn their own rules through 'training') are among the systems that have been used with some success in controlling the more complex bioreactors.

Bioreactor. A bioreactor is a vessel in which a biological reaction or change takes place, usually a fermentation or biotransformation. Bioreactors, and indeed fermentation and biotransformation, are central to much of biotechnology - everything from baking bread to producing genetically engineered interferon takes place in a fermentation, and hence uses a bioreactor. Bioreactor are conventionally divided up into three size classes. Laboratory bioreactors cover bench-top fermentors (up to 3 litres volume) and larger, stand-alone units (up to about 50 litres). These are used for research, and are usually used to create the fermentation process. Pilot plant fermentors are used to scale up a fermentation process, and to optimize it. They typically have a capacity of 50-1000 litres. Pilot plants have to be quite flexible to allow for process optimization. Production units can have any capacity, but usually hold at least 1000 litres, and can go up to the 1000000 litres of the ICI Pruteen plant. Generally, production units are much more specialized than pilot plants, being designed to operate one process with maximum efficiency. There are a number of separate entries about bioreactors. They cover different types of bioreactors: **tank bioreactors** (which is most of them), **immobilized cell bioreactors, hollow fibre and membrane bioreactors, digestors.** Other simpler types of reactor are not covered specifically. These include pond reactors and tower fermentors. Ponds are simply ponds: they are used mainly for growing algae. Often they are called oxidation ponds, because their large surface area allows for more rapid oxygen transfer, and so most of the degradation is by oxidation, rapidly reducing the BOD of waste. Tower reactors are relatively simple towers in which nutrient is injected at the base and product collected at the top. They are used typically for anaerobic fermentation, i.e. fermentation where no air in needed, as, for example, in brewing. Fermentation material is injected at the base of the tower, the flocculent organisms settle at the base of the tower, and the product is collected at the top through a series of baffles which separate foam from bulk liquid. Many of the classifications of bioreactors and fermentors depend on how the materials in them are cycled and recycled. One extreme is the completely mixed bioreactor, in which the contents are mixed uniformly during their passage through the reactor, and then pass en masse to the next stage. Most bioreactors take some of the output, usually some of the biomass, and feed it back into the reactor again. A general type of reactor is the plug flow reactor. Here a substrate flows past a plug of solid support materials, emerging from the end having been changed by the plug. Often the reactor is essentially a pipe, although it is sometimes a flat be. The solid support can contain an enzyme or an organism. This is, in fact, a bioreactor equivalent of column chromatography. Strictly speaking, a plug flow reactor is an unmixed system. In practice, many plug flow reactors have some stirring in the flow mechanism to even out the reaction across the 'plug'. In almost all bioreactors, mixing the contents is important to their efficiency. Often this is achieved through the use of stirring paddles or turbines. However, the inflow of liquid can be used to mix the content, as in the deep jet fermentor, where fluid is injected at high pressure at the base of the fermentor to stir the whole contents. The inflow of gas can be used to stir the fermentor contents too, as in the airlift fermentor or the deep jet fermentor, where an air jet, often starting above the liquid content of the reactor, plunges into the liquid and stirs it without the need for mixers. Other topics covered under fermentation include: **fermentation processes, gas transfer, sensor systems, fermentation substrates** (what the microorganism grows on)

Bioremediation. Bioremediation is the use of biological systems, usually microorganisms, to clean up a contaminated site (environment). Sewage treatment plants perform this in a limited way. Bioremediation covers the use of microorganisms to destroy more toxic materials than are usually found in sewage, and to destroy them in situ, usually in soil or in waste tips. The same approach (although using different technologies and organisms) is used for 'bioscrubbing' - removing waste materials from flue gases. The basic approach to most bioremediation projects is:

Selection of the microorganism. This is best done from soil that has been contaminated with the target chemical for some time. Often, such soil is found near a pipe junction or tank overflow valves in the plant that makes the chemical. Variants of an organism, which grow faster or degrade the target chemical more efficiently, are then created in the laboratory, by combinations of traditional microbial genetics, recombinant DNA methods, or selection. Typically, bioremediation methods use a consortium of organisms, rather than a single organism, that can catalyse the breakdown of different components of a pollutant or can perform different parts of the breakdown of a complex molecule. Even so, some molecules are quite hard to destroy: PCBs can be dechlorinated by obligate anaerobes (bacteria killed by oxygen), and the carbon skeleton broken down by aerobes (organisms needing oxygen); however, clearly these two cannot work together at the same site.

Development of organism physiology. Often, the right organisms are already present at the site, just not working fast enough. In this case, the scientist develops a cocktail of nutrients to boost the rate at which the soil organisms break down the target chemicals, a mixture often called an enhancer. Designs of enhancers are complex, since they must direct the bacterium's metabolism towards digesting the target chemicals, and not just feed the bug.

Inoculation of the environment. The microorganism is introduced into the site, usually with a nutrient mix to support its growth and encourage it to break down the target compound. Oxygen is usually a limiting factor, since most targets for bioremediation are complex hydrocarbon based compounds which must be metabolized by oxidation: nitrogen and phosphorus are also commonly added so that the bacterium's growth is limited by the availability of carbon. Thus, the bacterium is under continued selective pressure to use all the carbon available in the soil for growth, including that present in the target compound. This phase of bioremediation is as critical as identifying a suitable microorganism, and requires a substantial knowledge of microbial physiology and ecology. The main cause of failure of practical bioremediation projects is that the organism selected cannot perform the breakdown at a useful rate at the site, despite performing well in the laboratory. Clays, for example, are particularly poorly suited to bioremediation: because they are very densely packed, water penetrates them very slowly, and air hardly at all. Typical target compounds are chlorinated aromatics (although disposing of PCBs has met with only limited success), vinyl chloride, solvent residues, gasoline fractions, and crude oil. Alpha Environmental has hit the headlines on several occasions with its oil-eating bacterial preparation, used to digest oil spills at sea into soluble molecules that other bacteria can digest. Its most public application was in the Persian Gulf in 1991. This breakdown of compounds into biomass is a type of biodegradation. Other, non-organic, materials can also be metabolized if their end-product is non-toxic or volatile: selenium has been removed from soil by conversion to volatile compounds or elemental selenium, and nitrates have been removed from sewage waste by biological reduction to nitrogen gas for decades. If a target site is very highly contaminated, or too cold or dry for bacteria to flourish, then the soil can be placed in a tank bioreactor and the bioremediation carried our there. These bioreactors are essentially large, insulated tanks into which soil or waste is placed with a bacterial inoculum. Air is blown through the mass to keep it oxygenated. Peter Wilderer at Hamburg has used a biofilm-based tank reactor to remove aromatic hydrocarbons (specifically benzene, toluene, and xylene, the BTX mixture) from landfill site leachate. A film of organisms growing on a permeable membrane was used in order to capture the volatile hydrocarbons from the water.

Biosensors. Biosensors are devices that use a biological element as an intimate part of a sensor. For example, an electrode could have an enzyme immobilized on its surface so that it generated a current or voltage whenever it encountered that enzyme's substrate. There are several classes of biosensor: **ISFET** (ion-sensitive field-effect transistor)-based devices, **physical sensors** (including sensors for mass and thermal sensors), **enzyme electrodes, immobilized cell biosensors, immunosensors, optical biosensors.** Other biosensors use DNA probes as the biological element, or even multicellular organisms such as Daphnia (a small freshwater shrimp) or trout. Biosensors are also classified as potentiometric or ampometric. Ampometric sensors measure the current generated by a reaction, and are operated at as low a voltage as possible. Potentiometric sensors measure the voltage generated by the reaction, and are operated with as little current flowing through them as possible. Biosensors have the potential for being extremely sensitive and specific ways of detecting something. However, their practical application has been hampered by the biological element being very prone to destruction by whatever is meant to be being analysed. Thus, while for commercial applications, a sensor system should be either very cheap and disposable or able to operate continuously for some time, nearly all biosensors are very difficult to make in bulk and last for only a few measurements. The major problems found have been:

Biosorption. Biosorption is the sequestering (i.e. capture from solution) of chemicals (usually metals) by materials of biological origin. Biosorption is widely talked about and little used as a method for removing materials from waste or of purifying rare metals. Many organisms have components that bind metal ions: human bone matrix material, for example, binds strontium rather well. In some cases this is an active process (the organism uses energy to take the metal ions inside and trap them in an insoluble form) in others the process is passive (the metals stick of their own accord to a material that the organism makes). In both cases organisms can be selectec that can accumulate more of the 'target' metal, or that accumulate one metal specifically. For industrial use, bacteria or yeast are almost always the organisms used, although many other organisms, such as protozoa, simple plants, even trees, can accumulate substantial amounts of metals. Among the ways in which organisms actively accumulate metal ions is by precipitating them as phosphates or sulfides by 'pumping' them into special sections of the cell. 'Passive" systems include proteins that bind the metal specifically (metallothioneins, for example; sulfur-containing proteins found in many organisms), lignin (from wood), chitin, chitosan, and some cellulose derivatives. Biosorption is a biological phenomenon, and is interesting for its insight into how organisms cope with metal poisoning, lack of essential nutrient, etc. It can also be adapted to direct industrial use as a purification system, by immobilizing the organisms on filters or in pellets, by using a recycling reactor system that passes the water to be treated through a bed of bacteria in a fermentor, or by extracting the biosorptive material from the organism and using that on its own. This latter option allows non-microbial biosorption system: chitin, for example, absorbs a number of metal ions and is produced from waste prawn shells. One of the most common waste removal targets is removal of heavy metals from industrial waste water, especially nuclear waste streams, where the metals are present at low concentrations but are the most hazardous element in the water. There is also substantial interest in using biosorption to purify precious metals such as silver and gold from very low grade ores by washing the metal the metal out of the ore and then concentrating it from the leachate using biosorption. To be useful, biosorption must be specific and efficient. For metal removal from waste streams, removal must be at least 90% efficient to be of any industrial use, and the organisms or polymers must be able to remove at least 15% of their own weight in metal. Any less efficient system costs more to use than conventional removal systems (such as 'ion-exchange materials). The efficiencies for precious metal extraction can be lower, depending on how valuable the metal is, but must be very specific: there is no point purifying gold if you purify a lot of lead along with it. As well as being improved by breeding and selection systems, biosorption can be improved, in principle, by genetic manipulation, by altering the structure of metal-binding proteins such as the metallothioneins, or by altering the enzymes that make other materials such as the chitosans or lignins. However, although it is talked about a lot, biosorption is not understood well enough to make such genetic engineering feasible as yet.

Biostatistics. This is simply statistics applied to biology or biological research results. The term is often used in drug development, where it means using statistics to design a clinical trial properly and analyse the results.

Biosynthesis. The building up of biochemical compounds.

Biota. All of the organisms, including animals, plants, fungi, and microorganisms, found in a given area.

Biotechnology. Biotechnology is the application of knowledge of living systems in order to use those systems or their components for industrial purposes. The word biotechnology was first used by the Hungarian agricultural economist Kark Ereky in 1919, to mean 'all lines of work by which products are produced from raw materials with the aid of living organisms'.

Biotic. Pertaining to any aspect of life, especially to characteristics of entire populations or ecosystems.

Biotin. Biotin, a natural coenzyme, turns up in some unexpected places in biotechnology as a 'label' system. Biotin can be linked on to many different macromolecules by chemical reaction, a process called biotinylation. The protein avidin (usually made from egg white) or its bacterial counterpart streptavidin binds to biotin extremely tightly (far tighter than an antibody binds to its antigen). The avidin can be labeled with an enzyme, a fluorescent group, a coloured bead, etc. This will then seed out and recognize the biotinylated molecules, and will not stick to any others. This can be preferable to trying to link the enzyme, fluorescent tag, or other label on to the target macromolecule directly because: (i) you can get more biotin on to a macromolecule than enzyme molecules, and (ii) the biotin is very stable, and so can be treated with extreme pH, boiled, or irradiated, whereas an enzyme would be destroyed by these conditions.

Biotransformation. Biotransformation is the conversion of one chemical or material into another using a biological catalyst: a near synonym is biocatalysis, and hence the catalyst used is called a biocatalyst. Usually the catalyst is an enzyme, or whole, dead microorganisms that contain an enzyme or several enzymes. The advent of catalytic antibodies and ribozymes will broaden the definition somewhat. Conversion of one material into another using whole living organisms is usually called bioconversion. Biotransformation is one of the largest areas of applied biotechnology (as opposed to research technologies): around 5% by volume of the enzymes used industrially are used for biotransformation (nearly all the rest are used in the food industry, or in detergents). A wide range of materials are made by biotransformation, from commodity items such as high-fructose corn syrup, to speciality chemicals for the pharmaceutical industry. Some biotransformation processes, such as that producing vitamin C, produce thousands of tonnes of product per annum. The advantage of biotransformation over conventional chemistry is the specificity of the enzymes. Reactions can be: **stereospecific**; i.e. they produce only one optical isomer of a chiral compound, or **regiospecific**; i.e. they change only one part of a large and rather homogeneous molecule (analogous to only digging up one particular stretch of a motorway). A key use for biotransformation is in *resolution*. This is a biotransformation that takes a racemic mix of a chiral compound and converts one optical isomer into another compound. This means than conventional chemistry or separation techniques can now take what was a racemic mixture and produce an optically pure compound from it. The success of a biotransformation in making a chiral compound is measured by the enantiomeric excess of the product: the percentage amount by which one of the enantiomers (chiral versions) exceeds the other. The most commonly used biotransformations involve: **acylases** (to resolve chemically synthesized amino acids), **esterases** and lipases (to make a range of esters, lipids, and to resolve fatty acids and alcohols), **beta-lactamases** and penicillin acylas (to make penicillins and cephalosporins), **peptidases** and **proteases** (to make peptides), **steroid-transforming enzymes** (to make steroid derivatives). These are always used in whole organisms, as many enzymes are involved in each biotransformation. Biotransformations involving the use of proteases, glycosidases, and lipases are discussed separately.

Biotroph. A parasitic fungus which obtains its nutrients from living cells. Obligate parasites are characteristically biotrophic.

Biparental. Refers here to the progeny produced out of a cross.

Bipedal locomotion. (L. *bipes*: two-footed). Walking on two feet.

Birth rate. The number of births in a given year in proportion to the population of a given area.

Blastocoel. (Gr. *blastos*: sprout + *koilos*: hollow). The central, hollow cavity of a blastula.

Blastodisc. A disk of cells forming on the surface of a large yolk mass, comparable to a blastula, but occurring in forms in which the massive yolk restricts cleavage to one side of the egg only.

Blastomere. A cell produced by the division of a fertilized egg.

Blastopore. Asexualspore, produced by budding. The opening from the archenteron to the exterior or a gastrula.

Blastula. (Gr. *blastos*: sprout). An early stage in animal embryology; in many species, a hollow sphere of cells surrounding a central cavity.

Blocking antibodies. Antibodies which act to intercept the allergen from the IgE antibody. The allergen is neutralized by the IgE or IgA before it can react with the IgE antibody type. This term is also used to describe nonagglutinating antibodies that may be present in patients with brucellosis.

Blood group antigens. Genetically determined.

Blood products. Originally, these were biopharmaceutical products made from human blood, such as the blood clotting Factor VIII used to treat haemophiliacs. Such extracted products are usually made by a series of filtrations and solvent extractions. The major 'blood products' in this category are:

HSA (human serum albumin). This is the major human blood product by volume, and is used to produce blood substitutes and extenders for transfusion.

Human gamma globulins. These are antibody preparations, and are used medically to give people an extra high level of antibodies (immunoglobulins) when they might be exposed to specific, unusual diseases. The term blood products is also used to refer to biopharmaceuticals that act on blood or the cells that make blood. They are also usually made by those cells, but in such tiny amounts that extracting them from blood itself is impractical. So, they are made by genetic engineering instead. Among the 'blood products' category of biopharmaceuticals are:

Thrombolytics. Drugs such as tissue plasminogen activator (tPA) produced by Genentech, streptokinase, and eminase (made by Smith Kline Beecham). These dissolve blood clots in the arteries and therefore are used as treatments for heart attacks. An enormous clinical trial spreading over several years in the 1990s (GUSTO) showed that tPA gave a slightly better clinical outcome in heart attack patients than did streptokinase: however, it costs ten times as much. The protein hirudin, originally extracted from leeches, is also being developed as an antithrombin compound, which prevents further clotting rather than breaking up clots that already exist,

Clotting agents. Factors VIII and IX to treat haemophilia, a disease where these proteins are missing. Baxter Healthcare and Mile Inc. are developing recombinant Factor VIII.

Erythropoietin (EPO). This stimulates the bone marrow to make more red blood cells, and was the subject of a fierce patent dispute.

GCSF, GMCSF, etc. These are cytokines, materials made by the immune cells to regulate the immune system's function.

Animal blood products, notably fetal and newborn calf serum, are also used in the biotechnology industry: serums are used as a supplement for the media used to culture a range of mammalian cells.

Blood. The fluid circulated by the heart through the circulatory system. Carries oxygen, nutrients, hormones, carbon dioxide, wastes, and other substances.

Blots. A range of molecular biological techniques are called 'blots'. They all share a common appearance. At the start, biological molecules are usually present in a jelly-like matrix, often the result of separation by gel electrophoresis. The contents of the gel are then transferred on to a porous membrane, often a chemical derivative of paper or a nylon mesh. This technique was traditionally done by allowing liquid to seep through the gel, through the membrane, and into a pile of paper towels, which acted like blotting paper - the biomolecules travelled with the fluid until they stuck to the membrane.

Electroblotting (which uses an electric field to pull the molecules out of the gel) and vacuum blotting (which uses suction) are also used. Once on the membrane the molecules can be analysed by techniques that would not work in the original gel, such as antibody staining or DNA hybridization. The variations on this theme depend on the molecules:

Southern blot. Named after Professor Ed Southern, the gel here is a DNA electrophoresis system and so the molecules transferred are DNA molecules.

Northern blot. Almost the same as the Southern blot, but the molecules are RNA.

Western blot. Here the molecules are protein, also separated by gel electrophoresis.

Southwestern blot. This is a variant of the Southern blot used to find protein molecules that stick to DNA molecules. (Desperate attempts to get something (anything) named as the Eastern blot have not been generally successful).

Dot blot. Here DNA, RNA or protein is dotted directly on to the membrane support, so that they form discrete spots. Also slot blots (if the spots are not found).

Colony blot. Here the molecules (usually DNA) are from colonies of bacteria or yeast growing on a bacteriological plate. A variation (called the plaque lift) can also be used for viruses. With the advent of PCR and the fall in interest in Southern and Northern blots, the most commonly seen blot is now the Western blot, where proteins are separated according to size and then identified by reacting then with an antibody.

Bock beer. A dark-colored lager beer brewed with more malt and hops than light colored lager beers.

Bohr effect (boar). The reduction in affinity of hemoglobin for oxygen caused by acidic conditions, usually as a result in increased CO_2.

Boil. A localized abscess resulting from infection of a hair follicle, also called a furuncle.

Booster dose. Additional injection of an antigen to maintain antibody production at its peak.

Bordetella pertussis. Bordet-Gengou bacillus; gram-negative rod, the causative agent of whooping cough.

Borrella vincentii. A spirochete found in the mouth in cases of Vincent's angina. It occurs together with Fusobacterium fusiforme.

Botulism. Food poisoning, caused by the exotoxin of Clostridium botulinum.

Brachiation. (L. *bracchium*: arm). Movement by grasping branches above the head and swinging along beneath them.

Break tank. A tank that provides an air space in a water supply line in such a manner as to prevent reversal of water flow in case the supply pressure falls below the downstream pressure. It is considered more positive than the Backflow preventer or Vacuum breaker.

Breed count. Direct microscopic count of bacteria in milk.

Brewing. The brewing industry (and the associated distilled spirit industry) is both a major source of process biotechnology and a user of new techniques. Brewing and distillation is carried out on a large industrial scale. Wine production has been affected less by biotechnology than beer brewing, remaining a 'cottage industry' in many European countries. The two most common types of yeast used in beer brewing (*Saccharomyces carlsbergensis* and *Saccharomyces cerevisiae*) are mainstays of both molecular genetic research and biotechnological production processes. Brewing still goes through the same basic processes defined in pre Roman times, with correspondingly unscientific terms.

Malting. Barley is incubated in damp, warm conditions to make it sprout. The seeds make substantial amylase, which breaks the starch into simple sugars that yeasts can metabolize.

Mashing. After addition of hops (primarily for flavour), the plant matter is mechanically fragmented into a uniform suspension, a mash. The mash is then heated up to speed up amylase action. The result is called wort, which is filtered to remove most of the solids. Kettle boil heats up the ingredients to inactivate enzymes in the wort, precipitates a number of high molecular weight impurities (including a lot of proteins), and concentrates the sugars. It is then ready to be fermented.

Yeast performs fermentation. The culture in traditional brewing is not aseptic. Boiling kills off the endogenous microorganisms, and the yeast outgrows any newcomers, suppressing their growth by ethanol production.

Clarification. This is the removal of proteins and other high molecular weight material that can cause 'haze' in the beer. This material could be removed by filtering or centrifugation, but the particles are extremely small, and so these processes would be expensive. So they are usually removed by precipitation (started by adding a polymer that binds to them, making them stick together) or cooling (which can alter their chemistry, again making them sticky).

Potable alcohol is also made in many other fermentations. Many cultures produce low-grade fermentations from grains such as wheat or rice, or from potatoes, swede, or any other cheap carbohydrate source, and then distil the result. The fermentation processes are similar to those in beer. After the fermentation step, the entire broth is distilled to separate an alcohol-containing liquid from the yeast and remaining solids from the grain. In fact, the liquid distilled off is usually almost pure ethanol and water, with little flavour and a kick like a mule. In whisky distillation, this is called grain spirit or grain neutral spirit. Some drinks, such as vodka and tequila, use the primary distillate as a drink. Others make it more potable by 'ageing' it to diffuse woody flavours into it (whisky), or using it to extract the flavour from juniper berries (gin).

The solid material left after distillation of the alcohol from grain-based fermentations is called 'distillers grains', and is often used as an animal feed component. 'Dark grains' is the same thing, but with some of the spent liquor still absorbed into it.

New biotechnologies are applied to all aspects of this process. The use of recombinant yeasts with enzymes that degrade more of the sugars (producing 'lite' beers with little sugar left), or which can be separated more easily, is particularly high profile.

Bromthymol blue. A dye used as a pH indicator; pH range is 6.0-7.6.

Brownian movement. Random movement of bacterial cells and small particles due to bombardment by water molecules.

Brucella species. Small, gram-negative, coccoid to rod-shaped cells; causative agents of brucellosis in animals and humans.

Brucellergen. Concentrated culture medium after growth of Brucella; used to elicit skin reaction in diagnosis of brucellosis.

BST (bovine somatotropin). Bovine somatotropin is also called bovine growth hormone. The protein hormone is found naturally in cattle, and is the counterpart of human growth hormone, one of the earliest biopharmaceutical products. It has been cloned, expressed in large amounts, and is being marketed by Monsanto as an agricultural product to improve the growth rate and protein: fat ratios in farm cattle, and to improve milk yields.

There are animal welfare concerns about this, and health concerns about the possibility that BST will get into the milk or meat, and hence into people. In particular, the possibility that BST given to improve milk yield will end up in the milk given to children has proven a powerful weapon against Monsanto, one of the principal developers of BST for agricultural use. Monsanto has also been accused of reducing cows to unhappy milk-producing machines, as if 5000 years of breeding had not done that already. The debate has become highly polarized, with contenders on both sides seeing it as a trial case for the application of biotechnology to agriculture and food industries. The USSR, Czechoslovakia, Bulgaria, South Africa, Mexico, and Brazil have approved BST, but in many other countries the debate on its safety is holding up any approval. There is also a debate over whether BST will offer the consumer any advantage, especially in Europe where there is often a surplus of milk over the European Community's (EC) 'quota' for production. It would, however, allow the same amount of milk to be produced from fewer cows using less food.

Buboes. Enlarged lymph nodes seen in bubonic plague and lymphogranuloma venereum.

Bud primordium. (L. *primordium*: the beginning). In plants, a small mass of potentially meristematic tissue found in the angle between the leaf stalk and the shoot apex. Will give rise to a lateral branch under appropriate conditions.

Budding. An asexual reproductive process in which the parent cell retains its identity while the offspring (bud) is produced with a newly synthesized cell wall attached to parent cell; also a process in which an enveloped virus is released through the plasma membrane of a vertebrate cell.

Buffer. A substance containing the salt of a weak acid or a weak base which tends to resist pH change when either acid or alkali is added.

Buffering. A process by which a system resists change-particularly in pH, in which case added acid or base is partially converted to another form.

Bulk starter method. Starting with a purchased mother culture, the progressive inoculation and growth of starter cultures in increasing volumes of milk culminating with the inoculating of the cheese fermentation vessel.

Bursa of fabricius. A lymphoepithelial organ found at the junction of the hind gut and cloaca in birds which is the site of B cell maturation.

Burst size. The number of phage particles released from a bacterium after maturation of the phage. The burst size is usually in the range of 50 to 200 new bacteriophages.

Burst time. The number of minutes required from attachment of the phage to its bacterial host until the liberation of mature phages. The average burst time is between 20 and 40 minutes.

C

C_3 photosynthesis. The form of photosynthesis in which 3-phosphoglycerate is the first stable product, and ribulose bisphosphate is the CO_2 receptor.

C3b inactivator. A factor of the complement system recently renamed Factor 1.

C_4 metabolism. A strategy adopted by some higher plants which enables them to fix CO_2 into organic matter even when the CO_2 concentration is very low. Essentially a CO_2-concentrating mechanism, it is used in tandem with (and not in place of) the Calvin cycle.

C_4 photosynthesis. The form of photosynthesis in which oxaloacetate is the first stable product, and phosphoenolpyruvate is the CO_2 acceptor. C_4 plants also perform the reactions of C_3 photosynthesis.

Cabernet sauvignon. A variety of purple grape used in the production of high-quality red wines.

Cabinet array. A number of absolute barrier biological safety cabinets joined together. An array may be divided into two or more Cabinet systems by gastight doors or fixed partitions.

Cabinet system. A number of absolute barrier biological safety cabinets joined together to provide a single space, with a single inlet and exhaust for ventilation.

Caecum. (L. *caecus*: blind). A blind branch off the large intestine. In many nonruminant mammals, the caecum contains a colony of microorganisms that contribute to the digestion of food.

Callus. (L. *calleo*: thick-skinned). In plants, wound tissue, or relatively undifferentiated proliferating cell mass, frequently maintained in cell culture.

Calorie. (L. *calor*: heat). The amount of heat required to raise the temperature of one gram of water by one degree Celsius (1° C) from 14.5° C to 15.5° C. In nutrition studies, "calorie" refers to the kilocalorie (1 kcal=1000 cal), the amount of heat required to raise the temperature of one kilogram of water by 1° C.

Calvin cycle. The pathway which reduces CO_2 to carbohydrate (CH_2O) in the stroma of chloroplasts an the cytosol of autotrophic bacteria. Eukaryotes and blue-greens use $NADPH_2$ as a hydrogen donor; other auto-trophs use $NADH_2$. Some sulphur bacteria supplement the Calvin cycle by forcing the Krebs cycle to run backwards using reduced ferredoxin.

Calyptra. (Gr. *kalyptra*: covering for the head). A hood or cap found partially covering the apex of the sporophyte capsule in many moss species, formed from the expanded wall and neck of the archegonium.

cAMP (cyclic AMP). A compound, formed from ATP, that mediates the effects of numerous animal hormones. Also needed for the transcription of catabolite-repressible operons in bacteria. Used for communication by cellular slime molds.

Candida albicans. Yeast-like fungus; a member of the normal flora of humans, and associated with certain infections, e.g., thrush.

Canker. A disease that kills definite and relatively localized areas of bark on branches or trunks of trees.

Cannula. An artificial tube used for insertion into the body, for example, an artery.

Capillaries. (Gr. *capillaris*: hair). Very small tubes, especially the smallest blood-carrying vessels of animals between the termination of the arteries and the beginning of the veins.

Capillary zone electrophoresis. Also called simply capillary electorphoresis, this is an up-and-coming technique in many biochemical and biotechnological fields.

Gel electrophoresis is electrophoresis moving molecules around using electric fields, performed in a polymer material. The polymer does two things: it sieves the molecules by size, and it stabilizes the solution in which the electrophoresis is happening. Without it, any slight vibration or convection would stir all the molecules up, and the system's ability to separate very similar molecules (its resolution) would go down dramatically. However, the separation is a complex result of the molecule's shape, size, charge, and how it interacts with the polymer gel. This complexity can itself reduce the system's resolution. Electrophoresis without gels has been used. It is called free zone electrophoresis, and uses a flow of water, or sometimes a density gradient column (such density gradients are discussed further under centrifugation). However, the stirring effects in these systems can be substantial.

Capillary electrophoresis performs free zone electrophoresis is a very fine capillary tube (a tube with an internal diameter of less than 1 mm). Here, stirring effects undoubtedly occur, but they only stir up volumes of solution less than the tube diameter (i.e. less than 1 mm), and so the effect on resolution is very small. The electrophoresis can be 'run' much faster than conventional electrophoresis, where making the molecules go faster means putting a higher voltage across the gel slab, which means that more current flows through the gel, more heat is produced in the gel, and ultimately the biological molecules denature (or the gel tank cracks or bursts into flames). The mass of liquid in a capillary tube is so small that even very high voltages produce tiny currents, and the heat produced can be radiated away from the tube rapidly. So, the electrophoresis can either be run very fast or be run on a very long capillary tube, so increasing resolution. Capillary electrophoresis sometimes relies on electroendosmosis. This is the movement of a solution of ions in an electric field, when they are carried in a charged tube. It can be seen when salt solutions are exposed to an electric voltage in a charge gel (like agarose), or in a very small tube with charged walls (like silica). This latter is what happens in capillary electrophoresis, so that as well as the sample being pulled through the tube by the electric field, the whole liquid column in the tube is pulled through the tube by the same field. The direction of the flow depends on the charge on the tube walls, which itself depends on the chemical nature of the walls and the solution. Silica usually has a negative charge, but this can be 'converted' to a positive charge by coating the walls with positively charged detergent molecules.

Capillary electrophoresis system can also perform capillary gel electrophoresis. This uses a gel inside the tube. It has the same advantages of high power and good resolution, and separates molecules in a way more similar to other, conventional electrophoresis techniques. Capillary gel electrophoresis is gaining acceptance as a tool for DNA sequencing.

Capping. A process by which cell surface molecules are caused to aggregate (usually using antibody) on the cell membrane.

Capsomere. The individual protein units which comprise the capsid of viruses, particularly of the cuboidal ones.

Capsule. A mucilaginous envelope surrounding certain microorganisms, external to the cell wall; frequently consists of polysaccharide.

Carbohydrates. Organic compounds with the general formula CnH_2m Om. Common examples are sugars, starch, and cellulose.

Carbon budget. The amount of atmospheric carbon (from carbon dioxide) incorporated into organic molecules by a plant.

Carboxylic acid. An organic acid containing the carboxyl group, -COOH, which dissociates to the carboxy-late ion -COO.

Carbuncle. Enlarged skin abscess.

Carcinogen. An agent (chemical, physical, or biological) that can cause malignant tumor formation.

Carcinogenesis. The process of establishment of cancer.

Cardiac. (Gr. *kardia*: heart). Pertaining to the heart and its functions.

Caries. A disease of the calcified tissue of teethy tooth decay.

Carnivore. (L. *carn*: flesh + *vorare*: to devour). An organism that feeds on animal tissue. (Contrast with detritivore, herbivore, omnivore.)

Carotenoid. (L. *carota*: carrot). A yellow, orange, or red lipid pigment commonly found as an accessory pigment in photosynthesis; also found in fungi.

Carrier. A person who harbors and spreads disease-producing microorganisms without having the overt symptoms of the disease.

Carrier. An immunogenic molecule, or part of a molecule, which is recognized by T cells in an antibody response.

Carrying capacity. In ecology, the largest number of organisms of a particular species that can be maintained indefinitely in a given part of the environment.

Cartilage. In vertebrates, a tough connective tissue found in joints, the outer ear, and elsewhere.

Caseation necrosis. A type of necrosis in which the tissue resembles an amorphous mass of cheese.

Caseation. The process by which a tubercular lesion develops into a coagulated cheesy mass.

Casein. The principle protein (phosphoprotein) of milk; converted to calcium caseinate by renin, leaving the clear residual fluid, whey.

Casella slit sampler. A solid medium impactor consisting of a chamber in which an agar plate revolves beneath one or more slits. Tow models of the Casella sampler are commercially available, one that collects at 28 liters per minute, and one that collects at 700 liters per minute. Both models operate at a choice of three rotation speeds for the agar plate. One revolution per « minute, per two minutes, and per five minutes.

Cassava. Any of several plants of the spurge family grown in the tropics for their fleshy edible rootstocks, which yield a nutritious starch.

Caste. In social insects, any set of individuals of a particular anatomical type, or age group, or both, that performs specialized labor in the colony.

Catabolism. Negative metabolic balance; destructive metabolism; the breakdown of complex compounds, with the liberation of energy.

Catabolite repression. The decreased synthesis of many enzymes that tend to provide glucose for a cell; caused by the presence of excellent carbon sources, particularly glucose.

Catalase. An enzyme which catalyzes the breakdown of hydrogen perioxide to yield water and oxygen.

Catalyst. (Gr. *kata-*, implying the breaking down of a compounds). A chemical substance that accelerates a reaction without itself being consumed in the overall course of reaction. Catalysts lower the activation energy of a reaction. Enzymes are biological catalysts.

Catalytic antibodies can also work through reducing the entropy of reaction, i.e. bringing together two molecules in the right orientation to allow them to react. This can apply to two substrates for a reaction, or a substrate and a cofactor. Catalytic antibodies have been made that catalyse reactions through both these mechanisms. ('Entropy' in this case is chemical entropy, i.e. disorder. Two molecules aligned exactly right for a reaction represent a very ordered system - it is much more likely that they will collide in some unsuitable way, or, indeed, not collide at all. Thus the reaction has a high 'entropy barrier', which the catalytic antibody reduces by making the system more ordered - it brings the two reactants together in the correct way to react.)

Catalytic Antibodies, also called abzymes, are antibodies whose binding sites catalyse a reaction in their target 'antigen', rather than just passively binding to it. Antibodies do not normally possess catalytic activity.

Catheter. A tubular device used to withdraw fluids from, or introduce them into, body cavities.

Caudal. (L. *cauda*: tail). Pertaining to the tail, or to the posterior part of the body.

cDNA. cDNA is copy DNA (or complementary DNA). It is a DNA copy of an RNA, and is made from the RNA using reverse transcriptase. This is a gene cloning technology. There are two potential reasons for wanting to do this. **First**, the DNA gene itself may be unknown. In this case, the cDNA is made from a messenger RNA that codes for a known protein (or for a protein whose activity can be measured, by antibody reaction or because it is an enzyme). Then the DNA gene can be found using the cDNA as a *probe*. Secondly, the scientist might not want the 'original' gene. This is especially true if the purpose of cloning the gene is to express it in a bacterium. In this case, the scientist wants a section of DNA that codes for the protein concerned and nothing else. They do not want introns, other neighbouring genes, and so on in the gene clone. The cDNA is a much better approximation of this, consisting (in eukaryotes anyway) of a single mRNA with no introns that code for a single protein. Often, the cDNA can be spliced directly into an expression vector and some expression made to happen in bacteria. cDNA hit the headlines at the end of 1991 when Craig Venter at the US National Institute of Health (NIH) filed a patent claiming 337 new cDNA sequences that he had discovered using an automated DNA sequencer. The NIH's idea was to patent them as Venter churns the sequence out so that, if at any time in the future someone finds a use for those sequences, then the NIH can claim royalties on them. The Medical Research Council in the UK took the route of keeping its cDNA sequences generated by large-scale sequencing a secret, until the legality and acceptability of patenting cDNAs is established. It seems unlikely that the DNA patent will hold up, as Venter admits that he has no idea what the cDNAs do or what use the knowledge of their sequence could possibly be put to. cDNA is also made when you want to use PCR to amplify and RNA. PCR does not work on RNA, so you copy it to DNA using reverse transcriptase and then PCR amplify the cDNA. This is called RT-PCR (reverse transcriptase PCR).

Cell adhesion molecules (CAM). Also called ICAMs (intracellular adhesion molecules). These molecules are present on a wide range of human cells, and are part of the mechanism used by cells to recognize each other. They are glycoproteins, and the sugar residues can be crucial to their function: the differences between some blood groups, for example, are the result of variation in the sugar residues on some ICAM molecules. There are several families of mammalian cell adhesion molecules, with many variants of each family of each family being seen in different cell and disease states. The major families are:

Integrins. These are cell-surface receptors for a range of other cell adhesion molecules, and for molecules found in the intracellular matrix (which is the material that mechanically supports cells, and is secreted by them). Integrins bind to proteins with the three amino acid sequence RGD in them.

Immunoglobulin superfamily. These are molecules with a molecular similarity to the antibody (immunoglobulin) molecules. They include ICAM-1, VCAM-1, LFA-1, and Mac-1 (found on leucocytes only).

Selectins. They bind to carbohydrate cell surface molecules, and are important in the initiation of blood clots and inflammatory responses, among other processes.

Fibronectin is another of the cell surface molecules, which, in this case, is a glycoprotein that links the cell to other proteins in the intercellular matrix, notably to collagen. It is also sometimes called large external transformation-sensitive protein (LETS), because it is often absent from cancer cells.

Cell adhesion molecules are important to drug companies, and hence to biotechnology companies, because they are the molecules through which cells control their interactions with each other. There has been particular focus on ELAMs (endothelial lymphocyte adhesion molecules), because of their role in immune disease. These are proteins on the surfaces of lymphocytes and the endothelial cells (flat cells) that line the surfaces of blood vessels. During inflammation, white blood cells leave the blood and invade damaged tissue to engulf any invading organisms. They also release a range of chemicals that cause inflammation of the tissue. The invasion is partly controlled by the ELAMs, which allow lymphocytes to stick to and recognize the endothelial cells. Modulating this interaction, then, is a potential route to controlling inflammatory diseases. Cell adhesion molecules are important in many other processes in growth, development, and regeneration, and so are a hot research topic. They can also be the receptor that a virus binds to gain entry to the cells it infects. Research into cures for viral diseases can also be concerned with CAMs. The best known example is the CD4 receptor, a member of the immunoglobulin super family of receptors, which is the target for the AIDS virus.

Cell culture. This is the cultivation of cells from a multicellular organism, usually mammalian cells. The term 'tissue culture' is often used as a synonym, which strictly it is not: however, the intuitive nature of cell culture science does not lend itself to such pedantry.

Cell culture for production. This means the culture of mammalian cells (or occasionally the culture of isolated plant cells, as opposed to plant calluses), as opposed to bacterial or yeast cells, which is called fermentation. Cell culture is a difficult process, requiring sterility and careful control of the conditions. Large-scale cell culture has to take several additional factors into account.

Fermenter design. Confusingly, large-scale cell culture can take place in large vessels called fermentors or bioreactors. As mammalian cells are quite fragile, the usual stirring methods will destroy them, so gentle methods are used. For very large scales, airlift fermentation is established. Hollow fibre methods are increasingly used, not only for pilot scale, but also for smaller production runs.

Small-scale systems. The favourite small-scale system is the roller bottle. This can be used in a research laboratory, and scaled up to a dozen litres simply by adding more bottles. Roller bottle culture is often used to 'grow up' cells for use in a larger fermentor. Mammalian cells can usually only grow in relatively crowded conditions, so you cannot seed a 100-litre fermentor with a few cells and expect them to grow. You have to seed it with the output of a 10-litre fermentor.

Cell cycle. Cells grow in a carefully regulated cycle of events. In mammalian cells, the stages of the cycle are quite distinct. They are: G1, when the cell is making all the components for new cell, except DNA; S phase, when the DNA is replicated; G2, when the cell organizes itself to divide; and M phase (or mitosis), when the cell actually divides in two. M phase is the phase when DNA is organized into macroscopic chromosomes, which can be seen under the microscope. During G1, S, and G2 the DNA is contained in the nucleus, which looks like a relatively undifferentiated blob of DNA, RNA, and protein under the light microscope. Normal cellular metabolism is temporarily suspended during M phase, and in many cases cells that are usually flat, extended, and attached to the support they are growing on become much rounder and much less firmly attached. Cells in mitosis can be stained and examined under a microscope to count the number of chromosomes they have, a rough indication that they are the cells you thought they were: such a chromosome count is called a karyotype. In cell culture, cells are at a mixture of stages in the cell cycle. There are a range of chemical tricks to arrest them at one point, so that all the cells accumulate at, for example, the start of S phase. If you then release the cells, they all march through the cell cycle together: this process is called synchronization, and the result is a synchronized cell population. There is a lot known about the control of the cell cycle, which is important for the development of drugs that target abnormal proliferation of cells, such as cancer. The position of cells in the cell cycle is often one of the measures used to characterize cells. The position of individual cells is measured in a flow cytometer measuring the amount of DNA in the cell. If the amount of DNA in the cell after mitosis is 1 n, then cells in Gl have 1 n DNA, the cells in G2 have 2 n, and those in S phase have an intermediate amount of DNA. (Mitosis is so short in comparison to the other phases - minutes as opposed to hours - that it can be ignored here.) The amount of DNA is measured by staining the cells with a dye that binds to DNA and is flrorescent. In general, cells spend most of their time in G1. When the replication rate of mammalian cells speed up, it is usually because Gl gets shorter. FACS machines are also sometimes used to measure DNA content of individual cells. The FACS (fluorescent activated cell shorter, pronounced 'fax') separates cells according to their fluorescence. Cells are suspended in individual droplets, squirted past a laser that excites any fluorescent dye in them and past a detector that measures the fluorescence, and then into one of several receivers into which they are steered using electric fields. The detectors can detect one or two different colours of fluorescence (the latter being called two-colour FACS). Usually, one of the colours is provided by a flrorescently labelled antibody. Sometimes the other is a DNA-binding dye. Thus the experimenter can select for cells that are expressing a particular cell surface antigen (detected by an antibody) and are in S phase (detected by the dye). FACS machines are widely used in immunology, where they are used to identify and select populations of cells bearing specific combinations of cell surface markers. In this case, the cell cycle position is irrelevant, and is not measured. Some cells do not divide at all. These are usually said to be in G0, effectively a dead-end sideline off the main cycle at G1. After mitosis, cells that are not going to divide again go into this distinct G0 state and stay there as long as their developmental programme dictates. If they are dividing, they must first switch back to G1 before going on to S phase. This involves up-regulation of some aspects of cell metabolism, and switching on the signalling apparatus that tells a cell when to divide.

Cell disruption. Many fermentation processes produce products that are inside cells. Examples are many proteins produced by genetic engineering, enzymes, and large molecules such as the biodegradable plastic polyhydroxybutyrate. It is necessary to break the cells to get these products out. This process is called cell disruption. It is easy to break up mammalian cells, which spontaneously break open (lyse) if you put them in solutions of low osmotic strength, e.g. pure water. However yeast and bacterial cells are specifically 'designed' by evolution to be unbreakable. Thus, a lot of energy has to be put into breaking them, and there is the risk that energy will also disrupt the product inside the cell. Plant cells in culture are not usually as tough as yeasts, but are much stronger than mammalian cells. Methods used are:

Autolysis. This simply alters the conditions so that the cell digests itself. This is the simplest possible method, but tends to be useless for protein products since the cell digests itself from the inside out, so breaking down the product before breaking down the cell wall.

Enzyme action. This is very effective; the cells are treated with an enzyme that dissolves some key component of their cell wall, which then simply falls to bits. Typical enzymes used are lysozyme (for bacteria), chitinase or glucanase (for yeasts), and cellulase (for plant cells).

Detergents, alkali, osmotic shock (i.e. pure water), plasmolysis (treatment with high concentrations of salt), organic solvents. Any of these treatments will knock holes in the plasma membrane, the thin layer of lipids inside a cell wall which actually holds the cell's contents in (the cell wall, by contrast, is meant to keep the outside world out). If the product is small enough (as is true of small proteins and of metabolites) or if there is no cell wall (as is true of animal cells) then the product leaks out.

Freeze-thaw. freezing and thawing can break up any structure as ice crystals form inside the wet materials of which cells are made.

Mechanical methods. The most obvious method is to break the cells mechanically. There are many ways of doing this: **French press**, which forces the cells through a very small hole at high pressure. A large scale version of this is called a Manton Goulin homogenizer. **Mills**, in which the cells are shaken vigorously with abrasives, or metal balls (a ball mill), or rods (a cylinder mill). **Blenders.** Traditionally the laboratory uses a blender called the Waring blender (named after a 1930s New York dance band leader who invented or popularized it for making cocktails), but essentially this is a food processor with a powerful motor. Many of the mechanical techniques are direct translations from larger industrial processes. **Milling**, for example, is used in ore processing in the metals industry. A number of cell disruption techniques produce cells that are lysed, i.e.

broken open, but not otherwise disrupted. These cell suspensions can be extremely viscous, mainly because the cells' DNA has not been broken up, and so expands out of the cell to form a dense interlocking network of molecules. Thus many cell lysis treatments include a nuclease treatment step. Nucleases are enzymes that break down nucleic acids, and the object here is to find a very non-specific nuclease that will break down any nucleic acid into very small pieces, ideally only a few bases long. The solution's viscosity then drops dramatically. This also breaks up the RNA in the solution, which is present in much greater mass than the DNA (although it does not contribute to the viscosity problem), and can be a problem in other purification it is not broken down into small fragments.

Cell fission. Cell fusion. Asexual method of reproduction whereby a cell splits to from two cells. The fusion of two cells together results in a new cell that has all the genetic material of the two original cells, and hence is a type of cell. The ability to fuse different types of cell (from the same species or from different species) has been used widely in biotechnological research. Common methods used include: **Electroporation**. Fusogen-mediated fusion. Some chemicals, such as PEG (polyethylene glycol), can make cells fuse together: they are called fusogens. PEG is a polymer that binds into the lipid membrane of cells and fuses it with any other lipid membranes around. Thus is can mediate the fusion of any cells that are bounded by a lipid membrane (i.e. all animal cells, and plant or bacterial protoplasts). **Virus-mediated fusion.** Some viruses have lipid coats that fuse with the membrane of cells when the virus infects that cell. Thus they also act as fusogens. If the virus fuses with two cells at once, then it effectively joins the cell via a small bridge of membrane. Thus viruses have been used rather like PEG to fuse cells. Indeed, their cell-fusing capabilities were discovered before PEG's were, but PEG is preferred now because it is easier to get hold of and less potentially hazardous. Cell fusion is used in a variety of techniques. Making monoclonal antibodies relies on making a fusion between lymphocytes and an immortalized cell line. Some plant genetic engineering has used cell fusion to generate hybrid plants, i.e. plants with all the genetic material of two different plant types brought together into one species, by fusing the protoplasts of the two 'parental' species and then regenerating a plant from the result. (This is a difficult trick to achieve.) Polyploid plants, plants with abnormally large numbers of chromosomes, can also be made by fusing cells from the same plant together.

Cell growth. The growth of isolated cells in culture follows a characteristic curve. The phases of the curve are:

Lag phase. This occurs when the cells are introduced to their new growth medium, and is the time taken for them to adapt to it. If their new environment is identical to their old medium, then the lag phase can disappear: however, even the mechanical shock of moving some cells around can cause a lag phase.

Log phase. This is the main growth phase of the culture, when the cells are growing exponentially. When plotted on a logarithmic scale, the log phase shows as a straight line.

Transition. This is the period (which can be minutes to days) between log phase and

Stationary phase. Here the cells have stopped growing; they have reached the capacity of their growth system to sustain growth. In most bacterial systems, this is a balance between some cells dying and others continuing to grow. Mammalian cells usually stop dividing in stationary phase.

Death phase. If the cells are not then given fresh medium to start growing again, they will start to die off. The total mass of cells remains the same (top line) but increasingly few of those cells are alive (bottom line) in the sense that could start a new growth curve if they were given fresh growth medium. The length of the different phases varies enormously with different cell types. Thus, many common bacteria have a stationary phase lasting only a day or two before death phase starts. In contrast, mammalian nerve cells can last almost indefinitely in culture without dividing. A single mammalian cell isolated from skin or muscle and put into culture medium could take a week to divide, a single *E. coli* is unlikely to take more than 10 minutes to start growth. The other key idea in cell growth studies is doubling time. This is the time that is needed for the cell population to double in number, and is equal to the time an 'average' member of that population takes to go through one complete division cycle. The higher the doubling time, the lower the growth rate of the culture and the longer it will take for an inoculum to reach stationary phase ('growth rate' is sometimes defined as 1/doubling time). Doubling time depends on the growth conditions and on the organism being grown; some bacteria (notably Clostridium perfringens) can have a doubling time of 10 minutes in the right culture medium. Strictly speaking, the concept of doubling time only applies to organisms growing in log phase, i.e. growing exponentially. A related concept is generation time, which is the time that a single organism takes to go through one reproductive cycle. For organisms that grow by division, this is pretty much the same as doubling time. However, it can be applied to organisms with more complicated life cycles, such as flies and people, where you can get more than a doubling of the population every generation if each parent has enough offspring. Thus growth cycle is not the same as the overall life and senescence cycle of primary mammalian cells. Mammalian cells will stop dividing when they have exhausted one of the critical components in their culture medium, or when their surroundings are too crowded for their liking. However, if they are separated and put into new medium (a process known as 'splitting' the cells), then healthy cells will start to grow again. Senescence is what occurs when the cells have been split many times so that the total number of divisions they have gone through is 40-60. Then, they slowly cease to be able to divide any more, no matter how much new medium they are given.

Cell line. The term 'cell line' is usually applied to mammalian cells cultured in vitro, outside their original mammalian body. However, it can also be applied to plant cells. A mammalian cell line is a clone of cells, i.e. one that has been derived from one single cell. It is capable of being grown indefinitely, which mammalian cells taken straight from the body are not. Thus the cell has been 'immortalized', i.e. turned from a mortal cell (whose descendants are going to stop dividing after a few dozen divisions) into an immortal one. This can be by transformation of the cell with a virus, with DNA from an oncogene, or by mutagenesis of the cell and subsequent selection of anything that can continue growing. Cell lines should also be stable; i.e. they should not change their properties as they are grown. This can be difficult. Unlike normal cells, mammalian cells that have been immortalized often do not pass on their chromosomes very faithfully, and so can lose genes that are not essential to the survival of the cell. These may include genes vital to the biotechnologist, such as the genes making an antibody in a hybridoma cell. line. Before a cell clone is described as a cell line, its inventor has to demonstrate that it is stable in this sense. Favoured cell lines for general biological research and for production of recombinant proteins in cell culture include BHK cell lines (baby hamster kidney), CHO (Chinese hamster ovary). The names indicate where the cells to make these lines came from, and do not suggest that you have to cut up baby hamsters every time you need more cells.

Cell line rights. While a protein can be patented and its ownership is fairly clear, the ownership of a living system is rather more vague. In general, the rulings seem to suggest that any organism that is patentable at all can be patented if it has been manipulated to do something useful, almost no matter how the manipulation is done or how large or small it is. Thus the 'oncomouse' transgenic mouse has one new gene out of about 100000 but is still considered a 'new' entity. (For contrast, most mice (and people) probably contain at least half a dozen new, potentially physiologically significant mutations, never seen before, because of normal genetic change. Ownership of the new organism entity usually resides in who made it. It does not reside in the source of material for the new entity: the Moore case in the USA (when John Moore claimed that a cell line used in cloning interferon was derived from the hairy cell leukemia he was treated for in 1978, and hence was at least partly his) said that Moore had no rights to his own cell lines. In most countries, people do not have rights over organs removed in surgery: they only have the right to say what happens to their own, whole, body on death. (Perhaps the medics and lawyers have a point, as nearly everybody elects to have their body burnt or buried.). Interestingly, if the Moore decision had gone against Sandoz and Genetic Institute (who now own the cell line), then many other people would have rights to a wide range of cells in research and industry. The descendants of Henrietta Lacks, originator of the HeLa cell line 40 years ago, would now have rights to a substantial part of all molecular biology and a mass of cells that probably greatly exceeds her own weight in life.

Cell specific. A condition in which the effect produced is directed to the cells of a particular species; e.g., interferon inhibits viral replication most effectively in the species in which it was produced - it is cell specific.

Cell theory. The theory, well established, that organisms consist of cells, and that all cell come from preexisting cells.

Cell wall peptidoglycan. A structural component of the bacterial cell wall that provides the rigidity necessary to maintain the integrity of the cell. It is a very large molecule made up of N-acetylglucosamine and N-acetylmuramic acid to which is attached a pentapeptide. Bridges of amino acids cross-connect the pentapeptides.

Cell wall. A relatively rigid structure that encloses cells of plants, fungi, many protists, and most bacteria. The cell wall gives these cells their shape and limits their expansion in hypotonic media.

Cell-mediated immunity (CMI). T cell-dependent immunity cellulitis inflammatory condition of the subcutaneous tissue.

Cellular immune system. That part of the immune system that is based on the activities of T cells, Directed against cancer cells, parasites, fungi, intracellular viruses, and foreign tissues (grafts). (Contrast with humoral immune system.)

Cellular immunity. Immunity developed by certain cells that have acquired an increased ability to ingest or destroy specific foreign substances.

Cellulases. Hydrolytic enzymes which degrade cellulose to glucose; also called a α-glucosidase because it hydrolyzes the α-linkages of glucose which make up cellulose.

Cellulitis. Rapidly spreading infection, involving the skin and the subcutaneous and lymphatic tissues; can be caused by a α-hemolytic streptococcus.

Cellulose. A straight-chain polymer of glucose molecules, used by plants as a structural supporting material.

Central lymphoid organs. Lymphatic organs where lymphoid cell subpopulations become committed to specific functions; the thymus and bone marrow for T and B cells, respectively; also called primary lymphatic organ.

Centrifugation. This is one of the more common biochemical techniques, and it crops up very frequently in biotechnological purification schemes. Key terms are: **Zone vs. equilibrium centrifugation.** Zone centrifugation places the sample at the top of a tube, puts the tube in the centrifuge and spins it for a limited time, and then takes it out. What you want has then 'sunk' a certain way down the tube, separating it from what you do not want. If you run the centrifuge too long, everything sinks to the bottom of the tube. Zonal centrifugation separates things essentially according to their size. Equilibrium centrifugation runs the centrifuge until the contents have come to equilibrium, floating, for example, at their buoyant density. Further running will not alter the separation. This is related to:

Density gradients. Here, the solution in the centrifuge tube is arranged so that it gets more dense as you go down the tube. This is done by dissolving something in it: colloidal silica ('Percoll'), to separate live mammalian cells, sucrose to separate bits of cells, caesium chloride to separate nucleic acids, etc. When centrifuged to equilibrium, the sample will be separated according to its density, more dense parts floating further down the tube in more dense solution. This is also called 'isopycnic' centrifugation.

Density gradient stabilization is a also used in centrifugation, as well as in free zone electrophoresis and some other separation techniques. Here again the tube has an increasing density of solution in it, usually sugar solution. This, however, is not done to effect separation. Rather, it stabilizes the column of liquid against stirring. If a little bit of solution is stirred out of its 'proper' layer, then it will have a different density to the solution around it, and so will sink back to where it came from.

Rotors. Most centrifuges consist of a drive unit (which powers it, controls rotation speed, etc.) and a rotor in which the samples are placed and which goes round. The rotor is often removable, and fits in a bowl in the machine. In ultracentrifuges

(centrifuges capable of tens to hundreds of thousands of gravities) the bowl is armoured steel to protect the operator should a rotor 'fail' in a run. Legend has it that Svedberg, who developed ultracentrifugation for chemical and biochemical analysis, killed a couple of postdoctoral workers with bits of flying centrifuge. Some rotors are 'zonal' or continuous rotors. Liquid is fed into the middle of them, and bacteria or other particulate matter is centrifuged out to the outside. These are of obvious use in separating microbial cells from their culture medium, but are an expensive way of doing this for large volumes.

Centrifuge. (L. *fugere*: to flee). A device in which a sample can be spun around a central axis at high speed, creating a centrifugal force that mimics a very strong gravitational force. Used to separate mixtures of suspended materials.

Centriole. (L. *centrum*: center). A cylindrical organelle in animal and some protist cells, containing microtubules in a characteristic arrangement consisting of nine triplets A centriole is at each pole of the spindle apparatus in dividing nuclei (other than in vascular plants.)

Centromere. (L. *centrum*: center + Gr. *meros*: part). The point on a chromosome to which spindle fibers attach during nuclear divisions.

Centromere. The point on a chomosome that segregates during the first division of meiosis.

Cephalization. (Gr. *kephale*: head). The evolutionary trend toward increasing concentration of brain and sensory organs at the anterior end of the animal.

Cephalopod. A member of the mollusk class Cephalopoda, such as a squid or an octopus.

Cerebellum. (L. *diminutive of cerebrum*: brain). The brain region that controls muscular coordination; located at the anterior end of the hindbrain.

Cerebellum. The point on a chomosome that segregates during the first division of meiosis.

Chancre. Intial lesion in syphilis; primary stage.

Change room (s). Grouping of dressing rooms, locker rooms, lavatories, air locks, and showers to provide personnel access to and egress from contaminated areas without allowing escape of contamination.

Chaperones. Also known as chaperonins, these are a type of protein that help other proteins to fold up into their correct 3D structure. Some proteins will fold up correctly on their own as soon as they are made in the cell, forming a working protein molecule. Some, however, seen to do this very inefficiently, and to need other proteins to make them fold up properly. This is important to the production of 'foreign' proteins in bacteria. If a protein folds inefficiently or very slowly, then it will have a greater chance of aggregating into an insoluble, inactive mass, from which it may be very difficult to recover active protein. If the folding can be speeded up with chaperone proteins, then the amount of usable protein that can be got out of the bacterium (as opposed to the total amount of protein, usable or not) will be greater. Whether the role of chaperones in protein folding can be used, as opposed to describe, is still open to question. If it can, then it may be manipulated to produce protein complexes with functions we want that would not otherwise be produced. Some chaperones are also known as heat shock proteins (HSPs), because they are made when a cell is 'shocked', i.e. stressed by sudden heat or cooling, some chemicals, or other insults. It is believed that the HSP chaperones are made to help to refold proteins or remove those that have been damaged during the stress. Several groups are looking at how to harness this response so as to boost the cell's ability to repair itself after damage. Targets include reducing the damage caused by stroke, iscaemic heart disease, and in Alzheimer's disease. HSPs can also be used as diagnostic markers for environmental stress. If an organism is making HSPs then it's chemical environment is poor.

Character displacement. An alteration of the traits of a species as a result of competition or other interactions with associated species.

Chardonnay. A variety of green grape used in the production of white wine.

Chelate. Process of combining with a metal ion to keep in solution.

Chelating agent. A compound, usually organic, with two or more points of attachment at which an atom may be joined to form a ring-like structure.

Chemiosmotic hypothesis. According to this hypothesis, ATP formation in mitochondria and chloroplasts results from a pumping of protons across a membrane (against a gradient of electrical charge and of pH), followed by the return of the protons through a protein channel with ATPase activity.

Chemoautotroph. An organism which obtains its energy by chemical oxidations of inorganic substances, and which has the ability to fix CO2 as its sole source of carbon.

Chemokinesis. Increased random migratory activity of cells.

Chemolithotroph. Same as chemoautotroph.

Chemostat. A growth chamber for maintaining a bacterial culture in the logarithmic phase of growth by the continual addition of fresh medium.

Chemosynthesis. Synthesis of food substances, using the oxidation of reduced materials from the environment as a source of energy.

Chemosynthetic. Obtaining energy from chemical reactions.

Chemotaxis. Increased directional migration of cells particularly in response to concentration gradients of certain chemotactic factors. It is a movement of organisms in response to a chemical substance; the attraction of leukocytes to an area of injury is due to a release of a chemical substance from the injured tissue.

Chemotherapeutic. Chemical agents used in treatment of disease.

Chemotherapy. Treatment of disease using chemical substances.

Chiasma. (Gr. *cross*). An "x" -shaped connection between paired homologous chromosomes in prophase I of meiosis. A chaisma is the visible manifestation of crossing-over between homologous chromosomes.

Chimaerism. The situation in which cells from genetically different individuals coexist in one body.

Chimeric. In genetics refers to a DNA molecule derived from different sources.

Chimeric/humanized antibodies. A problem with using antibodies in medical therapy is that monoclonal antibodies are foreign proteins, and hence, when they are injected, the patient will have an immune response to them. This does not matter for a one-shoot therapy because the immune response in too slow to have an effect within hours of first encountering a foreign protein. But for longer term treatment it means that after a few days or weeks the patient will have their own antibodies that bind to and neutralize the immunotherapeutic one as soon as it is injected. This is known as the human anti-mouse antibody (HAMA) response (nearly all monoclonal antibodies are made in mice). It is extremely difficult to overcome this by making genuine human monoclonal antibodies as drugs: the technology of monoclonal antibody production works with mouse or rat cells, not human cells. To overcome this, scientists engineer the antibody so that is 'looks' like a human antibody to the immune system. The species-specific parts of the antibody to which the immune system responds are in the constant regions. Thus, by replacing the constant regions of a mouse antibody with those of a human antibody, a protein that binds to an antigen like the original monoclonal antibody, but which 'looks' to the human immune system like a human protein, can be made. This process is called humanizing the antibody. The fused protein is called a chimeric antibody. More sophisticated engineering can be done to produce a 'humanized' antibody. The ultimate engineering is to take only those parts of the antibody that determine the antibody's binding specificity (the complementarity determining regions - CDRs) and splice them into a totally human antibody. This is called CDR grafting. An alternative version is called 'antibody resurfacing' or 'antibody veneering'. This engineers all the surface residues on an antibody to 'look like' the human residues, so that the human immune system does not recognize them as 'foreign', but leaves the core residues and the binding site unaffected. Engineering antibodies in this way has an added complication. Antibodies consist of two protein chains (heavy and light chains) and so two genes must be engineered into the produced cell to make a finished antibody. While this is possible, and various technical tricks to make it easier have been developed, it would be much easier to handle only one protein chain. This is one advantage of Dabs and SCAs: they are antibody-based proteins that contain only one chain. The alternative approach is to make genuinely human antibodies by cloning them from humans and expressing them in bacteria or cultured cells.

Chiral chromatography. A racemic mixture of isomers is separated on a chromatographic column that is itself chiral, e.g. has a chiral compound linked to it or is made of a chiral material such as cellulose or protein. There are several routes to chiral synthesis that use biotechnology. The success of all of them is measured by the 'enantiomeric excess', the percentage by which one of the enantiomers outweighs the other in the preparation. An enantiomeric excess of 100% means that you have one of the optical isomers absolutely pure.

Biotransformation. This is the synthesis of a compound using enzymes. Since most enzymes produce only one enantiomer as a product, they can be used to take symmetric (i.e. not chiral) starting products and produce pure enantiomers from them.

Bioconversion. This is the same idea, but uses whole organisms to convert one chemical to another. This may be better than using isolated enzymes if the enzymes concerned are not very stable, or if a number of enzymes are needed to make one conversion. The chiral drug ephedrine has been traditionally produced by bioconversion.

Fermentation methods. If you can obtain the chemical from a fermentation culture, either of a microorganism or of a plant or animal cell, then that chemical will almost certainly be produced as one enantiomer. Many amino acids produced for animal feed supplements have been produced traditionally as single optical isomers by fermentation, especially in Japan. For all these processes, there are two approaches.

Stereospecific synthesis. This takes two, non-chiral starting materials and makes a chiral product from them. This has to be done using some third party to introduce the chirality into the system. This can be a third reagent, or a catalyst: often this chiral catalyst is an enzyme.

Resolution. This takes a racemic mixture (a 'racemate') of a chiral compound, i.e. a mixture in which all the various enantiomers are present as a mix, and removes one of them. A range of techniques can be used. One isomer may be bound to a material that is itself optically active (such as an optically active HPLC column, or an antibody), but because of their capacity to process only a few milligrams at a time these are usually used as analytical techniques rather than preparative ones. One isomer may be converted into another chemical (which can then be removed by conventional means) using another optically active chemical or, most effectively, an enzyme. The enzyme can either act on the compound you want (turning it into the product, or something more like the product) or on the one you do not want (turning it into something that is easy to remove). Often, chiral synthesis is not used to make the final chemical itself. Rather, it is used to make a precursor to it that is easier to make using the available enzyme systems. This precursor can then be turned into the final chemical using conventional chemistry.

Chiral synthesis. Chiral synthesis is the production of chiral compounds in only one enantiomer, or 'handedness'. Since chiral compounds can be made in two (or more) physical arrangements that are virtually indistinguishable chemically, this is a difficult task for conventional chemistry. Biological systems make just that sort of discrimination all the time, however, and so have great potential for making chiral compounds. To make chiral compounds in only one enantiomer, there are a range of chemical methods. These include: **Asymmetric catalysis.** A catalyst that is itself chiral is used in a key step in the reaction.

Chirality. Chirality is the chemical version of 'handedness'. Some molecules have distinct left-and right-hand forms, which, although containing the same atoms tied up in the same way, are not physically the same (just as your hands have the same number of fingers tied to the palm in the same way, but nevertheless are not the same). Such a chemical is called a chiral compound, and the two (or more) forms are called enantiomers (or optical isomers) of each other. Compounds that have two enantiomers are usually divided into L and D, or + and -, or S and R forms, so you have (L)-alanine or (+)-ephedrine. There are complicated rules about these nomenclatures for organic chemists. Usually there is no chemical difference between the enantiomers of a compound, or indeed between the pure enantiomers and an equal mixture of all of them (called a racemic mixture). The only detectable difference is that they interact with polarized light in slightly different ways. However, nearly all the molecules that make up living systems are chiral. Thus, all the amino acids in proteins are L-amino acids, not their chemically identical D forms. Because of this, all the chemistry of life is chiral, and so how other chemicals affect life depends on which enantiomer you have, just as it is easier to shake hands left-to-left or right-to-right, not left-to-right, because both hands are chiral, whereas it is as easy to pick up a light briefcase with left or right hands (because, although your hand is chiral, the briefcase handle is not). This has substantial implications for pharmaceuticals and agrochemicals. Different enantiomers of exactly the same drug can affect biological systems in quite different ways. Thalidomide is a case in point: an effective and safe antinausea agent, the teratogenic side-effects were not caused by the drug itself, but by its mirror opposite, the other enantiomer. However, the drug was given as a racemic mixture, so the patients got both therapeutic effects and side-effects. Clearly, as legislative pressures grow for chemicals used in agriculture and medicine to be more specific, there is increasing pressure for any chiral product to be manufactured by these industries as one enantiomer and not as a racemic mixture for those applications. Chiral synthesis is a major aspect of biotransformation and bioconversion technology. For biopharmaceuticals, of course, chirality is not a worry, since, as biologically derived proteins they all have the correct 'handedness' anyway.

Chitin. (Gr. *chitom*: tunic). The characteristic tough but flexible organic component of the exoskeleton of arthropods, consisting of a complex, nitrogen-containing polysaccharide. Also found in cell walls of fungi.

Chitin. Polysaccharide containing glucosamine; characteristic of cell walls of some fungi; also found in insects.

Chlamydiae. Obligate intracellular parasites which appear to be related to gram-negative bacteria, and which have a complex method of reproduction that sets them apart from Rickettsiae; they are not transmitted by an arthropod vector.

Chlamydospore. Resistant asexual spore characteristic of some fungi.

Chlorophyll. (Gr. chloros: green + phyllon: leaf). Any of a few green pigments associated with chloroplasts or with certain bacterial membanes; responsible for trapping light energy for photosynthesis.

Chloroplast. (Gr. *chloros*: green + plast. a particle). An organelle bounded by a double membrane containing the enzymes and pigments of photosynthesis. Chloroplasts occur only in eukaryotes.

Chlorosis. Refers here to failure of chlorophyll development because of nutritional disturbance or because of infection by virus, bacteria or fungi.

Chorioallantoic membrane. In embryos of birds, e.g., chick embryo, the chorion (outer covering of the embryo sac) is fused with the allantois (sac like out growth of ventral side of gut) forming a double membrane structure which has been extensively used for cultivating viruses in the laboratory.

Chromatic adaptation. (Gr. *chroma*: color). The ability of some algae to modify the kinds and amounts of different photosynthetic accessory pigments, producing those that absorb the wavelengths of light maximally available in their environment.

Chromatid. Each of a pair of new sister chromosomes from the time at which the molecular duplication occurs until the time at which the centromeres separate at the anaphase of nuclear division.

Chromatin (body). A network of deoxyribonucleic acid fibers and histones giving the appearance of a distinct body in the cell.

Chromatography. Any one of several techniques for the separation of chemical substances, based on differing relative tendencies of the substances to associate with a mobile phase or a stationary phase.

Chromatography. Many separation systems used in biochemistry, molecular biology, and biotechnological production are chromatography systems. Chromatography was originally developed as a way of separating pigments from plants by wicking them through paper, an experiment that many schoolchildren do today, and the same basic ideas apply to all chromatographic separations. A small sample is put on one end of a slab or wick of porous material. A solvent is then passed over the sample and up the wick or slab. Depending on whether the molecules in the sample stick to the solid wick or dissolve in the solvent, they either move up the wick or stay put. Most materials do a bit of both, and so move up the wick slowly: the exact speed varies with each component of the sample, and so they are spread out. The pattern of how materials are washed (eluted) off the end of the slab or wick is called the elution profile. This is, in fact, a two-phase separation, and so the two parts of the system are called the mobile phase (the solvent) and the stationary or solid phase (the solid material that the solvent moves over). There are a lot of variations of chromatography. The more common are:

Gel chromatography/gel exclusion chromatography/size exclusion chromatography. These sift by molecular size. The chromatographic material has small pores in it, into which small molecules can enter but from which large molecules are excluded. (Different materials have different pore sizes, so that the size limits 'small' and 'large' can be adjusted to suit whatever the scientist is trying to separate.) As the mixture of molecules passes down the column, small molecules diffuse into the pores, where liquid is stationary, and so spend some of their time standing still. Large molecules cannot enter the pores, and so spend all their time in the moving phase. Thus large molecules move down the column faster than small ones.

Affinity chromatography. Here a specific molecule is linked to the chromatographic material and molecules are separated by their ability to bind to it. If the bound molecule is large and the molecule to be separated is small, this is usually called affinity chromatography. If the bound molecule is small and the separated molecule large, the method can be called covalent chromatography, although this too is often called affinity chromatography.

Hydrophobic chromatography. This simply uses a hydrophobic material, such as untreated silica, as the stationary phase. Molecules stick to it depending on how hydrophobic they are, so it is an effective way of separating many metabolic products. Most HPLC systems are hydrophobic adsorption chromatography. So are adsorption fermentation systems, where a fermentation broth is passed continuously over some solid material that adsorbs the product of the liquid while leaving the microorganisms behind.

Gradient chromatography. Here all the molecules in a sample are bound to the solid support material, and then they are washed off one at a time with an increasing concentration of some solution, often increasing salt, acid, or alkali concentration. Chromatography also varies according to the physical arrangement of the solid material (the stationary phase):

Column chromatography. This is by far the most common; the solid phase is packed as small particles into a tube, and the liquid passes over it. Column chromatography methods capable of purifying kilograms of materials at a time have been developed. A variant is high performance liquid chromatography (HPLC; sometimes also called high pressure liquid chromatography) in which the liquid is pumped slowly over quite a small column at very high pressure. This greatly increases the resolution of method, i.e. how well it can separate similar materials.

Paper chromatography. This is essentially the same as the original method, and uses a paper wick as the solid phase. This is not as limited as it may sound since paper is a complex material and papers with widely varying properties are available.

Thin layer chromatography (TLC). Here the stationary phase is a thin layer of treated silica coated on to a glass plate.

Gas chromatography **(GC).** This is quite commonly used to analyse the volatile metabolites produced during fermentations, and also to analyse other biomolecules that have been made more volatile by chemical modification. GC passes a stream of hot gas through a very fine tube in a carefully thermostatically controlled oven. The volatiles partition between the gas phase and as liquid on the walls of the tube (hence this is sometimes called gas - liquid chromatography, GLC). The time that it takes them to get through the tube depends on this partition. Gas chromatographs often have ionization detectors at the end to detect the molecules coming off the tube. The gas is fed into a flame, and the ions produced as the target molecules burn is detected electrically. This is a sensitive and widely applicable detection method, but obviously not directly suited to collecting the molecules afterwards for further work.

Lastly, there are the different materials that can make up the and solid phase. In general the mobile phase is water, or some watery solution: this is because nearly all the materials that biotechnology is interested in are soluble to various degrees in water, and some, like proteins, are not stable in any other solvent. The solid phase gives greater flexibility.

Polysaccharides. Much loved by biochemists are polysaccharides such as cellulose (both as a granular material and as paper) or agarose. The popular Sepharose is a specific product of Pharmacia Biotech, but has almost achieved the status of a generic name like Hoover. They are used for gel chromatography and for affinity methods.

Synthetic polymers. Coming into increasing favour are synthetic polymers such as polystyrene, PMMA (perspex), and teflon, because they are easier to form into small, rigid, uniform spheres and are chemically more robust. Polyacrylamide is also used.

Silica. Chemically modified silica, especially silicas with chemically modified surfaces and silica materials with a porous structure (CPG, controlled pore glass) are used in many applications. For applications involving substantial pressure, such as HPLC (in which polysaccharide beads are likely to be squashed), silica is very useful. Generally, chromatographic methods are used to separate several different chemicals from a mixture at once. The pattern in which they emerge from the chromatographic column is called the 'elution profile'. The efficiency of a chromatographic system is defined in terms of 'theoretical plates'. The column can be viewed as a series of plates, each of which act like a cell with sample in one side and fresh solvent in the other. (Strictly, a theoretical plate is that bed length that brings about the same relative change in concentration of a solute as would be achieved on equilibrating with fresh stationary phase.) The efficacy of the column depends on how many theoretical plates it is equivalent to, and this can be calculated if you know the properties of the material involved. In practice, much biochemical chromatography development is developed on a trial and error basis.

Chromatophores. Discrete vesicles or special membrane systems which contain photosynthetic pigments in procaryotic cells.

Chromosomal aberration. Any large change in the structure of a chromosome, including duplication or loss of chromosomes or duplication of complete sets of chromosomes, usually gross enough to be detected with the light microscope.

Chromosome. (Gr. *chroma*: color + soma: body). In monera and viruses, the DNA molecule that contains most or all of the genetic information of the cell or virus. In eukaryotes, a structure composed of DNA and proteins that bears part of the genetic information of the cell.

Chronic. Of long duration; not acute.

Cide. Suffix, killing effect.

Cilia. Small hairlike projections found on the surface of some microorganisms and used for movement; found on respiratory epithelium, cilia are used to direct flow of various secretions and particles. Cilia have same internal structure as eukaryotic flagellum but are shorter.

Ciliate. A member of the protist phylum Ciliophora, unicellular organisms that propel themselves by means of cilia.

Circadian rhythm. (L. *circa*: approximately + dies: day). A rhythm in behaviour, growth, or some other activity that recurs about every 24 hours under constant conditions.

Cirri. Tuft of cilia used as organelle for motility and feeding in certain protozoa.

Cisternae. Stacked membranes associated with the golgi complex.

Cistron. A segment of the chromosome used as a pseudonym for gene; it codes for a single polypeptide.

Citric acid cycle. A set of chemical reactions in cellular respiration, in which acetyl coenzyme. A reacts with oxaloacetate to form citric acid, and oxaloacetate is regenerated. Acetyl-coA is oxidized to carbon dioxide, and hydrogen atmos are stored as NADH and FADH2.

Clade. (Gr. *Klados*: branch). A group of organisms descended from a common ancestor.

Cladistic systematics. Systematics based entirely on how long ago two species shared a common ancestor. (Contrast with phenetic systematics.)

Cladogram. Graphic representation of a cladistic relationship.

Class 1/2/3 MHC molecules. Three major classes of molecule coded within the MHC. ClasS 1 molecules have one MHC encoded peptide associated with α-2-microglobulin. Class 2 molecules have two MHC encoded peptides which are non-covalently associated, and Class 3 molecules are complement components.

Class. In taxonomy, the category below the phylum and above the order; a group of related, similar orders.

Clean change room. Dressing room for removal of street clothes and donning of laboratory-type clothing before entering contaminated change room, through an air lock.

Clean room. A clean room is a room that meets specific standards of cleanliness, especially with respect to what may go into and out of it and what the concentration of particles in the air might be. Clean rooms are central to pharmaceutical manufacturing, since it is by producing, formulating, and packaging drugs under rigorously sterile conditions that the sterility of the drug is assured. The same cleanliness conditions apply to a lesser extent to other biotechnological products, and also can apply to the research and development phase of recombinant DNA or plant or animal cloning work, where the objective is to stop contamination getting into the experiments. Clean rooms are classified in the USA according to US Federal Standard 209D. Roughly, clean rooms are classed by a number, which is the number of particles of more than half a micrometer diameter that are allowed per cubic foot of air. Thus a class 100 clean room would have around 100 0.5 μm particles per cubic foot. (The exact number varies slightly from this.) At the moment, class 100 rooms are the highest category of cleanliness required by the pharmaceutical industry. Other countries have different rating systems (notably, most are based on SI units), but the level of air purity required is similar. Clean rooms are kept clean by a variety of methods. Air going into the room passes through HEPA filters (high efficiency particulate air) to keep out even very small particles: ultra-clean rooms have several layers of filtering. Walls, floor, and ceiling are usually painted with materials designed not to stick any dust (and, obviously, not to peel or flake). People going into the clean room have to wear hats and overshoes, since hair and shoes are the most particle-laden parts of the worker, as well as the usual laboratory coat. For less rigorously clean areas there may be a 'sticky mat' just inside the door, which pulls loose dirt off the bottom of the shoes of everyone who walks in. To provide greater cleanliness within a clean room, a laminar flow hood are provided. These are benches either made of or surrounded by open meshwork, and covered by a hood.

Air flows up past the work-surface and into the hood, where it is filtered before being returned to the worksurface. Thus all the air passing over the workplace is in a separate flow from the air in the room, and has been extra cleaned. Clean rooms use much the same air filtration technology as containment laboratories, but for a different purpose. Containment laboratories are meant to keep possibly hazardous material in the laboratory, rather than external contamination out.

Cleaning-in-place. This is the cleaning (and sterilization) of a manufacturing system without dismantling it, so that the parts are cleaned as a whole: it also called 'in situ sterilization' or sterilization-in-place (CIP and SIP). This is a much easier operation to perform than cleaning and sterilizing all the components separately and then trying to assemble then under sterile conditions, or having separate cleaning and sterilizing operations. However, it requires some specialist techniques and equipment. In particular, bioreactor machinery much be designed so that there are no 'dead legs' (i.e. pipes that are blocked at one end), crevices, or 'shadowed' areas (i.e. areas where the bulk of some other piece of the apparatus prevents fluid from flowing) into which cleaning fluid could not flow. It is also useful for the equipment to be designed so that bits of it can be cleaned while the rest remains in operation. CIP systems are a combination of the equipment used and the processes used in it. Thus a full CIP system will include protocols for washing, usually with industrial strength detergents, and rinsing out the wash agents.

Cleavages. First divisions of the fertilized egg of an animal.

Climax community. The stable community characteristic of a site.

Climax. In ecology, a stable community that is the "end" point of succession. It is ore or less in equilibrium with the current physical environment.

Climograph. Graph relating temperature and precipitation with time of year.

Cloaca. (L. *cloaca*: sewer). In some invertebrates, the posterior part of the gut; in many vertebrates, a cavity receiving material from the digestive, reproductive, and excretory systems.

Clonal deletion. In immunology, the inactivation or destruction of lymphocyte clones that would produce immune reactions against the animal's own body. It is the mechanism by which exposure to antigen results in the activation of selected T-cell or B-cell clones, resulting in an immune response.

Clone. A clone is a collection of genetically identical individuals. They turn up in molecular biology and biotechnology in many contexts.

Clone. The aggregate of individual organisms produced asexually from one sexually produced organism. Here refers to the cells and tissuesdeveloped in culture.

Clostridium botulinum. Anaerobic spore-forming rod, gram-positive; causative agent of botulism.

Clostridium perfringens. Anaerobic sporeforming rod, gram-positive; the most common causative agent associated with cases of gas gangrene; frequent cause of food poisoning.

Clostridium tetani. Anaerobic spore-forming rod, with terminal spores, gram-positive; its exotoxin causes tetanus, or lockjaw.

Clostridium. A genus name, refers to anaerobic spore-forming rods.

Clutch. The number of offspring produced in a given batch.

CMI (cell-mediated immunity). A term used to refer to immune reactions which are mediated by cells rather than antibody or other humoral factors.

Coagulase. An enzyme which, in concert with certain serum factors, causes blood plasma to clot; produced by Staphylococcus aureus.

Coagulation. The formation of a mass, or clot.

Cobra venom factor. A cobra complement component equivalent to mammalian C3b.

Coccidioidin test. Skin sensitivity test used as an aid in the diagnosis of infections caused by Coccidiodes immitis. The injected material consists of dilution of culture filtrate of Coccidioides immitis.

Coccobacillus. Oval bacterium that is intermediate between the coccus and bacillus forms.

Coccus. A spherical bacterial cell.

Cochlea. (Gr. kokhlos: a land snail). A spiral tube in the inner ear of vertebrates; it contains the sensory cells involved in hearing.

Code date. A mixture of letters or numbers that signify the manufacturing place and date of a product; also called a manufacturing lot number.

Codon. A *triplet* of three nucleotides in messenger RNA that directs the placement of a particular amino acid into a polypeptide chain. (Contrast with anticodon.)

Coefficient of relatedness. The probability that an allele in one individual is an identical copy, by descent of an allele in another individual.

Coelom. (Gr. *koiloma*: cavity). The body cavity of certain animals, which is lined with cells of mesodermal origin.

Coenocytic. A filamentous cell or mycelium not divided by complete septa and containing more than one nucleus; also occurs in slime molds where there is no cell wall or mycelium.

Coenzyme. A nonprotein molecule playing a role in catalysis by an enzyme. The coenzyme may be part of the enzyme molecule or free in solution. Some coenzymes are oxidizing or reducing agents, others play different roles.

Coenzyme. The term cofactor is used almost interchangebly with coenzyme in most contexts. A coenzyme is a molecule that is needed by an enzyme to work, is part of the chemical mechanism of the enzyme, but which is not produced for its own sake. Rather, it acts as a shuttle molecule, taking groups between one enzyme and another or between one reaction and another. The most biochemically common coenzyme set is the NAD set. These molecules shuttle hydrogen atoms around the cell. There are two flavours (NAD (nicotinamide-adenine dinucleotide) and NADP (NAD phosphate)), and they come as hydrogenated (reduced) or nonhydrogenated (oxidized) molecules: NAD or NADP = oxidized, NADH or NADPH = reduced. Many cofactors and coenzymes are derived from vitamins. Thus NAD is derived from nicotinic acid. Some coenzymes are tightly, even covalently, linked to their enzymes: it is these that are often called cofactors. An example is the FAD (flavine-adenine dinucleotide) moiety that is needed by the common giagnostic enzyme glucose oxidase. If the FAD is removed the enzyme does not work. Such a cofactorless enzymes is called an apoenzyme. It contains all the protein of the intact, functional enzyme (the holoenzyme), but does not catalyse its reaction. Cofactors are important for the industrial application of enzymes. Many enzymes that perform industrially important reactions cannot be used in the appropriate industrial process because the cofactors they need are too expensive and are gradually lost from the reaction as the enzyme is used. Biotechnology therefore seeks either to replace them with other molecules that are cheaper or more stable (biomimetics) or to recycle the cofactors more efficiently. Both are difficult to achieve economically on an industrial scale. Coenzymes are also of concern to biotechnology in two other fields. The first is that they are often inconveniently complicated and expensive molecules to make and store, and so research is looking for synthetic alternatives. Secondly, some abzymes have been made that use coenzymes to catalyse reactions.

Coevolved relationship. Relationship between species that have influenced one another's evolution.

Cofactor. An inorganic molecule, such as a metal ion, that increases enzyme activity.

Cohort. (L. *cohors*: company of soldiers). A group of organisms, considered as it passes through time.

Coitus. (L. *coitus*: a coming together). The act of sexual intercourse.

Cold agglutinins. Antibodies which agglutinate at temperatures below 37øC only.

Coleoptile. (Gr. *koleos*: sheath + *ptilon*: feather). A pointed sheath covering the shoot of grass seedlings.

Colicin. A protein secreted by certain of Escherichia coli and lethal to other strains of the same species.

Coliform bacteria. Small, aerobic, gram-negative non-spore-forming bacilli; ferment lactose, with acid and gas production, normal inhabitants of lower intestine.

Colitis. Inflammation of the colon.

Collagen. (Gr. *kolla*: gluel). A fibrous protein found extensively in bone and connective tissue.

Collagenase. Enzyme which causes breakdown of collagen, the ground substance of bone, skin, and cartilage; may aid in the spread of the pathogen from the initial site of entry.

Collar. The thin disc which serves as an attachment area between the head and the tail of a binal virus.

Colon. (Gr. *kolon*: large intestine). The large intestine.

Colony stimulating factors (CSF). There are three varieties: G-CSF, M-CSF, and GM-CSF, which stimulate granulocytes, macrophages, or both, respectively. They stimulate the differentiation of some types of white cell. Ten companies are trying CSFs as biopharmaceuticals.

***Interferons* (IFN).** Well known as being one of the first proteins to be produced by the new biotechnology of the late 1970s, and touted as the wonder cure for everything, there are actually three classes of these cytokines. They are now consistently called interferons alpha, beta, and gamma. IFN-â is a potent stimulator of the activity of macrophages, encouraging them to kill tumour cells and intracellular parasites. Interferon à (from Biogen) has been approved as a treatment of hepatitis C. Interferon á has been approved for treating several cancers and for multiple sclerosis: it is produced by several companies including Ares-Serono and Biogen. Bovine interferon has also been shown to help improve the pregnancy rate in sheep, because it increases the process of 'maternal recognition' by which the ewe's immune system learns that the developing fetus should not be rejected. This unusual use of a cytokine could be as widespread as medical uses.

Tissue necrosis factor **(TNF).** Slows cell growth and kills some cancer cells and cell lines. It is therefore a hot candidate for an anticancer drug, and as the 'toxin' part of an immunotoxin. It is also involved in the cell destruction that can occur in some inflammations, so finding ways to block TNF's action is also a hot pharmaceutical topic. Several companies are developing genetically engineered cytokine preparations for drug use: Genentech (gamma interferon), Cetus (IL-2), Immunex (GM-CSF), Ares-Serono (beta interferon).

Colony. A discrete accumulation of microbial growth on the surface of a solid culture medium which is visible to the naked eye. A colony can be the result of the growth and multiplication of a single microbial cell, or of a clump of two or more cells. A colony formed from a single cell is referred to as a clone, but it is usually not possible to differentiate by mere visual examination a clone from a colony that may have formed from, for example, 100 cells.

Colostrum. The first milk from the woman's breasts after childbirth. Compared to later milk, it is higher in protein (chiefly immune globulins containing antibodies), betacarotene, riboflavin, niacinamide, and lower in fat and carbohydrate. It acts as a laxative and aids in the expulsion of the meconium.

Columella. The rounded apex of the sporangiophore within the sporangium of Fungus Class Zygomycetes.

Combinatorial chemistry. Combinatorial chemistry is a growing area of synthetic chemistry that provides new chemicals for drug discovery. Traditionally, new chemicals have been made by chemists, more or less one at a time. Thus the chemicals that can be screened for their potential as new drugs is limited to what happens to have been made before and stored in a company's chemical library. Combinatorial chemistry seeks to make many chemicals at once in small amounts, both as a source of new chemicals and to test which of many variants of a compound is the best drug.

Cometabolism. Metabolism of substrates incapable of being used as a carbon or energy source in the presence of alternative, usable substrates.

Commensal. A relationship between organisms in which one benefits but the other neither benefits nor is harmed.

Commensalism. The form of symbiosis in which one species benefits from the association while the other is neither harmed nor benefited.

Commercial sterility. The thermal processing of canned foods such that only thermophilic spores survive.

Communicable disease. A disease that is transmitted directly or indirectly from one person to another.

Communicable. Easily transmitted.

Communication. Action on the part of one organism (or cell) that alters the pattern of behaviour in another organism (or cell) in an adaptive fashion.

Community. Any ecologically integrated group of species of microorganisms, plants, and animals inhabiting a given area.

Companion cell. Specialized cell found adjacent to a sieve tube element in some flowering plants.

Compensation depth. Point where photosynthesis and respiration balance in an aquatic environment.

Compensation point. The light intensity at which the rates of photosynthesis and of cellular respiration are equal.

Competence. Transitory period during which the bacterium is capable of taking up transforming DNA.

Competition. In ecology, use of the same resource by two or more species, when the resource is present in insufficient supply for the combined needs of the species.

Competitive exclusion. Success of one species over another when they compete for a single limiting resource, so that only one species can continue to occupy that niche.

Competitive inhibition. Interference with a synthetic process due to an inhibitor which combines reversibly with the active site on the enzyme; e.g., sulfonamides inhibit folic acid synthesis.

Competitive inhibitor. A molecule that competes with the substrate for attachment to the active site of an enzyme.

Complement system. A group of eleven proteins that play a role in some reactions of the immune system. The complement proteins are not immunoglobulins.

Complement. A blood serum protein group involved in certain antibody-antigen reactions.

Complement. A complex system of serum proteins that are activated in a sequential fashion (in the complement cascade) by immune complexes or by other substances such as bacterial endotoxins.

Complementary base pairing. The A-T (or A-U), T-A (or U-A), G-C pairing of bases in double-stranded DNA, in transcription, and between tRNA and mRNA.

Complementary strand. Here refers to the newly synthesized of DNA utilizing a previously present strand as a template, thus forming a newly formed double stranded structure.

Complementation. (Latin: *complementum*, to fill). Here used to indicate a phenomenon relating to successful multiplication of a functionally incomplete bacteriophage only when it is associated with another functionally in complete phage. The latter acts as a `helper'.

Complex virus. Viruses with relatively complex architecture and also having a specific mode of development, e.g., poxviruses.

Compound microscope. A microscope that has two sets of lenses, one in the objective next to the object to be studied, and the other in the ocular next to the eye.

Compound. (1) A substance made up of atoms of more than one element. (2) Made up of many units, as the compound eyes of arthropods (as opposed to the simple eye of the same organisms).

Computational chemistry. This is a blanket term for using computers to predict or analyse the properties of molecules (as opposed to using computers simply to draw them, which is molecular graphics). Calculating the properties of molecules from first principles, which would be ideal, is impossible for practical purposes. Thus computational chemistry uses the known properties of chemicals to calculate the properties of similar molecules partly from empirical rules ('heuristics') and partly from 'rigorous' calculation. Typically, approaches use energy minimization, which calculates the total chemical energy of a molecular structure and seeks to adjust the structure to minimize it. Contributions to the energy include energy of solvation (binding water molecules), van de Waals forces between nearby atoms, and 'steric' constraints (which is a general term for how you can fit different parts of a molecule together without having two atoms in the same place at the same time).

Concatemer. A DNA molecule consisting of two or more separate genomes linked end to end to produce a linear structure.

Concentration. Biological products are usually produced in rather low concentrations, by fermentation methods or by extraction from animal or plant tissues. In order to keep the cost of purifying these materials down, it is useful to reduce the volume, i.e. to increase the concentration, as early as possible in the downstream processing stages of a biotechnological process. Many concentration methods also purify the product to some extent as well. The very best processes concentrate and purify in one step, but that is rarely possible. Concentration is inherently expensive. Preparing any material from a diluted solution costs more. This is why extracting gold from sea water is not economic, whereas extracting salt is: salt is so much more concentrated than gold that it tips the balance. Methods used in concentration are based on:

The size of the molecules. In this category come various filter methods are reverse osmosis. In reverse osmosis, the sample is placed on one side of a semipermeable membrane, i.e. one that will let through water but not other materials. A very high pressure then pushes the water through the membrane, giving water on one side and much concentrated product on the other. This can be a way of purifying water too: it is sometimes used to make drinking water from seawater. It is the reverse of osmosis, the process by which water moves from one side of semipermeable membrane to the other side if the concentration of soluble material is greater on the other side.

Ultrafiltration is a similar technique. Here the molecules are filtered through a membrane with pores of molecular size. Large molecules stay on the sample side, while water and small molecules, including salt, pass through. Again, considerable pressure is usually needed to make this happen.

Concurrent disinfection. Disinfection procedures carried out throughout the course of an illness with rigid aseptic disposal of contaminated materials.

Congenic. Animals which are genetically constructed to differ at one particular locus.

Congenital. Present at birth or acquired while in utero.

Conidiophore. A hyphal filament on which spores, called conidia, are borne.

Conidium (*pl.*: conidia). A spore produced by mitosis, externally on the tip of a more or less specialised hypha called a conidiophore. (synonym, conidiospore.)

Conjugate. A reagent which is formed by covalently coupling two molecules together such as fluorescein coupled to an immunoglobulin molecule.

Conjugation. (L. *conjugare*: yoke together). The close approximation of two cells during which they exchange genetic material, as in Paramecium and other ciliates, or during which DNA passes from one to the other through a tube, as in bacteria. It is a sexual method of reproduction involving contact between mating cells; in bacteria, genetic material is transferred from donor to recipient cell through actual contact of the two cells.

Conjunctivitis. Inflammation of the conjunctiva of the eye.

Connective tissue. An animal tissue that connects or surrounds other tissues; its cells are embedded in a collage containing matrix.

Constitutive enzyme. An enzyme that is present in approximately constant amounts in a system, whether its substrates are present or absent. (Contrast with inducible enzyme.)

Consumer. Organisms feeding on preformed organic material.

Consummatory act. One of a set of very specific, stereo-typed patterns of motor responses. The behavior of most animal species consists largely or wholly of such consummatory acts and is loosely referred to as "instinctive."

Contagious. Transmitted directly from one person to another.

Contagium vivum fluidum. Latin phrase coined by M.W. Beijernick describing diffusible, filtrable and infectious suspension of tobacco leaves infected with tobacco mosaic virus. Literally means living infection fluid.

Contaminated area. A building area with definite boundaries where hazardous biological work is being carried out, separated from noncontaminated and other contaminated areas by suitable barriers.

Contaminated change room. Dressing room for removal of laboratory-type clothing before entering clean change room, through a mandatory shower, to don street clothing.

Contaminated service. A service or utility, such as water or vacuum, which serves a contaminated area and is therefore segregated from similar services to noncontaminated areas, even though the service itself is noncontaminated.

Contaminated water. Water which contains either poisonous chemicals or pathogenic organisms.

Contamination. Undesirable organisms in a culture or other material.

Continental climate. A pattern, typical of the interior large continents at high latitudes, in which bitterly winters alternate with hot summers. (Contrast with meri-time climate.)

Continental drift. The gradual drifting apart of the world's continents that has occurred over a period of billion years.

Continental islands. Islands separated by rising waters from continents to which they were once attached.

Continuous cell line. Animal tissue cells that are propagated in a manner analogous to that used for a bacterial culture, and which can be used for virus replication.

Continuous disinfection. Concurrent disinfection; disinfection procedures carried out throughout the course of an illness with careful disposal of contaminated materials.

Contractile vacuole. An organelle, often found in protists, which pumps excess water out of the cell and keeps it from being *flooded* in hypotonic environments.

Convalescent. State of recovery from disease.

Coordinated enzyme repression. The process or condition whereby the synthesis of all of the enzymes involved in the biosynthesis of a particular molecule are simultaneously repressed.

Copy choice. Refers to the mechanism of genetic recombination in bacteriophages. Relates to exchange of fragments of one viral genome with that of another viral genome, both simultaneously infecting the same host cell. This exchange takes place during replication of the nucleic acid molecules.

Core. The central nucleic acid portion of a virus particle.

Co-repressor. The end product of a biosynthetic pathway that combines with aporepressor (a protein) to control an operon. It is a low molecular weight compound that unites with a protein (the repressor) to prevent transcription in a repressible operon.

Coriolus force. The force exerted on objects moving toward or away from the equator.

Cork. A waterproofing tissue in plants, with suberin-containing cell walls. Produced by a cork cambium.

Corn steep liquor. A waste product of the corn starch industry; it contains both carbohydrates and nitrogenous compounds and can serve as an inexpensive source of each in the cultivation of microorganisms.

Cortex. (1) In plants: the tissue between the epidermis and the vascular tissue of a stem or root. (2) In animals: the outer tissue of certain organs, such as the adrenal cortex and cerebral cortex. (3) A layer found inside the coat layer of the endospore.

Corynebacterium diphtheriae. Gram-positive, pleomorphic rod which shows granular staining; exotoxin produced by lysogenic cultures causes the symptoms of diphtheria.

Cost. See energetic cost, opportunity cost, risk cost.

Cotyledon. (Gr. *kotyledon*: a hollow space). A "seed leaf." An embryonic organ which stores and digests reserve materials; may expand when seed germinates.

Counterimmunoelectrophoresis (CIE). Electrophoretically directed movement of antigen and antibody toward each other through a support medium.

Covalent bond. A chemical bond that arises from the sharing of electrons between two atoms. Usually a strong bond.

Coxiella burnetii. A species of Rickettsia; the causative agent of Q fever.

Crista. A small, shelf-like projection of the inner membrane of a mitochondrion; the site of oxidative phosphorylation.

Critical night length. In the photoperiodic flowering response of short day plants, the length of night above which flowering occurs and below which the plant remains vegetative. (The reverse applies in the case of long day plants.)

Critical orifice. An opening through which air passes, the diameter of which is such that the velocity of the air is independent of the pressure differential across the opening, provided the differential is above a certain value. A number of aerosol samplers operate with a critical orifice.

Critical period. In birds, the age during which song learning must take place.

Cross (transverse) section. A section taken perpendicular to the longest axis of a structure.

Cross-flow filtration. This is a commonly used method of filtering the sorts of dense and thick fluids that have to be filtered in biotechnological separations in order to concentrate some material. If you try to filter (say) soup through a standard micropore filter in order to concentrate the particulate material, the pores will rapidly block up, and filtration will be brought to a halt. Cross-flow filtering does not seek to filter the liquid through the filter directly. Rather, it flows the liquid across the filter, allowing the carrier liquid to flow through. After it has passed over, the top (unfiltered) phase is more concentrated, and some of the fluid phase has passed through; meanwhile, the filter is not blocked up. The build-up of material on a filter, blocking the pores or gaps, is called fouling, and is a major problem for any filtration process. Cross flow filtration is one of the more effective ways of getting round fouling. Other approaches include scraping the filter clean or replacing it periodically, or back-flushing. In this last one, clean liquid is pumped back through the filter in the reverse direction, so that the accumulated material is pushed off the surface, and can then be cleared away. This is not as rough on the filter as scraping, and is simpler to engineer than cross-flow systems. However, it does not always work if the fouling material 'glues' itself to the filter.

Crossing-over. The mechanism by which linked markers undergo recombination. In general, the term refers to the reciprocal exchange of corresponding segments between two homologous chromatids. However, the reciprocity of crossing-over is problematical in prokaryotes and viruses; and even in eukaryotes, very closely linked markers often recombine by a nonreciprocal mechanism.

Cross-pollination. The pollination of one plant by pollen from another plant. (Contrast with self-pollination.)

Crude birth rate. the number of live births in a calendar year per 1000 enumerated or estimated population at midyear.

Crustacean. A member of the arthropod class Crustacea, such as a crab, shrimp, or sowbug.

Cryoglobulin. An antibody or immune complex which forms a precipitate at 4° C.

Cryopreservation. This is preservation of things by keeping them cold. There are several variations of relevance to biotechnology.

Freezing. The most obvious route. Just putting something in the fridge or freezer is fine for many biological materials, but not all, since the process of freezing sometimes destroys what you are trying to preserve. This is especially true of cells because ice crystals forming in the cells break the cellular structures up, leaving you with dead meat.

Freezing in mixed solvents. To prevent damage to cells on freezing, they are often frozen in a mixture of a watery material (their usual growth medium) and another material called a cryoprotectant, which mixes with water and stops it forming ice crystals. Glycerol is a favourite for bacteria, dimethyl sulfoxide (DMSO) for animal cells. Sucrose can also be used: this is why ice lollies are not crunchy with ice crystals.

Bacterial cells preserved in this way can be kept in a conventional freezer, but animal cells need to be stored at liquid nitrogen temperatures if they are to survive more than a few weeks. This is often called storing them in liquid nitrogen vapour phase since the tubes of cells are kept in a flask of liquid nitrogen above the nitrogen itself, so they are not actually immersed in the liquid, only exposed to its vapour. Apart from anything else, this stops the tubes filling up with liquid nitrogen and then exploding when you warm them up. Even at these temperatures, mammalian cells eventually 'go off'.

Antifreeze proteins. There are proteins known that prevent ice crystals forming that are found in some arctic fish. In principle they could be used to replace the glycerol or DMSO (which are somewhat toxic), but this is rarely done in practice.

Freeze-drying. This is not a cryopreservation method really since the dried sample is not stored cold, but it is often classed as one. An odd application of cryopreservation was suggested by Robert Ettinger in his book The Prospects of Immortality, published in 1964. He suggested that dying people could be frozen until medicine had discovered a cure for whatever was killing them, whereupon they could be thawed out and cured. Both ice crystal formation and the inevitable decay of mammalian cells even at liquid nitrogen temperatures means that this surely cannot work, but if you are dead anyway, why not try it?

Cryptic appearance. The resemblance of an animal to some part of its environment, which helps it to escape detection by predators.

Cryptococcus neoformans. Yeast-like organism; causative agent of cryptococcosis.

Cuboidal. Refers to the category of viruses having a cube like morphological appearance.

Culture. A medium containing living microorganism, It is meant to grow a particular strain of microorganism.

Culture collections. Many countries and institutions have set up places where samples of microorganisms and cell lines are stored. Other names are strain depositories and type culture collections, the latter because they are where the 'type specimens' (i.e. the definitive specimens, which describe that 'type' of organism) are kept. They have a triple function: they are a 'bank' for your valuable microorganisms (against the risk that your laboratory burns down); they are a centre from which other people can get samples of your organism (if you want them to) without bothering you; and they are somewhere where you can deposit an organism to prove you own it, a sort of biological patent office. Some patenting systems insist that you deposit a sample of any organism mentioned in a patent that cannot be created easily by someone else at a recognized depository so that, if there is dispute later, there is some way of saying what your original organism looked like. The best known depository is the American Type Culture Collection (ATCC), which collects all types or microorganism and cell line. ATCC is also the World Health Organization (WHO) international reference collection. There are a variety of other general depositories in other countries, some of which specialize in fungi, bacteria, or animal cells. There are also industry-specific depositories for the dairy industry, marine organisms, pathogens, etc. This can be confusing to someone trying to find a specific organism, so there are a number of centres and database to help track down organisms. Europe has a culture collection or purely mammalian cells: the European Central Animal Cell Culture facility (ECACC) at Porton Down, UK.

Culture media. A solid or liquid substance containing nutrient materials for the cultivation of microorganisms.

Cuticle. A waxy layer on the outer surface of a plant or an insect, tending to retard water loss.

Cutin. (L. *cutis*: skin). A mixture of long, straight-chain hydrocarbons and waxes secreted by the plant epidermis, providing a water-impermeable coating on aerial plant parts.

Cyanophage. Viruses which infect blue-green algae and destory them.

Cyclic photophosphorylation. The overall system in which an electron leaves chlorophyll and, after passing through a series of electron acceptors, returns to chlorophyll. During passage of the electrons from ferredoxin back to chlorophyll, two molecules of ATP are formed from ADP for each pair of electrons passing through the cycle.

Cyclodextrins. These are cyclic carbohydrates made of six, seven, or eight glucose molecules joined in a ring, to form alpha-, beta-, and gamma-cyclodextrin, respectively. The cyclodextrins all form cylindrical molecules with their water-soluble groups on the outside of the molecule and a relatively non-polar hole down the middle. This hole can accommodate another molecule, known as the 'guest' molecule. This allows them to be used in a wide range of applications, including improving the solubility of drugs and biopharmaceuticals, and selectively binding materials that 'fit' into the central hole in affinity purification and chromatography methods.

Natural cyclodextrins are not used extensively in drug applications because they are not very soluble and are rather toxic to injection. However, they may be modified by adding alkyl or hydroxyalkyl groups on to the hydroxyls of the natural cyclodextrin, which reduces toxicity and can enhance solubility.

Cyclophosphamide. A cytotoxic drug frequently used as an immunosuppressive.

Cyclosporin. An immunosuppressive drug which is particularly useful in suppression graft rejection.

Cyst (sist). (Gr. kystis: bladder, touch). (1) A resistant thick-walled cell formed by some protists and other organisms. (2) An abnormal sac, containing a liquid or semisolid substance, produced in response to injury or illness. It is encystment; the cell forms a heavy protective coating and becomes very resistant to adverse conditions; characteristic of many protozoa.

Cystitis. Infection of the bladder.

Cystoscope. An instrument for examining the interior of the urinary bladder.

Cytochomes. Protein molecules containing iron-prophyrin groups associated with electron transport. The iron ion playsa key role in passing on electrons:

Cytokinesis. (Gr. *kytos*: container + *kinein*: to move). The division of the cytoplasm of a dividing cell. (Contrast with mitosis.)

Cytolytic. Causing lysis and death of the cell.

Cytopathic effect. Destruction of tissue culture cells by an infecting virus.

Cytopathic. Characterized by pathological change in a cell.

Cytophilic antibodies. Antibodies that bind to cell surface receptors through the Fc portion of the lg molecule.

Cytophilic. Having a propensity to bind to cells.

Cytoplasm. The living cell substance within the cell membrane and surrounding the nucleus.

Cytoplasmic streaming. Continuous movement of cytoplasm within the cell which results in a constant distribution of intracellular contents, provides amoebic motility to some types of cells.

Cytoplasmic streaming. The flow of the cytoplasm in eukaryotic stationary cells that is used to distribute nutrients to all parts of the cell.

Cytosol. The fluid portion of the cytoplasm, excluding organelles and other solids.

Cytostatic. Having the ability to stop cell growth.

Cytotoxic reaction. Allergic reaction directed against antigens on the cell surface.

Cytotoxic. Having the ability to kill cells.

D

Dalton. (After J. Dalton, British physicist): A unit for measurement of molecular weight of complex macromolecules; is equivalent of 10-24 of molecular weight of oxygen.

Damping-off. Collapse of seedling near soil line from necrotic infection induced by water and fungus.

Darwinian cloning. This means selecting a clone from a large number of essentially random starting points, rather than isolating a natural gene or making a carefully designed artificial one. From this mixture you select, by whatever means at hand, those molecules that look more like those you want than the rest. (How you select them depends, of course, on what sort of molecules you want.) You mutate these to generate a new set of variants, and reselect, make more variants, and so on until you have the molecule you require. This is a process akin to natural selection, and so is sometimes called 'directed evolution' or (incorrectly) 'molecular evolution'. There are several classes of catalytic molecule suitable for this.

Catalytic antibodies. Indeed, all antibodies are evolved in this way: the body does the randomizing and selection procedures in the immune system.

Random proteins. In principle, you could clone a totally random bit of DNA in an expression vector, measure enzyme activity, make alterations in the DNA of the clones that show the best activity by random mutagenesis, select again, and so on. The most practical version of this is done using phage display technology.

Antisense. The word 'aptamer' has been coined for antisense RNAs and DNAs that have been selected by Darwinian cloning to bind very tightly to a particular gene or RNA. The starting point here is a random chain of bases, which is bound to the target molecule. Those that do not bind, or only bind weakly, are simply washed away and discarded. The few molecules (out of billions) that are left are then detached and amplified using PCR. These can then be altered by mutation, selected, amplified again, and so on.

Catalytic RNA. RNA can also be selected in this way, but with the added advantage that RNA can be a catalyst in its own right. This sort of Darwinian selection has been done to make RNAs that will bind specific low molecular weight chemicals very tightly. The next step is to find one that binds a transition state analogue for a reaction which would, plausibly, make a new catalytic RNA. **The advantage** of Darwinian systems is that they select a new catalyst from a vast number of possibilities. There are more possible 100-amino acid proteins than there are electrons in the universe, so screening them all is clearly impractical. However, this approach edges up to the catalyst you want one step at a time. If the catalyst you want has not been found in nature, then this could be an approach to getting it. A company, Affymax, has been founded specifically to take advantage of such technologies. Many other groups are using similar methods, all of which are still experimental.

Deaminase. An enzyme which catalyzes The removal of an amino group from a moleculewith the liberation of ammonia; the process is deaminations,

Death curve. A graph which shows the rate at which a microbial population dies. It can be plotted as a logarithm of the number of survivors at any given time. The curve shows that death occurs as a first order reaction.

Death phase. Phase in the growth of a culture when reproduction usually will have stopped. As the rate of death exceeds the rate of reproduction, the actual number of viable microorganisms declines.

Debilitated. Weakened.

Debilitation. Loss of normal function.

Decarboxylase. An enzyme which catalyzes the removal of carbon dioxide from the carboxyl group of an acid; the process is decarboxylation.

Decay. Aerobic breakdown of complex materials by microbial enzymes.

Deciduous. (L. *decidere,* fall off). Referring to a plant that sheds its leaves at certain seasons. (Contrast with evergreen).

Decomposer. An organism, such as a bacterium, fungus, or carrion beetle, that consumes dead organic matter, recycling it through the ecosystem.

Decontamination. The destruction or removal of living organisms to some lower level, but not necessarily to zero.

Deep-bed filter. Common form of high-efficiency filter for low-pressure use in ventilation systems.

Defective phage. Indicates functionally incomplete bacteriophages, those which cannot complete the replication cycle.

Defense guild. A group of plant species that interact with each other in their relationships with predators; the interactions are usually chemical. (Contrast with dependency guild.)

Degeneracy. The situation in which a single amino acid may be represented by any of two or more different codons in messenger RNA. Most of the amino acids can be represented by more than one codon.

Dehydration. Removal of water.

Dehydrogenases. Oxidoreductases, which act by removing hydrogen and electrons from a substrate.

Dehydrogenation. The usual type of biologic oxidation that occurs in microorganisms. The enzyme dehydrogenase causes the hydrogen donor to lose part of its hydrogen, and as hydrogen and electrons are transferred energy is released.

Delay hypersensitivity. An allergic or hypersensitive reaction to a foreign substance resulting from a cellular immunity to that substance. It is T cell mediated hypersensitivity in which the reaction reaches the maximum in about 24 to 48 hours after exposure to the antigen that elicited the original hypersensitive state.

Deletion (genetic). A mutation resulting from the loss of a continuous segment of a gene or chromosome. Such mutations never revert to wild-type. (Contrast with duplication, point mutation.)

Delfia. This is a trade name for Delayed Fluorescene Immuno Assay, marketed by Pharmacia. It is one application of a type of fluorescence detection called time-resolved fluorescence. The problem with fluorescence as a method of detection is that it is impossible to distinguish between the fluorescence of your 'marker' molecule (which you want to detect) and the fluorescence of everything else in your sample, including the sample holder (which you do not want to detect). One solution is to use a fluorescent material that has a long 'fluorescence half-life', i.e. one that goes on fluorescing for a long time after you have turned off the exciting light source. Then you look at the fluorescence after the exciting light has been turned off.

Deliberate release. This is putting something into the outside world ('the environment'), usually meaning putting a genetically engineered organism into field trials: they are often called GMOs (genetically engineered organism) or sometimes GMMO (genetically engineered microorganism). A wide range of such trials have been suggested, and some have been done: the first was probably the trial of a genetically engineered frost-proofing bacterial strain in California in 1986. By the end of 1989 there had been 140 deliberate release experiments in the USA, with about half that number in Europe. A wide range of social, scientific, and political pressure groups support and oppose such trials, on the basis that the organisms may be or are known to be dangerous. The biotechnology industry considers the fears to be widely exaggerated, and complains that every time they take a precaution to allay those fears, the opponents of the release experiments take it as proof that the organisms concerned really were dangerous after all.

Deliberate release experiments are the natural follow-on from laboratory trials, and then, for organisms involved in agricultural applications, greenhouse trials. Laboratories can include a range of barriers to stop genetically engineered organisms getting out: negative pressure rooms, sterilization procedures, and genetically engineering the organisms so that they cannot survive in the outside world. But, of necessity, none of these can apply to release into the outside world. Here, affected fields, animals, soil, etc. is kept isolated from surrounding farms, and affected material is destroyed after the trial (except for some Australian pigs, which accidentally got to market as human food in 1988). *Regulation about deliberate release vary widely.*

Deme. (Gr. *demos*: common people). Any local population of individuals belonging to the same species and among which mating is random.

Denaturation. Loss of activity of an enzyme or nucleic acid molecule as a result of structural changes induced by heat or other means.

Denature. To loss the natural configuration, for example, heating an enzyme in which there is change in conformation and usually biologic activity; also stripping a part of complementary strands of DNA.

Dendrite. (Gr. *dendron*: a tree). A fiber of a neuron which often cannot carry action potentials. Usually much branched and relatively short compared with the axon, and commonly carries information to the cell body of the neuron.

Denitrification. Metabolic activity by which inorganic nitrogen-containing ions are reduced to form nitrogen gas and other products; carried on by certain soil bacteria.

Density gradient. A solution, usually contained in a centrifuge tube, that varies continuously from a higher density at the bottom to a lower density at the top.

Dental caries. Dental cavities.

Deoxyribonucleic acid (DNA). The nucleic acid primarily found in the nucleus of cells; it contains-adenine, guanine, cytosine, thymine, deoxyribose, and phosphoric acid.

Deoxyribose. A five-carbon sugar found in DNA; it contains one oxygen atom less than ribose.

Dependency guild. A group of plant species that provide one another with support or nutrition. (Contrast with defense guild.)

Depurination. Removal of purine bases.

Dermatophytes. Fungi which infect superficial tissues, the skin, hair, and nails.

Desiccation. Drying, or removal of water.

Desquamation. Shedding of the superficial layers of skin.

Desulfurization. This is the removal of sulfur-containing materials, usually sulfides, from coal and oil. Sulfur residues in fuels end up as sulfur dioxide when the fuel is burned, causing acid rain. However, sulfur-containing fuels are often cheaper than 'clean' fuels. As a rough guide, a 'high sulfur' coal would contain 6% sulfur (mostly as pyrites), and cost $50-100/tonne less than a 'low sulfur' coal containing 1% sulfur or less. Thus there is an economic motive for removing the sulfur from oil and coal. The same types of bacteria that are used in microbial mining can be used for desulfurization of coal. They oxidize sulfides (which are insoluble) into sulfates (which are soluble). The sulfates can then be washed away, with the bacteria. This does not work well on lump coal since the bacteria cannot get inside the lumps fast enough to be economic, but it can be effective for treating pulverized coal such as that used in electricity generating stations. Crude oil also contains significant sulfur: between 0.1% (from the Far East) to up to 3% (for some Middle East oils). Usually, has its sulfur removed by hydrodesulfurization, a physicochemical technique, but work on using bacteria to remove the sulfur shows that it has some potential.

Detergent. A surface active agent used in cleaning. It is a synthetic substance that reduces surface tension of water.

Determinant group. The portion of an antigen which specifically combines with the antibody.

Determination. Process whereby an embryonic cell or group of cells becomes fixed into a predictable developmental pathway.

Detoxication. Biological destruction of a toxic chemical.

Detritivore. (L. *detritus*: worn away + *vorare* to devour). An organism that eats the dead remains of other organisms.

Deuteromycetes. Fungi imperfecti; fungi in which no sexual stage has been recognized.

Deuterostome. One of two major lines of evolution in animals, characterized by radial cleavage, enterocoelous development, and other traits.

Dextrin. A soluble polymer of glucose usually derived from starch.

Dextrorotatory. Capable of rotating plane polarized light to the right.

Diabetes. Disease in which the pancreas fails to produce insulin.

Diagnostics. This usually means chemically based diagnostic tests for disease. Although it has its roots in traditional medicine (medical student lore has some ancient clinicians tasting urine to see if it was sweet, and therefore if the patient was diabetic), chemical diagnostics has come of age with the development of precise chemical methods. Diagnostics test for chemical changes in the body that are characteristic of specific diseases. **Chemicals can be**: simple molecules or ions, like sodium or potassium (as checks of kidney function), or glucose (an excess of which is definitive of diabetes), **enzymes**, such as creatine kinase (characteristic of heart attacks) or a separate aminotranferase (released by damaged liver), or marker proteins, such as those made by viruses, bacteria, or cancers, or benign proteins such as the hCG hormone that marks pregnancy. This third group now covers more than 40% of all diagnostic blood tests, and is performed by immunoassay. Other technologies used to measure these chemicals are: **direct enzyme** activity assay (measuring the catalytic activity of an enzyme), and **enzyme assay** of another molecule (using an enzyme to measure the presence of another molecule, e.g. cholesterol or glucose). These assays can be colourimetric, can be linked to an enzyme electrode, or can be measured in a number of other, more complicated ways. Biotechnology helps develop diagnostics in two ways. If finds out more about living systems in health and disease, so we can discover the 'markers' that a test subsequently detects. It can then develop reagents to do the tests, notably monoclonal antibodies. DNA probe-based tests are another rapidly growing area of diagnostics that are based on biotechnology.

Diaphragm valve. Widely used in contaminated service because of zero leakage at the stem.

Diarrhea. Abnormal fluidity of fecal discharge.

Dicotomous. (Gr. *dis:* two + *kotyledon:* a cup-shaped hollow). Any member of the angiosperm class Dicotyledonae, flowering plants in which the embryo produces two cotyledons prior to germination. Leaves of most dicots have major veins arranged in a branched or reticulate.

Differential display. This is a group of techniques for showing the differences between the way that genes are working in two samples, for example between two related bacteria or between normal and cancerous tissue. There are lots of technical approaches. The original differential display technology amplified fragments of cDNA from the mRNA of the target tissues, and then separated then on a polyacrylamide get to detect the different sizes of fragments. The PCR primers have to be selected carefully so that they detect the majority of the mRNAs is without giving so many bands that the get that results is meaningless.

Virtual genomics. This is doing the comparison in a computer. You sequence every mRNA from two tissues. The mRNA from any one gene will be sequenced several times (depending on how much mRNA that gene is making). By comparing how many times you find the same mRNA in different tissues, you can estimate how often you find each mRNA in each tissue. Also called biology in silico or in machina, or the digital northern, this requires truly huge numbers of DNA sequencing reactions to work effectively.

Proteome. It is easier to separate a lot of proteins than a lot of mRNAs, so several groups are now looking at comparing the protein products of a cell's genes rather than the mRNAs. This general approach is called the proteome (contrasting with genome).

Differential medium. A medium which shows different color reactions due to the enzyme activity of different kinds of bacteria; e.g., Hektoen agar will be changed in color by lactose fermenters but unchanged by nonlactose fermenters.

Differential stains. Stains which reveal chemical differences in bacterial structures; more than one stain is employed.

Differentiation. Process whereby originally similar cells follow different developmental pathways. The actual expression of determination.

Diffusion. Random movement of molecules or other particles, resulting in even distribution of the particles when no barriers are present.

Digestion. Enzyme-catalyzed process by which large, usually insoluble, molecules (foods) are hydrolyzed to form smaller molecules of soluble substances. Large complex molecules of protein, carbohydrate, and lipid are hydrolyzed to simple molecules that go readily into solution and can enter the cell.

Digester. This is a bioreactor that uses microbial growth to break down waste. The most common use is in disposal of sewage or agricultural waste, where the digester is used to reduce the volume of biological matter to carbon dioxide or methane and compacted biomass (the cells).

Anaerobic digestors operate in the absence of oxygen, and generate methane gas, which can itself be used as a fuel. Anaerobic digesters for materials with high solids content, such as farm waste, can be very simple, and can generate useful amounts of methane. Typically 0.3-0.5 m3 of gas, which is 50-70% methane, can be generated from 1 kg of compacted material. Most of the rest of the gas is CO_2: the amount of CO_2 that comes out of the digester depends on the amount of liquid there is to dissolve this quite soluble gas. Digesters that handle material with more soluble organic matter in, such as sewage, need a more sophisticated design.

Anaerobic digesters can be batch fed or continuous. In the former, one batch of material is fed in, the unit sealed (to keep out air) and the digestion left to run to completion. In the latter, material is fed in continuously, and the flow must be balanced with the rate at which the bacteria in the digester can cope with new material. A variant on this is the anaerobic contact digester, where the living cells (active biomass) are separated from the material flowing out of the digester and fed back into it to digest more substrate. The IRIS (Institute de Recherche de l'Industrie) digester is of this design, but the separation system is integral to the digester itself.

Commercial anaerobic digesters of both designs are used to dispose of sewage sludge, and to produce biogas from purpose-grown energy crops. The biogas has to be 'scrubbed' before use because sulfur compounds in the feedstock are turned to hydrogen sulfide, which is smelly and, at moderate concentrations, poisonous. Farm waste and some industrial effluents can also be reduced using digestion, but usually have to be pretreated to break down the cellulose in them. Generally these digesters operate at mesophilic temperatures (around 35° C), although ones using thermophilic organisms operating at over 60° C are also sometimes used.

Aerobic digesters are also used, notably the activated sludge process in sewage disposal. This destroys the organic content of waste more completely, and produces fewer noxious side products (sulfur is oxidized to relatively harmless sulfate, for example). However, the growing microorganisms need lots of oxygen, which has to be pumped into the digester at considerable cost in energy. As well as the familiar trickle bed aerobic digesters, sewage waste and other organic materials are disposed in deep shaft systems. Here, oxygen or air is pumped into the bottom of a shaft at pressure, where it dissolves more rapidly that at atmospheric pressure. This allows high oxygen tension in the liquid, and so a high biomass content and fast metabolism of the waste.

Dikaryon. A mycelium in which each cell contains two genetically dissimilar nuclei. Dikaryotic mycelium is extensive in Basidiomycetes and forms the entire fruiting body (mushroom). It also leads to the formation of basidia and basidiospores. In Ascomycetes, only the ascogenous hyphae are dikaryotic.

Dimorphic. Occuring in two forms; characteristic of certain fungi, growing in the yeast phase under one set of conditions and filamentous under other conditions.

Dimorphism. The ability to exist in one of two morphological states; for example, some fungi may exist in filamentous or yeast-like state.

Dioecious. (Gr.: two houses). Organisms in which the two sexes are "housed" in two different individuals, so that eggs and sperm are not produced in the same individuals. Examples; humans, fruit flies, oak trees, date palms. (Contrast with monoecious.)

Diplococci. Cocci which occur in pairs.

Diploid cell strain. Cultures of cells or tissues derived out of primary cultures (q.v.).

Diploid. (Gr. *diploos*: double). Having a chromosome complement consisting of two copies (homologues) of each chromosome. A diploid individual (or cell) usually arises as a result of the fusion of two gametes, each with just one copy of each chromosome. Thus, the two homologues in each chromosome pair in a diploid cell are of separate origin, one derived from the female parent and one from the male parent.

Disaccharide. A sugar composed of two monosaccharide rings joined in glycosidic linkage; the most common are lactose, maltose and sucrose.

Disease. Any change from a state of health; interuption in the normal functioning of a body structure. It is an abnormal condition of the body, usually having characteristic symptoms.

Disinfectant shower. Unit at exit from ventilated suit area in which the suit is externally decontaminated by mist or spray of disinfectant, such as peracetic acid, before being removed.

Disinfectant. A chemical agent that destroys disease organisms; used on inanimate object, like vegetative bacteria, fungi, and viruses, but not spores.

Disinfection. The process of destroying pathogenic agents by the use of disinfectants.

Disjunction. (L. *disjungere*, to disunite or to separate). Refers to the dissociation of the component parts of a virus particle.

Dispersal stage. Stage in its life history at which an organism moves from its birthplace to where it will live as an adult.

Display. A behavior that has evolved to influence the actions of other individuals.

Disruptive selection. Selection in which extreme phenotypes are favored. (Contrast with stabilizing selection.)

Dissimilation. The act of breaking down a component into smaller parts.

Distal. Away from the point of attachment or other reference point. (Contrast with proximal.)

Disulfide bond. This is a form of chemical bond in proteins much talked about by biotechnologists because of its role in stabilizing the 3D structure, and hence the normal function, of proteins. They form when two cysteine amino acids in the protein react to form one cystine residue. They link via their sulfur atoms, which therefore form a bridge of two sulfurs between distant part of the peptide chain that fold close to each other in space. Once linked in that way, the chain is locked into that fold, since to unfold again would mean breaking the covalent bond.

Diverticulum. (L. *divertere*: turn away). A small cavity or tube that connects to a major cavity or tube.

Division. A term used by botanists and some microbiologists, corresponding to the term phylum used by zoologists.

DNA amplification. This is the use of enzymes to take a piece of DNA and multiply it in a test-tube into many thousands of millions of copies. It is of enormous potential use in detecting when specific genes are there without using radioisotopes. The best known and by far the most commonly used in the polymerase chain reaction (PCR) system invented at Cetus. Other systems announced or being developed include the following. (I will not attempt to describe them all in detail here!).

***Ligase chain reaction* (LCR).** Uses DNA ligase, the enzyme that joins two DNA molecules together, to link two oligonucleotides together if a target DNA is present. The linked molecules can then be templates for more molecules to link together, and so on. This technology is owned by Abbott Diagnostics, and was the first to be built into an automated test system reliable enough to be used in a routine medical diagnostic laboratory. There is a non-cycling version of LCR, called oligonucleotide ligation assay (OLA), which links the two probes together but does not then separate the product. If one probe is linked to a solid support and the other to a fluorescent label, then the hybridization and ligation links the fluorophore to the solid support. This is a version of a 'traditional' DNA probe assay.

***Nucleic acids sequence-based amplification* (NASBA).** This creates a new molecule of the DNA joined on to a promoter for RNA polymerase. The amplification cycle occurs when the RNA polymerase copies this DNA on to RNA, which is then turned back into DNA by reverse transcriptase. The advantages are that this all occurs at one temperature, and that the RNA polymerase creates many RNA molecules from one DNA molecule, so it has the potential for being very efficient.

***Branched multimer amplification*.** A system developed by Chiorn Inc. builds a branched tree of separate nucleic acids on a single probe. At the end of each branch is a label enzyme. This is a probe amplification system: the amount of target is the same as before, but each single probe hybridized gives rise to many signals, and hence a good chance that the detection method will find it. There is also an RNA-based system, the Q-beta system of Gene-Trak. The RNA of a small virus (Q-beta) is duplicated by the RNA polymerase enzyme, which the Q-beta virus carries. Add one molecule of Q-beta RNA to a tube of Q-beta replicase and the right chemicals, and the tube fills up with Q-beta RNA. The Q-beta replicase amplification system uses the enzyme to replicate RNAs that are related to the original RNA but have a probe sequence in them. Unlike the other systems mentioned above (which are target amplification systems), this is another probe amplification system. All these systems are being developed to be used in medical diagnostics. All suffer by a greater or lesser degree from the problems of their extreme sensitivity to contamination. Most DNA amplification systems are qualitative: if there is target DNA there, they will amplify it. It is hard to turn then into quantitative assays, which could tell you how much DNA was there. The relationship between how much DNA there was to start with and how fast it is amplified is complicated by many factors. The branched multimer technology is the only commercially used one that is relatively straight-forward to quantify. Other amplification techniques need to use reference samples with known amounts of target in to get a comparative measure of the target levels.

DNA fingerprinting. DNA (or genetic) fingerprinting, or profiling, is a way of making a unique pattern from the DNA of an individual, which can then be used to distinguish that individual from another. Most DNA fingerprinting systems rely on DNA probes, short pieces of DNA that hybridize to the genes from an individual to identify specific pieces of DNA within the total collection of DNA. The original DNA fingerprinting probes, discovered by Prof. Alec Jeffreys, used 'minisatellite' DNA, DNA that hybridized to short runs of bases called minisatellites, which vary greatly between individuals. Because there are 50-100 of each type of minisatellite in each person, the chances that any two individuals will have the same pattern of all the minisatellites is minuscule unless they are related. This type of repeat sequence is now usually called variable number tandem repeats (VNTRs).

Different DNA fingerprinting systems use different probes. Some also use PCR to amplify the target DNA, so that smaller amounts can be analysed. It is also possible to create 'single locus probes'. Whereas the normal DNA fingerprinting probes create a pattern like an irregular ladder to be compared between individuals, single locus probes detect just one DNA sequence, i.e. one rung on the ladder. This makes comparison between two individuals easier. DNA fingerprinting has been used extensively as evidence in paternity, rape, and murder cases to identify individuals. Until 1989 it was considered unassailable evidence, but since then several cases have questioned poorly collected or poorly analysed DNA fingerprinting data, starting with the case of State vs. Castro in New York, where supposedly water-tight DNA fingerprint evidence was refuted on technical grounds by the defence. This lead both to a better understanding of the strengths and weaknesses of DNA fingerprinting and to better quality control in DNA fingerprinting laboratories. DNA fingerprint evidence featured in the O.J. Simpson murder trial of 1996. The 'O.J. trial' showed that the value of DNA fingerprint evidence rests more on getting the samples carefully and agreeing what they mean than on the molecular genetic technology of actually making the fingerprint.

DNA probes. As well as being used as the genetic material to 'programme' cells to do things, DNA can be used as a reagent in its own right. DNA used in this way is almost always used as a DNA probe, also called a hybridization probe. One strand of the DNA double helix is used to bind to a target strand of DNA. If the base sequences are complementary (i.e. A matches with T, G with C), then the two strands will form a double helix. If they are not complementary, then no helix will form. Thus the DNA probe can be used as a reagent to detect when a specific DNA sequence is present among a mixture of sequences. This process of getting a DNA probe to bind to a target sequence is called hybridization, and can be used to detect DNA or RNA. DNA probes have been used in genetic research for over 30 years, but only became common when DNA cloning meant that pure DNA probes derived from just one gene could be made. DNA probes are still the standard method for finding a DNA sequence among a mixture, often allied with the 'blot' technology of analysing complex mixtures of DNA molecules. DNA probes are particularly used in medical genetics as a way of finding whether a person carries a particular gene or not (although in this application they are gradually being replaced by PCR-based techniques). They also have potential for use in detecting pathogenic bacteria, although this has not been realized as fast as was expected in the early 1980s. Probes are also the basis of DNA fingerprinting. One common use of DNA probes is to find a gene similar to one you already possess. Thus if I have a clone of a gene that performs a useful function in one organism, I can use the DNA from that clone to identify the similar ('homologous') gene in a range of related organisms, which may be better suited to my practical application. (Actually, purists insist that 'homologous' has a different definition, but few technologists are purists.) This can be useful for cloning, say, heat-resistant enzymes from thermophiles if you have already cloned the gene from an organism such as *E. coli* which is easier to grow and manipulate but not so biotechnologically useful. Traditionally, DNA probes have been made by cloning a gene and using its DNA as the probe. In the last few years oligonucleotides made on a DNA synthesizer have gained favour as probes. They react faster, so reducing assay time, they can be made more specific, so distinguishing between genes that differ by only one base, and they can be made in comparatively large amounts very cheaply. The primers necessary for such techniques as PCR could be considered as a form of probe. DNA probes are starting to be integrated into 'biochips', large arrays of probes immobilized on to very small devices for use as diagnostic assays. Many DNA tests actually need you to conduct dozens or even hundreds of hybridization reactions to get all the information you need. The biochip idea allows you to do all those reactions on one small sample (because the chip is very small) and one chip (because it has all the probes on it). So far only Affymetrix has succeeded in reliably building DNA biochips, by chemically synthesizing the DNA on a glass 'chip' using a process akin to photolithography.

DNA sequence patents. Biotechnology companies want to protect their intellectual property with patents, and since some of the leading edge of biology is concerned with gene sequencing, it is inevitable that researchers should try to patent DNA sequences that they have discovered. (Patenting invented DNA sequences, such as synthetic genes, has never been controversial.) The debate was kicked off in 1991, when the NIH in the USA filed a patent claiming as their invention the sequence of 337 cDNA clones, copies of naturally occurring RNAs, sequenced by Craig Ventor. This broke new ground in claiming that a newly discovered sequence was actually an invention. They also claimed it before utility had been proven - no one knew what the sequences could be used for. Several other blanket claims on EST (Expressed Sequence Tag) sequences from NIH and UK Medical Research Council (MRC) followed. The application was rejected in 1992, failing on the basis of novelty, utility, and obviousness. NIH and MRC dropped claims in 1994, but the debate over whether you can patent short DNA sequences of unknown utility still rages. Several companies, notably Human Genome Sciences and In Pharmaceuticals, they are basing their business in part on the patent protectable value of the sequences they discover. HGS owns the patent rights to Crag Ventor's work (now at The Institute of Genomic Research, TIGR), and was still pursuing the patents in mid-1996. The US pharmaceutical company Merck is deliberately putting all the DNA sequence they discover into the public domain as a 'spoiling action' to stop anyone else from patenting it (because once it is described in public, it is no longer novel). Meanwhile, several companies have applied for patents on genomic genes or gene variants. The first was Myriad Genetics, for the BRCA1 gene (a variant of which predisposes about 5% of women to breast cancer). Again, there is fierce debate between those that think that this is essential and those that think it immoral.

DNA sequencing. Determining the sequence of the bases in DNA (DNA sequencing) is one of the mainstays of gene cloning technology. The most common technology for this is the Sanger technique (dideoxy method, chain termination method). This uses enzymes to make a new DNA chain on the target you want to sequence, using the 'dideoxy' reagents to stop the chain randomly as it grows. A chemical technique invented by Maxam and Gilbert uses chemicals to break the DNA into fragments, but is not used much nowadays. In both cases the results of a series of reactions are analysed by polyacrylamide gel electrophoresis, to give information that can be read directly to give the sequence of the original DNA. DNA sequencing is starting to be used as a medical diagnostic technique as well as a research tool. Now, many genes are known, such as oncogenes, whose mutation is an important factor in disease. Because any mutation can cause disease, it is impractical to design a DNA probe to test for the disease-causing mutation. So clinicians would like to determine the whole gene sequence in their patients. At the moment this is very costly and complex, a situation the biochip might remedy.

DNA. Deoxyribonucleic acid. The genetic material of living organisms. Its double-stranded structure, the pairing specificity of the bases from which it is composed and its repairability enable it to carry out its own two main functions; (i) accurate transmission of heritable information from parent to daughter cell and (ii) determination of the amino acid sequence of proteins.

DNP. (Dinitrophenol). A commonly used hapten.

Dolipore. The circular perforation found in the septa of Basidiomycete hyphae. They have a characteristically thckened edge which is often associated with endoplasmic reticulum.

Domain. In immunoglobulin chemistry the folded regions or segments of H and L immunoglobulin chains that are stabilized with disulfide bonds. L chain have one domain in the variable region and one in the constant region. H chains have one domain in the variable region and three or four domains in the constant region. A region of a peptide having a coherent tertiary structure. Both immunoglobulins and MHC Class 1 and Class 2 molecules have domains.

Dominance distribution. Distribution of organisms resulting from territoriality or other defense of space, such that success is better for individuals with territories in better habitats. (Contrast with ideal free distribution.)

Dominance hierarchy. The set of relationships within a group of animals, usually established and maintained by aggression, in which one individual has precedence over all others in eating, mating, and other activities; a second individual has precedence over all but the highest-ranking individual, and so on down the line.

Dominance. In genetic terminology, the ability of one allelic form of a gene to determine the phenotype of a heterozygous individual, in which the homologous chromosome carries both it and a different allele. For example, if A and a are two allelic forms of a gene, A is said to be dominant to a if AA diploids and Aa diploids are phenotypically identical and are distinguishable from aa diploids. The a allele is said to be recessive.

Dormancy. A condition in which normal activity is suspended, as in some seeds and buds.

Dorsal. (L. dorsum: back). Pertaining to the back or upper surface. (Contrast with ventral.)

Downstream processing. This is a general term for all the things that happen in a biotechnological process after the biology, be it fermentation of a microorganism or growth of a plant. It is particularly relevant to fermentation processes, which produce a large amount of a dilute mixture of substrates, products, and microorganisms. These must be separated and the product concentrated and purified, and converted into a product that is useful. There are three general steps to downstream processing: **separation, concentration, and purification**. The first step separates the crude product from the microbial mass and other solid lumps, the second removes most of the water (and hence is often called dewatering), and the third takes the concentrated product and purifies it. The order of events can be different, but generally falls into this scheme. Separating the microbial mass is necessary whether the product is inside the microorganism or outside it; the difference is that in the first case you keep the mass, in the second you throw it away. This can be done by centrifugation (expensive but guaranteed efficient), filtration methods, especially cross-flow filtration, or by flocculation (adding something to the microbes so that they clump together and settle out on their own). If the product is inside the organism then separation also concentrates the product: however, you then have to break open the organisms to get it. Some of the same methods can also be used for concentration. Simply drying large volumes of liquid is usually too expensive, so again ultrafiltration or reverse osmosis (both membrane methods that keep the product on one side of the membrane while most of the water goes through it to the other side) are popular.

Product concentration. The result of the above steps is usually a rather dilute solution of the product, which must be concentrated. This can be by reverse osmosis, adsorption methods, or extraction with another liquid.

Purification. Most biotechnology products are produced as mixtures by cells, but are required pure. Purification methods include chromatography, affinity methods, and various specific precipitation methods. If the product is produced by genetic engineering, then is may be engineered to have a molecular 'hook' which makes it easier to isolate.

Droplet infection. Infection acquired from recently contaminated air via suspended droplets.

Droplet nuclei. Airborne particles that originated as droplets, but have dried and left behind mucus, microorganisms, and other material that may have been associated with the droplet. Because of their small size, they frequently remain suspended in the air for long periods of time.

Droplet. An airborne particle consisting primarily of liquid. Droplets are usually generated by talking, sneezing, or coughing, and while some settle out almost immediately, many dry to become droplet nuclei to add significant numbers of microorganisms to the air.

Drug delivery. This is the method by which a drug is delivered to its site of action. For traditional drugs this is a different name for formulation, i.e. in what form the drug is to be given to the patient (tablet, capsule, syrup, etc.). Drugs can also be made as a prodrug, a compound that is not itself a drug but which the body metabolizes into the drug. If the metabolism only occurs in one tissue or cell type, then the drug is only released there. Although a skilled area of pharmacology, this has little biotechnological content. However, two aspects of biotechnology have needed drug delivery technology. Biotechnology has allowed the development of a range of new drug delivery systems, such as liposomes and other encapsulation technologies, and antibody-based drug targeting mechanisms (such as immunotoxins) that target a drug to a particular cell or tissue. Secondly, biotechnology has created the need for new drug delivery systems to get biotechnologically derived drugs to their site of action. This is particularly acute in the case of biopharmaceuticals, protein drugs that cannot be taken by mouth since the stomach acid and gut enzymes would destroy them. Even if they could survive this digestive gauntlet, they would rarely get into the bloodstream, because the protein molecules are too big to diffuse through the intestinal walls. The current solution is parenteral delivery (i.e. injection): this works well, and has been the way of giving patients insulin (a protein drug) for decades. However, it is invasive, expensive, and carries a continued risk of infection or tissue damage. Thus, several biotechnology companies have set out to find a better way of getting proteins into the blood. There are several approaches:

Transdermal delivery. Using a method of getting the protein across the skin without knocking visible holes in it. Methods used include iontophoresis (using electric fields to push the drug across the skin) and high pressure jets of fluid. However, the skin has evolved to resist exactly this sort of attack, so these methods are not very effective for proteins.

Oral delivery. Taking the drug by mouth, but with some other materials which help it to survive the gut. These can include protease inhibitors (to block the digestive enzymes) or carrier materials, which protect the proteins but which dissolve at the right time to make them available for absorption. Other tricks include linking the protein to something (like vitamin B12) which is actively taken up from the gut, so the protein is absorbed along with it.

Nasal/pulmonary delivery. The cells lining the lungs and some of the nose (their epithelial cells) are very poor barriers to proteins compared with skin and guts, and so are potential 'weak points' for drug delivery. The nose is particularly attractive, because it has a large internal surface with lots of blood vessels, and is easily accessible.

Protein remodelling. This approach attempts to remodel the protein chemically to protect it against the difficulties of entering the body. This can be done by encapsulating it (as above), by embedding it in a variety of carrier materials such as dextrans, albumin, xanthan gums, or synthetic polymers such as polyethylene glycol (PEG), or chemically modifying it with these or other materials.

The blood-brain barrier. Many chemicals in the blood do not affect brain and spinal cord nerve cells. The nerve cells get their nutrients from surrounding cells and from the cerebrospinal fluid (CSF), which is not part of the circulatory system of the rest of the body. The cells form a barrier to penetration of drugs in the blood into the nerve cells of the brain. This can be a problem since taking drugs by mouth or even injecting them is much safer and easier than injecting them into the CSF. A substantial part of the effort on drug delivery is made directed at remodelling the drugs so that they can penetrate this blood-brain barrier. So far, protein drug delivery systems have been much hyped, but not very effective. It is not clear whether they ever will be, or whether redesigning biopharmaceutical drugs to be chemically more robust and better suited to getting into the body will be successful before the drug delivery systems are ever launched into general use.

Drug development pathway. ***Preclinical research.*** This means all the research that goes on before you try the compound out on people, but is often taken to mean animal studies of the drug. Studies using biochemical methods, receptor screening, or cell culture assays are usually considered just 'research' since most of the drug candidates they throw up will not make it even as far as clinical trials. It also means animal testing for toxicology and ADME (adsorption, distribution, metabolism, and excretion, i.e. pharmacodynamics).

Phase I trials. These are the first trials in which a drug candidate is given to people. The only permission needed to do Phase I trials is the approval of the local hospital ethical board or committee (which, however, has to be convinced that there is some point in you doing the trial). The people are normal, healthy volunteers (often medical students), and the purpose of the trial is to establish the pharmacokinetics of the drug, and to find the minimum dose that will have some effect: thus the trial starts out with a very small dose and works up. Usually only a small number of people (between 10 and 20) are involved. After Phase I the developer applies for an investigational new drug application (IND, in the USA), or the equivalent in other countries (e.g. the clinical trial exemption certificate (CTX) in Britain). This is the regulatory hurdle necessary to go on to Phase II trials, and at this point the developer must show that they have complied with a wide range of GLP practices in their preclinical and Phase I trials. For medical devices such as prostheses (whose development path is essentially the same), the IND is replaced by the 510(k) application in the USA.

Phase II trials. This is the first time the drug is tried on ill people. This trial is usually done at one hospital centre on a small number of patients, and looks for any evidence that the drug actually has a medical effect on the disease it is meant to treat. The drug is said to be being developed for one 'indication', i.e. one collection of symptoms or one disease type. The object of this and subsequent trials is to show that the drug has an effect for this indication. (Note that up to this point tests can be for any disease.) Again, the numbers of patients are quite small.

Phase III trials. This is where very big money is spent on drug development. The object of this phase is to see whether the drug is worth launching, because it is better that existing therapies, does not have severe side-effects, and so on. This requires hundreds or thousands of patients (all of which have to be followed in detail), usually at least six hospital centres. The trial is done 'double blind', so that neither the people giving the drugs nor the people analysing the results know who has received the drug and who has received a placebo until the study is over. It is also usually a cross-over trial, i.e. one in which, half-way through, those who had been getting placebo get given drug and those who had been getting drug get placebo. (This helps to avoid problems arising from the differences that people show in their response to drugs.). At the end of Phase III the drug is submitted for a new drug application (NDA, in the USA) or a product license application (PLA in Europe). (For a medical device, the equivalent is the premarketing approval PMA.) If this is approved, then the drug can be sold.

Phase IV trials. However, selling the drug does not mean that development is over. Phase IV trials, post-marketing surveillance or pharmacovigilance, then takes over to look for rare adverse reactions (adverse drug reactions, ADRs), to look for opportunities to decrease the dose (because the initial estimates, derived from Phase III studies, are often rather high), and to extend the range of indications for which the drug may be used. Extension of indications can come about because of off-label use, i.e. use of the drug by physicians for indications other than those for which the drug is licensed. There is nothing to stop people doing this, providing that they are very careful to emphasize to their patients that they are effectively doing an experiment on them. Successful experiments lead to new ideas for the use of the drug, and hence new clinical trials to see if the new indication is a suitable one for this drug.

Dry heat sterilization. Thermal sterilization carried out in the absence of added moisture. For the same sterilization time, dry heat usually requires a higher temperature than moist heat to achieve the same degree of sterilization.

DsDNA. Double-stranded DNA.

DTH (delayed type hypersensitivity). This term includes the delayed skin reactions associated with Type 4 hypersensitivity.

Dual host virus. Refers to viruses that can carry out the process of multiplication in the cells of two taxonomically distinct, unrelated and independent hosts, e.g., Kyasanur Forest Disease Virus (op. cit.).

Dunk bath. A tank device containing a liquid germicide that allows materials to be passed into and out of a gastight enclosure without transfer of air or micro-organisms.

Duodenum. The beginning portion of the vertebrate small intestine.

Duplication (genetic). A mutation resulting from the introduction into the genome of an extra copy of a segment of a gene or chromosome. (Contrast with deletion, point mutation.)

D-value. The decimal reduction time. The sterilization time necessary to reduce a microbial population by 90% (one log). It must be assumed that the rate of kill is linear with respect to the logarithm of the surviving population, then:

Dynein. Protein linkages found in the flagella of eukaryotes.

Dysentery. An intense diarrhea characterized by mucus and blood in the stool.

Dysmorphic. Structurally malformed, ill-shaped.

E

Early proteins. Proteinaceous substances, particularly enzymes, synthesised during the early stages of bacteriophage multiplication; are involved in various biochemical activities during phage multiplication.

Ecdysone. (Gr. *ek*: out of + *dyo*: to clothe). In insects, a pair of hormones inducing molting.

Echinoderm. A member of the phylum Echinodermata, such as a sea star or sea urchin.

Eclipse. (Greek *eklipsis*, failure to leave to come out). The period during bacteriophage multiplication which starts with the injection of nucleic acid into the host cell till the formation of first complete phage particle.

Ecological niche. (L. *nidus*: nest). The role of a species in relation to other species and its physical environment.

Ecology. (Gr. *oikos*: house + *logos*: discourse, study). The scientific study of the interaction of organisms with their environment, including both the physical environment and the other organisms that live in it.

Ecosystem. The organisms of a particular habitat, such as a pond or forest, together with the physical environment in which they live.

Ecto. (Gr.: *outer*, outside). A prefix used to designate a structure on the outer surface of the body. For example, ectoderm. (Contrast with endo-and meso-.)

Ectoderm. (Gr. *ektos*: outside + *derma*: skin). The outermost of the three embryonic tissue layers first delineated during gastrulation. Gives rise to the skin, sense organs, nervous system, etc.

Ectosymbiosis. A symbiotic state in which one organism attaches to the surface of another.

Eczema. Inflammatory skin disease characterized by scales and crusts.

Edema. An increase in the volume of interstitial fluid due to a disturbance in the mechanisms involved in fluid exchange.

Effector cells. A functional concept which is context means those lymphocytes or phagocytes which produce the end effect.

Effector T cells. T cells that have an effect on specific antigens (such as cytotoxic cells) as opposed to regulatory T cells that have regulatory effects on other lymphocytes.

Effector. Any organ, cell, or organelle that moves the organism through the environment or else alters the environment to the organism's advantage. Examples include muscle, bone, and a wide variety of exocrine glands.

Efferents. (L. *ex*: out + *ferre*: to bear). Neurons that carry impulses from the central nervous system. (Contrast with afferents.)

Egg. In all sexually reproducing, organisms, the female gamete; in birds, reptiles, and some other verterbrates, a structure within which early embryonic development occurs.

Elasticity. The property of returning quickly to a former state after a disturbance.

Electrochemical sensors. These are types of biosensor in which a biological process is harnessed to an electrical sensor system, making a sensor. The most commonly discussed type of electrochemical sensor is the enzyme electrode. Other types couple a biological event to an electrical one via a range of mechanisms. Among the more common are:

Oxygen electrode-based sensors. These are sensors in which an 'oxygen electrode' (Clark electrode), a standard electrochemical cell which measures the amount of oxygen in a solution, is coated with a biological material that generates or (more usually) absorbs oxygen. When the biological coating is active, the amount of oxygen next to the electrode falls and the signal from the electrode changes. Typical coatings might be an oxidase enzyme (which consumes molecular oxygen to oxidize a particular substrate) or a whole cell (which consumes oxygen when presented with a range of substrates). This latter type of biosensor (a microbial-or cell-based biosensor) can be used to detect poisons since poisons damage the cells and hence reduce the rate at which they consume oxygen.

pH electrode-based sensors. Again, a standard electrochemical pH electrode is coated with a biological material. Many biological processes raise or lower pH, and so can be detected by a pH electrode. Examples could include hydrolysis of an ester to acid and alcohol, or, again, the metabolism of neutral pH substrates by a bacterium. Indeed, in one study, which was meant to measure the pH inside a volunteer's mouth by placing a very small pH electrode there, what the electrode ended up detecting was sugar. Bacteria grew over the electrode and, every time the subject ate sugary food, the bacteria metabolized some of it to lactic and acetic acids and the pH next to the electrode fell from 7 to 4.5.

Electron. (L. *electron*: amber (associated with static electricity), from Gr. elektor: bright sun (color of amber)). One of the three most important fundamental particles of matter, with mass approximately 0.00055 amu and charge -1.

Electrophoresis. (L. *electrum*: amber + Gr. *phorein*: to bear). A separation technique in which substances are separated from one another on the bases of their electric charges and molecular weights.

Electroporation. This is manipulating cells by exposing them to a strong electric field. Initial studies (as one would expect) showed that while you could manipulate cells with strong electric forces, the cells never survived the treatment. However, by altering the conditions suitably, electroporation can be used t o transform cells with DNA and to fuse cells. Electric field induced fusion of cells is called electrofusion. Transformation of cells, getting DNA into them, can be achieved simply by exposing the cells to a suitable electric field while they are in a solution of the DNA. The electric field seems to modify the lipid membrane that surrounds the cells, and greatly increase the rate at which pinocytosis, a normal mechanism by which cells take up chemicals from solution, takes DNA into the cell. It is not widely used for animal or bacterial cells, where other methods have been developed which are quite reliable. However, electroporation is discussed fairly extensively when talking about getting DNA into plant protoplasts, and to a lesser extent into fungal cells. Some workers claim that electroporation or electrophoresis can even get DNA into intact plant cells (i.e. cells with their cell wall still there): the evidence for this is generally considered to be poor.

Fusing cells. This was the first application of electroporation. Protoplasts of plant cells or whole animal cells can be made to fuse by putting them next to each other and exposing them to a strong electric field. Typically, the cells are exposed to a low electric field, which induces a dipole in the cells and causes them to line up. They are then exposed to a quick pulse of very high voltage to make their membranes 'leaky' and so able to fuse. There seems to be no limit to the types of cells that may be fused together using this technology. Early studies uniformly resulted in dead cells: however, techniques have improved now so that cells may be fused to produce viable offspring using electrofusion. Application in plant genetics include the making hybrid plants and making polyploid plants.

Electrotonic potential. In neurons, a hyperpolarization or small depolarization induced by the application of a small electrical current. (Contrast with resting potential, action potential.)

Elephantiasis. A condition in which fluids accumulate in various tissues, causing grotesque features.

Embden-meyerh of pathway. A series of reactions for the conversion of glucose to pyruvic acid. There is a net yield of two ATP molecules per molecule of glucose degraded.

Embryo sac. In angiosperms the female gametophyte. Found within the ovule, it consists of eight or fewer cells, membrane bounded, but without cellulose walls between them.

Embryo. (Gr. *en*: in + *bryein*: to grow). A young animal, or a young plant sporophyte, while it is still contained within a protective structure such as a seed, egg, or uterus.

Emigration. The deliberate and usually oriented departure of an organism from the habitat in which it has been living.

Encapsulation. This is any method that gets something, usually an enzyme or bacterium, into a small package or capsule while it is still working (or alive). The capsule can be any size, but usually is no bigger than a few millimetres across. If it is too small to see fairly easily with the naked eye, the process is called microencapsulation. Encapsulation is one method of immobilizing cells for use in a bioreactor. Encapsulating agents can be anything that will form a shell around something, but usually they are polysaccharides such as alginate or agar, because they are inert, allow nutrients and oxygen to diffuse into and out of the sphere readily, and are easy to convert from a gel (solid) to a sol (liquid) or solution form by altering the temperature or the concentration of ions such as calcium. Proteins such as collagen (gelatin) are also used. Cell encapsulation is also used to try to develop artificial tissues. Enzymes may also be encapsulated, although they are more usually immobilized on the surface of polymer particles. Drugs are often encapsulated to help their survival or release in the patient. A variety of sustained-release cold cure medicines, which come as little particles inside a capsule, are actually encapsulated drugs: each particle contains a shell of a slowly dissolving material around a core of solid drug powder. Only after the shell has dissolved in the intestines can the drug get out to the body. By having a mixture of shell thicknesses the pharmacologist arranges for the drug to be released over a period of time. Similar approaches have been tried for biotechnological drugs, although not always very successfully. Encapsulation of drugs is also a methods of protecting them from, say, acid in the stomach so that they can be taken by mouth instead of having to be injected. This has been something of a holy grail of drug delivery technology for biopharmaceuticals, but has not been successful to date.

Encephalitis. Inflammation of brain.

Enclosure. A general term that includes cabinets and other devices designed to isolated or separate a working area for product protection or personnel protection reasons.

Endemic. (Gr. *endemos*: dwelling in a place). Confined to or native to a particular region, thus having a comparatively restricted distribution.

Endemic. Constantly present in a locality.

Endergonic reaction. One for which energy must be supplied. (Contrast with exergonic reaction.)

Endo- (Gr. within, inside). A prefix used to designate an innermost structure. For example, endocrine. (Contrast with ecto-, meso-.)

Endocarditis. Inflammation of the heart valves or chambers.

Endocrine. (Gr. *endon*: inside + *krinein*: to separate). Any gland, such as the adrenal or pituitary gland of vertebrates that secretes certain substances, especially hormones, into the body through the blood.

Endocytosis. A process by which liquids or solid particles are taken up by a cell through invagination of the plasma membrane. (Contrast with phagocytosis.)

Endoderm. (Gr. *endon*: within + *derma*: skin). The innermost of the three embryonic tissue layers first delineated during gastrulation. Gives rise to the digestive and respiratory tracts and structures associated with them.

Endodermis. (Gr. *endon*: within + *derma*: skin). In plants, a specialized cell layer marking the inside of the cortex in roots and some stems. Frequently a barrier to free diffusion of solutes.

Endoenzymes. Intracellular enzymes, concerned with cell synthesis and energy production.

Endogenous. Originating within the organism.

Endoplasmic reticulum. (Gr. *endon*: within + L. *plasma*: form; L *reticulum*: little net). A system of membrane-bounded tubes and flattened sacs, often continuous with the nuclear envelope, found in the cytoplasm of eukaryotes. Exists as rough e.r., studded with ribosomes,.and smooth e.r., lacking ribosomes.

Endorphins. Naturally-occurring opiate-like substances in the mammalian brain.

Endoskeleton. A skeleton covered by other, soft body tissues. (Contrast with exoskeleton.)

Endosperm. (Gr. *endon*: within + *sperma*: seed). A specialized triploid seed tissue found only in angiosperms; contains stored food for the developing embryo.

Endospore. Minute, highly durable body, developed within certain bacterial cells and capable of developing into new vegetative cells; characteristic of the genera *Bacillus, Clostridium, Sporosarcina, Sporolactobacillus*, and *Desulfotomaculum.*

Endothelium. Cells lining blood vessels and lymphatics.

Endotoxins. Large molecules of lipopolysaccharide that are normal components of cell walls of gram-negative bacteria; may cause fever and shock during infection.

Energetic cost. The difference between the energy an animal would have expended had it rested, and that expended in performing a behaviour.

Energy. Force, power, or the capacity for work.

Energy-gathering stage. That part of an individual's life history during which it feeds. (Commonly the entire life history.)

Enhancement. Prolongation graft survival by treatment with antibodies directed towards the graft alloantigens.

Enkephalins. (Gr. *en-*: in + *kephale*: head). Two of the endorphins.

Enteric. Relating to the intestinal tract.

Enteritis. Inflammation of the intestine.

Enterotoxin. Exotoxin released in food by certain microorganisms, which causes a severe reaction in the gastrointestinal tract.

Entropy. (Gr. *en*: in + *tropein*: to change). A measure of the degree of disorder in any system. A perfectly ordered system has zero entropy; increasing disorder is measured by positive entropy. Spontaneous reactions in a closed system. are always accompanied by an increase in disorder and entropy. Designated by the symbol S.

Envelop. (Old French enveloper, to cover). The outer membraneous lipo-protein structure covering the nucleocapsid particles of most animal viruses.

Environment. An organism's surroundings, including both living and nonliving things; includes such factors as temperature and light intensity.

Environmental biotechnology. Environmental biotechnology is a general term covering any biotechnological product or process that can be considered to be helpful to 'the environment'. Usually, this means control, reduction, or disposal of waste, removal of chemical pollutants, or reduction in power use, especially in industry. Because of the high political profile of 'the environment', a number of diverse biotechnology activities have been included under 'environmental biotechnology'.

Enzyme Commission (EC) number. All enzymes are given a systematic name and number which identifies them in the technical literature. (They may also have a 'common' name, like trypsin or rennin.) These names are given by the Enzyme Commission. The name and number are systematic descriptions of what the enzyme does. The number is a four-digit number. The first digit classifies the enzyme into one of six broad groups: 1. Oxidoreductases (transfer H atoms or electrons), 2. Transferases (transfer of small groups between molecules), 3. Hydrolases, 4. Lyases (addition to double bonds), 5. Isomerases, 6. Ligases (formation of bonds between C and another atom, using ATP as an energy source)

Enzyme electrode. A type of biosensor, in which an enzyme is immobilized on to the surface of an electrode. When the enzyme catalyses its reaction electrons are transferred from the reactant to the electrode, and so a current is generated. (This is distinct from other types of electrochemical biosensors, where the enzyme generates a distinct chemical product, for example an acid, which is then detected by a separate electrode system.). There are two types of enzyme electrode:

Ampometric. Here, the current generated by the reaction is measured, and the electrode is kept as near zero voltage as is practical. When the enzyme catalyses its reaction, electrons flow into the electrode, and so a current flows.

Potentiometric. Here, the voltage generated by the reaction is measured, and the electrode is held at a voltage that counteracts the voltage created by the enzyme's tendency to 'push' electrons into it. This may be done by actively adjusting the voltage or by not connecting the electrode to anything (as is the case in ISFET devices). The device's output is the voltage necessary to prevent any current flow through the electrode.

Enzyme immobilization. This is linking an enzyme to a solid support material, typically polymeric beads in a column or bed. Typical immobilizing materials could be silica or carbohydrate polymers such as Sepharose, although Altus Biologics has cross-linked enzymes in pure crystals to stabilize them (to form cross-linked enzyme crystals; CLECs). Immobilization has two general advantages: (1) It enables you to recover the enzyme quickly and easily from the reaction mixture, by separating the beads out. This is useful both for purifying the reaction mixture and for recycling valuable enzyme. (2) It alters the chemical properties of the enzyme. Many enzymes are more stable when bound to a surface, which appears to 'lock' their conformation into an active form, so slowing down the inevitable reduction in enzyme activity that occurs with use. (Although in theory an enzyme is a perfect catalyst, emerging unchanged from the reaction it speeds up, in practice, a wide range of side reactions slowly break up the protein's structure, rendering it useless as a catalyst. Sensitive enzymes can be inactivated in a few minutes. Robust industrial ones usually take hours or days.).

Immobilization is useful in industrial biotransformation processes. The synthesis of the anticancer drug Camptosar, for example, uses a lipase to resolve a chiral intermediate: the process in only economic if the lipase is immobilized, to give it added stability and efficacy. Enzymes are commonly immobilized in biosensor and bioassay applications. Here the main reason is to hold them in only place so that they do not wash off the assay strip.

Enzyme mechanisms. Since the use of enzymes is one of the most commercially important areas of biotechnology, understanding how they work is an important part of the research underpinning the technology. Indeed, one of the reasons that enzymes are so widely used is that their mechanism of action has been studied for most of this century, and the science of enzymology is correspondingly mature (as opposed to, say, the relatively new science of molecular genetics). Specific aspects of how enzymes work and how this may be improved for a particular application are dealt with elsewhere. The fundamental research involved is beyond the subject (and scope) of this book. In general, modification aims to improve the enzyme's specificity (i.e. what reactions it catalyses), its kinetics (how well it catalyses them), or its stability (how long it lasts). However, there are several lines of research that are relatively new technologies to *use in enzymology*:

Chemical modification. Changing an amino acid in the protein into another by chemically reacting it. This usually results in a change in the enzyme's activity, and if it does the change is almost always 'for the worse', i.e. it reduces the enzyme's catalytic effect, specificity, or both. Sometimes the change can result in a commercially more useful enzyme, in which case the modified protein is used commercially. However the protein is changed, the result is usually interesting to the enzymologist.

Site-directed mutagenesis. Changing an amino acid into another amino acid by genetic modification. This is more flexible than chemical changes, because an amino acid, perhaps identified from protein sequencing work or X-ray crystallography, can be changed specifically to another, closely related (or totally unrelated) into amino acid.

Enzyme production by fermentation. Industrial enzymes can be made by extraction from a naturally occurring source, often part of a large animal or plant, or by production from a microorganism in fermentation. The former requires less equipment, but is more prone to interruption of supply by seasonal variation, the vagaries of climate, international trade, and (in extreme cases) war. Fermentation has the potential to provide a more uniform and reliable source of material. The enzymes that account for the majority of production are essentially commodity products. Thus a substantial part of the cost of their production is the raw materials and power necessary to produce them. (This differs from enzymes used for research, such as restriction enzymes, which are produced in comparatively tiny amounts and whose production cost is dominated by the skilled labour necessary to make them.) Thus a successful fermentation process must use low-cost feed materials, an organism that does not require excessive heating or cooling, and one that makes a lot of the enzyme. Typical feedstocks are hydrolysed starch, molasses, whey, and cereals for carbon, and soy flour, fish meal, blood, and cotton-seed meal for nitrogen. For high value enzymes (for example, for use as drugs), some of these are inappropriate since they contain insoluble dirt, which will have to be rigorously removed from the final product. The fermentation conditions that must be monitored to optimize enzyme production include pH, oxygen, CO_2, aeration, temperature, agitation, and, since some enzymes are denatured on surfaces or may concentrate at them, foaming. In addition, the production of many enzymes by bacteria is induced and repressed by specific chemicals. Inducers must be present and repressors removed from the fermentation if the yield is to be satisfactory. Many industrial enzymes are sold as fairly crude preparations, with a mixture of proteins in them. These have been prepared by separating the cells from the fermentation broth, and then partially purifying the protein from the liquor by precipitation, ultrafiltration, or similar technique.

Enzyme repression. Inhibition of enzyme synthesis at the genetic level.

Enzyme stabilization using antibodies. This is a method of stabilizing proteins, usually enzymes, by binding antibodies to them. Some enzymes may be stabilized 200-fold by complexing them with an antibody, i.e. the 'half-life' of their enzymatic activity can be increased (from 5 minutes to 16 hours in the case of alpha-amylase). The antibodies have to be selected so that they do not block the active site of the enzyme, otherwise the protein is stabilized but becomes inactive as a catalyst: thus monoclonal antibodies, which bind to defined bits of the protein's surface, are usually used. The process works because the antibodies bind to the active structure of the enzyme. If the enzyme now 'tries' to unfold into an inactive structure, it must not only overcome its own binding energy but also throw off all the bound antibodies. This needs more energy to do, and so is a correspondingly slower process. Antibody stabilization is used routinely to stabilize enzymes used in medical diagnostic assays. The antibodies are too expensive for it to be routine for enzymes used in large-scale processes.

Enzymes. (Gr. *en*: in + *zyme*: yeast). A protein, on the surface of which are chemical groups so arranged as to make the enzyme a catalyst for a chemical reaction.

The core of traditional biotechnology, and a key feature of the new biotechnology of gene cloning, is the use of enzymes. For practical purposes these can be considered to be catalytic proteins, although recent work has shown that RNA can act as an enzyme. Enzymes are prepared from a huge variety of organisms, from viruses to whales. In general, they may be extracted from some organism that already produces the enzyme, extracted from a microorganism that is cultured under conditions where it produces the enzyme, or made from an organism that has been genetically engineered to produce the enzyme. Enzymes are so widespread in biotechnology that they crop up in many entries in this book. Specific classes of enzymes covered are glycosidases and glucose isomerase and invertase, proteases, and lipases. Enzymes also crop up in biotransformation, protein engineering, enzyme production by fermentation, and expression, as well as, en passant, in many other entries. The value of enzymes to the biotechnology industry can be estimated from the following table.

Epi- (Gr. upon, over). A prefix used to designate a structure located on top of another. For example, epidermis, epiphyte.

Epicanthus. A congenital downward fold of skin from the upper eyelid that hides the inner canthus and carnuele.

Epicotyl. (Gr. *epi*: upon + *kotyle*: something hollow). That part of a plant embryo or seedling that is above the cotyledons.

Epidemic. Disease found in large number of individuals in a community within a short period.

Epidemiology. Systematic study of the various factors that determine the spread of pathogenic organisms within a human community.

Epidermis. (Gr. *epi*: upon + *derma*: skin). In plants and animals, the outermost cell layers. (Only one cell layer thick in plants.)

Epigenetic. This is inheritance of biological traits outside the normal genetic mechanisms of DNA and genes. It is usually applied to biochemical aspects of a cell's metabolism. The first epigenetic traits known at a cellular level were the patterns of cillia (short 'tentacles') on the surface of pond-dwelling protozoa, which are passed down from mother to daughter organism even though they are not coded in any DNA: the pattern in one cell is copied on to its two daughter cells. Similar copying mechanisms also copy patterns from human mother cells to daughter cells, a form of cellular inheritance called epigenesis. The best-characterized epigenetic phenomenon is DNA methylation. A small fraction of the cytosine bases in DNA are altered by adding a methyl group on to them, using an enzyme called DNA methylase. This affects whether the genes containing those methyl cytosines are active or not. When DNA is made during cell growth, the enzyme has to 'decide' whether a particular cytosine should be methylated or not. It does so by methylating all the cytosines in the double helix that are next to another methylated cytosine. Thus the pattern of methylation is copied from mother to daughter DNA. (This only works because the methylated cytosines are always in CG pairs, called CpG dinucleotides.) Thus, although the pattern of methylation is not 'coded' in the DNA, it is nevertheless inherited from one cell to the next.

Epinephrine. (Gr. *epi*: upon + *nephros*: a kidney). A hormone produced by the adrenal medulla, also functions as a neurotransmitter. Also known as adrenalin.

Epiphyte. (Gr. *epi:* upon + *phyton*: plant). A specialized plant that grows on the surface of other plants but does not parasitize them.

Episome. (Greek *epi*, upon plus *soma*, body). Indicates the independently replicating genetical units that are often associated with the chromosome or the nuclear material of certain organisms. These are foreign in origin and are capable of imparting genetical changes in the host. e.g., temperate phages causing lysogenic conversion (q.v.)

Epistasis. An interaction between genes, in which the presence of a particular allele of one gene determines whether another gene will be expressed.

Epithelium. In animals, a layer of cells surrounding or lining an external surface or a cavity.

Epitope. A single antigenic determinant. Functionally it is the portion of an antigen which combines with the antibody paratope.

Epstein-barr virus (EBV). Causal agent of Burkitt's lymphoma and infectious mononucleosis. It has the ability to transform human B cells into stable cell lines.

Equilibrium constant. A particular kind of ratio of the concentrations of products to reactants when equilibrium has been reached in a chemical reaction. Designated by the symbol, Keq.

Equilibrium, chemical. A state in which forward and reverse reactions are proceeding at counterbalancing rates, so there is no observable change in the concentrations of reactants and products.

Erysipelas. Acute infection of the skin which results in characteristic red, edematous lesions; found most commonly on the face and legs; caused by a group a B-hemolytic streptococcus.

Erythrocyte (red blood cells). The biconclave, nonnucleated and agranular mature cell of vertebrate blood that transports oxygen to the tissues in combination with hemoglobin, the pigment responsible for the red color of fresh blood.

Essential amino acids. These are essential components of our bodies, but we cannot make them, so we must get them from food. They are important to biotechnology because many microorganisms can make them, and so can be used to produce them as food supplements. Other mammals, such as farm animals and pets, share much the same list, and much of the food supplement amino acids made are for farm animals rather than people.

Essential metabolite. A substance required by a microorganism, which must be supplied in a growth medium.

Essential nutrients. There are a range of nutrients that mammals need in their diet: these are called essential nutrients. As well as minerals, mammals need varying combinations of essential amino acids, fats, and vitamins. The particular combination depends on the species.

Estrogen. Collective term for the female sex hormones (natural; estradiol, estrone, estriol; synthetic: ethynylestradiol and stilbestrol) responsible for the development of the accessory sex structures and the secondary sex characteristics in the female. Estrogenic hormones also affect calcium and phosphorus metabolism, bone and lipid metabolism.

Estrous cycle. The repeated series of changes in reproductive physiology and behavior in female mammals (other than some primates), culminating in estrus.

Estrus. (L. *oestrus*: frenzy). The period of heat, or maximum sexual receptivity, in some female mammals. Ordinarily, the estrus is also the time of release of eggs in the female.

Ethanol. More ethanol is produced by biotechnology than any other chemical. Yeast fermentation of natural carbohydrate sources such as grain and grape has deep historical roots, and is nowadays a highly technological industry. Aspects of its use are discussed in separate entries on brewing and biofuels.

Ethology. (Gr. *ethos*: habit, *custom* + *logos*: discourse). The study of whole patterns of animal behavior in natural environments, stressing the analysis of adaptation and evolution of the patterns.

Ethylene oxide. A colorless gas having remarkable penetrating power that is frequently used as a sterilizing or decontaminating agent. It is effective at room temperature, although a certain level of moisture is required for optimum effectiveness. The flammability hazard is greatly reduced by diluting the gas with carbon dioxide or fluorocarbons such as Freon.

Ethylene. One of the plant growth substances, the gas $H_2C = CH_2$.

Etiolation. Growth in the absence of light.

Etiology. The cause of a disease.

Eukaryotes. (Gr. *eu*: ture + *karyon*: kernel or nucleus). Organisms whose cells contain their genetic material inside a nucleus. It includes all life other than the viruses and the kingdom Monera.

Eumycetes. True fungi; yeasts and molds.

Eutrophication. (Gr. *eu-*: well + *trephein*: to flourish). The addition of nutrient materials to water. This results in predictable consequences.

Evergreen. A plant that retains its leaves through all seasons. (Contrast with deciduous.).

Evolutionary biology. The collective branches of biology that deal with the evolutionary process and the characteristics of populations of organisms, as well as ecology, behavior, and systematics.

Evolutionary systematics. Systematics based on information from both cladistic and phenetic studies.

Excision repair. An enzyme system which can repair DNA by the elimination of ultraviolet light-induced thymine dimers.

Excretion. Release of metabolic wastes by an animal.

Exergonic reaction. A reaction in which free energy is released. (Contrast with endergonic reaction).

Exfiltration (Ventilation term). Ductless flow of air from a space to an adjoining space at lower pessure.

Exobiology. The study of the probable existence of life on other planets.

Exocrine gland. (Gr. *exo*: outside + *krinein*: to separate). Any gland, such as a salivary gland, that secretes to the outside of the body or into the gut.

Exocytosis. Extrusion of materials from the cell into the environment via vacuoles.

Exoenzymes. Hydrolases, or digestive juices, excreted into the environment.

Exon. A protion of a DNA molecule, in eukaryotes, that codes for part of a polypeptide. (Contrast with intron.)

Exoskeleton. A hard covering on the outside of the body; the exoskeleton of insects and other arthropods has many of the same functions as the bony internal skeleton of vertebrates.

Exotoxin. Toxin produced in the cell and released into the environment.

Exotoxins. Toxic substances excreted from the bacterial cell.

Exploitation competition. Competition that occurs because resources are depleted. (Contrast with interference competition.)

Expression compartment (inclusion bodies). Getting a protein made in a recombinant cell is relatively straightforward since a wide range of expression vectors exist to be used to clone the relevant gene. However, the protein is often produced in a form that is not useful to the genetic engineer. This is often a feature of where in the cell the protein is made.

Expression systems. Expression systems are combinations of vector and host that provide a 'genetic context' which makes the gene function in the host cell: usually, this means make a protein at high levels. (Usually a cloned gene will be inert, i.e. it will not perform its usual function in the host cell since it is outside its usual genetic context.) The proteins made in such systems are called foreign or heterologous' proteins, because they are not native to the cell in which they are being made. Thus, this is 'heterologous gene expression'.

Extinction. The termination of a lineage of organisms.

Extracellular virion. Refers to complete infective virus particeles as they exist prior to infecting a host.

Extremophiles. These are organisms that live in 'extreme' conditions, i.e. conditions substantially different from Western laboratories. They include:

Thermophiles. Organisms that live at high temperatures.

Psychrophiles. Organisms that live at very cold temperatures. They can be useful sources of enzymes that are active under cold conditions: most enzymes act only very slowly at temperatures much below the normal living temperature of the organism they came from. They can also be source of heat-unstable enzymes, enzymes that are destroyed by moderate heating to 40°C. This can be useful if you want to 'turn off' an enzyme without affecting other components of a complicated process. Another interesting target for biotechnology is fish that can live in arctic seas at temperatures that should freeze them solid. They are sources of 'antifreeze proteins', proteins that prevent their blood freezing.

Halophiles. Organisms that can live in concentrated salt solutions. Halophilic bacteria are interesting because of their metabolic curiosities, including the purple membrane protein of *Halobacterium halobium*. Biotechnology is also interested in engineering plants to be halophiles, so that they can be watered with sea water. Enzymes from extremophiles are potentially valuable in their own right because of their stability in industrial processes. They also have value for what they teach us about enzyme stability, and hence how we might make other enzymes more stable. They are sometimes called **extremozymes**. The centrifuge too long, everything sinks to the bottom of the tube. Zonal centrifugation separates.

Extrinsic isolating mechanism. Any barrier in the environment, such as a riverbed, desert, or ocean, that isolates one population from another and thus prevents mating between them. (Contrast with intrinsic isolating mechanism.)

Extrinsic protein. A membrane protein found only on the surface of the membrane.

Exudate. Fluid, often from formed elements of the blood, discharged from tissues to a surface or cavity.

F

Fab fragment. The fragment of the enzyme-digested antibody molecule that binds antigen, consisting of the combined amino terminals of the variable regions of an H and an L chain.

Fab. The part of an antibody molecule which contains the antigen combining site, consisting of a light chain and part of the heavy chain; it is produced by enzymatic digestion.

Facilitated diffusion. Passive movement through a membrane, involving a specific carrier protein. Does not proceed against a concentration gradient.

Facultative anaerobe. An organism that can utilize free oxygen or can grow anaerobically.

Facultative. Capable of occurring or not occurring, as in facultative aerobes. (Contrast with obligate.)

Familial route. Origin of sociality by sharing of care by older young. (Contrast with parasocial route.)

Family. In taxonomy, the category below the order and above the genus a group of related, similar genera.

Fatty acid. A molecule with a long hydrocarbon tail and a carboxyl group at the other end. Found in many lipids.

Fauna. All of the animals found in a given area (contrast with flora).

Faunal dominance. The ability of species originating from one part of the world to spread their range to other parts of the world and displace other species already living there.

Fc fragment. The crystalizable fragment of the enzyme-digested antibody molecule. It consists of the carboxy terminal half of the two H chains of an antibody molecule. Biological functions include attachment to appropriate cell-surface receptors, complement bindings, placental passage.

Fc. The portion of an antibody which is responsible for binding to antibody receptors on cells and the C1q component of complement.

F-duction. Transfer of genes from one bacterium to another, using the F-factor as a vehicle.

Feces (L. *faeces*: dregs). Waste excreted from the digestive system.

Feedback control. Control of a particular step of a multistep process, induced by the presence or absence of a product of one of the later steps. A thermostat regulating the flow of heating oil to a furnace in a home is a negative feedback control device.

Feedback inhibition. Temporary inactivation of an enzyme due to the end product of a series of reactions reacting directly with and inhibiting the activity of the enzyme of the first reaction.

Feeder root system. Fine root hairs of a tree that absorb nutrients.

Fermentation (L. *fermentum*: yeast). The degradation of a substance such as glucose to smaller molecules with the extraction of energy, without the use of oxygen (i.e., anaerobically). Involves the glycolytic pathway.

Fermentation processes. Strictly speaking, fermentation is microorganism metabolism under anaerobic conditions on a carbon substrate. However, it has been extended to mean growing microbes in liquid under any conditions. Culturing bacterial cells in small amounts on a Petri dish or analogous small-scale mammalian cell culture is called incubation, and takes place (unsurprisingly) in an incubator. There are three general ways in which fermentations are done, each with a variety of associated terms. In all cases there are come common terms in bacterial growth, such as the bacterial doubling time (the time needed to double the number of bacteria there). ***Common terms.*** For all fermentation processes, the first thing that happens is that the fermenter (or bioreactor) is sterilized. This can be done with stem, chemicals, washing, or some combination of these. The fermentation is then started with an inoculum, a small, actively growing sample of the organism to be cultured. Fermentation then proceeds according to one of the schemes below. Batch fermentation. Here the reactor is filled with a sterile nutrient substrate and inoculated with the microorganism. The culture is allowed to grow until no more of the product is being made, when the reactor is 'harvested' and cleaned out for another run. The culture goes through a lag phase (when the organisms adapt to their surroundings), exponential growth (when they grow in numbers), stationary phase (when they stop growing), and death phase. Depending on what the product is, the 'useful' part of the growth cycle can be any one of these four stages, although it is usually the growth or stationary phases. Fed batch fermentation. Here the batch culture is fed a batch of nutrients before it gets to the stationary phase, so that it never runs out of nutrients. At the same time some of the fermentation is removed and taken off for processing. Continuous culture. This is the logical extension of fed batch fermentation. The fermentor is fed continuously with nutrient and the culture medium removed continuously. This has some advantages over fed batch systems in that the culture conditions are always the same, but it is also harder to control. This is essentially a large-scale chemostat. Cascade fermentation. Here the fermenting 'liquor' is passed through a series of ferments, so that more and more of the product builds up each time. Each step can then be optimized for a specific condition. A typical example would be in brewing, where the beer would be fermented in several stages to increase the alcohol content, each stage using a yeast adapted to working in alcohol of that concentration. At the end of each stage the yeast is separated from the beer and used again in that stage, while the beer goes on to the next stage. Fermentations may also be classified according to when the product is made: **Type I fermentation:** product is made from primary metabolism. **Type II fermentation**: product is made from secondary metabolism at the same time as primary metabolism is going on (i.e. when the cells are growing). **Type III**: product is made by secondary metabolism at a different time from the primary metabolism (i.e. during stationary or death phases of the culture).

Lastly, fermentations may be classified according to how the culture is kept 'clean'. Aseptic/sterile fermentation. All other organisms are excluded by the biotechnologist. This is by far the most common class. **Consortium fermentations**. Here a group of organisms are growing together, rather than just one organism. For this to work, the organisms must be dependent on one another, otherwise one will outgrow the others and dominate the culture. **Protected fermentations**. Here the culture is not aseptic, but is performed under conditions under which only one type of organism will grow. Thus fermentations at extremely high temperatures, extremes of pH, or with very hard to metabolize substrates will tend to be able to support only the organism that the biotechnologist is after, thus removing the problem of having to keep contaminants out.

Fermentation substrates. Many materials are used as food for growing microorganisms. These are referred to as the substrate. The substrate and the trace materials needed, together with chemicals added to make the fermentation easier (such as anti-foam agents to stop froth forming) make the culture medium. The substrates can be divided according to the different essentials for life that they provide: a source of carbon, nitrogen, and (in the case of aerobic fermentation) oxygen. Usually carbon substrates cost the most, because you need the most of them. Among the common carbon substrates for large-scale fermentation are: **Molasses**. A side product of sugar refining that contains most of the material from sugar beet or sugar cane that is not sugar; molasses is one of the cheapest substrates available. There are several varieties with different properties: *Components and costs*. **Beet molasses**: the molasses left from processing beet. black strap molasses. *From an early stage in cane sugar processing*. **Cane refinery molasses**: from a later (more sugar rich) stage in processing. high test molasses: *Actually, concentrated cane juice treated with invertase*. **Malt extract**. Made from malted barley by soaking it in water. **Starch and dextrins**. Polysaccharides often made from cheap crops like potatoes. **Cellulose.** The world produces about 100 billion tonnes of cellulose a year, so it is a potential raw material for large-scale fermentation. But only a few organisms can degrade it. Sulfite liquor, a by-product of paper pulp production, which contains much of the fermentable sugars from wood without substantial cellulose. Whey. A side product of dairy processes, it is cheap but expensive to store or transport.

Methanol. A very cheap chemical from the oil industry, but it contains no nitrogen. Only a restricted range of organisms can grow on methanol. Similarly ethanol ('alcohol') can be used, but more usually ethanol is the product of a fermentation. Oil, gas. Some organisms can use natural gas or some components of crude or refined oil as carbon substrates. However, their commercial use depends critically on the price of oil. Nitrogen substrates include: **Ammonia.** A very smelly gas produced as a bulk commodity for the chemical industry. Most organisms can use ammonia. Sometimes it is converted into ammonium salts or into urea for ease of handling. **Corn steep liquor.** The liquid generated in the early stages of wet-milling maize to produce starch. Maize grains are submerged in water to soften them before milling. Low molecular weights sugars and peptides accumulate in the water. **Soy protein.** The protein left over when you have taken the oil out of soybeans. **Yeast extracts.** Made from waste yeast from industrial fermentations, they have everything necessary for microbial growth. **Peptones, casein hydrolysates**. These are partially digested meat or milk proteins, respectively. The proteins used are usually waste material from the food industry: nevertheless, this can still be an expensive source of nitrogen.

Fermentation. A central discipline of traditional biotechnology, and a key part of all biotechnology today, is fermentation. This is the growth of microorganisms, and encompasses a wide range of technologies, which are dealt with in different entries in this book. Fermentation includes: bioreactor design: the design of the container in which fermentation is to take place. substrates: what the microorganisms are to grow on. supports: whether the organism is to be on solid supports or in suspension. There is also a wide range of options in downstream processing.

Fermenter. A large cylindrical vessel, usually made of stainless steel, used to grow microbes on an industrial scale.

Fertility (F). Factor a genetic factor (circular double-stranded DNA) carried by some microorganisms that permits conjugation to occur with the same or related species.

Fertilization membrane. A membrane surrounding an animal egg which becomes rapidly raised above the egg surface within seconds after fertilization, serving to prevent entry of á second sperm.

Fertilization. Union of gametes. Also known as syngamy.

Fetal death rate. The number of fetal deaths as a proportion of the related total births (live births plus fetal deaths).

Fetal death ratio. The number of fetal deaths relative to the number of live births.

Fetus. (L.: *pregnancy*). The later stages of an embryo still contained in an egg or uterus; in humans, the unborn young from the eighth week of pregnancy to the moment of birth.

Fiber. An elongated and tapering cell of vascular plants, usually with a thick cell well. Serves a support function.

Fibrinogen. A plasma protein involved in clotting.

Fibroblast cells. (Latin *fibra*: thread plus Greek *blastos*, a sprout). A type of cells of irregular, branching shape found throughout vertebrate connective tissue; its function is to form and maintain collagen which is a kind of fibrous protein that acts as a cementing material for binding cells and tissues together.

Filter feeder. An organism that feeds upon much smaller organisms, that are suspended in water or air, by means of a straining device.

Filter. A device used for removal of undesirable particulates, including microorganisms, from air or other gases. Includes ultrahigh-efficiency filters and high-efficiency filters, as well as less efficient types.

Filtrable agents. Refers here to agents smaller than bacteria as these are able to pass through bacterial filters. Originally used as a criterion for defining viruses.

Filtrable virus. A virus that can pass through filters which do not allow bacteria to pass.

Filtration. In the excretory physiology of some animals, the process by which the initial urine is formed; water and most solutes are transferred into the excretory tract, while proteins are transferred into the excretory tract, while proteins are retained in the blood or hemolymph. It is removal of substances by filtering through any one of several types of filters. The pore size is usually small enough to prevent the passage of bacteria.

Fimbria. Fringe; in microbial genetics, a small nonconjugative projection on the surface of bacteria; term pilus has replaced fimbria.

Fingerprinting. In biotechnological terms, fingerprinting means making a characteristic chemical profile of something in order to identify it. There are many variants. DNA fingerprinting. Generating a profile of DNA fragment. Note that DNA fingerprinting can be applied to animals, plants, or bacteria in addition to it.

Protein fingerprinting. This generates a pattern of the proteins in a cell or organism, which provide a completely characteristic 'fingerprint' of that cell at that time. The technique usually used is 2D gel electrophoresis. This gives a complicated pattern of spots, which can be compared between cells. However, the comparison is difficult, and the pattern you get will depend on the cell's metabolic state as well as its origins, so this type of fingerprinting is now used only to compare the same cell at different times (e.g. with and without a drug added to its culture medium), and DNA fingerprinting is used to compare different cells.

Peptide fingerprinting. This can be one of two things. The first is use of peptides (i.e. the degradation products of proteins) to characterize a cell or organism, as above. The second is characterization of a single protein using its component peptides. Lots of peptide fragments can be generated from a protein, usually by partially digesting it with an enzyme. The peptides are easier to characterize and sequence than the whole protein. The pattern of peptides is used to identify the parent protein.

Chemical fingerprinting. Here, the low molecular weight chemicals in a cell are analysed, usually by GC or HPLC but increasingly by mass spectrometry, and the resulting pattern used to identify a cell or microorganism.

Footprinting. This is a distantly related idea. This is primarily applied to DNA, although the same logic can be applied to other molecules. It is a method of finding how two molecules stick together. In the case of DNA, a protein is bound to a labelled piece of DNA, and then the DNA is broken down, by enzymes or (as in the diagram) by chemical attack. This produces a 'ladder' of fragments of all sizes. Where the DNA is protected by the bound protein it is degraded less, and so the 'ladder' appears fainter. Footprinting is a common technique for homing in on where the proteins that regulate gene activity actually bind to the DNA.

Fitness. The contribution of a genotype or phenotype to the composition of subsequent generation, relative to the contribution of other genotypes or phenotypes.

Flagellate. A member of the phylum Mastigophora, unicellular eukaryotes that propel themselves by flagella.

Flagellin. The protein from which prokaryotic (but not eukaryotic) flagella are constructed.

Flagellum (pl., flagella). Slender, threadlike appendage, generally several times the length of the cell; provides for motility.

Flagellum. (plural: flagella) (L: whip). Long, whip-like appendage that propels cells. Prokaryotic flagella differ sharply from those found in eukaryotes.

Flavoprotein. A protein component of the electron transport system.

Flavour chemicals. An increasing use of biotechnology is to make flavour chemicals for the food industry. Flavour and fragrance chemicals are usually complex mixtures of very complicated molecules, and so are very difficult to reproduce with conventional chemistry. They arrive in food either as part of the ingredients or as chemicals created during cooking. Biotechnology can help both processes. (Most flavour chemicals are actually fragrances: we taste most of our food with our noses.) Flavour chemicals in ingredients are usually secondary metabolites from plants (especially spice plants) or bulk ingredients like salt and sugar. Often, plants are unreliable industrial sources of chemicals, so biotechnology has sought to grow their tissues in vitro to make the chemicals. Enzymes are used extensively to help to mature the flavour ingredients of food during processing. Biotechnologists can assist the development of new food enzymes by finding or engineering enzymes that fit better with the other processes that the food must undergo, like cooking or canning. These improvements could include making them more stable to acid or heat, or making them easier to remove once they have done their job, by, for example, immobilizing them on beads or columns so they can be separated from liquid food or food ingredients easily. Examples of the use of enzymes in flavour development include: Using endopeptidase (proteases) to break down bitter-tasting peptides in low-fat foods (fat usually dissolves these hydrophobic peptides and so disguises their taste).

Flight distance. Distance at which a prey organism initiates escape from a predator.

Flocculent. Masses of cells floating, or settled to the bottom, in a liquid medium.

Flora. All of the plants found in a given area. (Contrast with fauna.)

Florigen. A plant hormone (not yet isolated) involved in the conversion of a vegetative shoot apex to a flower.

Flower. The total reproductive structure of an angiosperm; its basic parts include the calyx, corolla, stamens, and carpels.

Fluctuation test. Test to determine the development of mutants in a culture. The test is based on the concept that mutation is a completely random event, and that spontaneous mutation would result in a large fluctuation in the number of mutants in a series of identical cultures.

Fluidized bed. Many types of bioreactor use particles to support the enzyme molecules or cells. These particles form a 'bed'. There are a range of ways of making sure that the reactor contents flow smoothly and uniformly past such a bed. Fluidized beds. Here the flow of material through a bed of particles is enough to shake them loose from their neighbors, but not enough to lift them. The bed of particles then behaves like a fluid, not a solid, and mixes very well with the material flowing through it. This technology is more widely used in the industrial processing of gases (such as burning coal dust in power stations), but can be used with liquids. Entrained bed reactors. Here the flow is higher, and the particles are kept in suspension by the fluid flow. Some cell culture systems use variants on this approach, with the cells growing on the surface of polymer particles that are kept suspended in a flow of growth medium. Note that this is not the same as keeping the particles in suspension by stirring. Packed bed reactors. Here the particles are packed into a bed, and remain essentially solid. This is the most common type of bed reactor, and is widely used in chromatography: a chromatographic column is in fact a columnar packed bed reactor. Fluid can flow unevenly through such a bed of solid particles, and so variants have been developed to encourage more even flow, such as the tapered bed reactor (with a conical base filled with particulate solids) and the disk bed reactor, where a disk of solid material has liquid fed in from the centre and flows out for collection at the outside.

Fluorescein. A fluorescing pigment produced by many *Pseudomonads* and observed by illumination with ultraviolet light.

Fluorescence microscopy. The staining of microorganisms with a fluorescent dye for microscopic study, using ultraviolet light as the illumination source.

Fluorescence transfer. An excited fluorescent molecule can give away its energy as light, or by donating it to another molecule. If the other molecule fluoresces, this is called fluorescence transfer. The second molecule glows (with a different colour to the first), where is would not on its own. If the second molecule does not glow, this is called quenching: nothing glows. Both instances can be used to see when two molecules have been brought very close together: in the first case, the light output changes colour, in the second it stops. The molecules needs to be within a few nanometres of each other for this transfer to work.

Fluorescence. (1) Light emitted by a substance after absorption of radiation of a shorter wavelength from another source. (2) Fluorescence is a very widely used way of detecting things in biology and biotechnology. A fluorescent chemical absorbs light of one wavelength are re-emits it at once at another wavelength. (A rarer phenomenon, phosphorescence, is when it re-emits it again later.) Usually this means shining an ultraviolet (UV) light (which we cannot see) at the chemical and having it emit visible light (which we can see). The molecule is said to be excited by the UV, and to emit the visible light. Because biological material is not generally very fluorescent, it is easy to see a fluorescent label molecule introduced to it. Fluorescent labels have been used to label a very wide range of biochemical reagents, including DNA probes, antibodies, and enzyme products. Common labels are fluorescein (which glows greenish yellow) and Texas red. Increasingly used are red and near-infrared (NIR) fluorescent dyes, which are excited by red light (e.g. from a laser) and emit deeper red or infrared light. They are attractive because biological material has almost no background fluorescence in the red, and because solid-state lasers are cheap, compact power light sources for these applications. DNA is very often stained with a UV dye, often ethidium bromide. This sits in between the bases of the double helix (it *intercalates*), where it glows much more brightly than on its own. If an agarose gel is soaked in a dilute ethidium bromide solution, the DNA shows up under UV light as bright bands on a dark background. An increasingly popular fluorescent label is green fluorescent protein (GFP), a protein that has a very strong green fluorescence. Since it is simply a protein, it can be made inside cells, and so one can engineer cells to make the GFP as a marker of gene activity. GFPs have been engineered to emit a range of colours between yellow and blue-green, so, in principle, one could have two or even three GFPs distinguishable within one cell. As well as simply tying a fluorescent molecule on to an otherwise non-fluorescent reagent, there are other tricks that use fluorescence.

Fluorescent antibody (FA). An antibody conjugated with a fluorescent dye that fluoresces on ultraviolet irradiation; used in an immunoidentification procedure; may be visualized by UV or fluorescent microscopy.

Fluorescent. Capable of emitting light of one wavelength when exposed to light of another wavelength.

Fluorochrome. A compound that releases its energy follows excitation in the form of light.

Fluorosis. Mottling of tooth enamel due to excessive amounts of fluoride in water.

Flying pin. Refers to the mechanism of transmission of myxomatosis virus causing myxomatosis of rabbits by mosquitoes. Refers particularly to the mosquito.

Focal infection. Local infection which acts as a nucleus for the spread of organisms to other sites.

Follicle stimulating hormone. A gonadotropic hormone produced by the anterior pituitary.

Follicle. (L. *folliculus*, little bag). In female mammals, an immature egg surrounded by nutritive cells.

Folliculitis. Inflammation of hair follicle.

Fomites. Inanimate objects or materials that act as intermediate carriers of microbial contamination. Contaminated tools used to assemble a sterile spacecraft could be fomites. (*Sing.* = fomes)

Food chain. A portion of a food web, most commonly a simple sequence of prey species and the predators that consume them.

Food infection. A gastrointestinal disturbance that results from eating food contaminated with certain living organisms, frequently Salmonella.

Food poisoning. A gastrointestinal disturbance due to ingestion of food containing toxins or poisonous substances which have been excreted into the food by certain species of bacteria, frequently *Staphylococcus aureus*, less commonly *Clostridium botulinum* and *Clostridium perfringens.*

Food subweb. A group of organisms united as a result of being food for the same or similar sets of predators.

Food web. The complete set of food links between species in a community; a diagram indicating which ones are the eaters and which are consumed.

Forespore. A structure that develops in spore-forming bacteria after the active growth period. It later becomes refractile, forms a thick wall, and becomes the resistant endospore.

Formalin. Formaldehyde in a 37 to 40 percent aqueous solution.

Fossil. Any recognizable structure originating from an organism, or any impression from such a structure, that has been preserved from prehistoric times.

Founder effect. Genetic drift resulting from establishment of a population by a very small number of individuals.

Fraction sterilization. Tyndallization sterilization of culture medium by exposure to steam for 30 minutes each day for three successive days to kill vegetative cells.

Frame-shift mutation. A mutation resulting from the addition or deletion of a single base pair in the DNA sequence of a gene. As a result of this, mRNA transcribed from such a gene is translated normally until the ribosome reaches the point at which the mutation has occurred. From that point on, codons, are read out of proper register and the amino acid sequence bears no resemblance to the normal sequence.

Free diffusion. Diffusion directly across a membrane without the involvement of carrier molecules. Free diffusion is not satiable and cannot cause the net transport from a region of low concentration to a region of higher concentration. (Contrast with facilitated diffusion and active transport.)

Free energy. That energy which is available for doing useful work, after allowance has been made for the increase or decrease of disorder. Designated by the symbol G (for Gibbs free energy), and defined by; **G = H - TS**, where H = heat, S = entropy, and T = absolute (Kelvin) temperature.

Freeze-drying. A technique in which microorganisms are frozen rapidly and then dehydrated in high vacuum directly from the frozen state.

Frei test. A skin test to determine sensitivity to the chlamydiae which cause lymphogranuloma venereum.

Frequency-dependent selection. Selection that is more intense when the individuals undergoing selection are present in greater abundance.

Freund's adjuvant. An emulsion of aqueous antigen in oil: complete Freund's adjuvant also contains killed Mycobacterium tuberculosis, while incomplete Freund's adjuvant does not.

Fruiting body. A specialized structure in some bacteria (Myxobacteria) and some fungi that produce spores.

Frustule. Another term of the cell wall of diatoms.

Functional genomics. This is performing genome-scale gene mapping or sequencing together with biological experiments that add meaning to the mass of data generated in a 'genome project'. Increasingly it is clear that automated sequencers and high-throughput mapping programmes can generate genetic data far faster than scientists can understand it or turn it into practical results. So the functional genomics movement is aiming to add high-throughput biology to the programme to identify what all these genes do.

Fundamental niche. The range of conditions under which an organism could survive if it were the only one in the environment. (Contrast with realized niche.)

Fungi. (*Sing.* = Fungus) Eumycetes; eucaryotic cells, frequently filamentous, which lack both a vascular system and chlorophyll; includes molds any yeasts. They are eucaryotes, members of Protists that lack chlrophyll. They share with plants the progression of cell wall, liquid-filled interacellular vacuoles, microscopically visible streaming of the cytoplasm, and lack motility. Many have cell walls composed of a very common plant material called **chitin**, rather than more common plant cell material *cellulose*. They do not contain photosynthetic pigments and are *chemo-organoheterotrophs*. They obtained energy by oxidation of organic substances. They grow aerobically. The common members are molds, yeasts, bracket fungi, and mushrooms. Non of these organisms has the plant-like characters like root, stem and leaves.

Fusiform. Spindle-shaped, or tapering at the ends; characteristic of Fusobacterium fusiforme.

Fusion biopharmaceuticals. Several biopharmaceutical proteins have been developed that are fusion proteins, i.e. they are the product of two genes, which have been fused together so that the proteins that they code for are joined end-to-end. The advantages of such proteins as drugs can be: They have two complementary or synergistic activities in one molecule. Thus, when the molecule binds to a cell, it does two things at once. To get the same effect with two molecules could need much more of both of them, to increase the chance that both would bind at once to one cell. The adverse effects or poor stability of one molecule are offset by the properties of the other. One molecule acts as a 'targeting' mechanism to bring the other to the site where it is meant to act.

Fusion protein. A fusion protein is a protein in which part of the amino acids come from one protein sequence and part from another. Fusion proteins are produced by splicing the gene for one protein next to, or into, the gene for another: the genetic apparatus reads the gene fusion as a single gene, and so produces a fusion protein. Fusion proteins are used in number of biotechnological applications. (1) To add an affinity tag to a protein. (2) To produce a peptide as part of a larger protein, which is then cut up after it has been made by cloning. (3) To produce a protein with the combined characteristics of two natural proteins (for example, a chimeric antibody). (5) To produce a protein where two different activities are physically linked (e.g. enzymes for substrate channelling or as a fusion biopharmaceutical).

G

Gall. Gross enlargement and deformity, usually globose, of plant tissue; caused by Agrobacterium tumefaciens.

Gametangium. (Gr. *gamos*: marriage + *angeion*: vessel or reservoir). Any plant or fungal structure within which a gamete is formed.

Gamete. (Gr. *gamete*: wife, *gametes*: husband). The mature sexual reproductive cell: the egg or the sperm.

Gametocyte. (Gr. *gàmete*: wife, *gametes*: husband + *kytos*: cell). The cell that gives rise to sex cells, either the eggs or the sperm.

Gametogenesis. (Gr. *gamete*: wife, *gametes*: husband + *gignomai*: be born). The specialized series of cellular divisions that leads to the production of sex cells (gametes). (Contrast with oogenesis and spermatogenesis.)

Gametophyte. In plants with alternation of generations, the haploid phase that produces the gametes. (Contrast with sporophyte.)

Gamma globulin. A fraction of globulin (a plasma protein); identified by electrophoresis; contains the majority of immuniglobulins (antibodies); also signifies the antibody containing fraction of plasma that is used for prophylaxis is therapy.

Gamma-richness. Addition of species on a broad geographical scale.

Ganglion. (Gr. *tumor*). A group or concentration of neuron cell bodies.

Ganglioside. Glycosphingolipids found in the neuronal surface membranes and spleen.

G-antigen. A particular kind of particles produced during multiplication of certain animal viruses; act as precursor to the complete virus particles. Consist of viral nucleic acid, haemagglutinin and lipids.

Gas sterilizer. An autoclave that is designed for or has been modified to permit optional use of a gaseous decontaminant instead of steam for sterilizing materials.

Gas transfer. One of the most important characteristics of a fermentation system is the rate at which gas can be transferred from the gas phase into solution. Often, the rate at which the organisms in the fermentor can metabolize is limited by how fast they can be provided with oxygen or have carbon dioxide, ammonia, or other `waste' gases removed. Oxygen is poorly soluble in water, and so the liquid itself holds very little, which the organisms in a dense culture can use up in a few seconds. Therefore they must be constantly supplied with oxygen gas, either as pure oxygen (efficient but expensive) or as air. Many fermentor design features are aimed at optimizing this transfer rate. There are several basic methods. Smaller bubbles of gas have a larger surface area per unit volume and greater internal pressure than large bubbles, and so gas diffuses out of them faster. Therefore the smaller the bubbles you can make, the faster oxygen diffusion will occur. However, creating small bubbles requires power, may cause disruption of the organism growing in the liquid, and can cause foaming, which fills the reactor vessel up with viscous foam. Anti-foaming agents can help this latter problem (which is also a problem when the organisms produce a lot of carbon dioxide gas). The sparger, the pipe system that delivers gas to the base of a tank fermentor, is responsible for breaking up the gas flow into the bioreactor into bubbles and making sure that they are evenly distributed in the reactor volume. There is a wide variety of methods for making small bubbles, keeping them suspended in the fermentation broth, and encouraging their content gases to dissolve. Among these are the deep jet fermentor, which injects the gas very fast into the bulk of the liquid, and the around the reactor loop at high hydrostatic pressure. At the base of the reactor the pressure is much higher, encouraging the gas to dissolve. The constant recycling ensures that bubbles stay in the liquid phase, and do not separate as forth. Other methods ensuring good gas transfer rely on increasing the surface of the liquid in contact with gas. Bubbling the gas through the liquid spreads the gas out: other methods spread the liquid out, for example in a thin sheet (in an oxidation pond), or in a thin permeable tube (as in a hollow fibre bioreactor).

Gas vesicle. A structure found in certain prokaryotes that contains gas and enables the organism to float or maintain certain levels in aqueous environment.

Gastroenteritis. Inflammation of stomach and intestine.

Gastrovascular cavity. Serving for both digestion (gastro) and circulation (vascular); in particular, the central cavity of the body of jellyfish and other cnidarians.

Gastrula. (Gr. *gaster*: stomach). An embryo forming the characteristic three cell layers (ectoderm, endoderm, and mesoderm) which will give rise to all of the major tissue systems of the adult animal.

Gel electrophoresis. Gel electrophoresis is one of the most common analytical methods in biochemistry and molecular biology. Electrophoresis is the separation of molecular according to how fast they move in an electric field. In gel electrophoresis this is done is a gel, a polymer matrix. Samples are put at one end of a slab of polymer gel (any jelly-like material). An electric field across the gel pulls the molecules through it: smaller molecules can pass through the gel more easily and so move towards the other end faster. Thus molecules are separated mainly according to size. A large number of materials are used to make the gel, but by far the most common are agarose (for DNA and RNA) and polyacrylamide (for DNA in DNA sequencing, and for proteins), which is often called PAGE (polyacrylamide gel electrophoresis). Various chemicals can be included in the gel to help the separation, such as the detergent sodium dodecyl sulfate (SDS) in protein gels, which unfolds all the proteins, and urea in DNA sequencing gels, which does the same to DNA. A recent variation in DNA gels is pulsed field gel electrophoresis (PFGE) and its variants. These also use electric fields to separate molecules, but with several sets of electrodes: the electric field is switchhed between them, which encourages the DNA to wiggle its way through the gel matrix, now headling one way, now headling another. This helps the separation of very large DNA molecules, up to the size of whole·yeast (but not human) chromosomes. Several terms such as orthogonal field gel electrophoresis and field inversion gel electrophoresis (FiGE) rafer to how the different fields are arranged. Variants on gel electrophoresis are isoelectric focusing (IEF) gels, which separate macromolecules on the basis of their isoelectric point (roughly the number of different charged groups they have) rather than on size. The gel has a pH gradient inside it, and the molecules migrate to the point where the pH of the gel is the same as their isoelectric point: at that pH they have no charge, so they stop moving and form a narrow band. O'Farrel gels (also called 2D gels) run an isoelectric focusing gel down one side of a gel slab, and then do a standard PAGE at right angles along the length: this produces a 2D pattern of protein spots which is as characteristic of a mixture of proteins as a fingerprint.

Gel. A mixture that is either semisolid or solid, but which has a large amount of liquid trapped within the solid component.

Gelatinase. A hydrolytic enzyme which degrades the protein gelatin.

Gene control. This is the genetic control of how a gene acts. Genes are pieces of information coded in DNA - on their own they do nothing. There must be a control system to make sure that their information is used at the correct time. Prokaryoyes (bacteria) and eukaryotes (yeasts, plants, animals) do this differently. In both types of organism there is a promoter region, which essentially says 'gene starts here' on the DNA. The promoter region contain DNA sequences that control when it is active, for example making sure that the gene is active under particular chemical conditions or (in eukaryotes) only in particular cells or at particular times during the organism's development. In prokaryotes such elements include operators and attenuators. In eukaryotes they include the hormone responsive elements (HREs) and a host of others. Eukaryotes also have enhancers, pieces of DNA that alter the activity of a gene but which are not next to the promoter. In general, in prokaryotes the promoter region is no more than 50 bases from the start of a gene. In eukaryotes it can be over 200 bases or more away. The enhancers can be up to several thousand bases away from the start of the gene in either direction, i.e. it is possible for a gene to contain its own enhancer. All these 'elements' are short pieces of DNA sequence, 5-30 bases long, each of which is recognized by a protein. The proteins are called regulatory proteins or transcription factors. They need genes to make them: in prokaryotes such genes are called regulatory genes, as opposed to the structural genes that code for the proteins that make the cell work. In eukaryotes this terminology is not used, because so many eukaryotic genes are to do with control at so many different levels.

Gene family. Many genes are not unique, but are closely related (anything from 30% the same to over 90% the same) to some other gene or genes. The whole collection is called a gene family. Gene families can be clustered into superfamilies, with even more members that are even less similar to each other. Gene families and superfamilies are often related in functions as well as DNA sequence, which is why people want to find them. Showing that a new gene is a member of the immunoglobulin-like super-family, for example, would strongly suggest that it has something to do with the immune system. There can be any number of members of a gene family, from three or four to thousands. Other DNA sequences that are not genes (i.e. they do not have a recognizable function) can also come in families of similar sequence. **Repetitive sequences.** Also 'SINES' in mammals, transposable elements in other organisms. These can be regarded as 'junk' DNA in any other than a genetic context. **Pseudogenes**. These are sequences that have nearly the same DNA sequence as a real gene, but do not work as genes because some critical bases are mutated. A variant is a processed pseudogene, which is a pseudogene that has lost all its introns. The *average sequence* of a gene family of any sort is called a consensus sequences. It is a sequence which, at each base, resembles the family members the most. Like that fiction 'the average man', a consensus sequence is an abstraction which represents the family as a whole: it may not represent any one member of it. You can also have a consensus sequence of any other bits of DNA, like promoters (the sections of DNA at the start of genes). The relationship of members of a gene family to each other is often shown as a phylogenetic tree, in which the length of the branches between each 'leaf' shows how different the sequences are.

Gene flow. The exchange of genes between different species (an extreme case referred to as hybridization) or between different populations of the species. It is the principal driving force of evolution.

Gene frequency. The proportion of a given allele in a population.

Gene library. A gene library is a collection of gene clones that contains all the DNA present in some source, but split up and joined on to suitable vector DNAs. It is also sometimes called a gene bank. If the original source of the DNA was the original DNA from a living organism, then the library seeks to include clones of all that DNA: it is called a genomic gene library, because it contains all the DNA from that organism's genome ('genome' simply means all the genes, or DNA, in an individual). If the DNA is from some other source such as copy DNA (cDNA) made by enzymatic copying of RNA, then the maker of the library seeks to include representative clones from all that source, and, in this case, it would be called a cDNA library. Gene libraries are not organized like book libraries, and can only claim to be complete because the number of clones in them is sufficiently large for us to be very confident that all the clones we want to be there are there, i.e. that there is a very small chance that anything has been left out. Usually genomic gene libraries are meant to be 95 or 99% complete, so that there is a 95 to 99% chance that the gene you are looking for is in there somewhere. The number of clones needed to get a complete gene library depends on how big the cloned pieces of DNA are. Thus, if you are using a lambda phage vector to make a genomic gene library from human DNA, you need about 500000 clones. However, cosmid cloning vectors can hold substantially more DNA: you would only need 200000 of them. YAC (yeast artificial chromosome) vectors hold 10 times as much DNA again, so you only need 10000 of them. This is why people use YAC vectors to make genomic gene libraries since screening 10000 clones for the one you want is almost invariably easier than screening 500000.

Gene pool. All of the genes in a population.

Gene synthesis. This is the complete synthesis of a gene using a DNA synthesizer ('gene machine'), rather than cloning it or assembling it from cloned fragments of DNA. Because most genes are longer than the maximum length of DNA that can be made conveniently on a DNA synthesizer, genes are usually assembled from a number of oligonucleotides. Each section of the gene is hybridized to the next section, and when the whole assembly has been hybridized together the DNA sections are joined together enzymatically to make one double helix. This requires the oligonucleotides to be designed carefully, so that they only hybridize to their correct partner and not to other oligonucleotides in the mix. Other concerns involve making sure that the same sequence is not repeated too often within the gene (repeated sequences can be targets for rearrangements of the DNA within bacteria) and making sure that the codons used are suitable. Different codons that code for the same amino acid are not used with equal frequency, and generally the more frequently used codons are translated faster than rare codons. However, which codon is used most frequently depends on the organism in which the gene is to be expressed. Other features of the gene are the presence or absence of restriction sites and suitable 'sticky ends' so that the final gene can be cloned into an expression vector easily.

Gene therapy. Gene therapy is changing the genetic make-up of a human. There are two approaches, germ-line gene therapy and somatic cell gene therapy. The former changes the 'germ cells', the cells that make sperm or ova. This has a permanent effect on all the individuals that are the descendants of whoever had the therapy. Somatic cells are the other cells in the body, i.e. the muscles, bone, nerves, etc. Changing them does not affect the germ cells, but does affect the engineered Person. Gene therapy of the germ cells of animals is called transgenic technology. Germ-line gene therapy of humans is considered unethical, and has not (legally) been tried. Somatic cell gene therapy can be aimed at correcting a genetic or a non-genetic defect. Current therapeutic targets include both categories. A relatively easy route to somatic gene therapy is bone marrow therapy since bone marrow is relatively easy to take out and replace, and generates itself inside the body. An engineered stem cell can reproduce itself inside the bone marrow, creating engineered blood cells as it does so. Targets for bone marrow therapy include SCID (severe combined immunodeficiency disease, a very rare genetic disease caused by a deficiency in the enzyme adenosine deaminase (ADA)). W. French Anderson and Michael Blaese used a gene therapy treatment for SCID on a four-year-old girl in late 1991, with reasonable success. Other targets include several cancer treatments, including introducing cells engineered to produce more tumour necrosis factor (TNF) or interleukin (IL-4) into cancer patients, where it is hoped they will be able to assist in destroying the cancer. There have been several proposals for ways of getting DNA into cells while they are still in the patient's body. Routes suggested include:

Using viruses. This is the most common approach.

Biolistics. As well as delivering DNA into isolated cells, biolistics can be used to put DNA into cells that are still part of an animal.

Injection. Simply injecting DNA complexed with calcium phosphate into liver or muscle causes some cells to take the DNA up and express the genes in it. This has attracted a lot of attention, because it offers the hope of a gene therapy treatment for muscular dystrophy, one of the most common genetic diseases.

Using lipsomes. DNA that has been encapsulated into liposomes and injected is taken up by the liver and, to a lesser extent, by the spleen, and any genes it carries are expressed briefly. Lipsomes can also be targeted to other tissues. DNA can also be complexed with negatively charged lipids the molecules form nanometer-scale clumps, which are taken up by cells.

Using cationic lipids. Positively charged lipids will form many-molecule aggregates with nucleic acids that are fairly lipid soluble and electrically neutral. These are taken up by cells much more readily than 'naked' DNA. The same approach has been to get antisense oligonucleotides into cells, using a lipid formulation called lipofectin.

Use of peptides. Rather than use whole viruses. one could complex DNA to proteins that bind to specific cells. For example, Genzyme are trying to link peptides with the RGD sequence in them to DNA. RGD binds to integrins.

Regulation. Applying gene transfer techniques to humans, usually called gene therapy, has been a major problem for legislators and regulators as well as for scientists. From the Martin Cline experience in 1980, there has been substantial reluctance to let anyone put genes into anyone else, no matter what for. Cline, a UCLA researcher, wanted to put the genes for β-globin into patients suffering from thalassaemia, a genetic disease caused by defects in the β-globin genes. Refused permission to do so in the USA, he performed the medical parts of his experiments in Israel and Sardinia (which have much higher incidences of thalassaemias). This brought universal disapproval and a determination to make sure that any future gene therapy experiments were tightly regulated. (The experiment was a complete failure.) Every agency or pressure group with an interest in biomedicine wants to have a say in whether gene therapy can take place. In late 1990 the first gene therapy experiment, giving as SCID patient the gene for ADA, took place. Before it could do so, the experimenters sought approval from the following agencies, any of which could have stopped the experiment.

Gene. (Gr. *gen*: to produce). A presumed unit of heredity. Used here as the unit of genetic function which carries the information for a single polypeptide. A particular segment of a nucleic acid (DNA usually) that codes for a single polypeptide or protein. A gene is a section of DNA that codes for a defined biochemical function, usually the production of a protein. DNA (deoxyribonucleic acid) is made up of repeating units which vary in their chemical details (as a magnetic tape is basically the same all the way along, but varies in the details of its magnetization depending on what has been recorded on it). The parts of the DNA that differ are the bases, so called because they are the chemically basic (alkaline) part of the overall acidic DNA structure. In DNA there are two strands winding round each other in a double helix, so the bases come in pairs. In RNA there is only one strand. Molecular biologists use 'base' and 'base pair' fairly indiscriminately to mean the length of a piece of DNA or RNA since RNA is copied off DNA base for base in the process of transcription. Genes are arranged on long DNA molecules called chromosomes, which may contain a few dozen genes in a few tens of kilobases (one kilobase = 1000 bases) in the chromosome of a virus, to tens of thousands of genes in hundreds of megabases (1 megabase = 1000000 bases) of DNA in the chromosomes of higher plants and animals. In eukaryotes (living things with cell nuclei, as distinct from prokaryotes, like bacteria) the chromosomes can be seen under the light microscope when the cell divides. Chromosomes can be chemically treated so that some regions take up dye more easily than others and so stain as darker 'bands', reflecting the genetic structure of the DNA. Each band can have hundred of millions of bases of DNA in it, so bands are not precise landmarks to genes. However, they form rough chromosomal 'addresses', which are often used to describe where a gene is on a chromosome. In addition, somewhere on a chromosome is a centromere, a waist-like constriction that is an essential part of the mechanism that makes sure each of two new cells gets one set of chromosomes each. The centromere divides the chromosome into two, usually unequal, parts. The short part is referred to as p, the long part q, so 16p2.3 refers to chromosome 16, long 'arm', band number 2, sub-band number 3. Preparing cells so that the chromosomes can be seen and hence counted is called karyology, and the resulting count of chromosomes is called the cell's karyotype. Preparing the more complex chromosomal band analyses is covered in the broader term cytogenteics. All the genes (and hence, inevitably, all the chromosomes) in an organism are called its genome. The human genome is about three billion bases long. The record size is that of the lungfish, which has a 140 billion base genome, nearly all of which appears to be 'junk' DNA. Animals and all larger plants have two sets of chromosomes in each cell, so that there are two copies of each gene there. Exceptions are the germ cells (sperm and ova in animals), which only have one chromosome set each. Thus, when a sperm and an ovum get together, the result is two chromosome sets again, a new genetic individual. Sometimes the two copies of a gene can be slightly different, in which case the copies are called alleles. If one of your genes is present as two different alleles in your cells, then you got one from your mother and one from your father. Organisms with two chromosome sets per cell are called diploid, and include animals, higher plants, and many fungi. Organisms with only one chromosome set per cell are called haploid, and include yeasts and bacteria. Some species have many chromosome sets: these are called triploid (for three), tetraploid (for four), and so on, or more generally 'polyploid'. In bacteria, genes that are regulated together (i.e. that are turned on at the same time and by the same stimulus) can be arranged in a tight cluster called an operon. This has one control region at one end, and then a series of coding regions, i.e. regions of DNA that code for single proteins. This whole cluster is transcribed as one RNA, which is then decoded into multiple proteins by the enzymes of the cell. Thiis operon structure is virtually unknown in higher organisms. So that all genes are not active all the time, genes need control regions attached to them to regulate their activity. In a bacterial operon, these regions are located at one end of the gene. In eukaryotes the control regions (or control elements, since they are usually very short sections of DNA) are mostly at the start of the gene, but can also be spread quite a way away from that start, both within the gene itself and away from it.

Generation time. Time required for a population of cells to double in number.

Genetic association. A term used to describe the condition where particular genotypes are associated with other phenomena, such as particular diseases.

Genetic disease diagnosis. A genetic disease is one that is caused by a gene, so we 'inherit' the disease from our parents. For a true genetic disease anyone with the correct genotype (collection of genes) will display the phenotype (the physical manifestations of the genes). In practice, quite a lot of genetic diseases have incomplete penetrance: i.e. the genes do not always cause the effect they are meant to. This makes detecting them rather complex.

Genetic drift. Evolution (change in gene proportions) by chance processes alone.

Genetic engineering. A general term for the directed manipulation of genes, and usually used synonymously with genetic manipulation or genetic modification. A wide range of technologies are involved in this, but most involve the recombinant DNA techniques. Genetic engineering falls into several different categories depending on what is being engineered. Bacteria, yeast. This is 'traditional' genetic engineering (i.e. over 10 years old). Using recombinant DNA techniques, genes are put into microorganisms to make them produce something we want, be it insulin, or better beer, or protein for food.

Genetic information. Projects such as the Human Genome Project and the development of tests for genetic predisposition to disease have lead to a lot of talk about how human genetic information should or could be used. This contrasts to genetic information about animals, plants, or microorganisms, which is not thought to have the same ethical status: debates about who 'owns' the human genome focus on high ethical philosophy, those about pig genomes centre around the patent courts.

Genetic map. A genetic map is a description of how genes lie along a chromosome. It looks rather like a ladder with very erratic rungs. There are several types. The map can show the distance between any two points in terms of how much DNA lies between them (usually measured in megabase, Mb). A genetic map of this sort is called a physical map, because it represents what the genes are physically like. A map can also represent the recombination distance between two genes. This is the chance that two genes at those two points will be separated by recombination during meiosis. (This is basic genetics. In summary, genes on the same chromosome are usually inherited together as a package between parent and child. When the sperm and ova are being made, a shuffling process called meiosis can shuffle these packages. The further the genes are apart on the chromosome, the more likely they are to be shuffled during meiosis. The shuffling process is called recombination. The chances of recombination between two genes can be described as a distance, and the distances used to build a map.) The resulting map is called a genetic map or a recombination map, with distances measured in centimorgans (cM), which is the distance apart genes have to be to have a 1% chance of being recombined during meiosis (named after American geneticist and Nobel Prize winner Thomas Hunt Morgan). The distances on the physical map and the genetic map are rarely the same, but the order of the genes is the same. (Just like a subway map vs. a town map: the distances on the subway map do not represent real-world distances, but the order of the stations is the same as the order you visit them in the real world.) Genetic maps are important for breeding experiments, where they allow us to predict how much effort it will take to get a particular combination of genes into one organism, and determine the best strategy for doing so. They are also important for gene cloning experiments, where they tell us how near one gene is to another, and hence how our cloning strategy should approach that gene. The landmarks on genetic maps are not always genes. They are more generally known as 'markers', and can be any piece of DNA, identified as a gene or by DNA probes or PCR amplification. The critical feature of a genetic map is its density: how many markers there are on it. The more markers there are, the more use it is in guiding our experiments, and hence the interest in the Human Genome Project in producing a high density genetic map with lots of markers on it. True genetic maps can only be made by genetic experiments, i.e. breeding. Physical 'genetic' maps can be made by a variety of DNA techniques, such as looking at RFLPs, making physical maps of a whole lot of cloned DNAs and then stitching them together, or in situ hybridization to chromosomes to see where a piece of DNA lies physically on these cellular structures. Ideally, the maps made by these different techniques should match. Usually, they do not, quite. The ultimate physical map is the DNA sequence of a chromosome. This tells where every single possible piece of DNA is. However, it may still not tell you what they do, and so is not the ultimate genetic map.

Genetic restriction. The term used to describe the observation that lymphocytes and antigen presenting cells cooperate most effectively when they share particular haplotypes.

Genoceuticals. A vogue term for a version of gene therapy where a gene is placed into a cell and there produces a pharmaceutically active protein. So far several studies have shown that DNA can be placed into the cells of adult mice and rabbits, and that the DNA can work there to produce proteins. This has two potential applications: however, both are speculative at the moment, not having been tried properly even in animals. **'Genetic antibodies'** are genes that have some antibacterial or (more usually) antiviral activity. The genes are placed into the cells that are the potential targets of the parasite. For example, a gene for a toxin could be linked to a controller gene that is activated by a virus: when the virus infects the cell the toxin gene is turned on, toxin is produced, and the cell dies.

Genome. The total genetic material contained within the cell.

Genotype. (Gr. *gen*: to produce + *typos*: impression). An exact description of the genetic constitution of an individual, either with respect to a single trait or with respect to a larger set of traits. (Contrast with phenotype.)

Genotype. The genetic material inherited from parents; not all of it is necessarily expressed in the individual.

Geographical (allopatric) speciation. Formation of two species from one by the interposition of a physical barrier. (Contrast parapatric, sympatric speciation.)

Geonome project. A genome project is a project to determine the exact genetic structure of an organism's genome, i.e. the DNA sequence of all its genes. There are quite a few. The Human Genome Project is a project to determine the base sequence of all the DNA in humans.

Bacterial genomes. Complete sequencing of the genome of *Escherichia coli* is well under way by a consortium of laboratories, and will probably be complete by the time this book is published. In 1995 two complete bacterial genome sequences were completed, for Haemophilus influenzae and *Mycoplasma genitalium*, both disease-causing bacteria. *Mycoplasma gentialium* has a simplified genome, because it relies on its host for much of its metabolic apparatus. Even so, it is 580067 bases round. *Haemophilus influenzae* is a more typical bacterium, and has a genome of 1830121 bases and 1749 genes. Scientists could assign a potential function to most of those genes by comparing them with genes from other species whose function was known. The complete genome sequence of the yeast Saccharomyces cerevisiae was completed in 1996, by a consortium of academic laboratories. The project took longer to complete than expected because the research groups held on to the sequence until the last bit was checked. Since the whole genome is 12146000 bases long, this took a while. Other genome projects include mice, cows, pigs, the weed *Arabidopsis thaliana*, the worm *Caenorhabditis elegans*, the fly Drosophila melanogaster and a number of crop plants: rice, wheat, maize, and apple among others.

Geophagia. The practice of eating dirt or clay.

Geosmin. Compounds produced by some microorganisms that contributes to "earthy" smell of soil.

Geotropism. (Gr. *ge*: the earth + *trope*: a turning). A directed plant growth response to gravity.

Germ cell. A reproductive cell or gamete of a multicellular organism.

Germ line. The genetic material which is passed down through the gametes before it is modified by somatic recombination or mutation.

Germ tube. A filamentous protuberance produced by yeast cells.

Germfree. Free of all microbial life detectable by examination.

Germicide. A chemical agent which kills bacteria but not necessarily endospores; generally considered synonymous with bactericide.

Germination. The process of vegetative cell formation from spore following a period of dormancy.

Gestational age. The age of a concepts computed from the first day of the last menstrual period to any point in time thereafter, but not usually calculated beyond the first few months of life after birth.

Giant cells. Large multinucleated cells of sometimes seen in granulomatous reactions and thought to result from the fusion of macrophages.

Gibberellin. (L. *gibberella*: hunchaback (refers to shape of a reproductive structure of a fungus that produces gibberellins)). One of a class of plant growth substances playing roles in stem elongation, seed germination, flowering of certain plants, etc. Named for the fungus *Gibberella*.

Gill. An organ for gas exchange in aquatic organisms.

Gingiva (gums). Soft tissue surrounding teeth.

Gizzard. (L. *gigeria*: cooked chicken guts). A very muscular part of stomach of birds that grinds up food, sometimes with the aid of fragments of stone.

Gland. An organ of secretion or excretion.

Glia. (Gr. *glue*) Cells, found only in the nervous system, which do not conduct action potentials.

Globulins. Class of proteins found in blood.

Glomerulonephritis. Disease of kidney, characterized by inflammatory changes in glomeruli; may occur as a result of an autoimmune disease in which the streptococci induce manufacture of antibodies against the individual's own kidney tissue.

GLP/GMP. These stand for Good Laboratory Practice and Good Manufacturing Practice. They are a code of practice that is designed to reduce to a minimum the chance of accidents that could affect a research project or a manufactured product. Sometimes referred to as cGMP (current GMP) or cGLP (clinical good laboratory practice). The GLP and GMP prescriptions are quite voluminous, but boil down to a few key points. The essential point of both GLP and GMP is that everything is recorded, and only established procedures are used by people who have been trained to use them. This may seem obvious, but extends to everything: in a true GLP laboratory, for example, only staff who have been trained to use a balance may use it, every weighing has to be checked by another person (who has also been trained for that specific balance) who has to sign to say that they have checked that the weight of material is correct, weighing has to be done according to a written standard operating procedure (SOP) for using that balance, the protocol used has to be noted in the record of the experiment, and so on. All records are kept, and have to be archived on microfiche, magnetic tape, or (more recently) write-once CD-ROM. Samples of every batch of material used in the experiment or the manufacturing process also have to be archived, so they can be referred to in future should this become necessary. Using these types of procedures, it is possible to trace exactly who did what at every stage of an experiment or a manufacturing process. Thus, if there is a problem afterwards, the GLP or GMP user can either point to a specific material or standard operating procedure that caused the problem, or can point to exhaustive documentation that shows that the problem is not their fault. This can be very important in pharmaceutical development and manufacture. (GLP was set up after some severe side effects of a drug were overlooked during pre-clinical research because the protocols for an experiment were faulty.) Many biotechnology companies claim to work to GLP or GMP (depending on whether they are concerned with R & D or manufacturing). In practice, many who claim to do so do not actually work to GLP. It is enormously difficult to do innovative research to GLP, where you have to define a set of standard operating procedures, train the staff formally, etc. just to do one experiment that might take half a day. GLP is more relevant to pharmaceutical development (where a large number of very similar experiments are performed). GMP is a standard requirement for pharmaceutical production, and for a number of other industries. GMP also stands for Good Microbiological Practice, a code of laboratory practice for performing basic microbiology safely. In this sense, GMP is simply a way of reducing the chance of contamination problems (contamination of the sample or of the laboratory) during a microbiological experiment of a minimum. The International Standards Organization (ISO) also has a 'good practice' standard, ISO9000. There are a number of substandards within ISO9000, dealing with 'good practice' in manufacturing, product development, provision of services, etc. Organizations can be certified to ISO9000 as a mark that they provide a consistent quality of result to their customers. Certification requires that the company be regularly inspected by external, qualified inspectors: it is not a 'once and for all' event. ISO9000 certification and GMP status are not formally linked, but have many of the same requirements for documentation and traceability.

Glucan. A term given to polysaccharides in which glucose is a component.

Glucoamylase. An enzyme that removes successive glucose units from the nonreducing ends of dextrin chains.

Gluconeogensis. Formation of glucose from noncarbohydrate precursors.

Glucose isomerase and invertase. Glucose isomerase is probably produced in larger amounts for industrial use than any other single enzyme (although the largest category of enzymes by far is the general class of alkaline proteases, used in detergents). It catalyses the interconversion of the two sugars glucose and fructose. Since fructose is slightly more chemically stable than glucose, a mixture of glucose and fructose with the enzyme will end up almost entirely as fructose. This is valuable for the food industry since fructose is substantially sweeter than glucose, and so you get more sweetness per gram than from glucose. The usual use for glucose isomerase is to take glucose made by hydrolysis of corn starch and turn it into a mixture of mostly fructose with some glucose. The corn starch is broken down using amylases. The result is called 'high fructose corn syrup' (HFCS). Invertase takes sucrose ('sugar') and turns it into glucose and fructose, a mixture called invert sugar or invert syrup. Invertase was the first enzyme to be used in industrial production, in the late 1940s. Thus, in conjunction with glucose isomerase it can convert sucrose into HFCS. Invertase can also be used in its own right to convert the easily crystallized sucrose into the less easily crystallized glucose-fructose mixture. 'After Eight Mints', for example, have invertase in their centres: it turns the hard sucrose core (which the chocolate coat was poured on to) into the soft centre we finally eat.

Glucose. (Gr. *gleukos*: sweet wine mash for fermentation). The most common sugar, one of several monosaccharides with the formula C6H12O6.

Glucosylation. Refers to the process of attachment of a glucose residue to the nucleic acid molecules of certain bacteriophage mutants. This renders the nucleic acid resistant to the influences of host cell nucleases, thus enabling the phage to infect hitherto resistant strains of the bacterium. Glucosylation is possible only when the phage previously infects a bacterium possessing the necessary enzyme system.

Glue. Biological glue is one of the many areas where biotechnology and medicine can overlap. Doctors are always interested in new medical techniques for repairing wounds. An obvious way is glue: however, the glue must have unusual properties. It must be able to set (cure) in a wet environment, not be broken down by watery liquids, not irritate or poison the body, not induce an immune or allergic response, and the body just be able to break it down after a time if its function is only temporary, like stitches. The most widely used and discussed 'glue' is the protein fibrin. The body itself produces fibrin, a component of the clotting proteins in blood: however, it is not a very strong glue, and, and, unless derived from human blood (with its concomitant risk of viral contamination), causes a strong immune response. However, it is a natural human product and is used in several commercial medical glue applications. Several marine organisms produce glues that could fit these criteria. Mussels and barnacles produce protein-based glue which could, in principle, be produced by more convenient organisms using biotechnology. Genex has produced a yeast that makes the protein (which has a very unusual amino acid composition, making it difficult for the yeast cell to make it efficiently). The protein also needs extensive and rather specific post-translational modification, which the yeast cannot perform. Thus, these proteins are some way from commercialization yet. Many other organisms make materials that glue them on to things or things (like eggs or nest material) on to other things. However, these have not been investigated nearly well enough to make them attractive targets for medical glues. Biotechnology is also investigating semi-synthetic glues made from natural materials. Proteins can be cross-linked into a glue-link material, which can be used in wound healing and surgery, being fairly rapidly readsorbed by the body.

Glycation. Glycation is the non-enzymatic reaction of sugars with proteins. Many proteins have sugars added deliberately by the body, and there are enzyme mechanisms to make this happen. However, sugars can also react with the amino groups in proteins in an uncontrolled, chemical reaction. Since every part of mammalian bodies has sugar in it, this means that all proteins get glycated after a while. This is much accelerated by heating or if the sugar levels are very high. Hence chemical glycation is important to protein processing, and hence flavour formation, in food. Chemical glycation is also very important in the damage done to diabetics when their sugar levels rise above normal, and to all of us as we age. Indeed, one school of thought holds that much of the damage that causes ageing is a result of the effects of glycation. In particular, glycated proteins can continue to react and form complex, stable cross-links with sugars and, through them, with other proteins. These complexes are called advanced glycosylation end-products (AGEs). The body seems to be unable to remove them specifically, and so they accumulate, cross-linking collagen into a rigid, inflexible net, damaging critical proteins in long-lived nerve cells, and maybe even directly mutating DNA.

Glycobiology. Glycobiology is the study of sugars and their role in biology. Usually this is taken to mean the study of complex sugars and what their functions are, and not the metabolism of how sugars are put together and taken apart. The twin thrusts of glycobiology are the study of glycoproteins, which are proteins with sugar residues attached, and the study of molecules that interact with sugars and affect sugar metabolism, especially the synthesis of those glycoproteins (i.e. glycosylation). Some glycoproteins have as much sugar in them as protein by weight, and the effects of this sugar on the protein can be substantial. Current theory suggests that the sugars on glycoproteins help in protein-protein binding (important for the mechanism by which cells recognize each other and by which viruses bind to and gain entrance to cells). From this, glycobiology is interested in how complex sugars on their own interact with glycoproteins, glycolipids (lipids with sugars attached), and each other. It is also interested in lectins (proteins that bind specific sugars) as tools. In living systems sugars, both as simple sugars and as blocks of sugar residues, are joined on to proteins at specific amino acid sites by glycosyltransferase enzymes (a process called glycosylation). Glycolipids can also be joined on to proteins by specific enzymes, a process called glypiation, producing glycolipoproteins. These complex entities are an important part of the surface membrane of cells, and so may be the 'docking molecules' that viruses use to attack cells: consequently, biotechnology researchers are interested in them because their study may lead to better antiviral drugs, and to markers for aberrant cells such as cancer cells. The application of glycobiology is sometimes called glycotechnology, to distinguish it from biotechnology, a discipline that concentrates more on nucleic acids and proteins. Companies such as Oxford Glyco-systems are developing a carbohydrate-based anti-AIDS drug (which acts by blocking HIV's mechanism for latching on to cells as it infects them), and Glycomed has a drug aimed at blocking the effect of the glycosylated ELAMS. Other uses of glycobiological expertise are in manipulating glycosylation in expression systems, and in analysis of carbohydrates and glycoproteins.

Glycogen. A branched-chain polymer of glucose, similar to starch (which is less branches and may be of lower molecular weight). Exists mostly in liver and muscle; the principal storage carbohydrate of most animals and fungi.

Glycoprotein. Proteins containing sugar sequences.

Glycosidases. A group of enzymes that break up complex sugars (such as starch or sucrose) into simple ones (such as glucose or fructose). About 12000 tonnes of glycosidases are made per annum, almost exclusively for use in the food industry. The major glycosidase enzymes are amylases (which break down starch) and glucose isomerase (which is used to turn glucose into the sweeter fructose). Amylases break the long chains of starch molecules and similar polymers into shorter segments, and ultimately into glucose. Amylases are commonly extracted from barley, beans, potatoes, and a variety of fungi. Other enzymes produced from bacteria and fungi for polysaccharide breakdown are isoamylases and pullulanases. These break off the side branches of starch, and are sometimes called 'debranching enzymes' for this reason. Since molecules that are single, unbranched strings of units have a quite different rheology from molecules that are branched like a tree, the debranching enzymes can be valuable to the food industry to alter the flow properties, or 'mouth feel', of food. A third group of these enzymes are the cellulases, which break down cellulose. Since cellulose is probably the most common biological material in the world, using it as a raw material makes economic sense. However, it is very difficult to break down into its glucose monomer units. A range of cellulases and hemicellulases are used to process food ingredients, for example, enhancing the maceration of fruit pulp to make fruit juices. Gist brocades a major provider of these enzymes, claim that enzymes can increase the amount of juice you can squeeze out of apples by 25%, through opening up the cell structure.

Goldberg-hogness box. Nucleotide sequences in eukaryotic promoter genes that bind RNA polymerase.

Golgi apparatus. A system of concentrically folded membranes found in the cytoplasm of eukaryotic cells. Plays a role in the production and release of secretory materials such as the digestive enzymes manufactured in pancreas. First described by Camillo Golgi (1844-1926).

Gonad. (Gr. gone: seed, that which produces seed). An organ that produces sex cells in animals either an ovary (female gonad) or testis (Male gonad).

Gonadotropin. A hormone that stimulates the gonads.

Graft. (Latin *graphium*: to write or to imprint). Here refers to the artificial union between tissues that are normally separate. Also refers to placing of a tissue from its normal position to another position of the same organism or to any position of a different organism. Grafting of a diseased tissue or a portion there form can help transmit the disease.

Gram stain. Differential stain which provides for grouping of bacteria as either gram-positive or gram-negative. If the organism retains the initial stain after contact with a decolorizer, it is gram-positive; if the organism loses the stain after exposure to a decolorizer, it is gram-negative.

Grana. Condensed bodies of chlorophyll within plant cell's chloroplasts.

Granulocyte. A type of white blood cell such as neutrophils, eosinophils and basophils containing granules of hydrolytic enzymes involved in phagocytic digestion.

Granuloma. A nodule formed in tissue as a result of the accumulation of fused macrophages, which try to destroy infectious agents.

GRAS. List an acronym for Generally Recognized as Safe. A list of chemicals that are allowed by law to be added to foods.

GRAS. This stands for 'generally' regarded as safe', and is an important category for acceptance of biotechnological products in Western countries, and especially the USA. It means that the product has a long history of use for a specific purpose. For microbial or genetically engineered products, regulatory approval for general release is much easier if the product is made in an organism that is GRAS, since the only unknown is then the new product, not the organism as well. For isolated materials, to be accepted as GRAS in one application (e.g. as a foodstuff) can greatly help getting approval for another application (e.g. cosmetics). The exceptions are usually pharmaceutical applications, where any new product, even if it is believed to be chemically identical to a previous product but made in a new way, must go though a full set of toxicological and clinical trials before it is released.

Gravity exhaust (Ventilation term). Discharge of air, resulting only from pressure differential, from a ventilated room to the outdoors through an exhaust duct.

Green beer. Freshly fermented beer before it has been aged.

Group transfer. The exchange of atoms between molecules.

Growth factor. An organic compound (vitamin, amino acid, purine or pyrimidine) required in small amounts to support growth of an organism.

Growth factors. Growth factors are materials (apparently invariably proteins in mammals) that stimulate growth. They are of great interest as potential drugs (biopharmaceuticals) because they could be used to assist wound healing or even encourage tissue regeneration. Growth factors not only stimulate cell division, but also influence differentiation of cells and, in some cases, select which cells in a mixed population divide or differentiate.

Growth stage. That stage in the history of an organism in which it grows to its adult size.

Growth. An irreversible increase in volume (probably the most accurate definition, but at best a dangerous oversimplification).

Growth and Development of Microorganisms. Growth is the self multiplication of living material, the protoplasm it self. Growth is an increase in size (volume or length) due to cell divisions and subsequent enlargement, It in an increase in dry weight of bulk of an organism associated with development.

Development is defined as an ordered change or progress, often towards a higher, more ordered or more complex state. Development may take place with growth and growth may take place with development. However, these are often quite integrated processes.

In most organisms, growth involves cell division, expansion, differentiation and morphogenesis. Increase in height and dry weight are the most obvious manifestation of growth, e.g., in forage and pasture crops. In grain crops dry weight increases in non-grain components are important only in producing a plant capable of high grain yield. Growth is also expressed as the advance of a plant from one development stage to another. A seed germinates, the seedling develops roots, leaves, and stem, and the plant grows to maturity. The plant flowers and produces seed to complete its life cycle.

Growth can be defined as *an orderly increases of all the chemical constituents of an organism.* An increase in total mass is not necessarily a reflection of growth; it may result from the synthesis and accumulation of a cellular reserve material (e.g. - glycogen), unaccompanied by synthesis of the major biopolymers (proteins and nucleic acids). Growth normally results in cellular multiplication, except in the special case of coenocytic organisms (coenocytes). In a unicellular organism, cellular multiplication leads to an increase in the number of individuals.

It is the addition of new material to the cell with the aid of enzymes catalysis. Others say it is just an increase in mass or size of a cell, even if the increase is due to accumulation of water or glycogen inside the cell. A change which shows an increase in mass or volume is positive; while a change showing a decrease in mass or volume is negative.

Population growth is the increase in the total mass of cells making the population, or the increase in the total number of cells of the population. Individual growth is the increase in mass of the individual cell.

Morphological and chemical differentiation takes place during the development. These processes of cells are complicated and involve synthesis of many organic compounds like protein, cellulose, and nucleic acids and requiring physical forces that cause cell enlargement.

Guild. A group of organisms that share a common food resource. Guild members have strong interactions with one another but weak interactions with members of other guilds.

Gumma. Granuloma found in tertiary stage of syphilis.

H

H antigen. Flagellar antigen of bacteria; (H form Hauch, Ger., Breath).

H chain (heavy chain). The larger of the two types of polypetide chains of the basic Ig molecule, there are two identical H chains per basic Ig molecule; H chain composition determines the class of the Ig molecule.

H-2. The mouse Major-Histocompatibility Complex.

Habit. The form or pattern of growth characteristic of an organism.

Habitat island. A patch of habitat, separated from similar patches by other habitats.

Habitat. The environment in which an organism lives.

Habituation. The simplest form learning, in which an animal presented with a stimulus without reward or punishment eventually ceases to respond.

Haemagglutinin (Gr. *Haima*: blood + L. *agglutinare:* to glue together). Refers to agents capable of inducing sticking together of red blood corpuscles. This happens when blood of incompatible group come together; viruses act as agglutinating agent.

Haemophilus. A *Blood-loving*, genus name.

Hair cell. A type of mechanoreceptor in animals.

Hairy root culture. This type of plant culture consists of highly branched roots of a plant. A piece of plant tissue (an explant, usually a leaf or leaf section) is sterilized to remove the bacteria on the surface, and then treated with a culture of the bacteria *Agrobacterium rhizogenes*. Like its cousin *Agrobacterium tumefaciens*, *A. rhizogenes* transfers part of its own plasmid DNA to the cells of an infected plant. This causes alterations in the plant's metabolism, including alterations in hormone levels. This in turn causes the explant to grow highly branched roots from the sites of infection. The roots branch much more frequently than the usual root system of that plant, and are also covered with a mass of tiny root hairs, hence the name of the culture system. The hairy root cultures do not require hormones or vitamins to grow, unlike explant cultures or cell cultures of plant cells, so they can grow on simple media of salts and sugars. They are also genetically stable, again unlike explant or cell cultures, so they can be cultured in bulk without the culture changing. Their most significant feature, however, is that they produce secondary metabolites at levels similar to those made in the original plant. Thus they can be used as a replacement for plants for making such compounds as food flavours or fragrances. As such they are the target of considerable interest and research, although there are no products yet. Hairy root cultures have been grown in several large laboratory fermentor systems as well as small pilot plants. They look like a mass of fibres when growing as an unstirred mass: they can be grown in a stirred tank reactor, but they are rather sensitive to being broken up by the stirring machinery. However, because they grow and metabolize more slowly than bacteria and do not have nearly such high oxygen demands, stirring is not usually necessary to obtain a successful culture.

Halophile. An osmophile that has a specific requirement for creased amounts of an inorganic salt (usually NaCl) for growth.

Halophillic. Salt-loving; growth is dependent on or enhanced by high salt concentration.

Hanging drop preparation. A method for observing bacterial motility in a liquid medium.

Haploid. (Gr. *haploeides*, single). Having a chromosome complement consisting of just one copy of each chromosome. This is the normal `ploidy' of gametes or of asexual spores produced by meiosis or of organisms (such as the gametophyte generation of plants) that grow from such spores without fertilization.

Haplotype. A set of genetic determinants located on a single chromosome.

Hapten. A substance that by itself does not elicit an immune sponse unless conjugated with an immunogenic carrier molecule; may react specifically with immune products in the sense of a carrier molecule.

Hardy-weinberg law. The law that the basic processes of Mendelian heredity (meiosis and recombination) do not alter either the frequencies of genes or their diploid combinations. The law also states how the percentage of diploid combinations can be predicted, on the basis of a simple formula, from a knowledge of the percentage of genes.

Harvesting. Harvesting in biotechnology terms usually means collecting cells or organisms from a growth system. If the organisms are very large (like, say trout) this is not difficult. However, most biotechnology uses single celled organisms like bacteria or yeast, which have to be actively collected. Among the ways of doing this are:

Centrifugation. Expensive, but guaranteed to collect any particle. It can be used in small volumes to purify viruses, and anything as large as a bacterium can be processed quite easily. There are two variants. The mixture can be centrifuged in a 'pot', where the biomass sinks to the bottom and forms a pellet. Alternatively, the mixture can be spun in a pot with holes in it, and the filtrate is spun out while the solid remains behind. This has the advantage that more liquid can be added, so the centrifuge can be run continuously. This is called basket centrifugation.

Filtration. There are a range of filtration systems. This is again cheap and effective, but usually they have limited capacity. The reason for this is that the filter is, of necessity, full of holes that are just a bit smaller than the cells you want to collect. So, after a while, the cells fill up all the holes, the filter is fouled, and filtration stops. Cross-flow filtration can help to solve this.

Flocculation. This is popular. By adding an additional reagent to the reaction mix, or by altering conditions, you get the cells to stick together and settle out like snow. (Any loose association of particles into an irregular clump is called a 'floc'.) This is often the only practical way of removing cells from really big fermentations, and is essentially how yeast is removed from brewing vats after fermentation is complete. Many types of cells form flocs spontaneously under the fermentation conditions being used: this can be valuable (to separate the cells from the liquor) or disastrous (because the cells will not stay in suspension depending on the needs of the fermentation.

Hydrocyclone. This is also called a whirlpool separator. Here the liquid flows relatively slowly round a bowl and the cells collect in the middle, like grounds in a stirred coffee cup. Cells are collected from the centre, usually to be recycled to the fermentation. Spent liquor, largely free of cells, is collected from the periphery. This is a highly energy efficient method of separation.

Haustoria. Hyphae that penetrate host cells for the purpose of obtaining nutrients.

Head. Refers to the cuboidal component of binal (q.v) viruses.

Healthcare reform. Reform of healthcare systems has been high on the political agenda of most Western countries in the last decade. This affects many aspects of biotechnology directly. The main concern is to reduce the rate of increase of the cost of healthcare. This has been a particular issue in the USA, where the Clinton administration promised a complete overhaul of the healthcare system, but was unable to deliver it. A particular target is Medicare, the government-supported healthcare in the USA for those unable to pay for the more widely used private healthcare. Biotechnology is affected in several conflicting ways. One of the aims of the reform programmes is to introduce price controls, limiting what companies can charge for therapies. Since biotechnology product development is very expensive, this may hit biotechnology companies hard. However, most of the cost of healthcare is the cost of hospitals and doctors. Therefore, if treatment were more effective then people would spend less time in hospital and the costs would go down, even if the cost of the treatment itself were high. The real target of cost reduction should be the overall cost of treatment (the argument goes), not just the cost of the pills. (This is the complex area of pharmcoeconomics.) This is an argument for expensive, highly targeted biotechnology drugs. Most of the cost of healthcare is directed at keeping old people alive for a bit longer, rather than at preventative healthcare. Biotechnological drug development is usually focused at the high-tech end of the therapeutic spectrum. Preventative medicine (basic nutrition, sanitation, vaccination programmes, and contraception advice) is not what the biotechnology industry usually provides. In an environment where everyone is talking about cutting healthcare costs, investors are less willing to provide money to biotechnology companies. In the early 1990s in particular, this and a series of very public failures of clinical trials of biotechnological products caused investors to flee biotechnology, with severe consequences for new companies struggling to raise capital. Other aspects of healthcare reform are a greater concern with respect to quality of life and ethical issues. The effect of quality of life can be estimated by discussing quality-adjusted years of life saved (QUALYs) rather than simply years of life saved (YLS), but, ultimately, what constitutes quality of life is a matter judgment. Biotechnology-based therapies are usually claimed to provide better quality of life than less well-targeted, conventional drugs. This is part of a range of social issues around the public acceptance of biotechnology, usually misnamed 'ethics', .which have been a significant brake on the industry.

Healthy carrier. A person who harbors and spreads pathogenic microorganisms without giving evidence of disease.

Heat shocking. A procedure used in bacteriological assays to eliminate vegetative cells from a sample, but to allow spores to survive. Usually carried out by heating samples to 80øC for 20 minutes.

Heat-labile. Inactivated or destroyed readily by heat, usually 56øC for 3 minutes.

Heat-stable. Heat resistant, or not readily inactivated by heat.

Heat-therapy. Refers to treatment of diseases by application of heat.

Helical. (Greek *helix*: a spiral). Indicates here the spiral arrangement of capsid subunits surrounding the nucleic acid of particularly the cuboidal viruses.

Helminth. Parasitic worm.

Helper cells. A functional subclass of T cells which can help to generate cytotoxic T cells and cooperate with B cells in the production of an antibody response. Helper cells usually recognize antigen in association with class 2 MHC molecules.

Helper T cells (Th). A subset of regulatory T cells that facilin the response of B and. other T cells to antigens.

Helper virus. A defective phage which can help another defective phage (q.v.) to make a successful infection, cf. Complementation.

Hemagglutination inhibition. Test is used in identification of viruses which cause red blood cells to agglutinate. If specific antibodies to the virus are present, agglutination is inhibited.

Hemagglutination. Clumping of red blood cells.

Hematocrit. The volume percentage of erythrocytes in the blood.

Hemoglobin. The oxygen-carrying pigment of erythrocytes, having the reversible property of taking up oxygen (oxyhemoglobin, HbO_2) or of releasing it (reduced hemoglobin, Hb), depending primarily on the oxygen tension of the medium surrounding it.

Hemolysins. Substances including antibodies (immune hemosins), which cause lysis of red blood cells; e.g., exotoxins or antibodies.

Hemosiderin. An iron-containing glycoprotein pigment found mainly in the liver, spleen, and bone marrow, representing colloidal iron in the form of granules larger than ferritin molecules.

Hepatic. (Gr. *hepar*: liver). Pertaining to the liver.

Hepatitis. Inflammation of the liver; frequently due to a virus.

Herbivore. (L. *herba*: plant + *vorare*: to devour). An animal which eats the tissues of plants. (Contrast with carnivore, detritivore, omnivore.)

Hereditary angioedema. A condition caused by congenital deficiency of C1 INH and a pathological activation of the classical complement pathway.

Hermaphroditism. (Gr. *hermaphroditos*: a person with both male and female traits). The coexistence of both female and male sex organs in the same organism.

Hertz (abbreviated as Hz)**.** Cycles per second.

Hetero- (Gr. *other*, *different*)**.** A prefix used in biology to mean that two or more different conditions are involved; for example, heterotroph, heterozygous.

Heterocysts. They are larger than vegetative cells and appear empty in the light microscope (akinetes appear full of storage products). They are produced from the vegetative cell after dissolution of storage granules, the deposition of multi layered envelope outside the cell wall, the damage and breakdown of photosynthetic thylakoids and the formation of new membranous structures.

The formation of heterocyst is inversely related to the amount of nitrogen in the alga. They fix atmospheric nitrogen because they contain enzyme- *nitrogenase*. They can not divide and have very limited period of physiological activity. They do not contain pigments associated with photosystem II (PSII) of photosynthesis. Heterocyst contains only half phycobiliproteins of vegetative cells (Kumar et al., 1982). It has high respiratory rate than the vegetative cell. The supply of respiratory substrate is made through cytoplasmic connections (*microplasmodesmata*).

Heteroduplex. DNA molecule in which one strand is from source while the other strand is from another source.

Heterogamy. (1) Unlike gametes. (2) DNA molecule in which one strand is from source while the other strand is from another source.

Heterokaryon. (Gr. *heteros*: different + *karyon*: kernel). A cell or organism carrying a mixture of genetically distinguishable nuclei. A heterokaryon is usually the result of the fusion of two cells without fusion of their nuclei.

Heterolactic acid fermentation. A series of reactions for glucose dissimilation with end products of lactic acid, acetic acid, ethanol, and carbon dioxide.

Heterologous. Refers to interspecies antigenic differences.

Heteromorphic. (Gr. *heteros*: different + *morphe*: form). Having a different form or appearance, as two heteromorphic life stages of a plant. (Contrast with isomorphic.)

Heterophile antibodies. Antibodies which combine with antigens from unrelated species; e.g., found in infectious mononucleosis.

Heterophile antigens. Antigens present in various organs or tissues of humans and animals which stimulate production of antibodies and which can react with antigens from unrelated species; e.g., Forssman antigens.

Heterotrophic. (Gr. *heteros*: different + *trophe*: food). Requiring organic materials for food. (Contrast with autotrophic.)

Heterozygous. (Gr. *heteros*: different + *zygotos*: joined). Of a diploid organism having different alleles of a given gene on the pair of homologues carrying that gene. (Contrast with homozygous.)

HEV (high endothelial venule). An area of venule from which lymphocytes migrate into lymph nodes.

Hexamer (Greek *hex*: six plus Greek *meros*: part). Here denotes the number of component protein units forming a capsomere.

Hexose monophosphate shunt. A series of reactions for glucose dissimilation with intermediates varying from two to seven carbons.

Hfr (for *high frequency of recombination*). Donor bacteria in which the F-factor has been integrated into the chromosome. This produces a bacterium that transfers its chromosomal markers at a very high frequency to recipient (F) cells.

Hibernation. (L. *hibernus*: winter). The state of inactivity of some animals during winter; often marked by a drop in body temperature.

High throughput screening (HTS). This is an approach to drug discovery. It searches for a chemical that will act on a defined target, such as a receptor or an enzyme. Scientific research will have identified that target as a part of the mechanism of a disease: for example, it may be a cell surface molecule through which cells of the immune system become activated to start an inflammatory reaction, and so be a target for anti-inflammatory drugs. There are two approaches to finding a chemical that will block that target. The first is a rational drug design approach, designing a molecule that you believe will interact with the target. This could be its natural ligand (insulin was 'designed' in this way) or it could be a synthetic molecule. Alternatively, you can search a huge range of essentially random molecules to find one that works fairly well as a blocking agent, and then optimize it. This is the HTS approach. HTS requires a large chemical library to screen, and a highly automated method of performing biological assays on each of the compounds. The chemical library can be a 'compound bank' (a collection of chemicals built up over years of chemical synthetic work), a natural products library (a similar bank of chemicals of natural origin, often not purified), or a combinatorial chemical library. Researchers here are seeking 'molecular diversity', a collection of molecules as different from each other as they can be. The assay can be anything a biologist would do in the laboratory, but must be amenable to being done by a robot, and hence usually it must also be done in small volumes of liquid. Performing HTS on half a million mice is not practical. Automation nearly always means using assays in the 96-well 'microtitre plate' format.

High-efficiency filter. Having a nominal efficiency of 99% for removal of 0.3 to 0.5 micron particles from air.

Hinge. The portion of an immunoglobulin heavy chain between the Fc and Fab regions. This allows flexibility within the molecule and allows the two combining sites to operate independently. The hinge region is usually encoded by a separate exon.

Histamine. A vasoactive amine released (especially in type I IgE-mediated hypersensitivity reactions) from mast cells and basophils; causes smooth muscle contraction, increased capilary permeability, and increased glandular secretion.

Histocompatability antigens. Tissue antigen whose similarine dissimilarities determine graft acceptance or graft reject. Also called transplantation antigens.

Histocompatibility. The ability to accept grafts between individuals.

Histology. The study of tissues.

Histone. Any one of a group of basic proteins forming the core of a nucleosome, the structural unit of a eukaryotic chromosome.

Histoplasmin. Filtrate from culture of Histoplasma capsulatum; used in hypersensitivity skin tests as an aid in diagnosis of histoplasmosis.

HLA. The human major histocompatibility complex. (Human Leucocyte-A locus gene).

hnRNA (heteronuclear RNA). The fraction of nuclear RNA which contains primary transcripts of the DNA prior to processing to form messenger RNA..

Holdfast. In many large attached algae, specialized tissue attaching the plant to its substratum.

Hollow fibre. Hollow fibres are tubes of a porous material. The tubes are very small, typically having an internal diameter of a fraction of a millimetre, and so their ratio of surface area to internal volume is very large. This has had two types of application. First, hollow fibres can be used as filters. Because they have a huge surface area, they take much longer to clog up than normal filters. The filters used in artificial kidney machines are often hollow fibre bundles. The second use is the hollow fibre bioreactor. This is a widely used type of bioreactor in which cells are kept inside the hollow, porous fibres, and the culture medium is circulated outside the reactor. The fibres have pores large enough to let the nutrients in and the product out, but not to let the cells out. The fibres are kept inside a shell: the space outside the fibres but inside the shell is called the shell space. Hollow fibre bioreactors enjoy very general use in some applications. They are very effective for maintaining mammalian cells in culture because they have a very large surface area for the cells to grow on, without needing a large reactor to hold them, and because the nutrient reaching the cells can be kept fresh: mammalian cells can be very susceptible to changes in the medium in which they grow. The reactor also provides an easy way of removing the product the cells are making: this has meant that hollow fibre reactors have been particularly useful for making large amounts of monoclonal antibodies. Hollow fibre reactors are less use when the cells themselves have to grow, because it is hard to get at the inside of the fibre to remove surplus cells, and it is hard to monitor exactly how many cells you have inside the fibres. This has meant that hollow fibre reactors have limited use for bacterial cultures. Hollow fibre bioreactors are versions of dialysis bioreactors or dialysis fermentors. This is any system where the cells in a fermentation are retained behind dialysis membrane (a membrane that lets small molecules pass through it, but will not let large molecules or organisms pass). The reason for using fibres as opposed to a single sheet across the bioreactor is that fibres have a very much larger surface area for fermentation substrates to diffuse across. A variant on the idea is the adsorption fermentor.

Holocarpic. A type of fungal species in which the thallus has no reproductive structures.

Holophytic. Nutrition typical of plants; nutrients are in solution.

Holozoic. Nutrition typical of animals; digesting solid food.

Holozoic. Nutrition typical of plants; nutrients are in solution.

Homeostasis. (Gr. homos: same + stasis: position). The maintenance of a steady state, such as a constant temperature or a stable social structure, by means of physiological or behavioral feedback responses.

Homeotherm. (Gr. *homos*: same + *therme*: heat). An animal which maintains a constant body temperature by virtue of its own heating and cooling mechanisms. (Contrast with poikilotherm.)

Homograft. Transplant of tissues from one member to another of the same species; does not refer to identical twins.

Homologous recombination. Homologous recombination is a biological process by which a living cell joins two pieces of similar DNA together. It is a version of the general genetic process of recombination, by which any two pieces of DNA are spliced together in a living cell. Recombination occurs in all living things: recombinant DNA technology was called that because of the similarity of its gene splicing technology to natural recombination processes. Homologous recombination is recombination between two pieces of DNA that are almost exactly the same, i.e. they are 'homologous'. This occurs much more readily than recombination between DNA that is completely different. It is used as a mechanism for ensuring that a cloned gene that the experimenter wishes to put into the chromosomes of a cell is inserted into those chromosomes at a specific point (i.e. at the point where the cell's DNA is the same at the cloned DNA). Because of this use, homologous recombination is sometimes called gene targeting. Homologous recombination is used in biotechnology in three areas. In generating new mutants of many organisms, but particularly yeast, homologous recombination is a method for targeting a specific bit of DNA. A piece of yeast DNA is joined into your plasmid: when the whole is introduced into a yeast cell, the piece of yeast DNA can undergo homologous recombination with the corresponding bit of DNA in the original yeast's DNA. The two are spliced together and, because the plasmid is all one piece, this means that all the other bits of DNA in the plasmid are also joined into the yeast DNA. This can be used to join a plasmid into the yeast chromosomes, or, if the yeast DNA is from a known gene, to disrupt that gene by putting a large piece of plasmid in the middle of it. The resulting mutant is called a 'knock-out' mutant. The second role is in manipulating large plasmids such as the Ti plasmid of *Agrobacterium tumefaciens*, which are too big to alter using recombinant DNA techniques. Genes can be spliced into them in exactly the same way as they are spliced into the yeast chromosome. The third application is in making transgenic animals (and potentially in gene therapy). Here, again, homologous recombination is used to carry a foreign gene into a cell's chromosome. However, the reason for doing it in this case is to avoid disrupting any genes in the target cell, and to make sure that the foreign gene arrives in a suitable chromosomal environment. The DNA that surrounds genes in mammalian cells (and many other types of cells) affects how that genes will be expressed. Thus it is important to target a foreign gene to a suitable place in the host cell's chromosomes, so that the gene functions correctly, and it is essential that the gene is not targeted to a site where it will damage the functioning of other genes. Homologous recombination offers a route to doing this, and hence makes the production of transgenic animals more reliable. It also offers the possibility of useful gene therapy in humans since one of the main problems with gene therapy ideas at the moment is the threat the 'therapeutic' gene introduced into a patient's cells will do as much damage as the original disease.

Homologue. (Gr. *homos*: same + *logos*: word). One of a pair, or larger set, of chromosomes having the same overall genetic set, of chromosomes having the same overall genetic composition and sequence. In diploid organisms, each chromosome and sequence. In diploid organisms, each chromosome inherited from one parent is matched by an identical (except for mutational changes) chromosome - its homologue - from the other parent.

Homozygous. (Gr. *homos*: same + *zygotos*: joined). Of a diploid organism having identical alleles of a given gene on both homologous chromosomes. An organism may be a homozygote with respect to one gene and, at the same time, a "heterozygote" with respect to another. (Contrast with heterozygous.)

Horizons. The layers making up the vertical soil profile.

Horizontal spread. The spread of infectious agents from one son to another through contact, insect vectors, or by continated food or fomities.

Hormone. (Gr. *hormon*: exicite, stimulate). A substance produced in one part of a multicellular organism and transported to another part where it exerts its specific effect on the physiology or biochemistry of the target cells.

Host induced modifications. Generally non-hereditary alterations in the nature of a virus induced by its host. For example, conversion of non-virulent phages into virulent ones after they pass through a host with glucosylating enzyme. The viral nucleic gets glucosylated (q.v) possessing bacteria.

Host. An organism that harbors a parasite and provides it with nourishment.

Humoral immune system. The part of the immune system mediated by B cells; it is mediated by circulating antibodies and is active against extracellular bacterial and viral infections.

Humoral immunity. Antibody-mediated (B cell) immunity bodies are primarily located and transferable in body (humors) such as serum.

Humoral. Pertaining to the extracellular fluids, including the serum and lymph.

Humus. (1) Organic fraction of the soil that is relatively stable. (2) Organic material in soil after decomposition by microorganisms.

Hyaluronidase. *Spreading factor*; enzyme capable of degrading hyaluronic acid, the intracellular material of connective tissues; may help the spread of the invading organism in the tissues.

Hybrid. (L. *hybrida*: mongrel). The offspring of genetically dissimilar parents.

Hybridization. Hybridization has several meanings in biotechnology and molecular biology.

DNA hybridization. This is the formation of a double helix of DNA from two DNA strands. The two separate strands of DNA will come together to make a double helix if their bases are complementary, so that whenever there is an A in one strand there is a T in the other, a G in one strand and a C in the other. (In fact there is a slight degree of laxity about this, and, depending on how long the DNA strands are, up to 10% wrong or 'mismatched' bases can be tolerated.) DNA hybridization is used as a method for using one bit of DNA (the probe) to find out if a complementary bit of DNA is present in some mixture of DNA species. It is used in 'blot' techniques, PCR, gene library screening, DNA fingerprinting, and a range of other techniques. DNA usually comes in a double helix. To make a hybrid molecule, the original DNA helix must be separated into two strands, a process called 'melting'. The temperature at which a particular helix 'melts' depends on: (1) how long it is (for helices shorter that about 50 base pairs, the shorter they are the lower the melting temperature), (2) how many G and C bases it has (more G + C means a higher melting temperature), (3) whether the helix is perfectly matched, or whether one or more of the bases in it is mismatched, i.e. is not opposite its complementary base in the helix, (4) a range of chemical conditions, like how much salt there is in the solution. The same 'melting temperature' governs the temperature at which a new hybrid helix will form. Thus it is often used as a measure of how stable a new helix would be.

Molecular hybridization. This is forming a new molecule that has functional parts from two different molecules. It could therefore have a combination of properties from its two 'parents'. Examples of this approach are the new antibiotics that can be made by combining the enzymes that make two old antibiotics in one cell, and making fusion proteins by joining two functional domains of other proteins together.

Cellular hybridization. This is essentially another term for cell fusion.

Species hybridization. This is forming a hybrid between two species. It can be done between closely related species by suitable breeding programmes: however, species do not hybridize readily, and, apart from a few closely related species such as donkey and horse, animals rarely form hybrids in this way. Alternative methods include making chimeras, cell fusion (for plants; this rarely works for animals) to produce a new species with all the genes of both the original, or using bacterial plasmids to shuttle genes between bacterial species.

Hybridoma. Cell lines created in vitro by fusing two different cell types, usually lymphocytes, one of which is a tumour cell.

Hydrocarbon. A compound containing only carbon and hydrogen atoms.

Hydrogen acceptor. A substance that accepts the hydrogen and electrons from the appropriate coenzyme in the dehydrogenation process.

Hydrogen bond. A chemical bond which arises from the attraction between the slight positive charge on a hydrogen atom and a slight negative charge on a nearby fluorine, oxygen, or nitrogen atom. Weak bonds, but found in great quantities in proteins, nucleic acids, and other biological macromolecules.

Hydrogen donor. The cellular substance that donates the hydrogen and electrons when it is oxidized.

Hydrogen swell. The nonmicrobial formation of hydrogen gas in a canned food product. usually resulting from a reduction of an acid food with the tin lining of the can.

Hydrolases. Enzymes which break chemical bonds by the addition of water; hydrolyze large molecules into smaller usable components.

Hydrolysis. Splitting a molecule by the addition of water. Like breakdown of a polymer into individual monomers by the addition of water.

Hydrolyze. (Gr. *hydro*: water + *lysis*: cleavage). to break a chemical bond, as in a peptide linkage, with the insertion of the components of water, -H and -OH at the cleaved ends of a chain. The digestion of proteins is a hydrolysis.

Hydrophilic. Readily attracted to water.

Hydrophobia. Fear of water.

Hydrophobic. (Gr. *hydro*: water + *phobia*: fear). Molecules and amino acid side chains, which are mainly hydrocarbons (compounds of C and H with no charged groups or polar groups), have a lower energy when they are clustered together than when they are distributed through an aqueous solution. Because of their attraction for one another and their reluctance to mix with water they are called hydrophobic. Oil is a hydrophobic substance; phenylalanine is a hydrophobic amino acid in a protein.

Hydrophobicity. A hydrophobic molecule is a molecule that dissolves very poorly, if at all, in water, but dissolves quite well in solvents such as butanol or toluene. They are non-polar molecules, i.e. they are essentially electrically neutral all over. The opposite is the hydrophilic molecule, which dissolves well in water or DMSO (dimethyl sulfoxide) but poorly, if at all, in toluene or long-chain alcohols. These molecules usually have partly charged groups on their surface, and often form ions when dissolved in water. Most biological molecules are to some degree hydrophilic, a major exception being the triglycerides ('fats'), which are hydrophobic. Hydrophobic molecules are therefore sometimes called *lipophilic*. When given a choice of environments (for example, a mixture of water and oil to dissolve in) hydrophobic molecules will chose a hydrophobic environment (in this case oil) and hydrophilic molecules a hydrophilic environment (water). Hydrophobicity and hydrophilicity are measures of how well molecules dissolve in non-polar solvents and water, respectively. There are degrees of hydrophobicity and hydrophilicity. Thus, among the amino acids, glutamic acid or lysine are highly hydrophilic, because they form ions easily and dissolve readily in water, while tryptophan has a large, uncharged side chain and is more hydrophobic in character. These differences in hydrophobicity can be used to separate the molecules. Hydrophobic chromatography uses this phenomenon: a mix of molecules is passed over a solid material that is mostly hydrophobic in character. The molecules that are more hydrophobic will stick to the material more strongly, and so will not be washed across the solid support as fast as the hydrophilic ones.

Many biological molecules are sufficiently large to have distinct hydrophobic and hydrophilic bits. These molecules are called amphipathic. If the two regions of the molecule are at opposite ends, then the result is a surface-active material: it will tend to congregate at the junction between a hydrophobic and a hydrophilic solvent. Phospholipids are of this type. Phospholipid membranes are arranged so that the 'tails' of the phospholiqids form a layer of hydrophobic 'liquid' that dissolves quite different chemicals to the watery phase around it. Proteins also almost invariably have a mixture of 'hydrophobic' and 'hydrophilic' amino acids. The protein folds so that most hydrophilic molecules are exposed to the waterery solution they are dissolved in, and the most hydrophobic ones tucked away inside the protein. Thus the distribution of hydrophobic and hydrophilic molecules along a protein (sometimes called a **hydrophobicity plot**) can be a clue to how the protein will fold up

Hydrosphere. Earth's squatic environment.

Hydrostatic skeleton. In annelids, not a chain of bones but rather the incompressible liquid trapped in each of the worm's body segments.

Hymeniun. Spore-bearing (sporogenous) layer of a fungal fruiting body.

Hymenopteran. A member of the insect order Hymenoptera, such as a wasp, bee, or ant.

Hyperbaric chamber. Altitude chamber containing more than the normal atmospheric pressure and an increased oxygen tension and used for the treatment of certain medical conditions.

Hyperplasia (Greek *hyper*: excessive plus Greek *plasis*: a forming or growth). An increase in the amount or mass of tissue due to an increase in the number of cells which individually keep their usual size. Often occurs as a result of viral infection.

Hypersensitivity. Increased sensitivity, or allergy, to foreign material. It is a state of overresponsiveness by the immune system to an antigen.

Hypertonic (Gr.: *higher tension*). Having a more negative osmotic potential, as a result of having a higher concentration of osmotically active particles. Side of one solution as compared with another. (Contrast with hypotonic, isotonic.)

Hypertonic. A solution in which there is a higher concentration of solute relative to that found in the interior of a cell (isotonic).

Hypha (*pl.* hyphae). One of the filaments that make up a fungal mycelium.

Hypocotyl. That part of the embryonic or seedling plant shoot that is below the cotyledons.

Hypolimnion. Deep waters of oceans or lakes below the thermocline.

I

Iatrogenic. A disorder or disease caused by the diagnosis or treatment by a physician.

Icosahedron (Greek *eicosei.* twenty plus Greek *hedra.* seat). A solid structure with twenty quadrangular faces. Refers here to twenty faceted virus crystals.

Icterus. Jaundice; yellow discoloration of skin due to bile pigments.

Id. An allergic reaction to fungi or fungal products.

Ideal free distribution. Distribution of organisms resulting from a process in which each newly arriving individual selects the habitat that is best for its success. (Contrast with dominance distribution.)

Idiotype. A single antigenic determinant on an antibody V region. Antigenic determinant associated with the antigen-binding site of an antibody molecule.

Imaginal disc. In insect larvae, groups of cells that develop into specific adult organs.

Imaging agents. A range of proteins are being developed as imaging agents or contrast agents. This means that they are for use with the various types of body scanners. The proteins (usually antibodies) are linked to a chemical group, which allows the scanner to see them very easily. The proteins bind to specific tissue types, usually tumours, and so allow the scanner to distinguish those tissues from the surrounding tissue. in the absence of a contrast agent the target cells look exactly like the surrounding tissue. Imaging agents can be made for any of the main imaging systems.

CAT scanning (computerized axial tomography). This techniques uses X-rays, and consequently the label tagged on to the antibody is an X-ray-opaque substance, usually something made from a 'heavy' metal like gold.

PET scanning (positron emission tomography). This technique injects a very small amount of a radioisotope into the body, and then tracks where it goes by following the path of the radioactive particles. The favoured element to tag on to your antibody for this is technetium.

NMR (nuclear magnetic resonance). This uses the way that the body absorbs microwaves when it is in a strong magnetic field. Chemical groups absorb microwaves differently depending on what sort of field they are in and what the group is. A wide range of materials can be used as 'contrast agents' for NMR scanning.

ESR (electron spin resonance). Little used, but of substantial potential, ESR detects 'unpaired' electrons that turn up in some types of compounds, usually those involved in energy metabolism. NMR and ESR use no radioactivity, and so are gaining favour as diagnostic systems because of widespread nucleophobia, which is especially prominent in the USA.

Imbibition (L. *imbibo.* to drink). The binding of a solvent to another molecule. Dry starch and protein will imbibe water.

Immediate hypersensitivity. Antibody-mediated (humoral) hypersensitivity reactions that occur within minutes after antigen is introduced into an individual two has been previously sensitized (i.e., has antibodies) to the antigen.

Immobilized cell bioreactors. Many of the plant and animal cells grown by biotechnologists are most effectively handled when the cells are immobilized on to a solid support. This helps to shield them against the stirring forces necessary to mix a bioreactor's contents, and makes them easier to move around and separate from the substrate. There are a wide range of immobilized bioreactors. They fall into three classes.

Membrane bioreactors. These grow the cells on or behind a permeable membrane, which lets the nutrients for the cell through but does not let the cells themselves out. Variations on this theme are the hollow fibre reactor, a common way of growing hybridoma cells for making monoclonal antibodies.

Filter or mesh bioreactors. Here the cells are grown on an open mesh of an inert material, which allows the culture medium to flow past it but retains the cells. This is similar to membrane and hollow fibre reactors, but can be much easier to set up, being similar to conventional tower bioreactors with the meshwork replacing the central reactor space.

Carrier particle systems. The term 'immobilized cells' is often taken to mean that the cells are immobilized on to something not all that much larger than they are, such as small nylon or gelatin beads. The reactor can then handle the beads in the same way as granular catalysts are handled in chemical reactions. There are several ways of doing this. Normal bioreactors of all sorts can be adapted to handle the larger particles. This works best if the particles are of neutral density, so that they do not sink or float. Alternatively, if the particles settle out fast, the bioreactor can be a fluidized bed or a solid bed bioreactor. In the former the particles are kept suspended in a dense, fluid mass by liquid pumped through them from the base. The mass behaves like a liquid, even though it is made of solid particles. In the latter the fluid flow is not fast enough to push the particles apart, and so they sit in a bed at the base of the bioreactor with the fluid flowing past them. Packed bed reactors come in various shapes (conical - 'tapered bed reactor'; disc-shaped - 'radial flow packed bed') to help fluid flow through them evenly.

Immobilized cell biosensor. These are biosensors, i.e. detector devices, that use a biological bit to allow them to detect only one thing at once, and use living cells as their detector system. They are often called microbial biosensors, since the cells are usually bacterial ones. As with any biosensor, there are two parts to immobilized cell sensors. the immobilized cell, which does the sensing and produces a very weak signal of some sort , and a detection and amplification system, which takes that very weak signal and amplifies it to something detectable. The cell used depends on what you want to detect. Some typical examples of analytes (*things to be analysed*) are. **amino acids** (using bacteria that metabolize them), **glucose** (using almost any cell), **toxic chemicals** (using any bacterium that is sensitive to the chemicals to be detected), **carcinogens** (using bacteria that are defective in DNA repair genes), **BOD** (biological oxygen demand, i.e. how much organic matter there is in waste waters), **heavy metals** (using metal-resistant bacteria), **herbicides** (using plant cells or blue-green algae), **toxicity** (using cultured animal cells). Only a few of these have been transformed into realistic sensor systems. The readout methods can be equally diverse.

Gas generation/depletion. A favourite method where the amount of oxygen burnt up or carbon dioxide produced by the bacteria is measured. This is rather non-specific since almost anything the bacterium does burns up oxygen and generates CO_2.

Light production. This uses luminescent bacteria, either ones that are naturally luminescent or ones that have the relevant genes (for luciferase, the light-generating enzyme) genetically engineered into them. Light production is either a measure of general bacterial well being (for toxicity sensors) or is coupled to the presence of a specific chemical.

Direct electrochemical coupling. Some groups are working on hooking the electrode directly into the bacterium's own electron transport system. This is a more sophisticated version of measuring oxygen uptake.

Bacterial biosensors are usually much less specific than other biosensors since bacteria are very diverse and complex things. However, they have the substantial advantage that they are very active, and so make a 'signal' that is easier to detect than that produced by antibodies or DNA probes. Of the few commercial 'biosensor' systems, several are bacterial biosensors. two luminescence-based bacterial biosensors (for toxicity and BOD measurement) are in use in the water industry, for example.

Immobilized enzyme. An enzyme that has been attached to or entrapped within an inert matrix.

Immortalization. Immortalization of a cell type is its genetic change into a cell line that can proliferate indefinitely. Cells taken from a mammal, called primary cells, will divide in culture for 20-60 divisions, but then stop dividing. This is not because they have run out of room or nutrients to grow, but rather because they have become incapable of dividing any more. They usually show characteristic changes in their structure, and reduce the amount of biotechnologically useful product they produce, be it metabolite or protein. This is called cell senescence, and severely limits the manipulations that primary cells can be put through. To avoid this cells are 'immobilized'. put through a treatment that allows them to overcome senescence and divide indefinitely, keeping whatever differentiated features they had to start with. This is done in a number of ways. Several oncogenes, when trans infected into a cell, will immortalize the cell. Some genes from oncogene (tumour-causing) viruses can also immortalize cells, notably the T antigen gene from the SV 40 virus. A third route is to look for a spontaneous mutant in the cells you to immortalize. This is done by growing a large number of the primary cells in culture and simply looking for any that keep on growing when the others have become senescent. The rate at which this happens varies between different organisms (mice seem to generate immortal cell lines much more readily than humans, for example). The last common route is via cell fusion. If a mortal, primary cell is fused with an immortalized cell line, the result is usually an immortalized cell. This is how the technology of making monoclonal antibodies immortalizes the one lymphocyte that is to make the antibody characteristic of a hybridoma. All the lymphocytes in a sample are fused with a suitable immortalized cell, so they are all then immortalized. The experimenter can then grow them indefinitely in the search for the hybridoma that is producing the antibody they want. A related concept is transformation of cells. This is their immortalization and conversion into a more aggressively growing form of the cell. Cells taken from advanced cancers are not merely immortalized, they are transformed. Typically, such cells will grow to form clumps of cells (colonies) in soft agar media, whereas normal, untransformed cells will not. They need a hard, charged plastic surface to grown on. You can measure the extent to which a chemical or a virus can transform cells by measuring its clonogenic potential in a clonogenic assay, i.e. exposing cells to the material and then seeing how many can form colonies in soft agar.

Immune complex. The product of an antigen-antibody reaction which may also contain components of the complement system.

Immune hemolysis. Red blood cell hemolysis in the presence of specific antibody and complement.

Immune modulators. Also immunomodulators. Cytokines come into this as part of the signalling system that collects at these cells together at the site of a potential infection. There are over 18 cytokines known, each with a slightly different message to carry. There are also lots of other cell-surface molecules involved. Here we will only mention the MHC (major histocompatability complex) which acts as a 'carrier wave' for the information that passes between cells about the epitope they are 'seeing'. Antigens that do not bind to the MHC properly are often not 'seen' by the immune system.

What can go wrong? If the immune system is constantly scavenging molecules, why does it not scavenge the body's own molecules, present them to T cells, and start of a process of self-destruction? Sometimes it does, and the resulting disease is autoimmunity. the molecules it recognizes as targets are called autoantigens. Autoimmune disease can range from minor ones such as eczema, where the body attacks itself to a small degree as a side-effect, through more severe attacks on particular tissues, as happens in rheumatoid arthritis, to the complete destruction of one tissue, as happens in insulin-dependent diabetes. It is also possible for the self control to break down completely, and the body to self-destruct massively. one example of such a disease is system lupus erythromatosis (SLE). Usually, however, the body prevents T cells that think that the body's own molecules are dangerous from ever reaching the blood. They are eliminated in the thymus, where T cells are made ready for action. T cells can also be made permanently unresponsive to a stimulus outside the thymus. They are then said to be in a state of energy, and the immune system is 'tolerant' of that antigen. One of the aims of biotechnological intervention in the immune system is to induce tolerance to selected foreign antigens, such as the foreign MHC of transplanted tissues. The immune system would then 'see' them as part of the body, and not as a potential invader. The only other route is to knock out those sections of the immune system with drugs that paralyse or kill the immune cells. This means that the body is now no longer able to fight off infection properly. The other type of disease of the immune system is inappropriate response. Some people's immune systems react far more violently to, for example, peanuts than to rabies virus, even though no one every died of an infection of peanuts. These people are allergic to peanuts. the immune system appears to be presensitized to reacting to some of the molecules in peanuts. Hypersensitivity of this sort is very common, and lies behind asthma, eczema, and probably many other diseases like arthritis. Asthma, in particular, is thought to be triggered by an allergy to the mites that breed in the warm fabrics of modern homes, especially mattresses. Disease like these are inflammatory diseases, where the immune system's actions cause inflammation that is not associated with fighting off an invader, so tissues get red, swollen, and ultimately attacked and even destroyed. Cytokines play a dominant role in this effect. As mentioned above, such diseases often have an autoimmune component, the body attacking not only the invading peanuts but also any other bystander molecule, which includes the body's own molecules. One hundred years ago, such immune disease were rare. Asthma, for example, was considered a mild disease that would be not be seen in most doctor's practices at all. Now between 2 and 5% of all children have asthma in the West, and the disease would be frequently fatal without controlling medication. Why this massive increase? Modern life styles may be a factor. only in the last 50 years have houses been warm, dry, and almost totally enclosed, favouring a population explosion in dust mites. Pollution may also be relevant. The immune system uses nitric oxide (NO) as a signalling chemical, and the NO in pollutants, particularly car exhaust, may be confusing the signals. Basically, however, no one knows.

Immune serum. Antiserum; blood serum containing antibodies.

Immune system. The immune system is the system of cells and molecules that gives us protection against invading organisms. The system cannot be preprogrammed to recognize pathogens, because the pathogen would simply alter its appearance at a molecular level to evade recognition, so the immune system has to 'learn' what each invader looks like, and then mount an appropriate response to it. Doing this takes a lot of different cell types and molecular interactions. These are important to biotechnology in two ways.

As a source of materials. Many materials made by the immune system are valuable biochemicals, and biotechnology has invested a lot of effort in making them efficiently. Leading the list are antibodies and cytokines.

As a target for therapy. The immune system can go wrong in many ways, and biotechnology has expended much ingenuity in trying to cure those disease without destroying the beneficial functions of the system.

How to immune system works. The immune system recognizes molecules on potential invaders. The molecules are called antigens, and the specific part of the molecule recognized by a specific cell is called an epitope. The immune system makes several families of molecules that fit these epitopes, binding to them very tightly. It does this by generating many random variants on a common structural theme and selecting the ones that bind. There are two classes of cells making such molecules. B cells (which make antibodies) and T cells (which make T cell receptors; TCRs). How do the B and T cells know which antibodies and TCRs to make? The key to the system are a group of T cells called helper T cells, or CD_4^+ cells. Potential antigens are continuously scavenged from the blood, broken down, and 'presented' to the T cells by antigen-presenting cells. If the fragments of antigen bind well to the TCR, then the T cell becomes 'activated' and passes on the news to other T cells and B cells. The B cells become activated in turn to make the corresponding antibody. Other types of T cells (CD_8^+ cells, called killer cells) become activated to identify that antigen when it is part of a cell, and attack the cell directly. This latter type of immune response is called cellular immunity or cell-mediated immunity, and is important for fighting off viral disease and immune attack on cancers. The antibody-type response is called humoural immunity, and is important for attack of bacteria. When the immune system experiences an antigen (usually called 'challenging' the immune system), it responds with antibodies and/or a T cell response, depending on the 'arm' of the immune system that has been stimulated the most. **Antibodies** can kill their targets in several ways. Antibodies stuck to a cell are a signal for macrophages to engulf the cell and digest it. Antibodies can also trigger 'complement', a complex set of proteins that puncture holes in cell membranes.

Immunity (Latin *immunis*. not serving or exempt). Refers here to the ability of plants and animals to resist infection by parasitic organisms including viruses. This ability could be genetical or be acquired. There are various mechanisms explaining immunity, some general and some specific.

Immunization. Immunization is the process by which we get an animal to make an antibody against something. The animal could be a human or a farm animal, in which case the purpose of immunization is to provide that animal with the ability to make that antibody so that they are protected against a disease. Or the animal could be immunized so that we could collect its blood and extract the antibody from it, thus providing us with a source or that antibody. There are a number of steps involved. The animal is injected with the antigen, i.e. the material that we want the antibody to react to. If that is a very small molecule (such as a steroid hormone or a short peptide) then it is usually chemically linked ('conjugated') to a much larger molecule, such as a protein. Favourite proteins are bovine serum albumin (BSA) and keyhole limpet haemocyanin (KLH). There are several terms for subsequent stages of the immunization process.

Adjuvants. The antigen can be injected together with an adjuvant, a material that increases the immune response. Common adjuvants are mineral oils and similar complex mixtures that causes inflammation. A common brew is 'Freunds complete adjuvant'. For human use, aluminium phosphate or hydroxide gels are used.

Boosting. The first injection will give rise to a primary immune response, the production of a relatively small amount of antibody. The antibody will be mostly IgM and its Ka will be low. If the same antigen is then injected again, a secondary response will occur, producing much more antibody, this time mostly IgG with a higher affinity. This subsequent injection is called boosting, and is usually done several times.

Titres. To test how the immunization is going, a small sample of blood is removed and the ability of the antibodies in it to bind to the antigen is tested. The blood is diluted until the antibodies in it are so dilute that they can no longer bind to the antigen to any detectable degree. This dilution is called the antibody's 'titre'. Therefore, when measuring the strength of an antibody preparation, people often cite a dilution figure (1 in 100000 being pretty good, 1 in 1000 being fairly hopeless). As immunization proceeds with additional boosters, the titre of the antibody should go up as the amount of antibody and its affinity goes up.

Immunization. Method of imparting or injecting an immunogen (q.v) into a susceptible host organism.

Immunoconglutinins. Autoantibodies which are formed against complement components.

Immunoconjugate. A compound that is a combination of an antibody molecule (or part of one) and another molecule. There are several types. Antibody contrast agents and tracers. These are used in conjunction with 'scanners'. CAT (computerized axial tomography, an X-ray technique), PET (positron emission tomography, a radioactive tracer system), or NMR (nuclear magnetic resonance) diagnostic devices. They all produce images of the insides of a patient, but these images can be much improved (in the case of CAT and NMR) or are only possible (in the case or PET) if some chemical is injected into the patient which the scanner can detect. If the chemical is linked to an antibody, then the scanner becomes a very sensitive way of tracking where the antibody ends up. Contrast agents are chemicals that increase the 'darkness' of the scanner image, and apply to NMR and CAT scanners (and to conventional X-rays, too). Tracers are materials that do something unique, so they 'light up' in the scan. some NMR reagent and PET scanner chemicals fall into this category.

Antibody-enzyme conjugates. These are complexes where the antibody has been chemically linked to an enzyme. They are widely used in immunoassays, where the enzyme acts as a 'flag' to mark the presence of the antibody. A few nanograms of antibody can easily be detected if a suitable enzyme is attached. Common ones are horseradish peroxidase (HRP) and alkaline phosphatase (AP).

Immunodiagnostics/immunoassays. One of the success stories of biotechnology, these are medical diagnostic methods that use antibodies. The antibody is used to detect the presence of something in a sample. The antibody latches on to its target very specifically, so is a very precise reagent. It can also latch on to the antigen at very low concentrations, and so can produce a very sensitive test. This combination has meant that monoclonal antibodies are used in about 35% of all medical diagnostic procedures. Exactly the same test technology can be used in other, non-medical applications, which are called immunoassays. The problem with immunodiagnostics is that the antibody does not do anything obvious when it latches on to its target, so we have to arrange the assay so that some other process detects that binding has occurred. There are various facets to this. The label. Antibodies can be labelled in various ways. As well as labels used for imaging agents,immunodiagnostics can use a variety of labels in 'in vitro' assays. These usually have different names. **ELISA.** enzyme-linked immunosorbent assay. Uses an enzyme label on the antibody. **RIA.** radioimmuno assay. Uses a radioactive label on the antibody or antigen. **FIA.** fluorescent immunoassay. Uses a fluorescent tag on the antibody or antigen. **CLIA.** chemiluminescent immunoassay. There are a variety of luminescent (light-generating) labels that can be used.

Immunogold. labelling the antibody with colloidal gold particles. This can be used as a label in a macroscopic assay (where is acts as a very intense dye) or in immunohistology (staining tissue sections with antibodies), where the gold particles show up very clearly in the electron microscope. The second faced is the physical format of the assay; what reagent is attached to what object. Common aspects of assay formats are.

Sandwich assay. Here, two antibodies are used that bind to different parts of the antigen. One is trapped on a solid surface (e.g. the bottom of the wells on a 96-well plate), the other has a label attached to it. If the antigen is present it links the two, and so the label stays in the plate.

Competitive assay (competition assay). This is like a sandwich assay, but the analyte is a small molecule that competes with the binding of an enzyme chemically linked to the analyte hapten (the hapten-enzyme conjugate) to the antibody. This is virtually the only way of making an immunoassay that can detect a small molecule. **Latex.** Latex particles are very small particles of plastic, usually coated with the antibody. typically they are polystyrene spheres 100 nm to 1 μm across. In the presence of antigen the particles stick together into larger lumps, held together by the antibodies that coat them, hence the name latex agglutination assay. The third aspect is the physical format of the assay.

Homogeneous, i.e.an assay that simply gives a result when the sample is added (together with suitable reagents), just as a colour pH indicator does. The chemistry of this can be very complicated. **Microtitre plate format**, i.e. done on microtitre plates (which have to undergo a series of washes between each reaction). Performing the assay on other surfaces, such as glass plates or silicon chips, can be essentially the same, but usually results in smaller tests. **Microparticle based**, i.e. the antibody is bound to tiny beads and these are moved between solutions by centrifugation, filtration, or other methods. (This is different from a latex agglutination assay, where the particles are the readout system as well). There are a range of semi-official trade names for more complex immunoassays (the competition for a good acronym for your immunoassay is intense). Among the more common are. **ARIS.** This uses a complex reaction in which the binding of an antibody to a synthetic target prevents glucose oxidase from working. This type of assay is now almost out of its patented period. It is a homogeneous assay (i.e. there are no washing or separation steps involved). It works for small molecule analytes. **EMIT.** This is another small molecule, homogeneous immunoassay, but one that is more sensitive than ARIS.

Other immunoassay formats fall into the biosensor category, which is considered to be more in the mainstream of biotechnology.

Immunofluorescence. A technique used to identify particular antigens microscopically in tissues or on cells by the binding of a fluorescent antibody conjugate.

Immunogen. An agent which can induce immunity in otherwise susceptible host; for example, inactivated virus particles.

Immunoglobulins (Igs). Plasma proteins of the gamma globulin class that are produced by or associated with B lymphocytes; the basic Ig molecule consists of two H chains and two L chains; antibodies are Igs.

Immunoglobulins. A class of proteins, with a characteristic structure, active asreceptors and effectors in the immune system.

Immunological tolerance. A mechanism by which an animal does not mount an immune response to the antigenic determinants of its own macromolecules.

Immunology. Study of reactions of an animal when foreign substances, antigens,are introduced into the body.

Immunosensors. Biosensors have a biological part and a detection part. The biological part confers selectivity on the sensor, the detection part detects the effect of whatever the biological part does and turns it into a recognizable 'signal' (usually an electrical signal). In immunosensors the biological part is an antibody. The physical part is usually a physical mass detection system or an optical device. The optical devices are discussed under optical biosensors. There are two groups of immunosensors based on mass detection. Both use extremely small mass detectors, usually manufactured on a **silicon chip** (and hence sometimes called 'microchip biosensors'), to detect the tiny changes in mass found when an antibody binds to an antigen. They are all resonance devices that measure the binding of the analyte to the probe. The simplest type is based on the tuning fork principle. The note a tuning fork sounds depends on the mass of the tines. If the mass increases, the note goes down. The sensors have the equivalent of a microscopic tuning fork with the antibody coated on the tines. The silicon surface from which the tines are made detects the frequency with which they vibrate. When something binds to the antibody, the note falls and the circuit picks this up. Surface acoustic wave (SAW) devices are a variation on this theme. Because the tuning fork is made of piezoelectric material, these are sometimes called piezoelectric sensors. The problem with such sensors is that anything landing them gives a signal. Thus, despite having a very specific antibody as a biological element, they are very prone to interference. So, while 'tuning fork' microdevices are well known in mechanical application such as strain gauges and gas sensors, they have not provided a reliable biosensor yet.

Immunosuppression. The reduction in the capability of the immune system to respond to an antigen; obtained through immunosupressive drugs,irradiation, and antilymphocyticserum (ALS).

Immunotherapeutics. These are drugs, usually biopharmaceuticals, that act on the immune system. Because the immune system regulates itself with a vast array of proteins that signal between the cells (including cytokines), most immunotherapeutics are proteins made by genetic engineering to boost some aspect of the immune system. the way the white blood cells grow, differentiate, or act. The cells of the immune system produce only minute amounts of these proteins, so the biotechnologist has to clone and express them in another organism to make them as drugs. Many of them have only been discovered by cloning their genes and then seeing what the resultant protein does. Among the ones being developed as drugs are.

Interferons. second oldest biotechnological protein, being used as an immune system booster for many diseases. There are three different interferons (alpha, beta and gamma) with different actions.

Interleukins. especially interleukin-2 (IL-2). **CSFs (colony stimulating factors).** These stimulate the growth of the cells that make the white blood cells, which are responsible for the immune response.

Immunotherapy. This is the use of antibodies or antibody-derived proteins in treating disease. The use of antibodies as targeting agents (in immunoconjugates or immunotoxins, for example) is usually not considered to be immunotherapy. Rather, immunotherapy means giving the patient an antibody that they cannot make for themselves, because their immune system cannot work fast enough, because their immune system is not working at all because of disease, or because the antibody is against an antigen that the body would not normally recognize as 'foreign'.

Antibody immunotherapy has been fairly spectacularly unsuccessful, costing much more to develop and being much less successful that originally hoped. It is now understood that you need to chose the target as well as the antibody very carefully. The molecular target must be highly specific to the disease you want to cure, and must be accessible to the antibody. most antibodies do not enter solid tumours, for example, so an antitumour antibody is not likely to have any effect. In addition, the antibody must be highly specific, and have the effect you want it to have. Antibodies can be used 'naked', using the body's immune system to attack antibody-coated cells. They can also be conjugated to a 'warhead', such as a radioactive atom, a toxin, or an enzyme. Bispecific antibodies can also be used, which bind the target cell with one arm and lock on to another cell with the other, bringing a killer cell to bear on your target. Immunotherapy can also mean using whole cells from the immune system as a therapy. This latter approach has been tried under the title of adoptive immunotherapy, where NK cells ('natural killer' lymphocytes, some of the white cells in the blood able to destroy other cells) were taken from terminal cancer patients, stimulated using cytokines to become more energetic, and then injected them back into the patient. The therapy had some effect, but severe side-effects. A variant of this is to use another class of white cells (tumour infiltrating lymphocytes (TILs)), which can target a cancer more specially. Again, they must be obtained from the patient first. TILs have also been tagged with foreign genes in a first approach to gene therapy in treating terminal cancer. The initial gene experiments put a 'useless' gene into the cells. the ultimate idea is to put a gene into TILs that will increase their effectiveness at killing tumours.

Immunotoxins. Immunotoxins are protein drugs. They consist of an antibody joined on to a toxin molecule. The toxins used (toxins from bacteria Diphtheria, Pseudomonas, or Shigella, or the castor bean toxin ricin) are extremely poisonous. Probably only a few molecules of ricin inside a cell can kill the cell. Thus they are of no use as systemic drugs. However, if they can be placed at a specific site, then they can be used to kill off one type of cell with very high efficiency. This is the idea behind immunotoxins. The toxin is joined to an antibody molecule that can bind specifically to one type of target cell. The resulting 'conjugate' is injected into the blood at extremely low concentration. When it encounters its target cell, the conjugate binds to it, concentrating the toxin there. The toxin then has a much higher chance of killing the cell. ImmunoGen has a ricin-based immunotoxin of this sort in clinical trials as a treatment for leukaemia. Refinements use parts of the toxin molecule, not all of it. Most toxins consist of a part that enables the toxin protein to enter the cell (the A chain) and a part that kills the cell (the B chain). The A chain is not toxic and the B chain on its own cannot get inside the cell to work, and so is much less toxic. Conjugating the B chain to an antibody makes a much less dangerous material. however, it can still kill cells if the antibody binds to them since the local concentration of B chains around that cell is so high that a few B chains get inside anyway. Immunotoxins have some limitations. As large molecules they cannot get inside solid tumours easily. They are also rapidly mopped up by the patient's own immune system unless the patient is on immunosuppressive drugs. Also, there are some cells that bind antibodies non-specifically as part of the normal immune reaction. They will bind the immunotoxin, and so be killed. Immunotoxins can be made by linking the toxin and antibody molecules chemically. They can also be made by fusing the genes for the toxin and the antibody. the resulting fusion protein is more stable, and can be smaller and less prone to binding to other tissues than a chemical conjugate. The antibody can also be 'humanized', reducing other complications. A related idea is to use the toxin themselves as biotherapeutics. A reverse version of an immunotoxin is to couple a non-antibody binding site to the constant region of antibody, producing an immunoadhesin.

Impetigo. Infection of the skin usually caused by streptococci.

Imprinting. A rigid form of learning, in which an animal comes to make a particular response only to one other animal or object.

IMVC. Biochemical tests (indole, methyl red, 'Voges-Proskauer, and citrate) used in identification of enterics.

IMVIC. Differential tests used to separate Escherichia and Enterobacter in the sanitary analysis of water.

In situ. The natural environment in which a microorganism resides.

In utero. Within the uterus.

In vitro. "In glass" as contrasted to in vivo; performed in a test tube or other glassware. It meanse outside of the body or performed in artificial environments.

In vivo (L.. *in the living state*). In a living organism. (Contrast with in vitro.) Many processes that occur in vivo can be reproduced in vitro with the right selection of cellular components.

Inapparent infections. Refers here to virus infections that remain hidden, i.e., are not expressed, at least for an indefinite period. Symptoms do not appear and the infective particles may be carried from one generation to the other.

Incidence. The number of new cases of disease in a particular population within a specified time period.

Incineration. The process of burning to ashes; e.g., sterilizing the inoculating loop in the Bunsen burner.

Inclusion bodies. General name given to the characteristic particulate structures present in certain virus infected host cells. For instance, the Negri bodies present in nerve cells infected by rabies virus. These bodies act as diagnostic features.

Inclusion bodies. In virus infections, the highly stainable components, usually virus, found in the cytoplasm or nucleus of the infected cell.

Inclusive fitness. The sum of an individual's own fitness (the effect of producing its own offspring; the individual selection component) plus its influences on fitness in relatives other than direct descendants (the kin selection component).

Incubation period. Time interval between infection and the appearance of the first symptoms.

Incubation. Maintenance of cultures under conditions favorable for growth.

Index case. The first case observed or detected in an epidemic.

Indicator. A substance which can be used to reflect pH changes by changing color.

Indigenous. Native to a particular place.

Individual distance. The fixed distance an animal strives to keep between itself and other members of the same species.

Individual selection component. That component of inclusive fitness that results from an organism producing its own offspring. (Contrast with kin selection component.)

Induced enzymes. Enzymes synthesized only in the presence of a specific substrate or inducer.

Induced fit. A change in the tertiary structures of some enzymes, caused by binding of substrate to the active site.

Induced mutants. Mutants resulting from the use of a mutagenic agent.

Inducer. In enzyme systems, a small molecule which when added to a growth medium, causes a large increase in the level of some enzyme. Generally it acts by binding to repressor and changing its conformation so that the repressor does not bind to the operator. In embryology, a substance that causes a group of target cells to differentiate in a particular way.

Inducible enzyme. An enzyme that is present in much larger amounts when a particular compound (the inducer) has been added to the system. (Contrast with constitutive enzyme.)

Induction (Latin inducere, to lead into). Refers here to the conversion of a temperate phage into a lytic phase (q.v.). It could be spontaneous or induced.

Induction. In biotechnological terms this means getting an organism to make a protein, usually an enzyme, by exposing it to some stimulus, usually a substrate for growth that requires the organism to have that enzyme in order to metabolize it. Induction involves the control of gene expression, but it is not a strictly genetic phenomenon since no new genes or gene rearrangements are involved. It is only the expression of genes already there. In general, an inducible gene, i.e. one that is capable of induction, can be induced by one or a few compounds. These are called inducers. These compounds (or sometimes their metabolites) affect how a protein binds to the promoter region of the gene concerned, and so affect the control of that gene. The exact mechanisms involve are as varied as they can be (as is the case for nearly all biology). Thus, in order to be inducible, a gene needs to have the right promoter region. Some expression vectors have inducible promoters in them. They must also carry the genes for any proteins involved, of course. the inducer does not bind to naked DNA on its own. A related term is repression, in which a compound has the opposite effect, i.e. reducing gene activity and so making the cell lose an enzyme activity. Such genes are called repressible, and the phenomenon repression. This can be very important in biotechnology since many genes for useful enzymes, such as those that make antibiotics and other secondary metabolites, are repressed by common substances such as glucose. A common mechanism is catabolite repression, where the presence of common metabolites in the growth medium of a microorganism represses the enzyme pathways that would otherwise make them. There are specific genes involved in catabolite repression, which can be removed by appropriate genetic manipulation in some cases. Induction also means a form of logic that reasons from specific examples of something to general rules about that something. This is something biochemists do a lot (despite the fact that it is logically indefensible, it seems to work), but that is rarely what they mean by induction. Induction is also a way of having babies - despite many women's pleas, there is no equivalent repression in this case.

Induration. A hardened area or lesion.

Infection. Invasion of the body by pathogenic micro-organisms.

Infectious microorganisms. The term is restricted to microorganisms capable of producing a disease process in man or domestic animals.

Infectious. Capacity to be transmitted and produce an infection.

Infestation. Attachment of parasites to superficial layers of the skin.

Infiltration (Ventilation term). Ductless flow of air into a space from an adjoining space at higher pressure.

Inflammation. Nonspecific response to irritants characterized by pain, heat, redness, and swelling. It is the body's response to injury resulting in a localized reaction.

Ingestion. Taking in of food by swallowing.

Inhibitor. A substance which binds to the surface of an enzyme and interferes with its action on its substrates.

Innate releasing mechanism (IRM). A system within the central nervous system that is especially responsive to a particular sign stimulus.

Inoculum. The transfer of microbial cells from inoculating needle or loop; also, the introduction of microorganisms into the animal body, as in a vaccine.

Insertion sequence. In genetics, the addition of gene or clustered genes to the chromosome and related to transposons.

Instinct. Behavior that is relatively highly stereotyped, more complex than the simplest reflexes, and usually directed at particular objects in the development of the bevior; the important point is that the behavior develops toward a narrow, predictable end product.

Insulin (L. *insula.* island). An animal hormone, synthesized in islet cells of the pancreas, which promotes the conversion of glucose to the storage material, glycogen.

Intention movement. The preparatory motins that animals go through prior to a complete behavior response; for example, the crouch before flying, the snarl before biting etc.

Intercalary meristem. A meristematic region in plants which occurs not apically, but between two regions of mature tissue. Intercalary meristems occur in the nodes of grass stems, for example.

Intercalary. Inserted between components.

Interdemic selection. Selection affecting the birth and death rates of demes rather than of individuals.

Interference (Latin *inter.* between plus Latin *ferire.* to strike). Here refers to the intervention of successful viral infection, occurring as a result of a variety of reasons; for instance, failure of infecting particles to get adsorbed to the specific receptor sites on the host cell wall.

Interference competition. Competition resulting from direct behavioral interactions between organisms. (Contrast with exploitation competition.)

Interferon. A specific proteinaceous substance, first discovered from influenza virus infected cells. These substances interfere with the process of multiplication of the virus in the host cell. These are host specific; different types of interferons are known to be produced when different host cell types are infected with the same virus; their mechanism of action involves prevention of protein synthesis.

Interleukins. The generic term for a group of peptides which signal between cells of the immune system.

Interphase. The period between successive nuclear divisions during which the chromosomes are diffuse and the nuclear envelope is intact. It is during this period that the cell is most active in transcribing and translating genetic information. However, because the nucleus of an interphase cell is less interesting to look at than the nucleus of a dividing cell, this phase has been misnamed "the resting stage", a term that is still sometimes used.

Intertropical convergence zone. The geographical area where the air rises most strongly.

Intoxication. State of being poisoned.

Intrinsic isolating mechanism. A genetically based trait that helps prevent members of other species. (Contrast with extrinsic protein.)

Intrinsic protein. A membrane protein that is embedded in the phospholipid bilayer of the membrane. (Contrast with extrinsic protein.)

Intron. A protein of a DNA molecule that, because of splicing out,is not involded in coding for part of a polypeptide molecule. (Contrast with exon.) It is a sequence in a eukaryotic gene that is transcribed but is excised from RAN before translation.

Invagination. Aninfolding.

Invasion. Able to invade or penetrate the body.

Invasiveness. The property that enables an organism to overcome the normal body defenses and causes an over infection.

Inversion (genetic). A rare mutational event that leads to the reversal of the order of genes within a segment of achromosome, as if that segment had been removed from the chromosome, turned 180° C, and then reattached.

Invertase. An enzyme that converts sucrose to a mixture of glucose and fructose.

Invertebrate. Any animal that is not a vertebrate, that is, whose nerve cord is not enclosed in a backbone of bony segments.

Invovulion forms. Abnormal forms which appear in old cultures of bacteria; they are likely to assume various odd shapes and may swell or show rudimentary branching.

Iodophors. Disinfectants consisting of iodine combined with a carrier molecule.

Ion (Gr. *wanderer*). An atom or group of atoms with electrons added or removed, giving it a negative or positive electrical charge.

Ionic bond. A chemical bond which arises from the electrostatic attraction between positively and negatively charged ions. Usually a strong bond.

Ionization. The separation of a molecule into atoms possessing a charge.

Ir gene. A group of immune response (Ir) genes determine the level of an immune response to a particular antigen or foreign stimulus. A number of them are found in the major histocompatibility complex.

Iron-binding capacity (IBC). A measure of the percent saturation of transferrin in the blood.

Irruptions. Large-scale movements of groups of animals; less regular than migrations.

Isoagglutination. Agglutination of blood cells following a blood transfusion by a different donor blood group. The result is agglutinatioins of the person's own blood cells by serum of the donor or agglutination of the donor's cells by the persons's own serum.

Isoagglutinins. Antibodies, found in the blood serum of all persons, against the major agglutinogens or antigens that the person does not have in his own red blood cells.

Isoelectric focusing. Separation of molecules on the basis of the charge. Each molecule will migrate to the point in a pH gradient where it has no net charge.

Isogamy. (Gr. isos. equal + gamos. marriage). A kind of sexual reproduction in which the gametes (or gametangia) are not distinguishable on the basis of size or morphology.

Isograft. Tissue graft between genetically identical animals, such as identical twins or inbred strains of animals.

Isokinetic sampling. The taking of air samples under conditions in which the velocity and direction of the air entering the sampler are the same as the velocity and direction of the ambient air.

Isolating mechanism. Geographical, physiological, ecological, or behavioral mechanisms that lead to a reduction in the frequency of hybrid matings.

Isolator. A barrier, usually flexible plastic, that operates at a positive air pressure to seal off the internal work space from the surrounding environment.

Isologous. Of identical genetic constitution.

Isomerases. Enzymes that catalyze the transfer of two hydrogen atoms from an adjacent carbon to an aldehyde group. This results in the reduction of the aldehyde group to a primary alcohol group, and the formation of keto group on the carbon atom that donated its hydrogen atoms.

Isomers. Molecules consisting of the same number and kind of atoms, but differing in the way in which the atoms are combined.

Isomorphic. (Gr. isos. equal + morphe. form). Having the same form or appearance, as two isomorphic life stages. (Contrast with heteromorphic.)

Isoniazid. Structural analog of pyridoxine; used in the treatment of tuberculosis.

Isotonic. (Gr. same tension). Having the same osmotic potential. Said of two solutions. (Contrast with hypertonic, hypotonic.)

Isotope. (Gr. isos. equal + topos. place). Two isotopes of the same chemical element have the same number of protons in their nuclei, but differ in the number of neutrons.

Isotype. Refers to genetic variation within a family of proteins or peptides such that every member of the species will have each isotype of the family represented in its genome (eg. immunoglobulin classes.)

Isozymes. Chemically different enzymes that catalyze the same reaction.

Iteroparous organism. An organism that reproduces more than once in its lifetime. (Contrast with semelparous.)

J

J chain (joining chain): Glycoprotein whose function is believed to be to join together basic IgM molecules to each other, and thus form polymers (IgM pentamer and IgA dimer and trimer).

J chain. Polypeptide chain (Mr = 35,000) linking, by means of disulphide bonds, the mu chains of the pentameric IgM molecule or the alpha chains of the dimeric secretory IgA molecule. It has a high cysteine content.

J gene. Gene (s) coding for the joining segment of polypeptide chain which links the *V* (variable regions) to the *C* (constant) regions of both light and heavy chains of immunoglobulins. During lymphoid development the DNA is rearranged so that the *V* genes are linked to the *J* region sequences.

J region. The polypeptide chains coded for by *J* gene.

J segment of immunoglobulin genes. One of three distinct gene segments encoding immunoglobulin *Light Chain* (the others are *V* and *C*) and of four gene segments encoding *Heavy Chain* (the others are *V, D* and *C*) . The *J* (joining) segments encode the last 10 amino acids of the variable region of light and heavy chains. Random genetic rearrangement between light chain *V,J* and *C*, helps to create diversity in antibody specificity.

Jarovization. Vernalization. The treatment of seeds, before sowing, by any technique designed to hasten the flowering period of the plants to which the seeds give rise (Also called *Vernalization*) .

Jaundice. Yellowing of body tissues.

Jerusalem artichoke. A plant (*Helian-thus tuberosus*) related to the sunflower capable of producing high levels of carbohydrate, in the form of the fructose polymer inulin, in its underground tubers.

Jet loop fermenter. A type of fermentation system in which agitation and gas absorption are achieved by recirculation of the broth with reinjection into the main fermenter vessel under pressure.

Jiffy pots. A brand name for peat pots that are sometimes used for ex vitro transplantation.

Joule: A unit of energy, equal to 0.24 calories.

Jumping gene. Transposon that carries genetic information coding for and controlling its own *Transposition.*

Junctional complex. A site at which there is direct contact between neighboring animal cells. There are at least four types of special contact between cells: an attachment plaque or desmosome; an adhesive zone; a tight or occluded junction; and a gap or intermediate junction.

Junctional proteins. Proteins specifically associated with *Cell Junctions.*

Junk DNA. DNA that has no genetic or structural function but persists within the genome since there may be no positive selection against it.

Juvabione (paper factor) . A JH-active chemical in paper products manufactured from American balsam fir trees: active against the European red bug, *Pyrrhocoris apterus* L. and accidentally discovered by Slama and Williams (1966).

Juvenile hormone (JH). A chemical produced by the corpora allata, one of the three major insect developmental hormones, and responsible for determining the type of molt which will occur when one is directed by ecdysone. At high JH blood titer, additional larval or nymphal molts occur: low titer or complete absence of JH causes the pupal or imaginal transformation to occur.

Juvenile hormone. In insects, a hormone maintaining larval growth.

Juvenile. The immature, usually non-reproductive [reproductively incompetent], sometimes phenotypically distinct phase of plant growth. **Juvenility** should not be confused with young in age although young plants may also be in a juvenile phase. Juvenile phase tissue has repeatedly been found to perform better in culture than adult phase tissue. It is more likely to grow and to form callus, embryoids, shoots etc.

Juvenoid (juvenile hormone analog: JHA). Any synthetic or natural chemical of plant or animal origin exhibiting biological activity similar to the true juvenile hormone.

J-value. The effective concentration of one end of a DNA molecule in the immediate neighbourhood of the other end of the same molecule. J-values and i-values (the total concentration of ends in the reaction mix) influence the formation of products in a ligation reaction. The J-value is a function of the length of a DNA molecule and is independent of the concentration of DNA.

K

K cell (killer cell). A group of lymphocytes which are able to destroy their targets by antibody-dependent cell-mediated cytotoxicity. They have Fc receptors. Weakly glass-adherent cell with an FC *Receptor* that potentiates antibody-dependent cellular cytotoxicity (ADCC). K cell may play a role in *Immune Surveillance* against tumours, and are arguably the same cells as NK *Cells.*

Kahn test: Precipitin test for diagnosis of syphilis using individual's blood serum.

Kairomone. A transpecific chemical messenger, the adaptive benefit of which falls on the recipient rather than the emitter; cf. *Allomone* (Brown et al. 1970).

Kakogenic. Dysgenic; tending to impair the hereditary qualities of the race, or strain.

Kallikrein plasma. Proteases normally present as inactive prekallikreins which are activated by Hagemann factor. Act on kininogens to produce kinins.

Kanamycin resistance. The ability to survive treatment with Kanamycin mediated by an enzyme that modifies the antibiotic and prevents its interaction with ribosomes.

Kanomycin. Antibiotic that operates through its ability to bind to bacterial ribosomes and cause misreading of the message.

Kapeller-adler test. A pregnancy test based on estimation of histidine in the urine.

Kappa (*Cuppa*) **Particles:** Self replicating (q. v) cytoplasmic particles present in **Paramecium**; these are instrumental in cytoplasmic inheritance and are through to be episomic (q. v.) in origin. Grows in the presence of gene K, producing a killer type.

Karyogamy. The fusion of two nuclei that exist within a common cytoplasm, as occurs in the formation of the zygote from two gametes. The process also occurs within the multinucleate plasmodium of fungi belonging to the Myxomycetes.

Karyokinesis. The division of only the nucleus during a cell division.

Karyological races. Race of species all having the same chromosome number which differs from that typical of the species.

Karyolymph. The nuclear sap enclosed by the nuclear membrane.

Karyolysis. The apparent dissolution of the nucleus during mitosis.

Karyomere. A small vesicle in the nucleus (after nuclear division) which usually encloses a single chromosome.

Karyomicrosome. A microsome located in the nucleus.

Karyomites. Chromosomes; chromophilic bodies, typically constant in number in any particular species, into which the nucleus resolves itself during cell division.

Karyomitome. The nuclear reticulum.

Karyomitosis. Nuclear division by mitosis.

Karyon. A nucleus; the most constant constituent of animal and plant cells which reproduces by mitosis and contains the chromosomes.

Karyoplasm. Nucleoplasm; the denser protoplasm of which nuclei are composed as distinct from the less dense cytoplasm outside the nuclear membrane.

Karyoplast. A cell nucleus surrounded by a narrow rim of cytoplasm and a plasma membrane which is obtained from the cell enucleation.

Karyorhexis. Nuclear fragmentation.

Karyosome. A vague term meaning a chromosome, a special form of nucleolus or even a nucleus.

Karyosphere. Part of the nucleus into which the chromosomes have contracted.

Karyota. Nucleate cells.

Karyotheca. The nuclear membrane; the membrane surrounding the nucleus, dividing the nucleoplasm from the cytoplasm.

Karyotin. Chromatin; a deeply staining substance in nucleus; the part of the chromosome that stains deeply with basic dyes during cell division.

Karyotype. The somatic chromosome complement of an individual, usually as defined at mitotic metaphase by morphology (including centromere location and often by special staining techniques, as fluorescence banding) and number, arranged in a sequence that is standard for that organism.

Karyotyping. The identification and classification of organisms, cells or tissue on the basis of their chromosome content.

Katachromasis. A general term for the changes which take place during telophase and the formation of daughter nuclei.

Katagenesis. Devolution; retrogressive evolution.

Kataphase. A collective term for all the stages of mitosis which follow anaphase.

Kb cells. Cultured cell line derived from neoplastic epidermal cells from the nasopharynx of humans; they were first propagated in fluid medium by Eagle in 1955.

Kbp. Kilobase pari.

Kefir. A fermented milk product produced in the USSR using a mixture of symbiotic yeasts, *lactobacilli* and *lactic streptococci.*

Kelp. A number of different genera of large brown seaweeds, some of which are used as a source of alginic acid. Kelp has also been suggested as a source of biomass.

Keratin (Gr. *keras:* horn). A protein which contains parallel polypeptide chains in α-helical or ß conformations, sulfur is part of such hard tissues as horn, nail, and the outermost cells of the skin.

Keratin gene. Since *Keratins* are, next to immunoglobulins, the most diverse proteins, the keratin gene family is very large. The genes range in size from 300 to 3000 bases, and some of the smaller ones have no intron. There is a tendency for some keratin genes, and some of these internal repetitious sequences consist of inverted repeats.

Keratin sulphate. Type of *Glycosamino glycan* found in the cartilage of cornea and the intervertebral discs.

Keratinization. The accumulation of keratin by a cell.

Keratins. Insoluble protective or structural proteins.

Kestose. A non-reducing trisaccharide formed by the condensation of sucrose and fructose with (2,6)-and (2,1)-linkages.

Ketal boli. The heating of wort to inactivate enzymes, precipitate impurities and concentrate the solution prior to inoculation with yeast.

Ketogenesis. The formation of ketone bodies.

Ketogenic amino acids. Amino acids whose carbon skeletons can be precursors of the ketone bodies.

Ketone bodies. Acetoacetate, D-ß-hydroxybutyrate, and oxidation of fatty acids.

Ketones: A collective term for the intermediate products of fatty acid degradation (acetoacetic acid, beta-hydroxybutyric acid, and acetone). The ketone bodies tend to accumulate when the rate of production exceeds the rate the organism can burn them.

Ketose. A simple monosaccharide having its carbonyl group at other than a terminal position.

Ketosis: The clinical condition in which ketones accumulate in the blood and appear in the urine. This condition is caused by a disturbance in carbohydrate metabolism (uncontrolled diabetes mellitus), dietary diet), or by a reduction in carbohydrate catabolism with resultant high mobilization of body fats (starvation).

Key factor. The hypothetical causative agent most responsible for population density change. In practice, a variable factor in the environment recognized as most closely linked to changes in population density, and measurable by continuous monitoring for use in predicting future population trends.

Key: In taxonomy, a device or quickly identifying a specimen down to the species (or a higher category) to which it belongs. The key consists of a series of choices made according to whether the specimen possesses one trait as opposed to another.

Kidneys: A pair of excretory organs in vertebrates.

Killer T cell (T): Sensitized effector T cells that are cytolytic for corresponding target cells.

Killer yeast. Yeast cells containing a double-stranded RNA *Plasmid* that encodes genes responsible for production of a killer substance which is lethal to other cells. The substance is closely similar in its action and synthesis to a bacterial *Colicin.*

Kilobase (*abbr.,* **kb.**). One thousand nucleotides or nucleotide pairs in sequence. May be used to pertain to either DNA or RNA.

Kilobase pair. One thousand base pair of DNA duplex. A convenient measure of DNA molecular weight.

Kimkap. A brand name for a plastic cap used as a test tube closure.

Kin selection component: The component of inclusive fitness resulting from helping the survival of relatives containing the same alleles by descent from a common ancestor. (Contrast with individual selection component.)

Kinase. An enzyme catalyzing phosphorylation of an acceptor molecule by ATP.

Kindred. A group of individuals each of whom is related in some way, either by blood relationship or marriage, to every other member of the group and, in so far as is or can be known at any given time from the available records, to no other person.

Kinesis: (Gr. Movement). Orientation behavior in which the organism does not move in a particular direction with reference to a stimulus but instead simply moves at an increasing or decreasing rate until it ends up farther from the object or closer to it. (Contrast with taxis.)

Kinetic body. A minute granular structure occurring in the immediate vicinity of the centromere.

Kinetic complexity. Complexity of a sample of DNA as determined by its reassociation kinetics in a renaturation experiment.

Kinetic constriction. Centromere; spindle attachment; insertion region; kinetochore; a non-staining, localized region in each chromosome which remains single for some time after the rest of the chromosome has divided, and which, at metaphase, appears to be attached to the spindle fibre.

Kinetic viscosity. A description of the flow behaviour of a fluid expressed in terms of the ratio of the dynamic viscosity to the density of the material measured at the same temperature.

Kinetics. Kinetics is the study of chemical reactions. In biotechnology, it can mean a variety of things.

Cell growth kinetics. This is the rate at which cells grow, and is discussed in more detail under cell growth.

Reaction kinetics. The speed with which chemical reactions can be made to happen is essential to many aspects to biotechnology. Among the commonly mentioned types of kinetics are:

Binding kinetics. How fast a molecule such as an antibody binds to its target. A broad use of the term includes considerations of how fast you stir the reaction, whether the molecules can physically get at each other, and so on. Usually the molecule doing the binding is called the ligand, and so these are ligand-binding kinetics. Related to this is ligand-binding thermodynamics, measuring how much energy is gained or lost when a ligands binds to its target, and hence how tightly it will stay bound. Values for this binding energy and usually expressed as ΔG, also called the `free energy of binding'. Basically, the more negative this number is, the tighter the ligand will bind to its target.

Enzyme kinetics. This is a rigorously defined area of science, and studies how fast enzyme-catalysed reactions can go. Key terms are the Michaelis-Menten equation, and its constants (Km and V_{max}). The latter is a measure of how fast an enzyme works, the former of how dilute a solution of its substrates it works on. (Of course, any enzyme can work as fast as you want it to, simply by adding more enzyme to the reaction. V_{max} measures how fast each unit of enzyme can work, when compared with other enzymes.) Another kinetic measure is turnover number, which is the number of substrate molecules each enzyme molecule processes each second. Yet another is Q10, which is the amount the rate of reaction is increased by increasing the temperature. You can guess from the Q10 whether a reaction is limited by how fast the molecules are getting to the enzyme (*diffusion limited*) or whether it is limited by how fast the enzyme works. Also part of the study of enzyme kinetics is the study of inhibitors. These are compounds that slow down the catalytic activity of the enzyme. They can act in many ways. Competitive inhibitors bind to the active site of the enzyme, and so stop the substrate getting in. Allosteric inhibitors bind to some other part of the molecule, and specifically alter the shape of the enzyme so that it is no longer as effective a catalyst. Both these types of molecule are reversible inhibitors: if they are removed, the enzyme can work again. Irreversible inhibitors chemically modify the enzyme so that it will not act as a catalyst ever again. End-product inhibition is a result of enzyme kinetics. It is an important effect controlling how much product an enzyme can make. In this case the end-product of a chain of enzyme reactions (a metabolic pathway) acts as an inhibitor of the first enzyme in the pathway. So when the pathway operates and the end-product builds up, the first enzyme is automatically `turned off' and the pathway stops again until whatever it is making is used up. This is a form of feedback control, also called feedback inhibition. End-product inhibition relies on one of the end-products of a pathway being an allosteric inhibitor of one of the enzymes.

Kinetin. (6-furfurylaminopurine). An artificial cytokinin found in extracts of denatured DNA; the first of the cytokinin to be isolated.

Kinetochore. The attachment region (within the centromere) of the microtublules of the spindle in cells undergoing mitosis or meiosis.

Kinetoplast. Specialized and very large mitochondrion associated with the basal body of the *flagellum* in some trypanosomes.

Kinetosomes. Granular or rod-like bodies often aggregated to form platelike bodies, which occupy the spindlypoles in the sporogenetic mitosis of mosses.

Kinety. Row of ciliary *Basal Bodies.* The kineties are composed of linked *Kintetosomes* and may be visualized by a silver staining procedure. Kinety pattern is independently inherited, and when artificially altered by surgery is passed on in modified form to progeny.

Kingdom. A major taxonomic category, consisting of several phyla or divisions.

Kinin. (1) A former name for cytokinin. (2) One of a class of polypeptides found in the blood (e.g. bradykinin). They mimic various stings and venoms in causing contraction of smooth muscle, dilating blood vessels, and irritating nerve endings. They are probably involved in blood clotting and inflammatory processes

Kininogen. Inactive precursor in plasma from which kinin is produced by proteolytic cleavage.

Kinoplasm. A modified type of cytoplasm which appears in the from of strands (e. g. spindle fibres) or of channels in the unmodified cytoplasm.

Kinosome. a minute body sometimes seen in the centromere in such cases there being one kinosome on each chromosome where it passes through the spindle attachment region.

Kirsten sarcoma virus. *Retrovirus,* originally derived from mice, but subsequently pass aged through rats. It contains the oncogene sequence (v-kiras). The human proto-oncogene sequence has not been definitely identified.

Kit. A package containing all the specialized reagents, enzymes, buffers cloning vectors, linkers, primers, etc. required to carry out a particular diagnostic procedure or technique in molecular biology. In general the kit provides the right proportions of the various components, packaged in a manner to make the best and most cost-effective use of expensive or difficult to obtain components.

Kleinschmidt spreading technique. Technique developed by A. K. Kleinschmidt in which nucleic acids are spread, in the presence of small amounts of protein, as extremely thin films on cleaned water surfaces. Such spreads may be picked up on copper grids (precoated) and subsequently examined by electron micorscopy.

Kleistogamous. Cleistogamous: with close fertilization, this taking place within the unopened flowers.

Klenow fragment. Larger part of the bacterial DNA polymerase 1 (76kD) which remains after treatment with subtilisin; retains some but not all exonuclease and polymerase activity.

Klimefelter's syndrome. A human syndrome that is due to the presence of one extra X chromosome in the male Karyotype (XXY).

Kline test: A slide precipitin test for the diagnosis of syphilis, using individual's blood serum.

Klon. A group of individuals all derived asexually from a single organism; clone.

Kluyveromyces. A genus of yeast used in the production of biomass (single cell protein).

Knight-Darwin law. The hypothesis that *nature abhors perpetual self-fertilization* and that *nature intended that a sexual itercourse should take place between neighbouring plants of the same species.*

Knock-outs and mutants. Knock-out organisms are ones where a specific gene has been inactivated, or knocked out, by recombinant DNA techniques. They are widely used as tools to find out what a gene does: knock it out, and see what effect this has. Knock-outs are made by using homologous recombination between the target gene and a piece of cloned DNA to insert a piece of *junk* DNA into the gene you need to disrupt. If the organism is haploid (i.e. only has one chromosome set) then this technique will result in that organism's only copy of the gene being knocked out. If it is diploid, then only one of the two alleles will be knocked out, and you will have to do some conventional breeding to make an animal that has two copies of the gene knocked out. Knock-out organisms, especially knock-out mice and yeast, have been used extensively to try to find out what the genes discovered by genome projects are actually for. Many of the genes discovered in even such a simple organism as yeast have no obvious function, so this is the only way to know what a new gene is doing. Knock-out mice can also be used as models of human disease (see Transgenic disease models).

Robert Koch was a German doctor who made several discoveries in the fields of microbiology. The main are:

- The bacteria can be isolated and cause diseases while working with anthrax disease of animals. He gave the final proof of these concepts.
- He developed the *Koch's postulates* which means that a specific microorganism is responsible for the cause of specific disease.
- He developed the technique of pure-culture.
- Robert Koch (1843-1910) discovered the bacterium that caused anthrax in cattle and demonstrated the progression of the disease.
- Robert Koch isolated and identified various kind of microbes, specially bacteria, therefore, he is regarded as first true bacteriologist. He is also known as founder of medical microbiology, bacteriology and virology. He was as a physician and practice as a doctor in small German town. He was interested in anthrax disease which begins with symptoms similar to a cold and cause itching skin, blisters or vesicles form which later turn black and swell. The bacteria may move into the blood and cause fever, shock and finally death. Koch isolated the bacterium that caused anthrax. He injected laboratory animals with the bacteria-laden blood to dead sheep. He then performed autopsies and recorded how the same symptoms appeared regularly. He also isolated few bacilli in a clear sterile vitrous humor of an ox's eye and observed their multiplication. He used house mice to show how the disease moved through animals and speculated on the formation of a resistant form of the bacterium- an *endospore*. He injected mice with a silver of wood containing endospores and found that the mice develops anthrax symptoms. On observing blood, several thread-like bacteria were present in it. He also studied the anthrax bacterium in frogs and horses, and demonstrated that how disease organisms became concentrated in the lymph sacs and spleen. In 1867, Koch presented *germ theory of disease* which was later on known as *Koch's Postulates*.

Koch's postulates state as follows: A definite sequence of experimental steps that shows the causal relationship between a specific organism and a specific disease.

- A particular organism can always be found in association with a particular disease, but not in healthy individual.

- The organism can grow in the laboratory itself.
- The pure growing culture will produce the same disease when placed back into a new susceptible animal.
- It is possible to recover the organism from the sick animal and grow them in pure culture.
- On the above basis it was concluded by the microbiologists that particular organism is in fact the cause of a particular disease. Koch isolated the bacillus- the causal pathogen of tuberculosis. He confirmed the theory of John Snow that water is the important cause for transmission of diseases. Koch was appointed to the University of Berlin (1885) and was made Director of Institute of Infectious Diseases in 1890. He continued his work on tuberculosis and was rewarded for this work with the Nobel Prize in 1905. He died in 1910.

Koji process. A process by which molds are cultured on the surface of moist bran.

Konberg enzyme (DNA polymerase l, pol l). DNA *Polymerase* discovered in 1956 by A. Kornberg and coworkers in cells of E. coli. It catalyses the addition of mononucleotides to the 3'-end of a growing DNA chain provided *in vitro* with a DNA template, the four deoxyribonucleoside triposphates and Mg^{2+}.

Koplik's spots. Small bluish red lesions appearing on the oralmucosa during-measles infection.

Kpn l. A type ll restriction enzyme from the bacterium *Klebsiella pneumoniae* OK8 which recognises the-DNA sequence.

Krebs cycle. An enzyme system which converts pyruvic acid to carbon dioxide in the presence of oxygen with concomitant release of energy which is captured in the form of ATP molecules. Also referred to as the *Citric Acid Cycle* or the *Tricarboxylic Acid* (TCA) *Cycle.*

Krill. Name applied to planktonic crustaceans.

Kupffer cells: Phagocytic cells which line the liver sinusoids.

Kurtosis. The departure of a symmetrical frequency distribution from the normal by excess (*Platykurtosis*) or deficiency (*Leptokurtosis*) in its shoulders as opposed to tails and centre.

Kuru. Chronic progressive degenerative disorder of the central nervous system caused by a virus and found among New Guinea natives.

Kwashiorkor: A nutritional deficiency disease of infants and young children mainly in the tropics and subtropics, occurring soon after weaning. It is caused by a deficient quality and quantity of dietary protein. The disease is characterized by retarded growth, edema, fatty liver, skin changes, severe apathy, and weakness.

Kynurenineless ephestia. A mutant type in the flour moth Ephestia incapable of producing a hormonelike substance (kynurenin) necessary for full pigment production.

L

L cells. Cultured cell line, originally described by Earle in 1943. He took normal subcutaneous connective tissue cells from healthy *C3H* mice and treated them with 20-methyl-cholanthrene. The *L* cells thus produced could be cultured and yielded sarcomas when injected into healthy *C3H* mice.

L region. Region of the major histocompatibility complex (H-2 complex) of the mouse.

Label (Radioactive). A radioactive atom, introduced into a molecule to facilitate observation of its metabolic transformations.

Labels and substrates. A range of standard methods for labelling or tagging macromolecules are used in the life sciences.

Enzyme labels. Many enzymes are used as tags in immunoassays. They are chemically tied on to the antibody, so that their catalytic properties can be used to track the antibody. Usually they are used to turn a colourless chemical (the substrate) into a coloured one, so they, can literally, be seen.

Fluorescent labels. The most common is fluorescein, which include. yellow-green, and rhodamine, which glows red, when illuminated with UV light. Fluorescein is usually linked on to a molecule by using the reactive derivative fluorescein isothiocyante (FITC). An issue of importance for all fluorescent labels is quenching, the reduction in the fluorescence output by other chemicals. Many fluorescent labels show self-quenching. when many fluorescent molecules are tagged on to one protein, they quench each other so that the whole does not fluoresce very well. Fluorescent proteins are becoming popular, because they are extremely efficient fluorophores and hence can be detected at very low concentrations. The most widely used are phycocyanins (from seaweed) and green fluorescent protein (GEP) (from the jellyfish *Acquorea victoria*). A variety of GFP mutants have been isolated, so you can get GFP mutants have been isolated, so you can get GFP labels that glow different colours.

Chemiluminescent labels. These are reacted with another chemical to generate light. There are two approaches. The luminescent chemical can itself be linked to what you want to detect. Alternatively, an enzyme can be linked to the target, and used to generate light in an enzyme-catalysed reaction (see Bioluminescence).

Radioactive labels. The ultimate label is a radioactive atom in your molecule, because its presence does not affect the chemistry of the molecule significantly. Radioactivity is used far less now for two reasons. Concerns over worker safety have made people use far less radiation generally. Additionally, more scientists use laboratory robots, and these are extremely difficult to clean up if they accidentally get contaminated with a long half-life isotope. Nevertheless, radioisotopes are still used in many applications. They are characterized by half-life (the time for the amount of radioactivity to decay to half its initial level, and not, as the press believes, the time taken for them to `become safe'), and the energy of the radiation they emit. Low energy labels like tirtium (3H) give radiation that does not penetrate far into solid matter, so they give finer `pictures' and are safer, but you cannot detect them so easily (or at all inside, for example, a human body). Higher energy labels (like 131I or 99Tc) are therefore used when you want to detect only tiny amounts of label, or label in an inaccessible site, such as inside the body.

Labile genes. Genes which are constantly mutating.

Labile. Easily destroyed usually by heat.

Laboratory scale. It is a process carried out using only small amounts (i. e. milligrams or grams) of reagents in order to validate the nature of the products and the characteristics of the reaction.

Lac. The three letter symbol for the genotype concerned with lactose utilisation and/or uptake. In *E. coli* the lac genes are organised into an operon of *lacZYA* which encode respectively, the ß-galactosidase, lactose permease and lactose transacetylase enzyme.

Lactate dehydrogenase (LDH). Enzyme that catalyses the interconversion of lactate and pyruvate in anaerobic glycolysis. Lactate dehydrogenase is a tetrameric molecule, each subunit of which can be of either the *H* (heart) or *M* (muscle) type. There are five isoenzymes of lactate dehydrogenase (i. e. four *M*-type subunits, three *M*-type subunits plus one *H*-type subunits, etc.).

Lactation. The period of maternal milk secretion; the secretion of milk by the mammary gland. The amount of milk produced is affected by the maternal nutritional state, frequency of infant sucking, ingestion of medicines, and by hormones.

Lactic acid bacteria. A group of bacteria that ferment carbohydrates in the presence or absence of oxygen, with lactic acid always a major end product. They have a high tolerance of acid conditions. Lactic acid bacteria are involved in the formation of yogurt, cheese, sauerkraut, and silage. They can occur as spoilage organisms and some are pathogenic, causing infections of the nasopharynx.

Lactic acid. A syrupy liquid occurring in sour milk as a result of fermentation by *Lactobacillii.* It is produced (L-form only) during anaerobic respiration in animals as the end product of glycosis.

Lactic dehydrogenase (LDH). The enzyme which catalyses the formation and removal of lactate according to the equation : pyruvate + NADH = lactate + NAD. The appearance of LDH in the medium is often used as an indication of cell death and the release of cytoplasmic constituents.

Lactic heterofermentation. A type of fermentation occurring in some microorganisms, the end products being lactic acid, ethanol and carbon dioxidentation.

Lactobacillus. A genus of rod-shaped, Gram-positive bacteria characterized by the production of lactic acid as an end product of metabolism, through a heterofermentative or a homofermentative route.

Lactoferrin. It is iron-binding protein of very high affinity (K_d 10^{-19} at pH 6. 4;26-fold greater than that of transferrin) found in milk and in the specific granules of neutrophil leucocytes. A protein found in external secretions of vertebrates that binds (chelates) iron.

Lactoflavine. Part of the vitamin B_2 complex.

Lactoglobulin. A crystalline protein obtained from milk.

Lactoperoxidase. Peroxidase enzyme from milk which finds an important use in generating active iodine as a nonpermanent radiolabel for membrane protein.

Lactose carrier protein. The best known example is the product of the lacy gene, coded for in the lactose operon and responsible for the uptake of lactose by *E. coli.*

Lactose repressor. Protein (tetramer of 37kD subunits) which normally binds with very high affinity to the operator region of the lactose operon and inhibits transcription of the downstream genes by blocking access of the polymerase to the promoter region. When the lactose repressor binds allolactose, its binding to the operator is reduced and the gene set is dipressed.

Lactose. A disaccharide carbohydrate (disac. charide) also known as milk sugar. On hydrolysis it is split into glucose and galactose.

Laemmli loops. Loops of DNA visualized when chromosomes are viewed by electron microscopy following removal of histone; named after originator of the technique, U. K Laemmli. The loops all extend out from a central matrix or scaffold of non-histone protein, and each loop represents a distinct DNA domain.

Laevorotatory. Describing compounds that rotate the plane of polarized light to the left (anticlockwise as viewed facing the oncoming light). Compare *Dextrorotatory.*

Lag phase. The period of slow, orderly growth when a medium is first inoculated with a culture. Used to describe the first of five growth phases of most batch propagated cell suspension cultures in which inoculated cells in fresh medium adapt to the new environment and prepare to divide. (see growth)

Lager beer. Beer that has been aged to develop its characteristic taste.

Lager. It is a type of beer produced using a bottom yeast (*Saccharomyces uvarum*).

Lagging strand. Strand of DNA that is synthesized discontinuously during DNA replication, and consists of a series of okazaki fragments lighted together. Compare *Leading Strand.*

Lagging. Delay in moving from the equator to the poles at anaphase, of one or more chromosomes so that these become excluded from the daughter nuclei.

Lamarckism. The theory of evolution by acquired characteristics, as propounded by Jean Baptiste de Lamarck. The hypothesis of evolution based on inheritance of acquired characters; `use' and `disuse' of organs play an important role in this. Now, this concept is not accepted (cf. *Darwinism*).

Lambda (virus). A bacterial virus of *E. coli* noted for its ability to integrate into bacterial chromosome.

Lambda chain. One of two types of light chain found in immunoglobulins (for the other) which differ by about 60 percent in their constant region amino acid sequence.

Lambda phage. Double-stranded DNA phage that infects *E. coli* and is widely used as a cloning either a lytic growth or lysogenic growth cycle depending on a balance of genetic and molecular factors within the cell.

Lamella. A thin, plate-like structure. In plants, lamellae are the layers of closely appressed membranes that occurs in the chloroplasts.

Laminar air flow cabinet or hood. Layered monodirectional airflow. In a laminar airflow (LAF) cabinet air is forced through filters and specially designed ducting to produce an airflow with a minimal lateral disturbance. This has the effect of producing `layers' of air where the adjacent layers do not mix except on the molecular scale. In the vertical LAF cabinet this airflow is in the form of a narrow `curtain' of air passing down the inner front surface of the cabinet: in a horizontal LAF cabinet the air forms a `block' the size of the front opening blown towards the operator. The air is filtered through prefilter (furnace-filter quality), then a high efficiency particulate air (HEPA) filter that strains out particles greater than 0. 3 m. These hoods are commonly employed for aseptic plant tissue culture manipulations and are designed to accommodate one to the several persons, depending on the model.

Laminaribiose. A disaccharide of glucose.

Laminarin. The chief carbohydrate food reserve of the brown algae. It is a polymer of glucose, is soluble, unlike starch, and closely resembles the callose of higher plant sieve tubes.

Laminin. Glycoprotein found as a major component for the basal lamina, which forms the extracellular matrix of many cells.

Landfil site. An area of land where domestic solid refuse and similar waste is discharged and buried. Landfill gas is produced in this area. The gas is a mixture of methane (50-65 percent) and carbon dioxide (30-40 percent) together with traces of nitrogen, hydrogen and a wide range of volatile organic compounds generated by anaerobic bacteria present in landfill waste disposal site.

Langmiur-Blodgett films. These are thin films of molecules formed on the surface of water. The original Langmiur-Blodgett film was a lipid layer on top of the water, but the term is often used to describe lipid films in which both sides are in water, or these films when they are transferred to a solid surface. Lipids have a polar, water loving (hydrophilic or lipophobid) `head' and a water hating (hydrophobic or lipophilic) `tail'. Thus, half of the molecule is soluble in water, half is not. The most stable arrangement of such molecules is to have them in clusters with the tails on the inside, away from the water, and the head on the outside. One such cluster arrangement is a flat sheet, with the tails in the middle and the heads on either side. This is the Langmiur-Blodgett film, or lipid bilayer. It is the basis of the membranes that surround living cells and some of the organelles inside cells. Lipid bilayer films or membranes are only one example of `liquid membranes', in which a thin layer of liquid is stabilized so that it can last for a long time in water. They all have to be stabilized by some chemical means, otherwise they collapse into little globules of liquid, or dissolve in the water. Lipid bilayer membranes have applications in drug delivery systems (as liposomes), in biosensors, in separation processes, and in some bioreactors. Nearly all of these applications are still only laboratory demonstrations. Liposome and liquid membrane separations are described separately. Biosensor applications rely on the high electrical resistance of a Langmiur-Blodgett film, or on its optical properties. Electrical sensors are based on the ability of some proteins to carry ions across a lipid membrane. Some antibiotics, proteins from nerve cell membranes, and a variety of *transport* proteins that allow cells to get materials from outside the cell into the cell without making holes in the membrane, can all be inserted into the lipid membrane. The protein can allow one material or type of material (an amino acid, a metal ion, or maybe simply protons) to cross the membrane. in the presence of that material the membrane will conduct electricity. In its absence, the membrane will have a much higher resistance, because there will be no path for any other charged species to cross it. Thus the membrane could be a very sensitive detection system. The problem with this is that the membranes are mechanically and chemically unstable, as are most of the proteins we want to put in them. Thus, while a sensor system may be improvised in the laboratory, none have worked `in the field'. A related use of Langmiur-Blodgett films is as switching elements in computer-like circuits. An alternative sensor system based on Langmiur-Blodgett films is an optical sensor system. Because the films are extremely thin, they cause interference effects when light is shone through them or reflected off them, and these effects depend critically on how thick the film is. If antibodies are immobilized on the surface of the film, then when they bind to their antigen the total thickness of the assembly will change (from being film+antibody to film+antibody + antigen), and so the colour of reflected light will change. Again, this can be demonstrated for some very simple, model systems in the laboratory, but not for any realistic sensor application.

Lanthanium. Element used, usually as its hydroxide, as an electron-dense tracer in electron microscopy. The small size of the molecule allows it to be used to indicate sites of membrane permeability.

Large scale. Descriptive of a process carried out using the large amounts of material (i.e. kilograms or tonnes) that are similar to those required for commercial production of the product.

Larva (L. *ghost*: early stage)**.** An immature stage of any invertebrate animal that differs strongly in appearance from the adult.

Lassa virus. Virulent and highly transmissible arenavirus whose normal host is a todent (*Mastomys natalensis*); first recorded from Nigeria.

Lata-type. A mutant individual possessing one or more supernumerary chromosomes in its nuclei.

Late genes. Genes that are expressed late in the life cycle of a virus. These genes usually code for proteins such as capsid subunits. Compare *Early Genes.*

Latency (L. *latens*: to be hidden)**.** Here refers to the state of in-apparent infection of a virus. A state of inactivity.

Latent image. Image produced on photographic film following exposure, but prior to development.

Latent learning. Learning that is not put to immediate use but is "stored" for possible future use.

Latent period. The time that elapses between stimulation of an irritable tissue (e.g. a nerve or muscle) and the production of detectable response. Compare *Reaction Time.* This may be regarded as the penultimate stage of the replication cycle of the bacteriophages. This period begins with the formation of the first new phase particle in the host cell and ends with the lysis of the same (q. v.).

Latent virus. A virus that remains within the host without producing any obvious effects.

Latent. Not obviour or manifest, as a carrier of a disease, that shows no symptoms, has a latent infection. The noun is latency.

Late-proteins. Indicates the proteins synthesized during the later stages of the replication cycle. These are usually the structural proteins necessary for the formation of capsids.

Lateral gynandromorph. An individual, half of whose body is male and the other half female.

Lateral inhibition. A simple form of information processing. The classic example is found in the eye, whereby ganglion cells are stimulated if photoreceptors in a well-defined field are illuminated, but their response is inhibited if neighboring photoreceptors are excited (an *on field/off surround* cell), or vice versa an *off field/on surround* cell. The effect of lateral inhibition is to produce edge-or boundary-sensitive cells, and to reduce the amount of information which is sent to higher centers; a form of peripheral processing.

Lateral. Pertaining to the side.

Late-replicating X-chromosome (Barr body). Condensed copy of the *X* chromosome of mammals. As compared with the non-condensed copy of the *X*, this condensed copy, which is composed of facultative heterochromatin, is replicated late in the S phase of the cell cycle.

Latex. It is a complex organic liquid found in certain plants which contains a variety of secondary products, including termens, terpenoids and alkaloids. Its high molecular weight products found in rubber.

Lauter tub. In the brewing process it is the tank where insoluble spend brewers grains are removed after mashing is completed.

LAV (lymphadenopathy virus). Causative agent of human AIDS, and now more properly known as HIV (human immuno-deficiency virus).

Law of homologous series in variation. The principle enunciated by Vavilov. In general, closely allied Linneon species are characterised by similar and parallel series of variations; and as a rule, the nearer these Linneons are genetically, the more precise is the similarity of morphological and physiological variability; generally nearly related Linneons have consequently similar series of hereditary variation.

Lazy leucocyte syndrome. A rare and possibly apocryphal human complaint in which neutrophils display poor locomotion towards sites of infection. LD. (1) Lethal dose. The number of pathogenic microorganisms required to cause death in a given species of animal or plant. (2) Lethal dose. Radiation required to kill within a specified time all the individuals in a large group of organisms.

LCM (lymphocytic chriomeningitis). Aviral disease of mice.

LD_{50}. (1) The dose) that will kill 50 percent of the animals in a test series. (2) The dose of radiation to kill 50% of the individuals in a population.

Leachate. An aqueous solution containing inorganic or organic material. It results from the percolation of water through various types of solid.

Leaching. Microbial leacing, or bioleaching, is the use of microorganisms, usually bacteria, to isolate metals from mineral ores by solubilizing them and allowing them to be washed (`leached') ou of the ore. Thus it is method of mining, and is a major component of microbial mining (biohydrometallurgy) technology. Many ores cannot be processed economically because the concentration of metal in them is too low. Some of these ores are low-grade ores which are discarded as waste during mining operations aimed at higher grade ores. (The `grade' of an ore depends mainly on how much metal there is in it, but also on how accessible that metal is. Clay has a very high content of aluminium, but it is extremely expensive to extract aluminium from clay.) However, if the metal can be released as a soluble salt, then it can be washed out and collected without the ore having to be mined, crushed, and smelted, as in a normal mining operation. Leaching is also used to extract gold and uranium from normal ores (see Gold and uranium extraction). Leaching can be done in three physical arrangements. Slope or dump leaching is where a pile of the metal ore on the side of a hill is sprayed with the bacterial culture at the top ad the eluant with the metal is collected at the bottom. Heap leaching is similar, but the material is in an isolated heap, which is more common in most mining sites. In situ leaching pumps the bacterial culture into the centre of the ore body along pipes or tunnels and then allows it to filter down to the base, where it is collected. Leaching is a chemical process. In some cases bacteria such as *Thiobacillus ferrooxidans* or *Thiobacillus thiooxidans* oxidize sulfur compounds in the mineral to sulfuric acid, deriving metabolic energy. The sulfuric acid solubilizes the metal (copper sulfate is soluble, while the sulfide is not, for example), and so the metals wash out in an acid solution. In others the bacteria act directly on the metal ion, for example oxidizing uranium (IV) (insoluble) to uranium (IV) (soluble). The ore to be leached is sprayed with the bacteria in a suitable nutrient mix, which supplies all the other chemicals it needs for growth. Thus the bacterium is limited by the energy it can obtain from digesting the mineral, and thus digests the ore as fast as it can. Optimizing the nutrient mix is a critical factor in making a bioleaching process work at a commercially useful rate.

Lead citrate. Heavy metal salt used as a stain in the preparation of specimens for transmission electron microscopy.

Leader peptide. A short peptide synthesized *in vitro* by translation of the leader sequence of some bacterial mRNAs. These peptides are not formed *in vivo.*

Leader region. A region of regulatory DNA involved in amino acid synthesis, within the promoter-operator region, which is transcribed even in the presence of a repressor.

Leader sequence peptide. Short length of amino acid sequence at the amino end of some protein molecules, which, as a result of hydrophobic amino acids, carries the protein through the membrane of the endoplasmic reticulum and so ensures its eventual secretion. The leader sequence peptide is commonly removed before the protein becomes active.

Leader sequence. It is the length of mRNA, extending from the 5'-terminal to the beginning of the coding region of the first structural gene, which is not translated during protein synthesis. It may contain ribosomal binding sites or an attenuator.

Leading strand. Strand of DNA that is synthesized continuously during DNA replication. The leading strand is synthesized towards the replicating fork in a 5'-to 3'-direction. Compare *Lagging Strand.*

Leafhopper. Delicate, leaf-sucking, homopterous insect; often a disease vector.

Leaky mutant. Mutant gene that shows some residual activity.

Leaky protein. A protein coded by a mutant gene that shows some residual activity.

Leavening agent. A compound or system used in baked food products to produce carbon dioxide. Production of the gas causes the mix or dough to rise.

Leavening. The production of gas in the manufacture of a baked food product.

Lecithin. A naturally occurring, choline-containing phospholipid present in animal and plant tissue. Chemically lecithins are similar to fats, but they also contain phosphorus and nitrogen.

Lecithinase. Alpha toxin; toxin which hydrolyzes the phospholipid lecithin, a component of cell membranes; produced by Clostridium perfringens. Also a component of many snake venoms.

Lectins. Glycoprotein found on the surface of plant cells. It is cell-agglutinating proteins. Most so far studied have been isolated from plant seeds.

Leghaemoglobin. A component of the nodules in nitrogen-fixing plants. It occurs in the bacteriod membrane, although it is synthesized on the plant genome. It is haemoglobin-like molecule with a high affinity for oxygen.

Legume. A group of plants such as peas, beans, and clover, in which nodules form on the roots and contain bacteria that fix nitrogen.

Legumin. A storage protein (along with vilivin) found in seeds of legumes. The protein has 12 components, each with an average molecular weight of 33,000. It contains more acidic and basic residues than vicilin, and three times as many tryptophan residues.

Leishman's stain. A stain used in haematology. It stain some blood cells red-pink.

Lek. A traditional courtship display ground, where males display to females.

Lepidoptera. Moths and butterflies; an order of insects, characterized by a covering of bright scales over the wings and bodies.

Lepore haemoglobin. Variant haemoglobin in a rare form of thalassemia; there is a composite chain as a result of an unequal crossing over event. The composite chain is functional but synthesised at reduced rate.

Leprosy. Disease caused by *Mycobacterium leprae,* an obligate intracellular parasite which survives lysosomal enzyme attack by possessing a waxy coat. Leprosy is a chronic disease associated with depressed cellular (but not humoral) immunity; the bacterium requires a lower temperature than 37°C, and thrives particularly well in peripheral Schwanna cells and macrophages. Only humans and the nine-banded armadillo are susceptible.

Leptospira. Spirochaete bacterium which causes a mild chronic infection in rats and many domestic animals. The bacteria are excreted continuously in the urine and contact with infected urine or water can result in infection of humans via cuts or breaks in the skin. Infection causes leptospirosis or Well's disease, a type of jaundice, which is an occupational hazard for sewerage and farm workers.

Leptotene (Leptonema). An early stage in meiotic prophase in which individual chromosomes are first visible as long threads.

Lesch-Nyhan syndrome. A sex-linked recessive inherited disease in humans which results from mutation in the gene for the purine salvage enzyme HGPRT, located on the *X* chromosome. Leads to severe mental retardation and distressing behavioural abnormalities, such as compulsive self-mutilation.

Lesion. An area of injury, or a circumscribed pathologic tissue change.

LET. Linear energy transfer. The energy released by the radiation per unit length of absorbing material. RBE increases with LET.

Lethal (semilethal) gene. A gene that, when expressed before (after) age 1 year, results in the eventual death of the organism. A gene whose phenotypic effect is sufficiently drastic to kill the bearer. Death from different lethal genes may occur at any time from fertilization of the egg to advanced age. Lethal genes may be dominant, incompletely dominant, or recessive.

Lethal Factors. Genes which, in the homozygous state, have such a marked deleterious effect that such homozygous organisms are inviable. Where the homozygote is capable of survival but with impaired efficiency the factors responsible are termed Semilethal.

Lethal mutation (gene). A gene which renders a gamete nonviable (gametic lethal) or one which prevents the complete development of a zygote (zygotic lethal). Zygotic lethals may produce their effects at any time from the fertilized egg on.

Lethal zygosis. A term borrowed from classical genetics used to describe the formation of pocks in a lawn of Actinomycetes. Certain plasmids of *Streptomyces* species contain a gene or genes responsible for pock formation and these are known as *ltz* for lethal zygosis.

Lethal. Capable of causing death.

Lethality (adj. lethal). Mortality owing to genetic causes or environmental actions. In the first case, it may be due to the presence of lethal genes or other alterations in the hereditary constitution, especially structural chromosome changes.

Leu enkephalin. A natural peptide neurtransmitter;

Leucine (leu). One of the 20 common amino acids. It carries a non-polar side-chain.

Leucine aminopeptidase. An exopeptidase which removes neutral amino acid residues from the N-terminus of proteins.

Leucocyte interferon (alpha-interferon, INF-Â). An antiviral protein isolated from blood. Normal leucocytes cannot be cultivated, hence mass production is hampered unless gene manipulation techniques transformed cells or hybridomas are used.

Leuconostoc. A genus of microaerophilic, spore-forming, Grampositive cocci that produce lactic acid. *Leuconostoc* species also produce large amounts of extracellular dextran. Species of *Leuconsostoc* are used in the production of dextran and a number of fermented dairy products.

Leucoplast. A colourless plastid, i.e. one not containing chlorophyll or any other pigment.

Leucosin. One of the structurally stable scleroproteins found in wheat.

Leucoviruses. Now little used name for retroviruses.

Leukemia. An excessive increase in number of leukocytes due to malignancy. It is a form of cancer characterized by extensive proliferation of nonfunctional immature white blood cells (leukocytes).

Leukocidin. A substance produced by some pathogenic bacteria and toxic to white blood cells. It kills leukocytes.

Leukocyte (white blood cell). The nonpigmented nucleated cell of the blood essential in the body's defense mechanisms. Those found in normal blood are divided according to their staining reaction into granular leukocytes. neutrophils, eosinophils, basophils; and nongranular leukocytes, lymphocytes and monocytes.

Leukocytosis. An increase in the number of leukocytes caused by the host's body response to an injury or infection.

Leukopenia. A smaller than normal number of circulating leukocytes.

Leukosis. Uncontrolled proliferation of mature forms of white blood cells.

Leukotrienes. A collection of metabolites of arachidonic acid which have powerful parmacological effects.

Levan. A polysaccharide formed chains of ß- (2,6)- linked fructofuranose units branched through ß- (1,2)-links.

Level of significance. Risk of incorrectly rejecting an appropriate model.

L-form. A prokaryotic form lacking a cell wall, developing from a normally wall-containing organism. For example, bacterial forms devoid of a cell wall; some can revert to normal cells.

Lichen. A mutualistic association of a fungus and either algae or cyanobacteria, resulting in a new structural form. The new formation is morphologically distinct from either of its microbial components and possesses separate chemical and physiological properties.

Life cycle. The entire span of the life of an organism from the moment of fertilization (or asexual generation) to the time it reproduces in turn.

Life history. The stages an individual goes through during its life.

Life table. A device for expressing in an orderly fashion, observations on the changing density of an insect population in time and space, and the processes which direct those changes especially in relation to the age-specific distribution of mortality and its causes. It is the table showing, for a group of equal-aged individuals, the proportion still alive at different times in the future and the number of offspring they produce during each time interval.

Ligament. A band of connective tissue linking two bones in a joint. A tough band or capsule of connective tissue that connects two bones together at a joint. It has a high proportion of elastic fibers and white collagen fibers; therefore it combines strength with elasticity to control movement at the joint. The fibers of ligament penetrate the tissue of the bones, making a very strong connection.

Ligand. A ligand is anything that binds to another molecule. Usually it means something that bnds to a receptor or enzyme molecule, such as a hormone or protein growth factor. Thus the antigen molecule that binds to an antibody is an example of a ligand. Enzyme substrates can be considered as ligands, although they are not usually called this because, soon after they bind, they are converted to something else. Discovering the natural ligands for receptors is a substantial part of the research into what receptors are doing. Often a ligand will bind to more than one receptor, and a receptor will bind more than one ligand. These networks are important in the control pathways in many living things. Ligands that bind receptors on cell surfaces can be agonists or antagonists. The former trigger the receptor to perform its function, i.e. signalling to the cell, opening an ion channel, or whatever. Antagonists block that function. The strength of the ligands binding is measured by its affinity constant (see Binding), or the reciprocal of the affinity constant, the dissociation constant.

Ligases. Enzymes that catalyze the linking together of two molecules. They take part in many of the steps involved in the synthesis of macromolecules, such as proteins and nucleic acids. (They are also known as synthetases).

Ligation. The process of joining two linear nucleic acid molecules together via a phosphodiester bond. In a cloning experiment a restriction fragment is often ligated to a linearised vector molecule using T4 DNA ligase.

Light chain (L chain). The smaller of the two types of polypeptide changes (the other is the H chain) that compose the basic Ig molecule; each basic Ig molecule contains two identical light chains. Generally refers to the light polypeptide components (MW-23,000) of the immunoglobulin.

Light compass reaction. A reaction of many invertebrates in which the angle between the direction of movement and the direction of the sun is kept constant.

Light reactions. The light-dependent reactions of photosynthesis that convert light energy into the chemical energy of NADPH and ATP.

Light-harvesting system. Set of photosynthetic pigment molecules which absorb light and channel and energy to the photosynthetic reaction center, where the light reactions of photosynthetic occur. Higher plants, contains chlorophyll and carotenoids, and is present in two slightly different forms in photosystems I and II.

Lignification. Hardening of tissue from wall deposition of lignins.

Lignin. Common constituent of plant cell walls; major component of wood. It is a complex polymer of somewhat variable composition, made up from alcohols, sugars, aromatic amino acids and phenolic substances.

Lignocellulose processing. It is a technique or combination of techniques, used to upgrade the value of lignocellulose by breaking it down to simpler molecules of value.

Lignocellulose waste. Forestry and agricultural waste with a high lignocellulose content.

Lignocellulose. It is the complex mixture of lignin, cellulose and hemicellulose which occurs in wood and woody parts of herbs and trees. Most intractable wastes are lignocellulose.

Liguleless seedling. A seedling in maize without a ligule, one of the distinctive features of the family Gramineae. Several recessive genes cause liguleless seedlings.

Likelihood function. The function relating an unknown parameter to observations from which it can be estimated. Cf. *Probability Function.*

Limited Chromosomes. Chromosomes which occur only in nuclei of the germ-line and never in somatic nuclei.

Limiting factor. An environmental variable whose absolute level at a given time limits the growth or other activity of an organism. Any factor in the environment that alone governs the behaviour of an organism or system by being above or below a certain level. In general, the behaviour of a system depends on number of different factors; under certain conditions, one of these can limit the behaviour. For instance, plant growth is limited by low temperature and increases with rising temperature to an optimum, beyond which growth rate decreases.

LIMS. Laboratory information management systems (LIMS) are the computer systems used to manage the information that a laboratory needs and uses. They have become topical in biotechnology because of the large amounts of experimental results produced by high throughput, automated laboratories in genome projects and high throughput screening. LIM systems need to solve several related problems. 1. How to get lots of different machines to communicate with one system. This sounds trivial, and is not. 2. Capturing data directly off instruments, so a scientist does not have to read it off a screen or printout and then type it in to the system. This is particularly important for high throughput discovery mated microtitre plate readers that are part of HTS systems (see High throughput screening). DNA sequencer LIMS systems, for example, will read the gel outputs and turn them directly into DNA sequence data suitable for input into analytical programs. 3. Tracking samples consistently, so that each sample, analysis, or run is identified, its results linked to it, and the protocol used in that run is known. All this data should be linked up, not stored on separate (often incompatible) systems. 4. Appropriate data storage and presentation. Data should be summarized to a suitable level. A chemist may want to see detailed NMR spectra, but a molecular biologist will not, so the details must be summarized in the presentation software. In theory, a good LIMS system can result in a completely paperless laboratory, but that ideal is rarely achieved.

Lincomycin. Antibiotic active against Gram positive bacteria. Acts by blocking protein synthesis by binding to the 50S subunit of the ribosome and blocking peptidyl transferase reactions. Clin-damycin, a derivative of lincomycin, is used as an antimalarial drug.

Line of breeding. A rather indefinite term used somewhat loosely, applied to a group of individuals which have descended from one individual; used more frequently and correctly with reference to male lines of descent.

Line. Group of homozygous individuals that form a pure breeding population distinct from other individuals of the same species. It is the collection of cells produced by continuously growing a particular cell culture *in vitro*. Such cell lines will usually contain a number of individual clones.

Lineage. In man a section of a tribe or mating group forming a unit of common descent, with a name inherited in either the male or female line, within which mating is tabu, i. e. forbidden by custom or law.

Lineal consanguinity. Marriage of persons one of whom is descended from the other.

Linear phase. The constant increase in cell number following the exponential growth phase and proceeding the deceleration phase in most batch suspension cultures.

Linear. A term used to describe the physical state of a nucleic acid molecule. The two strands at the end of a linear molecule of double-stranded DNA can be free, as in restriction fragment, bound to a specific protein, as in some eukaryotic extrachromosomes. Eukaryotic chromosomes are linear and their ends, the telomeres, have special sequences which promote their application.

Linearise. To introduce a single double-strand break in covalently closed circular DNA molecule and so convert it to a linear molecule. The cleavage of a unique restriction site within a plasmid will linearise it.

Lineweaver-Burk plot. It is a mathematical analysis used to tread data obtained from simple enzyme reactions.

Linin. A vague term designating one or more particular structural components of the nucleus.

Lining epithelium. An epithelium lining a duct, cavity or vessel, that is not particularly specialised for secretion or as mechanical barrier. Not a precise classification.

Linkage (genetic)**.** Association between markers on the same chromosome such that they do not show random assortment. Linked markers recombine with one another at frequencies less than 0. 5; the closer the markers on the chromosome, the lower the frequency of recombination.

Linkage analysis. The determination of the frequency of crossing-over or recombination of DNA sequence. It is used as means of establishing the position of a gene, or sequence, in a chromosome or length of DNA.

Linkage disequilibrium. (1) A condition where two genes are found together in a population at a greater frequency than that predicted simply by the product of their individual gene frequencies. (2)Nonrandom association of alleles at different loci in, a population.

Linkage group. A group of characters all showing linkage with others in the same group. The genes for these characters are arranged linearly in one chromosome, the genes close together showing close linkage.

Linkage map. A representation of a linear array of genes with appropriate distances between them. Linkage maps are usually constructed from the proportions of different types of off-spring from a cross.

Linkage value. A recombination percent expressing the proportion of cross-overs to non crossovers (the parental type). The linkage value may vary from zero to 50 percent, which is the value for independent assortment.

Linked enzyme assay. It is a test procedure in which two or more enzymes are used to catalyze a series of reactions enabling identification and/or qualification of a specific substrate (or levels of a specific enzyme) and the visualization of the result. A common technique is to link one enzyme assay with a second enzyme reaction which results in either the oxidation or reduction of NAD or NADP. The rate of oxidation or reduction by the coenzyme is then followed by measuring the change in absorbance at 320 nm.

Linked genes. Genes that are located on the same chromosome and that therefore tend to segregate together.

Linker DNA. (1) Segment of DNA between adjacent nucleosomes. Its length is variable, averaging about 50 base pairs, and it is a probable attachment site for histone H1. (2) DNA sequence used in recombinant DNA technology to join together two other DNA fragments or the ends of one fragment.

Linker. A synthetic oligodeoxyribonucleotide which contains a restriction site. Linkers may be blunt end-ligated onto the ends of DNA fragments to create restriction sites which can be used in the subsequent cloning of the fragment into a vector molecule. It may be necessary to protect the fragment to be cloned from the action of the restriction endonucleases by treating them with DNA methylase.

Linking number. Number of times that two strands of double-helical closed-circular molecule cross one another. Molecules of DNA that are identical in all respects other than the linking number are termed topological isomers.

Linoleic acid (Vitamin F). Unsaturated fatty acid required by various insects and mammals, probably including man.

Lipases. Lipases are enzymes that break down lipids into their component fatty acid and `head group' moieties. The lipases used in biotechnology are almost invariably digestive lipases, meant to break down the fats in food. However, they can be turned to a number of different uses. They can be used to break down complex fats into their components, which are then uses to make other materials. This, however, is a relatively prosaic use. Two more valuable uses are lipid synthesis and transesterification. Lipid synthesis aims to make the lipase put a lipid back together from a fatty acid and an alcohol. This replaces conventional acid-catalysed chemistry, which, as well as using high temperature and pressure, produces a product contaminated with side products, which must then be purified. Unilever produce isopropyl myristate (an ingredient of cosmetics such as moisturizing creams) using a lipase based biocatalytic synthesis. In order to reverse the normal reaction of lipases (which is to use water to split lipids), you have to be able to remove the water from the reaction. This means that the reaction must be carried out in non-aqueous solvents, such as benzene (see Organic phase catalysis). A related reaction is transesterification, a process where a lipase is used to swap the fatty acid chains between lipids without ever releasing significant amounts of fatty acid. This is a useful thing to do since it enables the biotechnologist to take a saturated fat (with a high melting point) and an usaturated one (with a low melting point) and produce mixed molecules with intermediate properties. depending on how the ingredients are mixed, the properties can be determined fairly accurately. This application also requires that the lipase work in organic solvents. Transesterification of triacylglycerol fats (the normal `fat' in animal tissue) which is specific to the 1 and 3 fatty acids is a relatively specific and widely used transesterification, and is called Interesterification.

Lipid annulus. Ring of lipid molecules bonded by hydrophobic interactions around an intrinsic membrane protein. This lipid may have a different composition from the rest of the membrane lipid.

Lipid bilayer. An early model for the structure of cell membranes based upon the hydrophobic interactions between phospholipids. The polar head groups face outwardly to the solvent, with the hydrophobic tails clustered in the interior.

Lipid micelle. Amphipathic lipid molecules will not only self-assemble into lipid bilayers in aqueous solutions, but also into micelles which are small (colloidal-size), spherical, molecular aggregates in which the hydrophobic tails projects inwards and the surface comprises the hydrophilic poles.

Lipid monolayer. A film of lipids, one molecule thick, that forms on aqueous surfaces due to the orientation of the lipid molecules. The polar groups are in the liquid and the fatty acid chains are above the surface.

Lipid. A group of molecules containing esters of fatty acids. Typically, lipids contain long chain aliphatic `tails' linked to water-soluble `heads'. Lipids can be broken down into their component water-soluble heads and sparingly soluble tails by alkali (a process called saponification) or by lipase enzymes. Common tails are fatty acids, of which the more common saturated ones (fatty acids with no carbon-carbon double bonds) are.

Lipidoses. Storage diseases in which the missing enzyme is one which degrades sphingolipids (sphingomylin, ceramides, ganglioside). In Tay-Sachs disease the lesion is in hexosiminidase A, and enzyme which degrades ganglioside Gm_2; in Gaucher's disease, glucocerbrosidase; in Niemann-Pick disease, sphingomyelinase.

Lipoamide dehydrogenase. An enzyme which regenerates lipoamide from the reduced form dihydrolipoamide.

Lipoamide. The functional form of lipoic acid in which the carboxyl group is attached to protein by an amide linkage to a lysine amino group.

Lipofuscin (age pigment). Pigment that tends to accumulate as brown granules in the cytoplasm of long-lived, non-dividing cells; granules are often bounded by a single membrane. Lipofuscin, derived from lipoproteins, may be a residue from lysosome activity, particularly under hyperoxic conditions. Excessive accumulation of lipofuscin may impair cell function and contribute to cell ageing.

Lipogenesis. Production of fatty acids in biological systems.

Lipoic acid. A sulphur-containing fatty acid found in a wide variety of natural materials. It is an essential component in metabolism, although it is active in extremely minute amounts. It has not yet been demonstrated to be required in the diet of higher animals. Lipoic acid is classified with the water-soluble B vitamins.

Lipold. (1) Substance resembling fat in solubility, but not containing fatty acid (e. g. steroid, carotene, terpene). (2) Substance, resembling fat in solubility, which may contain acid (e.g. phospholipin); but excluding neutral fats. (3) Occasionally, fat meaning (1) or (2).

Lipolysis. The splitting of the component fatty acids from a lipid; i. e. part of the process of catabolism of lipid molecules. Lipolysis is effected in the body, largely in the gut, by the lipase enzymes.

Lipolytic enzyme. An enzyme that hydrolyzes fats and lipids.

Lipopolysaccharide (LPS). The lipid and sugar complex that makes up part of the outer membrane of the gram-negative cell wall.

Lipoprotein. A conjugated protein formed by the combination of a protein with a lipid. Lipid material in mammalian blood is associated with plasma proteins, forming lipoproteins. Some lipoproteins are used to transport fats to metabolic sites whilst others are used to form the structures of cell membranes. They are found in mitochondria, nuclei, and microsomes.

Liposome. A liposome is a small capsule made of lipids. Lipids can form stable structures in solution, in which the polar `heads' point outwards into the watery solution and the apolar `tails' stick together into the middle. If this is a droplet of lipid, it is called a miscelle. If it is a sheet of two layers of lipid molecules, then it is a Langmiur-Blodgett film. If such a film closes up into a ball, the result is a sphere with watery solution outside and inside separated by a lipid `bilayer'. This is a liposome. Liposomes can have multiple layers stacked inside each other, but are often considered as if they were single (unilammellar) bags. Some lipids will spontaneously form miscelles or liposomes if shaken up in water at their `critical miscelle concentration'. (At higher concentration the miscelles just lump together into a bulk of lipid, and at lower concentration the lipid dissolves to form a true solution.). Liposomes have been suggested as the basis of several methods of drug delivery, especially for delivering peptide drugs. This is because they could protect their contents from digestion in the stomach and so deliver them to the intestine, where they would be absorbed, or they could allow them to be injected into the bloodstream and be carried around to a specific organ. Here the organ would recognize the lipids and absorb them specifically (this would work well for the liver, which tends to absorb liposomes from the blood spontaneously). Alternatively, antibodies linked on to the outside of the liposome would bind it to the specific cells that the drug is meant to affect (a combination sometimes called an immunoliposome). Liposomes tend to accumulate at sites of inflammation, and in some tumours, and so they are potential delivery vehicles for anti-inflammatory and antitumour drugs. Liposomes are also being as adjuvants in immunization. they can be used to carry the proteins of a synthetic vaccine to an antigen-presenting cell in the immune system and stimulate it to produce an effective immune response. Liposomes are considered particularly useful for this sort of application as they are made of the same materials (lipids) s the outside of cells, and so are less `alien; to the body. Trapping things inside liposomes is a form of encapsulation, and as such can be used in many other areas. Here, however, liposomes are less favoured because they can be rather less stable than other, polymerbased encapsulation methods.

Lipotrophin (LPH). A product of the mammalian pituitary gland which contains peptides (Â-and ß-LPH) structurally related to the melanocyte-stimulating hormone. They stimulate lipolysis.

Lipoxygenase. Enzyme which catalyzes the oxidative conversion of arachidonic acid to the hydroxyeicosatetraenoic acid (HETE) structure in the synthesis of leucotrienes.

Liquefaction. (1) A series of biological and/or physicochemical processes that result in the conversion in insoluble solid organic material into a liquid or soluble form. (2) Transformation of a gel to a liquid.

Liqueur. An alcoholic beverage produced from aromatic extracts of fruit or other plant material to which sugar is added.

Liquid chromatography. A separation technique in which a mixture of compounds in solution is passed through a column packed with a solid matrix.

Liquid culture. The culture of plant cells on liquid medium, in suspension or on supports. Cultures are either held steady (stationary culture) or are shaken (agitated culture or shake culture).

Liquid manure. Animal waste from intensive animal husbandry units where no bedding material is used. It is usually collected as a slurry.

Liquid medium. Medium not solidified with a gelling agent. Liquid media are used for suspension cultures and for a wide range of research purposes. They are also useful for Stage I cultures in some micropropagation protocols. Usually a support structure or wick is used to hold the tissue above the nutrient medium.

Liquid membrane separations. Liquid membranes are thin layers of a liquid that does not mix with the water on either side. (In principle, they could also be thin layers of water with some other liquid on either side.) If something can dissolve in the liquid, then it can pass through the membrane. This can be the basis for separating materials that do dissolve in the liquid from those that do not, by putting the mixture on one side of the membrane and pure water on the other. the soluble component diffuses across the membrane, the contaminants cannot. More sophisticated separation mechanisms can be based around this idea. The membrane can be impregnated with a carrier molecule which can carry across one type of molecule but not others. These usually bind to target molecule, making it soluble in the lipid (as a `complex') whereas it usually would not be. Chemicals that can do this could include some peptide antibiotics, clathrins, crown ethers, or cyclodextrins. The transport of the molecule you want can also be linked to the transport of another molecule (for example, a proton). this is called `coupled transport', and is the way that living cells concentrate many molecules inside themselves. Ion exchange systems can also be used with a supported liquid membrane, in an ion exchange membrane (IEM).

Liquid membranes. These are thin films made up of liquids (as opposed to solids) that are stable in another liquid (usually water). Thus the liquid must not dissolve in the water, but nevertheless must be prevented from collapsing into a lot of small droplets. There are several types of liquid membrane.

Langmiur-Blodgett films. These are `true' liquid membranes, in that there need be nothing in them except the liquid (see Langmiur-Blodgett films).

***Immobilized or supported membranes (immobilized liquid membranes*; **ILM).** Here the liquid is a thin film trapped in some solid material. This might be a porous polymer (such as scintered glass) or a fibrous one (such as cellulose). The liquid fills up the pores in the material, and so forms a series of minimembranes.

Supported membranes can also be ion exchange membranes (IEMs), if the supporting material is something that binds ions strongly. When something dissolves in the liquid part of the membrane, it is held there by the solid part. This can be the basis of separation methods.

Emulsion liquid membranes **(ELMs).** Here the watery part and the non-watery liquid are mixed up with a detergent. This makes small droplets of water in the other liquid (or the other liquid in water) stable. The result is a mix of water inside liquid droplets, themselves inside water. This is a membrane since it is a liquid barrier between two volumes of water. **Liquid membranes** can be used in a number of applications. (The specific applications of Langmiur-Blodgett films are described separately.) Their main potential is as the basis of separation systems.

Liquid nitrogen. Nitrogen that has been liquefied using cryogenic techniques. It is used as a refrigerant to keep lyophilized stock cultures at -130° C or as a coolant for instruments such as electron microscopes.

Liquid scintillation counter. Electronic instrument used to measure quantities of radioactive substances by assay of the emission of radiation during radioactive decay. The sample containing the radioactive compound is placed in a fluid containing a fluorescent chemical. This chemical emits photons of visible light when exposed to beta particles or gamma-radiation. Photomultiplier tubes measure the light emitted and so determine the radiation levels involved.

Liquor folliculi. An albuminous, alkaline, yellowish fluid filling the space between the ovum and the epithelial cells of the Graafian follicle.

Listeria monocytogenes. Rod-shaped Gram positive bacterium. It is wide-spread and can grow over an unusually wide range of temperatures (0-45°C); normally saprophytic but is an opportunistic parasite, in that it can survive within cells (particularly leucocytes) and can be transmitted transplacentally. It has caused number of serious outbreaks of food poisoning with a high mortality rate in recent years.

Liter. Metric unit of volume containing 1,000 milliliters (ml).

Lithotroph. A microorganism that obtains its energy from the oxidation of reduced inorganic compounds.

Litmus paper. Red or blue acid-alkali indicator papers. In acids blue litmus paper turns red and in alkalis red litmus paper turns blue. The paper is prepared by soaking absorbent paper in litmus solution and then drying it. Litmus solution has a pH range from 4. 5-8. 3. It is obtained from a lichen (*Lecanora tartarea*).

Litmus. Plant extract, used as a pH indicator, also an oxidation-reduction indicator; it turns blue when alkaline and red when acid in reaction.

Littoral zone. The coastal zone from the upper limits of tidal action down to the depths where the water is thoroughly stirred by wave action.

Live vaccines. Live vaccines are vaccines containing living organisms or intact viruses, rather than inactivated (killed) organisms or extracts of them. They can cause better immunity in patients, but have the potential drawback that, unless they are thoroughly `crippled' in some way, they may cause disease. They can also be harder to transport and store, because you have to avoid killing the organism. Biotechnology has generated ideas and research products for live vaccine development in a number of areas. Viral vaccines are dealt with elsewhere. Bacterial live vaccines can be developed in a number of ways.

Attenuation. Bacteria need a number of specific genes (virulence genes) to be able to cause disease, but these genes are not essential for growth in the `test-tube'. When pathogenic bacteria are grown outside their host, they tend to lose their virulence genes by mutation. The result is an attenuated bacterium, which will cause an immune response similar to the original but which is harmless. Usually, several mutations are needed to make sure that a strain is really attenuated. If the nature of the virulence genes is known, then conventinal and `molecular' genetics can be used to select for mutatins in or loss of those genes.

Gene cloning. An alternative is to place some key genes from the pathogenic bacterium into another, harmless organism. These may be the genes for surface parts of the pathogenic bacterium, such as pili or transport proteins, that are seen by the immune system. The degree to which an antigen, or a particular part (epitope) of an antigen is detected by the immune system, and hence the amount of antibody response that the immune system makes against that antigen, is called its immunogenicity. A key part of designing a good vaccine is deciding how to make the vaccine highly immunogenic, so that it is `seen' by the immune system very clearly. On vaccination with such a material, the immune system `learns' what those parts of the pathogen look like without ever encountering the whole organism. This is similar to cloning the prtein as a vaccine, but has the advantage that, as part of a living organism, it can stimulate the immune systems to greater prodigies of invention in generating good antibodies against it. Live bacterial vaccines are generally suggested being used for enteric infections, including tooth decay, and some parasitic diseases.

Liver. A large digestive gland. In vertebrates, it secretes bile and is involved in the formation of blood.

Load. Total count of bacteria in a consumable product.

Loading. The introduction of a feedstock into a reactor or a sample for analysis into an analytical instrument, or onto a gel or similar support material in a chromatographic or electrophoretic technique.

Local anaphylaxes. A type I hypersensitivity reaction that primarily affects a given organ or tissue (such as skin or nasal mucosa).

Localization. The genotypic property of restriction of crossing-over and chiasma formation to certain corresponding parts of all the chromosomes. It may be procentric or proterminal according to whether the Contact Points are near the centromere or near the ends. (Darlington, 1931).

Localized (L. *localis*, a place). A kind of infection which is localized or restricted to a particular tissue or organ of an organism.

Lock and Key Theory. The theory, propounded by Dufour, that the female and the male genitalia of the same species (at least in insects) are so exactly fitted to each other that even slight deviations in the structure of either make copulation physically impossible.

Locullicidal (Bot.). Describing dehiscence of multilocular capsule by longitudinal splitting along dorsal suture (midrib) of each carpel, e. g. iris. Cf. Septicidal.

Locus. In genetics, a specific location on a chromosome. May be considered to be synonymous with 'gene'. It is the position of a gene on a chromosome. Alleles of the same gene occupy the equivalent locus on homologous chromosomes.

Lod score. Measure of the statistical significance of a departure from expected values; logarithm of the odds of such a departure on the basis of chance alone.

Loeffler's medium. Coagulated-serum medium used to culture *Corynebacterium diphtheriae* in diagnostic bacteriology.

Log phase. The period of most rapid reproduction in the growth phases of a culture; the generation time is constant when plotted on a graph, and the log of the number of organisms appears as a straight line.

Logistic growth. Growth, especially in the size of an organism or in the number of organisms that constitute a population, which slows steadily as the entity approaches its maximum size (contrast with exponential growth).

Long terminal repeat (LTR). Sequence of several hundred bases found at either end of some DNA molecules, especially the genomic DNA of retroviruses. The sequences at either end are identical to one another or, in at least some cases, are inverted repeats.

Longistylous. With a long style, in plant species that are heterostylous.

Long-period interspersion. Pattern of distribution of DNA sequences found in the genomes of some organisms such as Drosophila. It involves an alternative distribution of non-repetitive DNA sequences between long stretches of moderately repetitive DNA.

Loop bioreactors. Also loop fermentors, these are bioreactors in which the fermenting material is cycled between a bulk tank and a smaller tank or loop of pipes. The circulation helps to mix the materials and to ensure that gas injected into the fermentor (usually oxygen or air) is well distributed amongst the liquid. The reactors are also very useful for photosynthetic fermentations, where they allow for the photosynthesizing organism to be passed along a large number of small pipes, where the light can get to them easily, rather than inside a single volume, where only the organisms near the edges get much light. There are lots of types of loop bioreactors, but they break down into those that have an internal loop (e.g. a stirred tank bioreactor with an internal draft tube) and those that have an external loop. Some airlift where air or oxygen is injected into the riser half up, so pushing the flow round the vessel. A variant that is found in all of them is the jet loop reactor, in which the recycled liquid is injected with considerable force back into the main tank. This means that it does not merely circulate the re-injected liquid around, it stires the rest of the tank contents up as well. This has the advantage that the recycling mechanism is also the stirring system, removing the need for stirrers and baffle plates. One popular type of loop bioreactor is the airlift fermentor (or bioreactor). See also tank bioreactor, gas transfer.

Loop fermenter. A bioreactor in which the culture medium is continuously recycled through two chambers, or through a side arm attached to a single vessel, as an aid to mixing and mass transfer.

Lophotrichous. Having a polar tuft of flagella.

Lost gene. A situation in which all the genes at a certain locus are the same and the other allele has been permanently eliminated from the population.

Low pressure liquid chromatography. An automated system for carrying out liquid chromatography.

Low-density lipoprotein. Plasma lipoprotein fraction that is the chief source of cholesterol for cells outside the liver and intestine. Cells bind the lipoprotein by means of specialized receptors, and there is evidence that individuals suffering from the inherited disorder hypercholester-olaemia have absence or deficiency of these functional receptors. This deficiency produces abnormally high levels of circulating cholesterol and low-density lipoprotein, with consequent deposition of cholesterol in arterial plaques and greatly increased risk of heart disease.

Low-gelling-temperature agarose, LTG-agarose, low-melting-point agarose. An agarose derivative which will remain fluid at temperatures around 37° C. It also melts at low temperature, **ca.** 60-65°C and this means that DNA fragments can readily be recovered in their native form following electrophoresis.

Lowry method. The most widely used chemical method of protein analysis. A sample containing up to 500 μg of protein is dissolved in a solution of sodium carbonate, copper suplhate, sodium hydroxide and sodium potassium tartrate, followed by addition of the phenol reagent. The dark blue colour produced by interaction of the reagents with tyrosine in the protein, is compared colorimetrically with the colour generated using standards, usually bovine serum allumin (BSA).

LPR (lymphoproliferation gene). A gene found in MRL mice which is involved in the generation of autoimmune phenomenon.

LPS (Lipopolysaccharide). A produce of some Gram negative bacterial cell walls and which can act as a B cell mitogen.

LSD (lysergic acid diethylamide) Hallucinogenc drug that acts by increasing the concentration of serotonin in the brain.

Luciferase. Enzyme involved in biological photoluminescence.

Luciferin. A complex compound responsible for emitting light in bioluminescence.

Lucigenin. Compound used as a bystander substrate in assaying the metabolic activation of leucocytes by chemiluminescence. When oxidise by superoxide it emits light.

Lucke virus. Herpes virus that causes adenocarcinomas of the kidney in frogs.

Lumen (L. *light*). The cavity inside any tubular part on an organ, such as piece of gut or a kidney tubule.

Lumicolchicine. A derivative of colchicine produced by exposure to ultraviolet light and which does not inhibit tubulin polymerisation, although it has many of the non-specific effects of colchicine.

Luminescence. Luminescence, the production of light by chemicals, is gaining increasing use as a labelling system for antibody-and DNA-based tests. Luminescent tests are of interest because, if carried out in a rigorously light-proof box, they can be extremely sensitive. a photomultiplier tube can detect when only a handful of photons have been given out by a reaction, so offering the potential of detecting only a handful of DNA or antibody molecules. There are two ways of generating light using chemicals.

Chemiluminescence. This uses specific chemical groups which, when reacted, give out light. They can be attached to many other chemicals (e.g. proteins, DNA). There are also chemiluminescent groups that are `blocked' (they do not themselves react to give out light). However, when the blocking group is chopped off, they react rapidly to generate light. A phosphate group is a common blocking group, which allows a chemiluminescent reaction to be used to detect an enzyme that cleaves phosphate groups, such as the widely used alkaline phosphatase (AP). Adding chemiluminescence to such an assay enhances its sensitivity greatly.

Bioluminescence. Some specialized enzyme systems can also generate light, using the energy of ATP (adenosine triphosphate) to do so. The most commonly used is luciferase, an enzyme from bacteria. Enzymes from fireflies have also been used. As well as being used as a label, these enzymes can be used as exquisitely sensitive detectors for ATP. a few hundred molecules of ATP, the amount in a few bacterial cells, can produce a detectable glow. This makes the enzymes useful for rapidly detecting bacteria for hygiene monitoring. (However, the enzymes are also subject to interference from many of the products they are meant to test.)

Luminescent labels are widely replacing radioactive ones in some applications. Thus the scientist will chemically link a luminescent group on to antibody or a DNA probe where before they would have used a radioactive group. luminescent labels can be as sensitive as radioactive ones. The light can be detected by a sufficiently sensitive camera, or by exposing the gel or blot directly to photographic film, as is done in an autoradiograph. (Confusingly, this is sometimes called an autoradiograph as well, although it has nothing to do with radioactivity.) This approach has the advantage that the active luminescence of living cells can be measured (it is usually too weak to be seen by the naked eye). This has been used to monitor cellular processes as cells metabolize. A commercial example is the Microtox environmental monitoring system, which uses living luminescent bacteria to detect if an environmental sample contains any toxic material. If it does, the bacterial glow fades.

Luminol. Compound used as a bystander substrate in assaying the metabolic activation of leucocytes by chemiluminescence. When oxidised by myeloperoxidase hydrogen peroxide system, it emits light.

Lungs. A pair of sac-like chambers within the bodies of some animals, functioning in gas exchange.

Luria and Delbruck experiment. Experiment in genetics designed to show whether a specific mutation could be environmentally induced or occurred randomly. It involved the use of a fluctuation test, in which numerous cultures of bacteria were simultaneously screened for resistance to lysis by bacteriophage. If mutation to phage resistance were random, then considerable fluctuation in the numbers of mutant bacteria would be expected in the parallel cultures, depending on precisely when the mutation had occurred (if it occurred early in the life of the culture, many progeny would have accumulated, if late, no progeny would have accumulated). Since marked fluctuations in the numbers of surviving resistant colonies were obtained, it was concluded that mutation was occurring at a constant random rate and was not being induced by exposure to the phage or other environmental factors.

Luteal hormone. Progesterone; a sex hormone which has a specific action on the uterine endometrium.

Luteinisation. The formation, in large amount, of corpora lutea in the ovary.

Luteinizing hormone (LH, interstitial-cell-stimulating hormone, ICSH). A glycoprotein hormone secreted by the anterior pituitary lobe under regulation of the hypothalamus. In female mammals it stimulates secretion of estrogen, ovulation, and formation of corpora lutea. In male mammals it stimulates interstitial cells in the tests to secrete androgens.

Lutel tissue. Tissue which fills cavity of ruptured Graffian follicle constituting a corpus luteum which is derived from follicle cells. Secretes progesterone.

Luxuriance (adj. luxuriant). The high degree of vegetative development that is often observed in species hybrids.

Luxury gene (tissue-specific gene). DNA sequence that is not expressed in all cells. Such genes include many which have been intensively researched, since luxury genes often code for products which are made in very large amounts in particular tissues. The products of luxury genes are sometimes referred to as Luxury Proteins. The non-luxury genes, the so-called house-keeping genes, include sequences coding for histone or lactate dehydrogenase, which occur in all cells.

Luxury protein. A term sometimes used to describe those proteins which are produced specifically for the function of differentiated cells and are not required for general cell maintenance (the so-called "house-keeping" proteins).

Ly antigen. A group of cell surface markers found on murine T cells which permit the differentiation of T cell subpopulations.

Ly antigen. A group of cell surface markers found on murine T cells which permit the differentiation of T cell sub populations.

Lyases. Enzyme that (i) remove groups from substrates nonhydrolytically, usually bearing double bounds, or (ii) add groups to double bonds. This most commonly involves the removal of water, ammonia, or a carboxyl group.

Lyb antigen. Surface antigens of mouse ß-cells.

Lymph (L. *lympha:* water). A clear watery fluid that is formed as a filtrate of blood; it contains white blood cells; it collects in a series of special vessels and is returned to the bloodstream.

Lymphadentis. Inflammation of lymph node.

Lymphatic system. A closed system of vessel through which the lymph is transported from the interstitial fluid to the blood starch. The flow of lymph in mammals is maintained by muscular activity.

Lymphatic tissues. Those tissues, including the lymph nodes and vessels, the thymus, the spleen, and the Bursa of Fabricius (in birds), which produce and contain the lymphocytes.

Lymphoblasts. The cytoplasm-rich cells differentiated from antigenically stimulated T Lymphocytes.

Lymphocyte activation (lymphocyte transformation). The change in morphology and behaviour of lymphocytes exposed to a nitrogen or to an antigen to which they have been primed. The result is the production of lymphoblasts, cells which are actively engaged in protein synthesis and which divide to form effector populations. Should not be confused with transformation of the type associated with oncogenic viruses, and *activation* is therefore perhaps a better term.

Lymphocyte. A major class of white blood cells. Includes T cell, B cell, and other cell types important in the immune response. It is devoid of cytoplasmic granules, associated with immune response and chronic inflammation.

Lymphocytic leukemia. A form of blood cancer in which a clone of white blood cells continuously increases in number.

Lymphokines. A generic term for molecules other than antibodies which are involved in signalling between cells of the immune system and are produced by sensitized lymphoid cells (classically T cells) in response to antigenic contact; such factors modulate the biological activity of other cell populations, especially macrophage cells.

Lymphoma. Any neoplasid (abnormal growth) associated with lymphatic system. It is the malignacy of lymphoid cells resulting in a solid tumor.

Lyon hypothesis. That is any given cell of a female one *X* chromosome is active and the other is inactive.

Lyonization. Term based on an interpretation of certain genetic traits expressed in the female as if she were carrying an *X*-linked recessive gene in a homozygous condition (e. g., a woman who suffers from haemophilia yet carries this normally recessive allele on only one *X* chromosome). The phenomenon is described as lyonization on the assumption that one of the two *X* chromosomes is preferentially condensed in most or all somatic tissues.

Lyophilization. Freeze-drying, or dehydration from the frozen state.

Lyophilize or freeze dry. To freeze rapidly then dehydrate under high vacuum. and process is **lyopholization.**

Lyophilized culture. A preparation of microorganisms produced by freeze drying.

Lys. An abbreviation used to denote the amino acid lysine in protein sequence and elsewhere.

Lysate. Product of cell lysis; often used to refer to a suspension of phage particles released by the lysis of host bacteria, or to a cell-free homogenate following the action of detergent or mechanical means to break the cells.

Lysine. An enzyme or other substance capable of disrupting cells.

Lysine production. Lysine is produced by fermentation, generally using molasses as substrate.

Lysis (Gr. *a loosening*). Bursting of a cell. It is a process of disintegration or destruction of host cells (e. g., bacterial) during and on the completion of bacteriophage multiplication.

Lysogen. A strain of bacteria carrying a prophage.

Lysogenic bacteria (Gr. *lysis* + new Latin *genic*, to give rise to). Here used to indicate bacteria carrying a phage in its temperate state; the phage in its temperate state; the phage may eventually lead to the destruction of the erstwhile host bacterium itself.

Lysogenic bacteria (Gr. *lysis* + new Latin *genic*, to give rise to). Here used to indicate bacteria carrying a phage in its temperate state; the phage in its temperate state; the phage may eventually lead to the destruction of the erstwhile host bacterium itself.

Lysogenic conservation. (phase conversion) The influence of the prophage DNA on the host bacterium carrying it. Such influences include change in pathogenicity and in morphological nature of the host.

Lysogenic pathway. The sequence of development of a phage that results in the production of a lysogen. In this form of development, almost all of the genes carried by the phage are repressed. This requires the insertion of the phage genome into the host genome, as well as activation of repressor synthesis. This repressor represses the synthesis of the products of other genes and regulates its own synthesis, with the concentration of repressor controlling the rate of synthesis of its own mRNA.

Lysogenic phage. A bacteriophage which, instead of always lysing its host, can sometimes enter into a stable relationship enter into a stable relationship with it. Its genes being replicated in phase with the bacterial chromosome. To achieve this, the phage DNA is either integrated into the host chromosome or exists as a plasmid. A bacterium harbouring a phage genome in this state is said to be a lysogen.

Lysogenic viruses. Viruses that can become prophages.

Lysogenic. The term applied to bacteria which are carrying prophage.

Lysogenization. Production of lysogenic bacteria by infecting them with a temperate phage.

Lysogeny. A condition in which the nucleic acid of the bacteriophage exists inside the host bacterium and replicates along with the bacterium's nucleic acid but does not produce mature phage particles. It is the state of a bacterial cell in which a bacteriophage genome is in association with the bacterial genome.

Lysophosphatides. Mono-acyl derivatives of diacyl phospholipids which are present in membranes as a result of cyclic deacylation and reacylation of phospholipids. Membranolytic in high concentrations, and fusogenic at concentrations which are just sub-lytic. May have important modulatory roles.

Lysosomal diseases. Diseases (also known as storage diseases) in which a deficiency of a particular lysosomal enzyme leads to accumulation of the undigested substrate for that enzyme within cells. Not immediately fatal, but within a few years leads to serious neurological and skeletal disorders and eventually to death.

Lysosomal enzymes. A range of degradative enzymes (hydrolases), most of which operate best at acid pH. The best known marker enzymes are acid phosphatase and ß-glucuronidase, but many others are known.

Lysosome (Gr. *lysis:* a loosening + *soma:* body). A membrane-bounded inclusion found in eukaryotic cells (other than plants). Lysosome contain a mixture of enzymes that can digest most of the macromolecules found in the rest of the cell. They are very small organelles, within the cytoplasm of animal cells, which contain a large number of hydrolytic enzymes, capable of breaking down all the major constituents of living cells.

Lysosome-phagosome fusion. A process which occurs after the internalisation of a primary phagosome. Fusion of the membranes leads to the release of lysosomal enzymes into the phagosome. Some species of intracellular parasite evade intracellular destruction by interfering with this process.

Lysozyme. Enzyme which hydrolyzes the bond between N-acetylglucosamine and N-acetylmuramic acid of the peptidoglycan of the bacterial cell wall; this causes the cell wall to break, and lysis of the cell usually follows.

Lysyl oxidase. Extracellular enzyme which deaminates lysine and hydroxylysine residues in collagen or elastin to form aldehydes, which then interact with each other or with other lysyl side chains to form crosslinks.

Lyt antigen. A set of plasmalemmal surface glycoproteins on mouse T-lymphocytes. Possession of Lyt 1 partly defines a T-helper cell, and of Lyt 2,3 suppressor and cytotoxic cells, Formerly known as Ly antigens.

Lytic complex (membrane attack complex). The large (2000kD) cytolytic complex formed from complement C5b6789.

Lytic cycle control. A mechanism whereby the pathway of transcription of phage DNA is controlled in such a way that either the lytic pathway or the lysogenic pathway is formed if a protein (*cl* repressor) is formed the lysogeny results, if cro protein is formed progeny are formed.

Lytic cycle. The sequence of events that occurs when a phage infects a bacterium. It includes adsorption to a receptor site on the bacterium's cell wall, injection of nucleic acid, replication of phage nucleic acid, the assemblage of complete phage particles, and the breaking open of the bacterium with the release of mature phage particles. The sequence of events terminates with the lysis and death of the bacterial cell.

Lytic growth. Developmental phase of a virulent phage during which the viruses multiply vegetatively and kill the host cells by lysis at the end of the growth cycle. Lytic growth occurs also following induction of a temperate phage.

Lytic infection. Viral infection leading to lysis of cell and "burst" of progeny virus.

Lytic pathway. The sequence of reactions, involved in the expression of the phage genes that results in the completion of a lytic cycle. The genes of the phage are regulated to permit controlled DNA replication, recombination, and synthesis and assembly of the progeny phage. In the prophage stage, all the genes required for forming the lysogemy results, if corporation is formed progeny are formed.

Lytic phase (*vegetative phase*). The phase of the viral life cycle during which viral replication and host-cell lysis occur.

Lytic viruses. Viruses whose multiplication leads to lysis of the host cell.

Lytic. Condition in which a bacterium lyses shortly after infection by a virus; the viral genome does not become stabilized within the bacterial cell. (Contrast with lysogenic.)

M

M_1, M_2 etc. Symbols used to designate the first, second etc. generations following exposure to mutagenic agents (ionizing radiations, chemical mutagens, etc.).

M13. Widely used cloning vector; single-stranded bacteriophage with a closed circular DNA genome. Since the product of replication of this phage, even when it carries an insert of some other gene, is a single-stranded DNA molecule, this implies that the molecules are readily used as template for sequence analysis.

Macerase. A brand name for pectinase. Pectinase is useful in the isolation of intact cells and protoplasts of higher plants.

Macerate. To disintegrate or separate tissues through cutting, soaking enzymatic or other action, resulting in cell dissociation.

Machupo virus. A member of the Arenaviridae that may cause a severe haemorrhagic fever in humans. The natural hosts are rodents and transmission from human to human is not common.

Macro- (Gr. *makros.* large long). A prefix commonly used to denote something large. (Contrast with micro-).

Macrobead. A large particle used as stationary phase in affinity chromatography where large molecules or whole cells are being separated.

Macrodescriptor. An index or term describing some aggregate feature of groups of organisms.

Macroelement. One of the main plant nutrient elements essential for plant growth, and required in greater amounts than the others (macroelements or macronutrient elements). The microelements are carbon (C), oxygen (O), hydrogen (H), nitrogen (N), phosphorus (P), potassium (K), calcium (Ca), magnesium (Mg), and sulphur (S).

Macroevolution. A large amount of evolutionary change, involving many elementary changes in gene proportions or chromosome structure. (Contrast with microevolution.)

Macrolide. An antibiotic inhibiting protein synthesis, for example erythromycin.

Macronucleus (meganucleus). Larger of two distinct sizes of nuclei occurring in ciliated protozoa. These nuclei are highly polyploid and are essentially responsible for vegetative growth, whereas the much smaller micronuclei (absent in the cells of some ciliate strains) have important sexual functions during conjugation. Compare Micronucleus.

Macrophage activation factor (MAF). A lymphokine that increases the capacity of macrophage cells to destroy antigens.

Macrophage colony stimualting factor (M-CSF). Growth factor which stimulates committed cells of bone marrow into differentiating towards the production of monocytes (mononuclear phagocytes)

Macrophage inhibition factor (MIF). A group of lymphokines (including a 14kD glycoprotein) produced by activated T-lymphocytes which reduce macrophage mobility and increase heterotypic-homotypic macrophage adhesion.

Macrophage. A large mononuclear cell derived from the mononuclear phagocytic system and having phagocytic activity.

Macro-pyrenic. (i) Having nuclei which are considerably larger than the average for the species, individual, etc. (ii) An individual of this type.

Macroscopic. Visible to the naked eye, without the use of a microscope.

Macrotype egg. A type of tachina fly egg which is broadly oval in outline, with a thick, tough dorsal and lateral chorion, and a flat, membranous ventral surface by which it is glued externally to the integument of the host. cf. Microtype Egg.

MACS, BACS, etc. MACS are mammalian artificial chromosomes. BACS are bacterial artificial chromosomes. Both are complex pieces of DNA that can be used as cloning vectors, to allow a foreign piece of DNA to replicate in a cell or organism. Together with YACs they can be used to clone very large pieces of foreign DNA, in theory tens of millions of bases long. Most cloning vectors are based on plasmids or viruses; they allow our cloned DNA to grow rapidly in their host cells. However, these are not genetically stable; the cells will lose the cloned DNA easily if we do not continuously elect for it to be there. The DNA will either be rearranged by the cell, usually removing part of the gene we want to clone, or will be lost completely. Chromosomes, by contrast, fit in with the normal gene-building machinery of the cell which ensures that every new cell gets an exact copy of the genetic material of the old cell. BACs and MACs are synthetic versions of chromosomes that use this machinery. All chromosomes need to include an origin of replication, the signal that DNA polymerase uses in the cell to start replicating the DNA before the cell divides. Thus MACs and BACs must provide these. Chromosomes also need genetic elements that make sure that each cell contains one copy of the chromosome, and that when they are duplicated during cell division that exactly one copy goes to each new cell. Some BACs use the F1 plasmid to provide origins and other necessary sequence; others (sometimes called PACs) use the bacteriophage P1. Mammalian chromosomes need several origins. They also need a centromere somewhere in the middle (this attaches to the spindle during cell division, so the two new copies of the chromosome end up in the two daughter cells, not both in one), and telomeres at the ends. Telomeres round off the ends of the chromosome and make sure that they are genetically stable. In theory, many cloning vectors can handle very large bits of DNA. In practice, YACs, MACs, etc. are the only convenient way of handling very long pieces of DNA, i.e. over 100000 bases long. MACs are being explored as ways of putting very large gene clusters into transgenic animals, and as vectors for gene therapy. BACs are used in cloning large stretches of bacterial DNA into other bacteria. Both are also used to study the control and function of genes, and of telomeres, centromeres, and origins of replication.

Macule. Small, round, colored area on the skin.

Maculopapule. Raised, small, round, colored area on the skin.

Magainins. Peptides of about 20 amino acids residues with antimicrobial activity, found in amphibian skin. Probably have membrane insertion and lytic properties. Sequence related to melittin.

Magnesium porphyrin. A substituted tetrapyrrole in which the four nitrogen atoms are coordinated to a magnesium atom.

Magnesium. An element essential for plant and animal growth. It is contained in the chlorophyll molecule and is thus essential for photosynthesis. In animals it is found in bones and teeth. As magnesium carbonate it is found in large quantities in the skeletons of certain marine organisms, and is found in smaller quantities in the muscles and nerves of higher animals. It is essential cofactor for certain phosphate enzymes, e.g. phosphohydrolase and phosphotransferase. High concentrations of magnesium ions, Mg^{2+}, are need to maintain ribosome structure.

Magnetic immunoassay. An analytical technique in which an antibody is immobilized on a negative microcarrier.

Magnetic microcarrier. A type of microbead which has magnetic properties and can be used in magnetic immunoassay.

Magnetic stirrer. An apparatus used for stirring solutions. A magnetic stir bar is placed at the bottom of the container and is twirled in the solution that requires mixing by an electrically driven, rotating magnet below the platform of the apparatus.

Magnetosome. Enveloped compartment in magnetotactic bacteria containing magnetite particles. By using this organelle to detect the vertical component of the Earth's magnetic field, the bacteria swim towards the bottom of aquatic habitats.

Main band. In density gradient separation of genomic DNA, a broad peak containing most of the DNA and excluding any visible satellite DNA bands.

Maintain. To sustain, provide for or support; as cultures are kept maintained on fresh medium in incubators. The process is maintenance.

Maintenance energy. Energy required by an organism or a cell culture in order to survive in a viable condition, but without further increase in size or cell number. This may be expressed in terms of the weight of substrate, or energy in joules, required per dry weight of cells or body mass.

Major element. An element which is essential for life and is required for the formation of the major structural parts of living organisms. They include carbon, nitrogen, oxygen, hydrogen, sulphur, phosphorus, calcium and, in instances, silicon.

Major genes. Genes whose differences or mutations are large enough to be identified by their individual effects in individuals, and which are therefore presumed to have a large effect on the selective advantage of the individuals. Cf. Polygenes.

Major histocompatibility complex. A complex of linked genes, with multiple alleles, that control a number of immunological phenomena; it is important in graft rejection.

Major histocompatibility complex. A complex of linked genes, with multiple alleles, that control a number of immunological phenomena; it is important in graft rejection.

Malachite green ($C_{22}H_{25}N_2Cl$ mw 364.66). A chemical used as a seed disinfectant in agriculture. It is included in some tissue culture media as an antiviral agent but has received mixed reviews.

Malaise. General discomfort, uneasiness.

Malaria. Disease caused by the single celled parasite *Plasmodium falciparum* Geographical areas of major incidence of the disease correspond closely with the distribution of sickle cell anaemia.

Malate dehydrogenase. An enzyme which catalyzes the reversible reduction of oxaloacetic acid to malic acid in a reaction linked to the oxidation-reduction of NADP.

Male gametophyte. In highest plants, the pollen grain plus the pollen tube, containing three haploid nuclei.

Male haploidy. A system of sex-determination in which the haploid individuals are males whilst the diploid condition determines femaleness.

Male sterile maize (genic). A plant producing no viable pollen because homozygous for a recessive pollen lethal usually designated *ms/ms.*

Male sterile or cytoplasmic male sterile. Plants incapable of viable pollen production.

Male sterility. A condition in which the pollen is not produced or is non-functional in flowering plants.

Male-limited chromosomes. Chromosomes which occur only in the nuclei of the germ-line, never in somatic nuclei of either sex, and never in nuclei of the female germ-line; androsomes.

Malic acid fermentation. A metabolic process that results in the degradation of malic acid to ethanol. There are two possible pathways based on malic enzyme or malate dehyrogenase. In the former, malic acid is decarboxylated to pyruvic acid, and then to acetaldehyde and ethanol. In the latter, the malic acid is oxidized to oxaloacetic acid, decarboxylated to pyruvate and then metabolized to ethanol as above. These pathways are characteristic of yeasts such as *Schizosaccharomyces prombe.*

Malic enzyme. An enzyme that catalyzes the decarboxylation of malic acid coupled to the reduction of NAD(P).

Malignant. Threatening to produce death.

Malnutrition. Poor health state due to an excess or lack of nutrients; may be primary, caused by poverty, inadequate food supply, ignorance, indifference; or secondary, caused by disease, drug abuse, or mental illness.

Malnutrition. Poor health state due to an excess or lack of nutrients; may be primary, caused by poverty, inadequate food supply, ignorance, indifference; or secondary, caused by disease, drug abuse, or mental illness.

Malo-lactic fermentation. A number of different metabolic pathways by which microorganisms convert malic acid directly to lactic acid plus carbon dioxide. It is of importance in the production of wine, where it is catalyzed by lactic acid bacteria.

Maloney leukemia virus. Mouse retrovirus that produces lymphocytic leukemia. Newborn mice can be infected with the virus via the milk from an infected female mouse.

Malpighian tubule. A type of protonephridium found in insects.

Malt extract. An extract from dried and ground, germinated grain seed (usually barely, sometimes rice or corn). It is an undefined constituent of some plant tissue culture media.

Malt. A mixture of starch hydrolysis products produced from germinating barley. Barley (*Hordeum distichum* or *H. hexastichum*) is sprouted under controlled conditions and then dried in a kiln.

Maltase. An enzyme that hydrolyzes maltose. the product of hydrolysis is glucose.

Malthusian parameter (of population increase). A measure of the relative rate of increase or decrease of a population when in the appropriate steady state (*Fisher*, 1930).

Maltose binding protein. Protein of the bacterial (*E. coli*) surface which links with *MCP-II* and is involved in the chemotactic response to maltose; probably derived from a similar protein which links with a trans-membrane transport system.

Maltose. A carbohydrate; a disaccharide produced by the enzymatic hydrolysis of starch by diastase.

Mammal (L. *mamma*. breast, test). Any animal of the class Mammalia characterized by the production of milk by the female mammary glands and the possession of hair for body covering.

Mammal (L. *mamma*: breast, test). Any animal of the class Mammalia characterized by the production of milk by the female mammary glands and the possession of hair for body covering.

Mammalian cell cloning. The gene manipulation of mammalian cells. One vector that has been widely investigated is the papovaviurs SV-40. This is suitable since it consists of a single small, covalently closed circular DNA molecule, the entire nucleotide sequence of which has been determined. The genomic regions responsible for the viral functions have been established, and information is available on the replication and expression of the viral genome. The virus can replicate either vegetatively (in the cytoplasm of the host) or as an integral part of the cellular chromosome. Improved SV40 vectors, which retain the regions implicated in initiation, termination, splicing and polyadenylation, have been constructed.

Mammary tumour virus. Retrovirus causing mammary tumours in mice and transmissible from mother to off-spring via milk.

Man made pest. A species which is a pest only because of human interference with natural control processes that normally regulate its density to non-pest levels. Most commonly generated through pest upsets, i.e. the unintended destruction of the natural enemies of a non-pest species with a chemical insecticide, or through monocultural practices.

Manifestation. The noticeable effects of a gene. Manifestation may change depending on different external influences or different gene environments, and may sometimes even be entirely lacking.

Mannan. A major constituent of the hemicellulose of soft (coniferous) woods. Its hydrolysis yields mannose, glucose and galactose in the ratio of 3.1.1.

Mannitol ($C_6H_{14}O_6$ mw 182.17). A sugar soluble alcohol (carbohydrate) widely distributed in plants and forming a characteristic food reserve of brown algae and often employed as a nutrient and osmoticum in plant tissue culture work; as in suspension medium for plant protoplasts.It is hexahydric alcohol i.e. each of the six carbon atoms has an alcohol (hydroxyl) group attached

Mannose ($C_6H_{12}O_6$ mw 180.16). A hexose component of many polysaccharides and mannitol. It is occasionally employed as a carbohydrate source in plant tissue culture media.

Mannose-6-phosphate. Mannose derivative which is of particular interest when formed by phosphorylation, in the Golgi complex, of certain mannose residues on N-glycan chains of lysosomal enzymes. Believed to function as targetting signal which causes entry of these enzymes to the lysosomes.

Mannosidase. Enzyme catalyzing hydrolysis of the glycosidic bond between mannose residues and a variety of the hydroxyl residues and a variety of the hydroxyl containing groups.

Mannuronic acid. A uronic acid structurally related to the hexose mannose. It is a constituent of alginate and other plant gums, as well as occurring in plant cell wall matrix polysaccharides.

Manometry. It is a gas exchange in small pieces of tissue or cell suspension.

Mantoux test. Test for tuberculin reactivity in which tuberculin PPD (purified protein derivative) is injected intracutaneously. The injection site is examined after 2-3 days. A positive reaction, indicating current or previous exposure to *Mycobacterium tuberculosis* (in an uninoculated individual), is a firm red nodule, a chronic inflam matory granuloma. The classical delayed-type hypersensitivity reaction.

Mantoux test. Tuberculin skin test. A dilution of purified material prepared from culture filtrates of tubercle bacilli is injected into the skin. An inflamed area at the site of injection that reaches a maximum in about 48 hours provides a positive test indicating that the person has been previously infected with the tubercle bacillus.

Map distances. Distance between two gene loci expressed as Morgan units or some other representation of a map unit.

Map units. A number proportional to the frequency of recombination between two genes. One map unit corresponds to a recombination frequency of 1 percent.

Map. *As a verb.* to determine the relative positions of genes or restrictions sites on a DNA molecule. Genetic mapping is done by carrying out mating experiments to determine the linkage relationship between genes and the frequency of recombination between them. The further apart two genes are on a chromosome, the greater will be the frequency of recombination between them; if they are on different chromosomes, i.e. they are on different chromosomes, i.e. they are unlinked, they will show the maximum recombination frequency of 50% Physical mapping is usually performed by carrying out a number of single and double digests with a range of restriction enzymes. *As a noun.* a diagram showing the relative positions of, and distances between, genes or restriction sites.

Maple syrup disease. Inborn error of metabolism affecting the degradation of valine, leucine and isoleucine by oxidative decarboxylation. Since degradation is blocked, blood and urine have markedly higher levels of these three amino acids and a corresponding increase in the ß-keto acids derived from them. The deficiency gives the urine of affected patients an odour of maple syrup. Unless affected individuals are placed on a diet low in these amino acids early in life the disease normally proves fatal.

Mapping. A process used in the production of map. These processes may include breeding experiments to determining the order of genes on a chromosome and the distances between them and to produce linkage maps, the use of endonucleases to produce restriction maps, DNA sequencing or observations on dividing cells made under the light microscope.

Marasmus. An extreme form of chromic undernutrition caused by a lack of calories and protein. Seen mostly in children, it results in loss of subcutaneous fat, inelastic wrinkled skin, loss of muscle tissue and strength. It may be due to chronic malnutrition or defective absorption or utilization of an adequate food supply.

Marburg virus. A Filovirus that causes Marburg disease, a severe haemorragic fever developed in many people who work with African green monkeys.

Marfan's syndrome. Disorder of connective tissue in the human, inherited as an autosomal dominant. The disease is characterized by extremely long fingers and toes, and has a prevalence of approximately one in 20000.

Marginal meristem. A meristem along the leaf primordial margin that forms the mesophyll and epidermal tissues of the leaf blade.

Marine (L. *mare*: sea, ocean). Pertaining to or living in the ocean. (Contrast with terrestrial.).

Maritime climate. Weather pattern typical of coasts of continents, particularly those on the western sides at mid latitudes, in which the difference between summer and winter is relatively small. (Contrast with continental climate).

Marker rescue. Phenomenon in which a population of bacteria is infected with two distinct `marked phages (i.e. two types that can be genetically distinguished), but following the irradiation of one of the two phage strains. Because of recombination, some of the surviving phage possess properties of the irradiated strain.

Marker. One of the most flexible words in microbiology, this can mean many things in different contexts.

Assay markers. The system that allows you to seed the result of a chemical assay or test. Also called labels; examples include enzymes, radioactive atoms, and luminescent groups.

Genetic markers. Landmarks in a genetic map.

Biological markers. These are chemical signposts that something is happening in a cell or tissue. Often they can be produced by reporter genes, genes we introduce into a cell to show when something has happened. Markers can also be chemicals produced by the cell, even things such as cell shape. `A marker for' an event is usually shorthand for `something that is convenient to measure to show that event has happened'.

Disease markers. These can mean genetic markers in the sense above for genetic diseases. Usually this means a chemical `signpost' that someone has a disease. Thus the presence of some heart-specific enzymes in the blood (such as CK-MB) is a marker for heart attack. it shows that the patient has recently had a heart attack. Biotechnology contributes to these by finding them, and by developing tests to measure them.

Marocytes. Abnormally large red blood cells, numerous in pericious anaemia.

Marsupial. A mammal belonging to the subclass metatheria, such as opossums and kangaroos. Most have a pouch (marsupium) that contains the milk glands and serves as a receptacle for the young.

Masked mRNA. Messenger RNA present in a cell, but not being actively translated into protein. Masked mRNA is associated with particular types of proteins as messenger ribonucleoprotein (mRNP) particles and can be activated to function in protein synthesis by specific physiological triggers (e.g. hormones).

Mass meristem. A meristematic tissue which increases in volume through cell divisions in various planes.

Mass selection. Artificial selection practiced by choosing in each generation, individuals with maximum (minimum) expression of a given trait as parents of the following generations.

Mass spectrometry. Mass spectrometry (MS) has been adopted by a wide range of biological researchers in the last five years, as the instruments and methods have made it more routine. A mass spectrometer measures the exact molecular mass of a molecule by measuring its flight path through a set of magnetic and electric fields. More exactly, it measures the mass/charge ratio. To allow the molecules to fly uninterrupted, the operating chamber of the mass spectrometer contains a vacuum. In vacuum nearly all biological molecules have no charge, but the method relies of them being charged, so that the magnetic and electric fields can affect them. So, mass spectrometers include a method of bombarding the target molecules with radiation to charge them up. This can fragment the molecule. This is a benefit because then you can measure the mass/charge ratio of bits of the molecule as well as the whole thing in one experiment. Piecing together these data allows the full structure of the original molecule to be deduced.

Mast cell. A cell found in the ground substance (matrix) of areolar (loose) connective tissue, which it is thought to secrete. Mast cells are also found in the lining of blood vessels, and some occur in the blood, as basophils. They have lobed nuclei and cytoplasmic granules that are sustained by basic dyes. Most cells secrete heparin (an anticoagulant) and histamine.

Master gene. Gene that controls other genes, particularly when the genes controlled are all expressed characteristically in a particular type of differentiated cell.

Master-slave hypothesis. Proposal made by H.G. Gallan that most eukaryotic genes existed in multiple identical copies (the slaves) and that the sequences were all kept identical by frequent correction, using a `single master' copy of the gene as a template for correction of the `slave' copies. Apart from possible correction mechanisms in ribosomal gene copies, the hypothesis has not been substantiated by the evidence.

Maternal antibody. Any antibody transferred from a mammalian mother transplacontally into the foetus.

Maternal chromosomes. In bisexual organisms the chromosomes derived from the mother.

Maternal effects. Effects of the phenotype or genotype of the mother on the phenotype of immediate off-spring, generally of an impermanent nature. Thus maternally derived protein or messenger RNA stored in the egg leads to maternal effects in some early embryos. Compare Maternal Inheritance.

Maternal influence (Ephestia). The influence of the genotype of a hetrozygous mother on the eggs and larvae. The dominant colour of the mother persists for a short time in eggs and larvae, even with recessive genotype.

Maternal inheritance (cytoplasmic inheritance). Inheritance in which the phenotype of the offspring depends on factors, such as mitochondria or chloroplasts, that are inherited from the female parent through the cytoplasm of the female parent through the cytoplasm of the female gamete.

Maternal mortality rate. The number of women dying from causes directly related to childbearing per 10,000 live births during a calendar year.

Maternal mRNA. Messenger RNA found in oocytes and early embryos which is derived from the maternal genome during oogenesis.

Maternal. From the mother. Some characters, for various reasons, tend to follow a matrilineal inheritance; these are known as maternal characters.

Mating Group. A group of individuals, haploid or diploid, within which mating is favoured at the expense of mating outside the group, by genetic or environmental conditions characteristic of the group.

Mating system. The pattern of mating in sexually reproducing organism; two types of mating systems are Random Mating and Assortative Mating.

Mating type gene. Sequence that controls the choice between two mating types known as Â and a; best known in the yeast *Saccharomyces.* Haploid yeast cells occasionally switch their mating type. At a genetic level such a switch has been shown to involve excision of a particular mating type gene sequence from the genome, with its replacement by a gene of the alternative mating type.

Mating type. In some bacteria, fungi, and protists, sexual reproduction can occur only between partners of different mating type. *Mating type* is not the same as *sex*, since some species have as many as 8 mating types; also, as in *Paramecium aurelia*, mating may be between hermaphroditic partners of opposite mating type, with both partners acting as body "male" and "female" in terms of donating and receiving genetic information.

Mating. The assortment of individuals, usually zygotes, in couples leading to sexual reproduction; by extension includes self-mating. *Mating continuum* is an aggregate of individuals whose genes systematically recombine. *Illegitimate mating* is a kind of mating normally precluded by the breeding system of the organism (e.g. self-pollination in incompatible or heterostyled plants, or the crossing of species).

Matric potential. A water potential component, always of negative value, resulting from capillary, imbibitional and adsorptive forces.

Matrix. A substance that apparently accumulates around the chromosomes during the last stage of the prophase of meiosis. The matrix causes the chromosomes to appear thicker and tends to conceal their spiral structure.

Matroclinal. A case where the offspring resembles the mother more than the father.

Matroclinous. With reference to hybrid or to an offspring that is similar to the mother. The hybrid or offspring is said to be patroclinous, if similar to the father. Matrocliny is usually caused by cytoplasmic factors.

Maturase. Protein that catalyses the removal of introns. It is likely that such proteins are not enzymes, but act as factors that modify the activity of the splicing enzymes.

Maturation (L. *maturus*, ripe). Here indicates the process during which the individual components assemble and form into mature particles; an essential feature of virus multiplication. It is the automatic development of a pattern of behavior, which becomes increasingly complex or precise as the animal matures. Unlike learning, the development does not require experience to occur.

Maturation (L. *maturus*, ripe). Here indicates the process during which the individual components assemble and form into mature particles; an essential feature of virus multiplication. It is the automatic development of a pattern of behavior, which becomes increasingly complex or precise as the animal matures. Unlike learning, the development does not require experience to occur.

Maxam-Gilbert chemical sequencing. A technique for sequencing nucleic acids based on the differential ability of the four organic bases to react with various reagent such as a piperidine and hydrazine.

Maxicells. Maxicells are bacterial cells that have a mutation in the genes that control how they divide. Under the `right' conditions (usually when the temperature of the medium is raised) they simply stop dividing. However, they do not stop growing, so the result is a huge bacterial cell. This can be helpful, in that such large cells are much easier to separate from their medium than the relatively small normal cells. they will, for example, settle out of solution under their own weight in a fairly short time. A related idea is the minicell. This is another cell division mutant. Here, under the `right' conditions the cells divide, but they do not divide in the middle. Rather, a small `minicell' splits off from one end. All the bacterial DNA stays in the main cell, so the splits off from one end. All the bacterial DNA stays in the main cell, so the minicell has no DNA of its own. Therefore it cannot make any new RNA, and as soon as the RNA that happened to be in the cell has broken down it cannot make any new protein either. However, this rule is broken if the cell contains some types of plasmids. a few plasmid molecules can end up in the minicell. Thus, when all the trapped RNA has broken down, the only proteins being made by the minicell are those coded by genes on the plasmid. This is very useful for studies of gene expression, since, by isolating minicells, the proteins being made by the plasmid can be studies without having the purify them from all the other proteins being made by a normal bacterial cell.

Maximum enzyme velocity (V_{max}). It is the theoretical maximum rate at which a growing enzyme can catalyze a specific reaction when supplied with substrate at an infinite concentration. V_{max} is determined experimentally using, for example, Lineweaver-Burk kinetics.

Maximum oxygen transfer rate. It is a measure of the fastest possible rate of uptake of oxygen by a growing culture expressed in terms of the volumetric mass transfer coefficient times the oxygen solubility.

Maximum permissible dose. Maximum level of ionizing radiation that present safety standard permit over a unit period.

Maximum specific growth rate (μ max). The theoretical maximum rate of growth of a culture under optimal conditions of nutrient supply. μ max is determined experimentally from a double reciprocal derivation of the Monod equation analogous to the Lineweaver - Bruk plot for enzyme kinetics.

MDA. Microdroplet array technique.

Mean (arithmetic). The arithmetic average of a series of observations, or quantities.

Weighted Mean, a mean obtained when different classes of observations or quantities are given different weights in the calculation.

Working mean, a value approximating to the mean used for the purpose of lightening the calculations of the mean, the sum of squares and other quantities.

Mean corpuscular hemoglobin concentration (MCHC). An expression absolute terms, of the average hemoglobin concentration per unit volume (per 100 ml) of packed erythrocytes, calculated from the equation.

Mean corpuscular volume (MCV). An expression, in absolute terms, of the average volume of the individual erythrocyte, calculated from the equation.

Mean detention time. The average time that each unit volume of mixed liquor is retained in a bioreactor, fermenter or other vessel. The hydraulic mean detention time refers to the liquid portion of the liquor and the solid means retention time refers to the solid portion.

Mechanist. One who regards the phenomena of nature as the effects of merely mechanical forces.

Median. (1) The middle value of a series of readings arranged serially according to magnitude. (2) The value of a variate on each side of which lie equal numbers of observations.

Medium formulation. One of many available, usually derived formulas that are used in plant tissue culture. The formulas commonly contain the macroelements and microelements (high and low salt formulas are available), some vitamins (B vitamins, inositol), hormones (auxin, cytokinin and sometimes gibberellin), a carbohydrate source (usually sucrose or glucose) and often other substances such as amino acids (the most common addition is glycine) or complex growth factors. Media may be liquid or solidified with agar, the pH is adjusted (ca. 5-6) and the solution is sterilized (usually by filtration or autoclaving). Some formulations are very specific in the kind of explant or plant species that can be maintained, some are very general.

Medium. The substrate for plant growth; as nutrient solution, soil, sand, etc. This is a general term for the liquid or solidified formulation upon which plant cell, tissues or organs develop in plant tissue culture.

Medulla. The inner part of an organ.

Medusa. The tentacle-bearing, jellyfish-like, free-swimming sexual stage in the life cycle of a cnidarian.

Mega- (Gr. *megas.* large, great). A prefix often used to denote something large. (Contrast with micro-.)

Megaloblastic anemia. Type of anemia caused by a deficiency of folic acid; characterized by an increased level of megaloblasts, the primitive nucleated red blood cells which are much larger mature erythrocytes.

Megasporangium. The special structure (sporangium) that produces the megaspores.

Megaspore (Gr. *megas.* large + *spora.* seed). In plants, a haploid spore that produces a female gametophyte. In many cases the megaspore is larger than the male-producing microspore.

Megaspore mother cell (megasporocyte). A diploid cell ($2n$) that undergoes two divisions, one a reductional division, to produce four megaspores.

Megaspore. The larger of the two kinds of haploid spores produced by vascular plants; the smaller kind is called a microspore. In seed plants, the megaspore develops into the embryo sac (the female gametophyte), while the microspore gives rise to the pollen grain (the male gametophyte).

Megasporogenesis. The process of the production of the megaspores from a megasporocyte.

Meiocyte. Any cell that undergoes meiosis.

Meiosis (Gr. *diminution*). In eukaryotic cell division of a diploid nucleus to produce four haploid daughter cells. The process consists of two successive nuclear divisions with only one cycle of chromosome replication. It is a reduction-division process nuclear division, daughter nuclei have half the chromosome number of the parent nucleus; occurs in the formation of sex cells.

Meiospore. In plants, one of the asexual reproductive cells produced by meiosis from a meiocyte.

Meiotic analysis. Used to analyze chromosome-pairing relationships.

Meiotic drive. Any alternation of the process of meiosis with the consequence that a parent heterozygous for two genetic alternatives produces an effective gametic pool with an excess of one type.

Melanic. Describing an individual that, because of increased melanin production, is much darker than normal Melanism is common in birds, mammals and occasionally occurs in insects and other invertebrates.

Melanin. A dark brown to yellow pigment found in cells of the skin, hair, and iris of the eye; also in many plant species.

Melanism. Genetically controlled darkening of colour, *cf* Industrial Melanism.

Melanocyte. Pigment cell containing Melanin granules.

Melanocyte-stimulating hormone (intermedin). A polypeptide hormone produced by the anterior pituitary gland in man. Its role is uncertain although its blood level is often disturbed in patients with endocrine disorders, accompanied by pigmentation abnormalities. It has a marked action on pigmentation in the skin of amphibians. Compare Melatonin.

Melanoma. Tumour caused by excessive growth of Melanocytes. Non-invasive tumours of melanocytes include moles, whereas malignant melanomas are noted for their efficient metastatic spread.

Melanosphore. A cell containing high levels of melanin.

Melatonin. A hormone, produced by the pineal gland that has no known action in man but produces marked lightening of the skin in embryonic fish and larval amphibians.

Melting curve. A graphical plot of the change in absorbance at 260 nm, as a function of temperature that occurs when a solution of DNA is heated. Heating results in denaturation due to breaking of the hydrogen bonds between the bases and dissociation of the two strands of the double helix. The absorbance of double-stranded DNA is always lower than that of a single-stranded DNA which is lower than that of an equivalent amount of free nucleotides in solution.

Melting out temperature. Temperature at which nuclei acid duplexes of DNA-DNA or DNA-RNA melt by breakage of the hydrogen bonds between the complementary molecules.

Melting profile. Profile, commonly represented on a graph, showing the percentage of a given sample of DNA that melts in a specific time on a gradually increasing temperature scale. Since some DNA, such as imprecisely base-paired sequences, low-molecular-weight material, and molecules rich in AT bases, will melt out at lower temperatures than other DNA, some useful information can be derived from melting profiles.

Melting. The separation of the two strands of duplex DNA to form single strands by disruption of hydrogen bonds between the duplex strands.

Membrane filter bioreactor. A bioreactor in which the fermenting broth is circulated continuously through a membrane filter. The product diffuses from the reaction zone through the membrane and is withdrawn continuously. This method is also applied to enzyme and resting cell reactors. The hollow fibre fermentor is one variation of this type of bioreactor.

Membrane filter. A filter medium made from various polymeric materials, such as cellulose, polyethylene, or tetrafluoroethylene. Membrane filters usually exhibit very narrow ranges of effective pore diameters, making the filters very useful in the collection and sizing of microscopic and submicroscopic particles and in the sterilization of liquids.

Membrane fluidity. Biological membranes are viscous two-dimensional fluids within their physiological temperature range.

Membrane potential. More correctly, transmembrane potential difference; the electrical potential difference across a plasma membrane.

Membrane transport. Transport across biological membranes. Since the lipid bilayer of the membrane is inherently impermeable to most polar molecules, cells have evolved special ways of transferring such molecules across their membranes. These include the use of specific membrane transport proteins, some acting as channel proteins and others as carrier proteins, the latter being involved in facilitated diffusion. Sodium pumps are also an important aspect of membrane transport.

Membrane. In most plants and animals, a bilayer of lipids (e.g., phospholipids, glycolipids and sterols) in which globular proteins are embedded.

Memory cell. A lymphocyte of either B or T cell lineage that was formed during a primary immune response and that is capable of synthesizing specific immune products at an increased rate during an anamnestic response.

Mendelian Character. A character the inheritance of which follows Mendel's Laws i.e., one showing allelomorphic inheritance.

Mendelian inheritance. The inheritance of genetic characteristics according to the laws proposed by Gregor Mendel (1822-84).

Mendelian population. A local population of individuals belonging to the same species, in which mating is random.

Mendelism. A field of genetic research that is especially concerned with the analysis of gene effects and segregation ratio.

Mendel's law. The principle that hereditary characters are determined by discrete particles (genes) that segregate at random in gamete formation.

Meninges. Protective coverings of the brain and spinal cord.

Meningitis. Inflammation of the meninges (membranes that surround the brain and spinal cord).

Meningococcus. *Neisseria meningitides;* Gram negative non-motile pyogenic coccus which is responsible for epidermic bacterial meningitis.

Meniscus. The curved surface at the top of a liquid column. The bottom of the meniscus is used as a guide when liquid volumes are measured in calibrated containers.

Menopause. The period during which the "change of life" occurs in woman. During this period the ovaries atrophy and oestrus cycle is terminated.

Mercaptoethanol. (ethanethiol, CH_3CH_2SH). Strong reducing agent that is often used for the reduction of disulphide linkages in proteins. The reaction converts a cystine which bridges two parts of the polypeptide chain into two chains cysteine residues having free thiol groups.

Mercuric chloride or mercury bichloride ($HgCl_2$ mw 271.52). This compound was once commonly used as an antiseptic and fixing agent (0.05-0.10%) for plant material. However, it is highly toxic and is now little used for this purpose.

Mericlinal chimera. An incomplete periclinal chimera.

Mericlone. A meristem tip source clone. This term can be synonymous with calliclone or may not involve callus.

Mericloning (Meristemming). A population term, not in scientific usage, referring to the *in vitro* vegetative propagation of orchids from excised shoot tips, axillary buds, or floral organs.

Meristem (Gr. *Meristos*: divided). Plant tissue made up of actively dividing cells.

Merocytes. The nuclei derived from the supernumerary sperm-nuclei which have failed to conjugate with the egg nucleus in cases of physiological polyspermy.

Merodiploidy. Temporary partial diploidy in bacteria due to the presence of transferred genes.

Merogamy. A condition in which the gametes are smaller than the negative cells, often of different structure, arising by division from the gametocytes.

Meromyosin heavy. Portion of the myosin molecule with ATPase activity and Ca-binding properties produced by trypsin digestion of myosin.

Merostathmokinesis. Incomplete inhibition of the spindle (usually as a result of treatment with colchicine or some substance with similar effects). This may occur either through the failure of the cytoplasm to behave normally, or through the failure of the nucleus, this latter effect being termed hemikinesis.

Merozoite. A plasmodium trophozoite resulting from multiple fission (for example, malaria).

Merozygote. A partially diploid bacterium (homogenote or heterogenote) containing a complete recipient genome and a partial genome (merogenote) derived from a donor.

Meselson and Radding model. Modification to the Holliday Model of chromosome recombination. A twisting and isomerization of strands in the cross-over intermediate is proposed, avoiding the requirements in the original Holliday model for nicking in both strands.

Meselson and Stahl experiment. Experiment in molecular genetics first published in 1858. It proves the semi conservative replication of DNA. The protocol used involved labelling *E. coli* DNA with ^{15}N, then abruptly transferring the bacteria to a medium containing only ^{14}N. The distribution of DNA that had incorporated the isotopes was followed by caesium chloride density centrifugation, and the appearance of a band partially labelled and therefore of intermediate density constituted the evidence of a semiconservative pattern of DNA replication.

Meso- (Gr. *middle*). A prefix often used to designate a structure located in the middle, or a stage that appears at some intermediate time. For example, mesoderm, mesozoic.

Mesocotyl. The internode between the coleoptile and the scutellum in grass embryos and seedlings.

Mesoderm (Gr. *mesos.* middle + *derma.* skin). The middle of the three embryonic tissue layers first delineated during gastrulation. Gives rise to skeleton, circulatory system, muscles, excretory system, and most of the reproductive system.

Mesophile. A microorganism that grows best at temperature between 20ø and 45° C.

Mesophilic. Designating microorganisms with an optimum temperature for growth between 25-45° C. Compare Psychrophilic, Thermophilic.

Mesosome (Gr. *mesos.* middle + *soma.* body). A localized infolding of the plasma membrane of a bacterium. It is relatively large, irregular invaginations of the plasma membrane.

Messenger RNA (mRNA). A kind of ribonucleic acid formed by the copying of certain segments of DNA complementary to it. This type of RNA carries the instruction or the codes, in the form of bases, necessary for the synthesis of a specific protein. It acts as an assembly line or a template in which the amino acids transported by transfer RNAS are arranged in an orderly fashion for the formation of a particular protein.

Messenger splicing. The `editing' of the primary transcription of a gene from a higher organism to excise the no coding intervening sequences. This leaves a continuous coding sequence specifying the protein in question.

Meta- (Gr. between, along with, beyond). A prefix used in biology to denote a change or a shift to a new form or level; for example, as used in metamorphosis.

Metabolic block. A block in a metabolic pathway that prevents the completion of a product.

Metabolic cooperation. Transfer between tissue cells in contact of low-molecular weight metabolites such as nucleotides and amino acids. Transfer is via channels constituted by the *connexons* of gap junctions, and does not involve exchange with the extracellular medium. First observed in cultures of animal cells in which radiolabelled purines were transferred from wild-type cells to mutants unable to utilize exogenous purines.

Metabolic differentiation, Theory of. The theory that sex determination is conditioned by the degree of metabolism of the developing individual.

Metabolic engineering. The manipulation of cells to make them produce more of a product is sometimes called metabolic engineering. This focuses on altering the enzymes present in a cell, rather that just altering their growth conditions. Typically, a cell will be genetically engineered to produce more of a specific enzyme, or to replace one enzyme by another with more desirable properties. The difficulty in metabolic engineering is that just replacing one enzyme by another usually does not have much effect, because other control systems in the cell stop the new enzyme from working effectively. It used to be thought that the amount of chemical being processed by a specific pathway (the `metabolic flux') was determined by flow through one key `rate-determining' step in the pathway. If you doubled the amount of the enzyme that catalysed this step (the rate-determining enzyme), then you would double the amount of stuff made by that pathway. Many experiments have shown that this is not so. All the enzymes of a pathway contribute to controlling how fast that pathway works to a greater or lesser degree, and if you double the amount of one enzyme it has little effect overall. Thus, metabolic engineering is about predicting changes in the whole of metabolism, and altering a number of enzymes and physiological conditions to bring about those changes. Applications are generally in fermentation, where metabolic engineering seeks to improve the yield of primary or secondary metabolites. However, it can also be applied to plants that produce valuable chemicals, and even to human cells where the same analysis can be used to discover what goes wrong with metabolism in diseases such as diabetes and obesity. Transgenic plants have been suggested as practical ways of making a variety of biomolecules such as modified starch and oils through manipulating their metabolism. The degree to which a single enzyme affects the flux through a pathway is called its control coefficient, and is given as a number between 0 (has no effect at all) and 1 (is the only control on that pathway).

Metabolic heat. Energy lost as heat during the growth of an organism. In aerobic biocatalysis and fermentation processes, knowledge of the amount of heat produced is important in bioreactor design since provisions have to be made to remove the heat. Heat production may be calculated from an enthalpy balance for a growing culture. However, for aerobic systems the amount of heat released can be derived on the basis of oxygen consumption. In general, 460 kJ of heat are released per mole of oxygen consumed. In fermentation, to produce ethanol, around 70 kJ are produced per mol of glucose converted to biomass or to ethanol.

Metabolic mutant. Mutant having a lesion in a gene coding for an enzyme involved in metabolism. Non-emtabolic mutants mist (for example) have alterations in parts of their DNA not expressed as enzyme proteins.

Metabolic nucleus. A `resting' nucleus, *i.e.* one which is not dividing.

Metabolic pathway. The chemistry that turns an organism's food into more organism is collectively known as `metabolism'. The whole of metabolism is a very complex network, but specific series of reactions can be found in which chemicals (`metabolites') are converted one into another to get from a starting material to an end material in a series of relatively small steps. Such series of reactions are called metabolic pathways. Metabolism is usually divided into anabolic pathways (pathways that build up molecules for use by the organism (e.g. those pathways that make amino acids)) and catabolic pathways (ones that break down molecules, either for energy or simply to get rid of undesired materials (e.g. breaking down hydrocarbons to obtain energy)). Some pathways, especially those at the centre of metabolism (e.g. those that break down glucose) perform both functions, and are called amphibolic. Pathways can also be divided into `primary' and `secondary'. Primary metabolism is making the basics of life that are common to nearly all living things. Secondary metabolism is making specialist chemicals such as hormones or toxins. Mutant organisms that lack enzymes of the anabolic pathways of primary metabolism are called auxotrophs. they require that the end-product of that pathway be provided as food since they cannot make it themselves. Mutants of the pathways that make secondary metabolites are called idiotrophs. Many anabolic pathways are end-product inhibited. This means that the end-product of a pathway is an inhibitor of the enzymes in the pathway, usually the first enzyme that is specific to that pathway. So, if a lot of the end-product builds up, the pathway automatically shuts down the synthesis of more of that product. This can be very irritating for the biotechnologist, who therefore has to remove the end-product of the pathway very fast, so that it does not build up, or find some way of knocking out the inhibition.

Metabolic rate. A measure of the rate of chemical reactions in an organism. For instance, in an aerobic organism it may be expressed as the rate of oxygen consumption.

Metabolic stage. Resting Stage.

Metabolism (Gr. *metabole.* to change). The sum total of the chemical reactions that occur in an organism, or some subset of that total (as in *respiratory metabolism*). It is the sum of all chemical changes in the body; the two phases are anabolism and catabolism.

Metabolite. A product of metabolism. Addition of certain metabolites may hasten or retard (repress) biological reactions.

Metacentric. With the centromere approximately in the middle of the chromosome.

Metachromatic granules. Cytoplasmic inclusions of concentrated cellular material.

Metachromatic stains. Stains that colours in cells or tissues which are different from that of the stain solution. For example, aqueous methyl violet, differentiated with acetic acid, stains amyloid pink/red.

Metafemale. An individual that has a greater dose of female determiners than to normal females; also called Superremale.

Metagenesis. Alternation of generation; an alternation of a sexual with an asexual form, or, cytologically, the alternation of a haploid with a diploid stage.

Metalloenzyme. Enzyme consisting of a protein complexed with one or more metal atoms. Carboxypeptidase A, for example, contains an atom of zinc.

Metalloprotease. An enzyme, containing an atom of zinc as·a prosthetic group, that hydrolyzes proteins. They are divided into neutral and alkaline enzymes. Neutral proteases of microbial origin are specific towards hydrophobic amino residues on the amino side of the cleavage point. Metalloproteases are inhibited by metal-chelating agents.

Metallothionein genes. Small gene family of some 11 separate sequences (in the mammal, including human); the number of copies in the family is less in lower eukaryotes. The genes code for the metal-binding protein metallothionein and are inducible by heavy metals such as cadmium and zinc, and corticosteroid hormones.

Metamale. Abnormal males in Drosophila with an overabundance of autosomes to *X* chromosomes; *X/A* ratio less than 0.5.

Metamorphosis (Gr. *meta.* between + *morphe.* form, shape). Aradical change occurring between one developmental stage and another, as for example from a tadpole to a frog or an insect larva to the adult.

Metanephridia. Nephridia (excretory organs) with open inner ends.

Metaphase (Gr. *Meta*: between). The stage in nuclear division at which the centromeres of the highly supercoiled chromosomes are all lying on a plane (the metaphase plane or plate) perpendicular to a line connecting the division poles.

Metaphase pairing index. The proportion of metaphase cells of which two particular chromosomes have paired, to the total number of cells examined.

Metaphase plate (Equational plate). The group of chromosomes arranged in the equatorial plane of the spindle in the mitotic metaphase and in first metaphase of meiosis.

Metaplasia. The production by the cells of one tissue, of a different tissue.

Metaplasm. A general term for non-living inclusions in the protoplasm.

Metapneustic. An insect respiratory (particularly in some dipterous larvae) in which only the last abdominal pair of spiracles is functional. cf. Amphipneustic.

Metastasis. The spread of cancer cells from their original site to other parts of the body.

Metasyndesis. Parasynapsis; the side-by-side conjugation of chromosome at zygote.

Metatonin. A hormone produced by the pineal gland of vertebrates that regulates the dispersal of melanin in the skin.

Metaxenia. Immediate effects of the pollen upon the development of the mother tissue of the fruit.

Metazoa. Animals whose bodies consist of many cells.

Methaemoglobin. Oxidized haemoglobin, containing iron (lll), produced by oxidizing poisons (e.g., nitrite, chlorate). If a cell's reductive capacity is exceeded, methaemoglobin builds up and dissociates into its heam and globin components.

Methane digester. A synonym for anaerobic digester.

Methanobacterium. A genus of strictly anaerobic bacteria that reduce CO_2 using molecular hydrogen, H_2, to give methane. They show a number of features which distinguish them from bacteria, and are now classified as a separate group within the *Archaebacteria.* Methanobacteria are found in the anaerobic sediment at the bottom of ponds and marshes (hence marsh) gas is the common name for methane) and as part of the microflora of the rumen in cattle and other herbivorous mammals.

Methanogen. A strictly anaerobic bacterium that is capable of producing methane. Methanogens include the genera *Methanobacterium, Methanobacillus, Methanococcus* and *Methanosarcina. Methane bacteria* are found in anaerobic soils (marshes and bogs), sewage sludge, landfill sites and anaerobic digestion systems. They also play an important role in the metabolism of ruminants.

Methanogenesis. The final step in the production of methane during which acetate, hydrogen and bicarbonate are converted to methane due to the activity of methanogens.

Methanol dehydrogenase. Enzyme responsible for the oxidation of methanol to formaldehyde in methylotrophic bacteria.

Methionine (Met, $C_5H_{11}NO_2S$ mw 149.21). An amino acid precursor in ethylene synthesis and occasionally added to plant tissue culture media.

Methotrexate. Resistance to the drug methotrexate (amethopterin) is a popular selectable marker in cloning experiments with mammalian cells. The drug is an inhibitor of the enzyme dihydrofolate reductase (DHFR) and one route to resistance is the over-production of this enzyme. Thus the inclusion of a DHFR gene in a cloning vector permits the selection of methotrexate transformants.

Methyl blue. Dye, normally dissolved in ethanol, that is used in histology for staining cell nuclei.

Methyl green. Basic dye used in histochemistry to stain DNA. It is often used in conjunction with pyronin.

Methyl methanesulfonate (MMS, $C_2H_6SO_3$ mw 110.13). A frequently used, very potent chemical mutagen which acts by adding methyl groups to guanine and subsequently causes base pairing errors as it binds to adenine.

Methyl violet. Histological stain, soluble in water or ethanol, the is used to stain cell nuclei.

Methylase. A DNA modification enzyme which covalently attaches a methyl (CH_{-3}) group to specific nucleotides base within a DNA molecule.

Methylated cap. Methylated guanine nucleotide that is added to the 5'-end of eukaryotic hnRNA and mRNA after transcription but prior to later processing and translation.

Methylation. Attachment of a methyl group to some other molecule, such as a histone RNA or DNA.

Methylotroph. Microorganism able to grow at the expense of reduced carbon compounds containing one or more carbon atoms, but not carbon-carbon bonds.

Methyltransferase. Enzyme which transfer a methyl group from S-adenosyl methionine to a substrate. Encountered in post-translation modification of proteins and nucleic acids, in the removal of methyl groups added to DNA by alkylating carcinogens, and in bacterial chemotaxis where the methyl-accepting chemotaxis protein (MCPs) become methylated in the course of *adaptation.*

Metric character. A trait that varies more or less continuously among individuals, which are therefore placed into classes according to measured values of the trait; also called Quantitative Character.

Metric system. Standardized decimal measurement based upon the meter as a unit of length, the gram as a unit of weight, and the liter as unit of liquid measure.

MHC (the major histocompatibility complex). A genetic region found in all mammals whose products are primarily responsible for the rapid rejection of grafts between individuals, and which function in signalling between lymphocytes and cells expressing antigen.

MHC restriction. Restriction on interaction between cells of the immune system because of the requirement to recognize foreign antigen in association with MHC antigens (major *histocompatibility* antigens). Thus, cytotoxic T-cells will only kill virally-infected cells which have the same class I antigens as themselves, whereas helper T-cells respond to foreign antigen associated with Class II antigens.

Micelle. A colloidal particle formed by the reversible aggregation of dissolved molecules. Charged micelles form colloidal electrolytes such as soaps and detergents.

Michaelis constant (K_m). Concentration of substrate that permits an enzyme reaction rate to be half-maximal, or at which half of the substrate molecules are engaged with enzymes molecules.

Michaelis-menten constant. A constant that specifies the quantitative relationship between substrate concentration and maximum velocity for different enzymes.

Michaelis-Menten kinetics. A generalized description of enzyme kinetics based on a model that assumes the formation of an intermediate enzyme-substrate complex, which has a greater tendency to dissociate.

Micro- (Gr. *micros.* small). A prefix often used to denote something small. (Contrast with macro-, mega-).

Microaerophile. An organism that grows in air but at a concentration of oxygen less than our atmosphere. They grow in reduced oxygen tension.

Microbial film. A thin layer of microorganisms absorbed onto a supporting surface.

Microbial floc. Particles formed by the aggregation of bacterial, yeasts or microalgal cells.

Microbial insecticide. A pathogenic microorganism or its products (e.g. toxins) when used by man to suppress an insect population. use of the word *insecticide* is properly reserved only for chemical agents of insect mortality, and Microbial Pathogen is a preferable term for materials whose active agent is a microorganism.

Microbial leaching. A method of recovering such metals as copper and uranium from ores of low content. This is carried out using bacteria of the genus *Thiobacillus,* such as *T. thiooxidans.* These organisms grow chemoautotrophically on energy released by oxidation of sulphur compounds, generating sulphuric acid and copper ions.

Microbial metal accumulation. All microorganisms accumulate metals from their environment since many are essential components of enzymes, pigments or intermediates of electron transport. Such process may be applied to the removal of toxic heavy metals from dilute waste streams or used as the basis for the recovery of precious metals from dilute solutions.

Microbial mining. This is the use of microorganisms to remove minerals, and particularly metals, from rocks. It is a specific application of biohydrometallurgy. Microbial mining is related to the use of microbes in desulfurization, and for bioremediation. Microbial mining falls into two areas.

Leaching. This is using bacteria to process ores to make the metals in them more accessible. Usually it involves using bacteria to release the metals as soluble salts, to be washed out for subsequent recovery. However, it can also involve preprocessing of the ores which, while not releasing metals directly, allows them to be separated more easily by washing out, flotation, or other `traditional' methods in a further processing step.

Purification. Using microorganisms or microorganism components to separate and concentrate metals from very dilute solutions. This is also called biosorption.

Biohydrometallurgy is used commercially to recover copper and uranium from low-grade ores, especially chalcopyrite ($CuFeS_2$), covellite (CuS), chalcocite (Cu2S), and uraninite (UO_2). A number of other metals (antimony, arsenic, molybdenum, zinc, cadmium, cobalt, nickel, and gold) can be extracted using bacteria, but these are not used to a significant extent. Plants can also be used to extract metals, although this is not used as an extraction technology commercially to any significant extent. Phytoextraction is the concentration of metals from the soil into the plant, usually the leaves. Some soils can produce quite toxic crops because plants concentrate elements such as selenium and cadmium in their leaves. The roots of water-growing plants such as the water hyacinth can also concentrate metals from their water supply, a process called rhizofiltration, which is used in reed bed biofilter technology. The only commercial use of such plants is to use metal-tolerant plants to stabilize waste tips (and to make then less ugly).

Microbial pathogen. Generally, a microorganism which causes disease in its host; more specifically, a term used in preference to *microbial "insecticide"* to denote a microorganism used by man to suppress insect pest population.

Microbial pest control. A technique in which microorganisms (bacteria, fungi or viral culture) or derivatives from them, such as the spore protein of *Bacillus thuringiensis,* are used in place of chemical pesticides to kill harmful insects in particular.

Microbial physiology. Physiology is the description of the chemical processes that happen in a living organism. The physiology of microorganisms is important because many biotechnological processes such as fermentation, biosorption, and even bulk growth, depend on the right chemistry happening under the right conditions. So microbial physiology is critical to controlling a range of processes. Aspects of microbial physiology that scientists seek to control include. ***Overall thermodynamics*** (where the energy goes to) and mass balance (where the mass goes to) in the organism.

Specific substrate utilization and growth kinetics. how fast microorganisms grow on a particular substrate, and hoe much of it they convert to the product you want.

Growth energetics (where the specific energy sources go to in the bacteria, and how energy is used in different parts of metabolism).

Redox state or redox balance. the balance of oxidizing and reducing molecules in the bacterial cell. This is important for the energy balance, and also for what chemicals the cells can produce. A related concept is proton motive force (pmf) which is the chemical `push' behind the central reactions of energy metabolism.

Transport mechanisms. how materials get to the cell, and then into or out of it.

Heterologous protein production. how the chemical conditions inside the cell affect how much of a `foreign' protein is produced, and whether it is produced as a soluble material, as an inclusion body, or as denatured, useless material.

Microbial polysaccharide. A simple or complex polysaccharide that is formed as a storage product, cell wall constituent or extracellular gum or slime by a bacterium, alga or fungus.

Microbial sensor. A biosensor in which the catalyst is a microorganism. For instance, a suspension of anerobic bacteria, such as Pseudomonas fluorescence, is immobilized in a collagen membrane and attached to detect changes in rates of oxygen gas exchange, which in turn are proportional to the glucose concentration of any solution in which it is placed.

Microbial thin film. A layer of active biomass formed by the growth of microorganisms on a solid support such as in a biological filter or anaerobic filter.

Microbiological barrier. A material object or set of objects that separates, demarcates, or serves as a barricade to prevent the passage or migration of microorganisms.

Microbiological tests. Methods of examination of specimens, objects, or materials to determine the presence or absence of microorganisms, their taxonomic identification, and/or their relative frequency and types.

Microbiology (Gr. *micros:* small + *bios:* life + logos. discourse). The scientific study of microscopic organisms, particularly bacteria, unicellular algae, protistans, and viruses.

Microbes are the part of our lives in more ways than most understand. They have shaped the present environment and their activities will greatly influence our future. Microbiology is one of the most applied of all the biological sciences. Although microbiology did not exist as a true science before the latter part of 19th century, humans have undoubtedly been practitioners of microbiology since prehistoric times. Knowledge gained in this field has lad to the development of many concepts, which in turn resulted in action. Since the prefix *micro-* means *small.* microbiology must be the biology of small living things. In fact, *microbiology* is the branch of biology that deals mainly with the study of microscopic organisms called *microbes,* which are composed of only one cell. They are typically uni- and multicellular microscopic organisms, widely distributed in air, water, soil, sea matter and with the bodies of the living plants and animals. They are included in the group of lower plants. They should not be considered apart form humans, but should be considered a part of our world. Microbes are present in both natural and man made worlds. Every microbe has special qualities which enable it to survive in such unique places as the soil, oceans, rivers, and streams, ice, water pipes, concrete, hot springs, the human intestine, root of plants, and even oil wells. They are omnipresent and are characterized by very high degree of adaptability. They constitute a world of their own; full of uniqueness from different biological standpoints. Microbes are spread over the entire biosphere- the lithosphere, the hydrosphere and the atmosphere.

The science of microbiology is subdivided into a number of diverse fields: ***Bacteriology*** is the study of the bacteria, ***Virology*** the study of viruses, ***Protozoology*** the study of protozoans, ***Phycology*** the study of algae, and Mycology is the study of fungi. These fields centre around the kinds of organisms under study. Other fields of microbiology centre around where the microbes grow and have the most significant effects like medical microbiology, food microbiology, agriculture, waste treatment, industrial microbiology, etc.

Microbody. A small organelle, bounded by a single membrane and possessing a granular interior. Peroxisomes and glyoxysomes are types of microbodies.

Microcarrier culture. A technique in which animal cells are grown on the surface of small beads that are kept in suspension in a nutrient medium by gentle stirring.

Microcarriers. A small beaded material used as a support for the culture of anchorage-dependent animal cell lines. The beads are derived from silica, glass, dextran or similar materials.

In microbiology, microcarriers are generally small particles used as a support material for cells, are particularly mammalian cells, in a large-scale culture. Mammalian cells are too fragile to be pumped and stirred in the same way that bacterial cells are, but still need to be fed with oxygen and nutrients and must still be separated from their culture medium when the times comes to collect the product. In mammalian cell culture, microcarriers are particularly useful for culturing cells that would normally grow attached to a solid surface (attached or surface culture, as opposed to suspension culture). Rather than have to have acres of flat plastic surface, the cells are grown over the surface of small polymer spheres, especially, polystyrene, gelatin, collagen, or polysaccharides like dextran or cellulose. The surface area for growth is huge, and the spheres can be treated like bacterial cells for filtering and for (gentle) centrifugation, and they protect the cells from the shear forces involved in pumping and aeration. Some microcarriers are simply solid spheres, some are porous. Porous ones have a larger surface area for cells to grow on, since the cells can grow inside them as well as on the outside, so giving them greater protection. However, it is harder to see the cells on the carriers, which can be important if you want to know how well your culture is growing. An alternative to growing cells on microcarriers is to grow the cells as aggregates. Cell aggregates have some of the mechanical robustness of cells on microcarriers, and have much higher cell content for a given amount of solid matter. However, getting cells to grow in aggregates can be much harder than getting them to grow on suitably treated polymer surfaces.

Microcephaly. A condition characterized by a small head whose circumference is less than two standard deviations below mean for age and sex; may be due to congenital hypoplasia of the cerebrum or be the consequence of severe brain damage in early life.

Microchromosome. In reptiles and especially in bird, the chromosome complement is comprised of large chromosomes and many quite small chromosomes, called microchromosomes. Only recently has it been possible to count their number accurately.

Microclimate or microhabitat. The climate in the immediate vicinity or surrounding an organism or its parts.

Microcutting. A tiny cutting; as an explant such as a meristem tip removed for culture with the use of a dissecting microscope.

Microdissection (micromanipulation). The technique of dissecting under a microscope using the mechanically manipulated instruments. Such techniques are often used when dealing with living organisms at the cellular level, and have been applied successfully in transferring nuclei between species of Amoeba.

Microdroplet array or multiple drop array (MDA) or hanging droplet technique. Introduced by Potrykus, Harms and Lorz (1979), this technique is used to evaluate large numbers of media modifications, employing small quantities of medium into which are placed small numbers of cells. Droplets of liquid culture (medium and suspended cells or protoplasts) are arranged on the lid of a petri dish, inverted over the bottom half of the dish containing a solution with a lower osmotic pressure, and the dish is sealed. The cells or protoplasts form a monolayer at the droplet meniscus and can easily be examined.

Microeinstein (μE). A photometric measurement consisting of 0.000001 mole photons.

Microelectrode. Electrode made of glass, pulled out so that the end tip has a diameter of a fraction of a micrometre, and filled with a conducting electrolyte such as potassium chloride. Such electrodes are used to study the electrical properties of cells and cell membranes.

Microevolution. A small amount of evolutionary change, consisting of minor alterations in gene proportions, chromosome structure, or chromosome number. (A larger degree of change would be referred to as macroevolution or simply as evolution.)

Microfibrill. Threadlike structure found in the cell walls of filamentous fungi. Chemically, microfibrils consist of chitin.

Microfilament. Minute fibrous structure found in the cytoplasm of eukaryotic cells. They play a role in the motion of cells.

Microfilaria. Microscopic larvae of certain nematodes, e.g. *Wuchereria banocrofti,* the causative agent of elephantiasis.

Microglobulin. Small protein, with some homology to immunoglobulin, which binds to glycoproteins of the major histocomapatibility complex.

Micrognathia. Abnormal smallness of the jaws, especially of the lower jaw.

Micrograft. An *in vitro* graft. This procedure involves placing a meristem tip or shoot tip explant onto a decapitated rootstock, that has been grown aseptically from seed or micropropogated. The purpose may vary, from virus elimination, using meristem tips and virus-free rootstocks, to viral assays or to expedite grafting that is usually done in the greenhouse. The explanted section is placed directly onto the cambium of the severed stock or inserted into a T or inverted T-shaped epidermal incision on the stock. This technique has been used to eliminate viruses, viroids and mycoplasmas from *Citrus* species.

Micrograph. A photograph taken with the aid of a microscope. *Photomicrographs* and *electron micrographs* are produced using optical microscopes and electron microscopes respectively.

Microhabitat. The microscopic environment that is important to a population or community in nature.

Microinjection of cells. Procedure whereby molecules (e.g., antibody, mRNA, DNA) or organelles (e.g., nuclei, mitochondria) are introduced into cells by direct injection using ultrafine glass needles. Although relatively large tissue culture cells can be injected, favoured cells are the eggs of eukaryotic animals, since they are often very large; in some cases direct injection into the cell nucleus is possible. Injection of DNA may lead to permanent incorporation of some of the DNA into the nuclear genome of the cell. If the cell is a fertilized egg, this may achieve genetic transformation of the organism and its progeny.

Micromanipulator. A device for manipulation of macroscopic specimens under a microscope.

Micrometer (μm). A commonly employed unit of measurement in microscopy, it equals 1 x 10^{-6} meter or 1 x 10^{-3} millimeter. *Synonym, micron* (μ) in older usage.

Micromole (M). One millionth (0,000001) of mole (10^{-6} M). The unit of concentration used for microelements, vitamins, hormones and some other organic addenda used in plant tissue culture media.

Micro-mutation. A mutation, or sudden heritable change, at a single gene locus.

Micron (μ). A unit of length convenient for describing cellular dimensions; it is equal to 10^{-3} cm or 10^{5}, Å.

Micronucleus. (1). The small permanent reproductive nucleus of the Infusoria, as opposed to the short-lived multiple macronucleus. (2) A nucleus separate from the nucleus formed at telophase by one or more lagging chromosomes or fragments.

Micronutrient. A nutrient required in more than trace amounts by an organism.

Microorganism safety classification. One major concern about biotechnology is whether it is safe. As much of biotechnology involves the genetic manipulation, selection, or physiological manipulation of microorganisms and their subsequent production in large amounts, some of this concern translates into a concern about the safety of industrial-scale microbiology. Most guidelines and codes of practice in handling microorganisms are aimed at medical microbiologists and microbiologists handling pathogens to produce vaccines. Thus most of the guidelines on how microorganisms should be handled in biotechnology are derived from these medical examples. The World Health Organization has no evidence that genetically manipulated organisms are inherently more dangerous than any others, and has found no case where a laboratory or industrial worker has contracted an infection as a result of contact with a genetically engineered organism. The principle of classifying the danger from a microorganism, and so of deciding how to contain that danger, is to classify the organism according to how likely it is to escape, how likely it is to survive if it does escape, and how much damage it could do if it does survive. Different countries have different rules about how this is done. the table summarizes a few of these. If an organism is outside the Class 1/Grou 1 area, then it can be contained by a variety of physical or biological methods (see Clean room, Physical containment). A range of national safety committees ensure that appropriate containment is used for organisms in each class (even if, in other industries, no containment is needed for those same organisms at all).

Microorganisms. Form of life of microscopic dimensions. There are a very wide range of microorganisms used in biotechnology. A few are mentioned elsewhere in this book under specific heading. Some are used for more than one thing, and so crop up in several biotechnological contexts. Microorganisms, indeed all life, are classified into prokaryotes (organisms without a cell nucleus) and eukaryotes (organisms with a cell nucleus). Animals, plants, and fungi are all eukaryotes, bacteria and archebacteria are prokaryotes. Bacteria are classified into Gram-positive and Gram-negative. These names reflect whether their cell walls will absorb Gram's stain, but the division they represent is quite a substantial one, and Gram-positive and Gram-negative organisms are biochemically and genetically quite different. However, they may look much the same under the microscope. Microorganisms may be ball-shaped (cocci), rod-shaped, or made of very long `strings' called hyphae. Hyphae may be branched or unbranched. in either case they are often harder to grow in bulk because the stirring needed to get nutrient to all the hyphae can break them. Organisms that grow as long strings or filaments are called `filamentous. Microorganisms are also classified into aerobes (grow in the presence of oxygen) and anaerobes (do not use oxygen). These can be facultative or obligative.

Microperoxidase. Part of a cytochome *c* molecule which retains its haem group and has *peroxidase* activity.

Microphage. A granulocytic phagocyte.

Microphyll. A foliage leaf that, even if large, has a single unbranched vein running from base to apex (some fossil lycopods are an exception). It is the typical leaf of lycopods, horsetails and psilophytes. The stele remains entire when a leaf trace branches off to a leaf. Compare megaphyll.

Microprocessor. An electronic device that accepts and processes coded instructions. Microprocessors are now being widely used in the fermentation industry; uses include monitoring routine process variable, data logging, planning maintenance schedules, monitoring inventories to maintain delivery and shipping schedules, processing documentation, generating process status information and sending appropriate alarms to operators, and automatic control of fermenter functions.

Micropyle. The canal formed by extension of the integuments of the ovule and through which the pollen tube grows into the embryo sac. In the mature seed it appears as a minute pore in the seed coat through which water is imbibed when the seed begins to germinate.

Microscope. An instrument used to produce a highly magnified image of a sample. In biology, electron microscopes and light microscopes are used.

Microscopic. Anything so small that is cannot be seen without the aid of a microscope.

Microscopy. Microscopy is a standard method of examining biological systems, both to determine their structure and to examine their chemistry. Two more specialist techniques are.

Fluorescence microscopy. This version detects the light given out by from a fluorescent molecules in a sample. Almost always the fluorescence comes from a fluorescent label attached to an antibody or DNA probe, so we can tell what part of the cell the fluorescent reagent has bound to. Several colours of label can be used, and the results are very pretty. Techniques using fluorescence microscopy include FISH and immunocytochemistry.

Confocal microscopy. Here the microscope effectively focuses on only a very thin slice of the sample at a time. So a picture is built up, not as if you were looking down through a thick sample, but as if you had sliced it into layers. Often, confocal microscope systems are connected to computers which can reassemble the slices into a 3D picture of the original object.

Microsequencing. A technique used to increase the sensitivity of detection of products of the Edman reaction during amino acid sequencing of proteins.

Microsome. Largely submicroscopic particles in the cytoplasm characterized by the presence of ribose Nucleic Acid and phospholipids. (*Claude*, 1943).

Microspikes. Narrow (0.2μ) cytoplasmic projections which can reach 20μ in length. They extend and retract rapidly from the cell surface, perhaps playing a sensory role.

Microsporangium. In heterosporous plants, the sporangium that produces the microspores, located on the microsporophyll. The microsporangium wall splits to disperse the mature microspores. Microsporangia are found in some pteridophytes (e.g. Selaginella and many ferns) and are represented by the pollen sacs in gymnosperms and angiosperms.

Microspore (Gr. *mikros.* small + *spora.* seed). In plants, a haploid spore that produces a male gametophyte. In many cases the microspore is smaller than the female-producing megaspore.

Microspore mother cell. A diploid cell that goes through two divisions, one a reductional, producing four microspores. It is the same as the pollen mother cell.

Microsporocyte. In plants, a cell that is destined to produce microspores by meiosis. Synonyms, Microspore Mother Cell and Pollen Mother Cell.

Microsporogenesis. The process in which microspores are produced. Meiosis produces four cells that, when mature, are the microspores (pollen grains).

Microsurgery by radiation. The breaking of a chromosome in two places and the insertion of a fragment into another chromosome that has likewise been broken.

Microtome. An instrument for making thin sections of tissue or cells.

Microtrabecular lattice. A system of strands (about 6 nm in diameter) forming a meshwork that connects and supports the organelles of a eukaryotic cell.

Microtrabecular lattice. A system of strands (about 6 nm in diameter) forming a meshwork that connects and supports the organelles of a eukaryotic cell.

Microtubule. A hollow tubular cytoplasmic component, outside diameter about 15 to 30 nm, found in the cytoplasm, especially of motile cells, and forming the spindle of the mitotic apparatus.

Microtubule-organizing centre. Region of the cell cytoplasm in which Microtubules are chiefly synthesized or organized. Many microtubules continue to associate with these centres after growth, although some become linked to other cellular structures.

Microtype egg. A minute type of egg produced in large numbers by some Tachinidae, and by the Acroceridae and Trigonalidae. The ventral or posterior end is membranous and used for attachment to a leaf or other surface. whereas the rest of the chorion is thick, tough, and sculptured. In most tachinid microtype eggs, hatching takes place only after ingestion by the host; cf. Macrotype Egg.

Microvacuole. A minute vacuole which may be invisible to the light microscope but observable with the electron microscope.

Microvilli. Long, narrow cytoplasmic projections which increase cellular surface area and form the brush borders of certain (e.g., intestinal) cells.

Microwave oven. An oven in which heating results from microwave radiation of the food or contents. Such ovens are commonly employed to melt agar in making up plant nutrient media and may be useful to sterilize liquids and some plastic or glass items.

Mid-body. The first stage in the formation of a cell-wall between daughter cells; the cell plate.

Middle lamella. Intercellular layer occurring between adjacent plant cells. It is secreted outside the cell wall and is composed of magnesium and calcium pectates.

Middle piece. A vague term signifying the middle region of a sperm.

Mid-prophase. The stage following zygotene when paired chromosomes are somewhat shortened and thickened, a stage at which chromosomes have been studied extensively. Also called Pachytene.

Migration coefficient. The constant proportion of individuals migrating from one population to another.

Migration inhibition factor (MIF). Soluble substance released by sensitized lymphocyte in the presence of its specific antigen; this inhibits migration of macrophages; **in vitro** system for measuring cellular immunity. It is a group of peptides produced by lymphocytes which are capable of inhibiting macrophage migration.

Migration inhibition factor (MIF). Soluble substance released by sensitized lymphocyte in the presence of its specific antigen; this inhibits migration of macrophages; in vitro system for measuring cellular immunity. It is a group of peptides produced by lymphocytes which are capable of inhibiting macrophage migration.

Milk factor in cancer transmission. A factor present in the milk of mice that transmits cancer.

Millimeter (mm). A unit of distance measurement equal to 1 x 10^{-3} meter.

Millimicron. 0.000001mm.

Millimole (mM). One thousandth (0,001) of a mole (10^{-3} M). A unit of concentration used for macronutrient elements and organic nutrients in tissue culture media.

Millipore filter. The brand name for a type of membrane filter. These filters come in a range of pore sizes and are used for filtering gases or liquids.

Millon's reagent. A mixture of mercury, nitric acid, and water that is used in testing for the presence of proteins. The reagent is heated with the test material and the production of a brick red precipitate indicates the presence of protein.

Mimic genes. Two or more non-allelomorphic genes which produce similar or identical effects.

Mimicry. The resemblance of one kind of organism to another; serves the function of discouraging potential enemies or attracting potential prey.)

Mineral. Any of many homogeneous, usually solid inorganic substances present in the soil, occurring as the result of natural processes.

Mineralization. Conversion of the organic form of an element to the inorganic state.

Mineralocorticoid. A corticosteroid hormone, such as aldosterone. It stimulates the re-absorption of salts and water in the distal tubules of the kidneys.

Mini Ti plasmid. A derivative of the wild-type tumour-inducing plasmid of *Agrobacterium tumefaciens,* in which part of the DNA sequence that is not needed for infection and replication is excised. This leaves a smaller plasmid that may be used as a cloning vehicle for recombinant gene technology in higher plants.

Minicell. Small, spherical bacterial cell that is produced during the culture of some mutant strains of bacteria.

Minichromosome. Chromatin complex of DNA and Histone which is the form adopted by the genome of certain eukaryotic viruses when present in cell nuclei. Viruses such as Papovaviruses replicate in cell nuclei and have double-stranded DNA as their genetic material, and this viral nucleic acid adopts a nucleosomal organization by combining with histones of the host cell nucleus.

Minimal essential medium (MEM). Medium that contains only the minimum ingredients to sustain growth of a particular cell line in culture.

Minimal inhibitory concentration (MIC). The lowest concentration of a drug that will prevent growth of a standard microbial suspension.

Minimal medium. A medium containing certain inorganic salts, a suitable carbohydrate, such as sucrose, and the vitamin Biotin, which will support the growth of wild-type cells.

Minimum effective cell density. The inoculum density below which the culture fails to give reproducible cell growth. The minimum density is a function of the tissue (species, explant, cell line) and the culture phase of the inoculum suspension. Minimum density decreases inversely to the aggregate size and division rate of the stock culture. It is important to know this ratio when plating suspension cultures for selection of single cell lines so that colonies will not overgrow one another prior to obtaining readily manipulable size.

Minimum Inoculum size. The smallest inoculum that can be successfully used for subculture.

Minimum turbulence air flow. Unidirectional air flow within a work space where undisturbed air moves in parallel lines and where the location of people and objects within the work space is controlled to allow reasonable recovery from turbulence downstream of the people and objects. The rate of air movement is usually such that surface contamination from air-borne particulate fallout is minimized.

Minor base. It is a nucleotide which differs in structure from the four primary building blocks of nucleic acid (*A, C, T, G*) and carries a substituent (usually a methyl group), or possesses a saturated bond or an isomeric form of the glycosidic linkage.

Minute. Dominant mutation of Drosophila, lethal when homozygous, and thought to involve the coding sequences for Transfer RNA. Affected heterozygotes develop slowly; the adults are small, sterile and possess an abnormal pattern of wing veins.

Miscegenation. Interbreeding as between different varieties.

Misdivision. Spontaneous crosswise (instead of lengthwise) division of the centromere on the spindle especially of univalents at MI and daughter univalents at M II.

Mismatch. Incorrect base pairing in DNA or RNA. mismatching can occur during renaturation or Molecular Hybridization experiments *in vitro* if conditions in the assays are not carefully optimized, or as a result of DNA damage or mutation *in vivo.* A-A, A-G and A-C are examples of mismatched base pairs in contrast to the normal pair A-T.

Mispairing. The presence in one nucleotide chain of a DNA molecule of a nucleotide which is not the complement of that at the corresponding position in the other chain.

Missense mutation. A mutation by which a particular amino acid, often resulting in an inactive protein. A Missense Codon results when one or more bases of a sense codon are changed so that a different amino acid is coded for.

Mistranslation. Incorrect placement of an amino acid residue in a polypeptide chain during Translation.

Mitochondria (*Sing.* Mitochondrion, (Gr. *mitos*: thread + *chondros*: cartilage, or grain).). A kind of cytoplasmic organelles carrying most of the respiratory enzymes. The usual shape of mitochondrion is cylindrical consisting of a double layered membrane, the inner one of which projects as cristae into the cavity of the chamber dividing it into a number of incomplete regions.

An organelle that occurs in eukaryotic cells and contains the enzymes of the citric acid cycle, the respiratory chain, and oxidative phosphorylation. A mitochondrion is bounded by double membrane.

Mitochondrial ATPase. An enzyme found on the inner mitochondrial membrane that catalyzes the formation of ATP from ADP and orthophosphate.

Mitochondrial DNA (mtDNA). DNA of mitochondria. It consists commonly of a closed cricle of double-stranded DNA, of between 5 and 30μm in length, comprising 15-20 kilobases in most animals (longer in higher plants) and including about 50 gene sequences, of which about one-half code for transfer RNA.

Mitochondrial enzymes. The mitochondrial matrix contains all the enzymes of the TCA cycle, with the exception of the succinate dehydrogenase which is built into the inner surface of the inner membrane. Mitochondria also contain enzymes for fatty acid metabolism and interconversion of some amino acids.

Mitochondrial membranes. The two membranes of a mitochondrion differ in structure and composition. The outer membrane has a higher lipid content and a distinct enzyme complement.

Mitochondrial replication. The formation of new mitochondria by division of existing ones. Division is accomplished by invagination of the inner membrane.

Mitochondrial respiratory chain. A sequences of enzymes and proteins present in the inner mitochondrial membranes of eukaryotic cells and in the cell membranes of prokaryotes. The chain receives electrons or hydrogen atoms derived from reactions of the TCA cycle.

Mitochondrial ribosome. A ribosome associated with the mitochondrion having a sedimentation constant of 70S.

Mitogen. Substances which cause cells, particularly lymphocytes, to undergo cell division.

Mitogenetic radiation. Radiation supposed to emanate from certain growing tissues and to induce mitosis in other tissues.

Mitomycin C. Antibiotic, derived from **Streptomyces caespitosus,** that inhibits DNA synthesis and mitosis by cross-linking DNA, being particularly potent against gram-positive, gram-negative and acid-fast bacilli. It is used in immunology to prevent immunologically reactive cells from dividing, and its use has also been suggested for cancer therapy.

Mitosis. The term proposed by Flemming (1882) for the process of nuclear division, during which the chromosomes become evident as they shorten and thicken (prophase), and then take up position on the equatorial plate of the spindle (metaphase), to be followed by separation of the two daughter-chromosomes of which each is composed to opposite poles of the spindle (anaphase), and their gradual return to a slender elongated condition (telophase). (Gk. *mitos* = thread).

Mitotic analysis. Examination of the mitotic division cycle. The purpose might be to count the number of chromosomes (at metaphase) or to calculate the rate of cell division, etc.

Mitotic apparatus. The collection of cytoplasmic structures present in mitotic cells and consisting of the asters (if present) surrounding each centrosome and the spindle of microtubules.

Mitotic chromosome. Chromosome occurring at the mitotic stage of cell division.

Mitotic death. Cells fatally damaged by ionising radiation may not die until the next mitosis, at which point the radiation damage to the DNA becomes evident, particularly when there is fragmentation of chromosomes.

Mitotic index (MI). Fraction of a sample of cells that are undergoing mitosis. MI is commonly expressed as values below 1, so that 0.1 represents 10 percent of cells in mitosis. Percentages of cells in mitosis can vary greatly, from more than 10 percent in tissue culture to much less than 1 percent in relatively slowly growing tissue *in vivo.*

Mitotic nondisjunction. Occurs when sister chromatids fail to migrate to opposite poles of the cell during mitotic anaphase. The result is daughter cells with hyperploid and hypoploid chromosome counts.

Mitotic poison. Compound that specifically prevents mitosis, often by blocking spindle formation or operation. Colchicine is a widely used mitotic poison in cytogenetic studies. Some mitotic poisons are used as cytotixic drugs in cancer chemotherapy.

Mitotic recombination. Crossing-over between homologous chromosomes during mitosis, which leads to the segregation of heterozygous alleles.

Mixed culture. A culture of microorganisms containing a number of species or genera. Many traditional food fermentation processes are dependent on the activities of mixed cultures.

Mixochimaera. An individual obtained by mixing the protoplasm of two individuals. (Such individuals have been formed artificially in certain fungi).

Mixochromosome. A newly fused chromosome pair produced by syndesis (zygotene).

Mixoploid. Refers to cells with variable (euploid, aneuploid) chromosome numbers; as mosaic or chimaeral components that differ in chromosome number, or the result of a variety of mitotic irregularities.

Mixovariation. A variation due to genetic segregation or to recombination.

M-N blood groups. Oligosaccharides present on human erythrocytes that form an antigenic system separate from the ABO blood group system and rhesus system. Individuals are typed as M, N or MN, since M and N are inherited without dominance. A further antigen, S, may be closely linked with M-N blood groups.

Mob. The genetic notation for gene(s) involved in the mobilisation of a plasmid from cell to cell. Certain broad-host-range cloning vectors are mob to enable them to the mobilised into a large variety of Gram-negative bacteria.

Mobbing. Gathering of birds around a predator, calling vigorously and thus reducing the probability that the predator can hunt successfully in the area.

Mobilisation. (i) The transfer, between bacteria, of a non-conjugative plasmid by a conjugative plasmid. (ii) The transfer, between bacteria, of chromosomal genes by a conjugative plasmid.

Modal bulk. A modal bulk is produced by determining the mean values for the most important characters, then discarding all material deviating from any one of these values plus or minus, an amount exceeding the standard deviation and repeating this process several years.

Mode. The value of a variate shown by the class most frequently observed.

Model organisms. Many of the experiments that biotechnologists might conceive of doing to develop new products or processes cannot be done for ethical, moral, legal, financial, or practical reasons. Breeding experiments on people or trees, studies of individual cells in a whole cow or a pig, such experiments may give valuable knowledge to agriculture, healthcare, or environmental studies, but we cannot sensibly do them. So, instead we do a model experiment, one that seeks to duplicate some critical aspect of the experiment, one that seeks to duplicate some critical aspect of the experiment we want to do, but avoids the impracticalities. Some organisms have become standards for such model experiments, and so are called model organisms. They are models for some aspect of other organisms.

Moderately repetitive DNA. Fraction of DNA obtained by renaturation of DNA fragments from a sample of eukaryotic DNA of reduced molecular weight. Such DNA hybridizes weight. Such DNA hybridizes fairly fast because many copies of the same DNA sequence are present in the solution, but less speedily than does Highly Repetitive DNA or Fold Back DNA.

Modification methylase. An enzyme which catalyzes the chemical alteration of certain bases of DNA by methylation of nucleotide within restriction recognition sites corresponding to their own restriction enzymes. This protects their own DNA against attack.

Modification. A temporary change or variation in the characteristics of an organism.

Modified base. Base other than those bases from which DNA and RNA are normally synthesized (i.e. thymine, cytosine, adenine, guanine and uracil). Modified bases result from postsynthetic changes in nucleic acid and include 5-methylcytosine and 6-methyladenine.

Modifier complex. An assemblage of genes which conjointly affects the degree of expression of a given pair of alleles.

Modifier gene. A gene that interacts with other genes by modifying their phenotypic expression (cf. Regulatory Gene, Structural Gene).

Modifying factor. A gene that affects the expression of another nonallelic gene. Often without other known effects.

Modifying gene. A gene which alters the expression of another gene that is non-allelic, e.g., genes affecting the amount of spotting in Holstein cattle, a spotted breed.

Modulating codon. DNA coding triplets (Codons) that code for rare transfer RNA (tRNA) molecules. Since the tRNA which carries the appropriate amino acid to the mRNA is in short supply in the cell, the presence of such codons tends to slow down translation of an mRNA containing them.

Modulator. The regulatory metabolite which binds to the allosteric site of an enzyme and alters the maximum velocity (also called an effector or modifier).

Moiety. A part of a molecule having a characteristic chemical property.

Molal. Descriptive of a solution that contains one mole of a compound dissolved in one litre of water or 1000 grams of solvent.

Molasses. A concentrated syrup that contains a significant amount of low molecular weight fermentable sugars. Molasses is produced as a by-product of the sugar, paper pulp, fermented beverage and fuel alcohol industries, but the term is mainly applied to by-products of the sugar industry.

Mold. Another name for fungi that exhibit branching or mycelium formation.

Mole. A quantity of a compound whose weight in grams is numerically equal to its molecular weight expressed in atomic mass units. Avogadro's number of molecules. 6023 x 10^{23} molecules.

Mole. A quantity of a compound whose weight in grams is numerically equal to its molecular weight expressed in atomic mass units. Avogadro's number of molecules. 6023 x 1023 molecules.

Molecular biology. A modern branch of biology concerned with the understanding of biological phenomena in molecular terms. Molecular biology arose largely from the fusion of genetics with biochemistry.

Much of microbiology is based at least in part on molecular biology. Molecular biology, and its twin science molecular genetics, started in the late 1940s around a group of biologists, and physicists turned biologists, who were looking for a new way to attack the fundamental problems of life. Traditional biochemistry at that time (and much of biochemistry today) sought to attack complex systems by taking them apart and analysing all the bits as carefully as possible in terms of chemicals and biochemicals. The `new' science instead chose the simplest system they could and sought to analyse it using genetics as their primary tool. The system they chose was the bacteriophage, and hence many of the founders of molecular genetics were members of a semi-official `phage group'. The genetic approach paid off handsomely in three ways. First, it opened up whole new areas of genetics (genetics at a molecular level rather than the whole organism genetics that had characterized previous work on *Drosophila*, mice, and plants, or the biochemical genetics of bacteria and yeast). This in turn allowed the investigators to start to decipher the genetic code and deduce some of the mechanisms of protein synthesis. But probably more importantly, it gave credibility to a new way of thinking about biology. This is now entrenched in the normal way of thought, and envisages the molecular basis of biology as being made up of quite understandable building blocks that bump into each other and latch on to and off each other in defined ways. Whereas in 1950 an enzyme was a squiggle in an equation, in 1990 it is a coloured `blob' on a computer graphics display. The molecules at the basis of life became much more real and much more important. Life became a discrete machine, and the instructions for that machine lay in DNA. Hence the centrality of DNA to much of biology today. This approach to living systems, as discrete blocks labelled `protein' and `gene', has been called **molecular Lego**. As well as this philosophical change, which enabled us to imagine a `biotechnology' of engineered genes and molecules, the phage group work gave us the basic tools of recombinant DNA technology. Restriction enzymes, DNA ligase, and many cloning vectors all come directly from bacteriophage genetics. Thus, molecular biology is not a science in the sense that it studies molecules or biology-biochemistry, physiology, pathology, microbiology do that too. It is more a way of doing biology, both a way of thinking about it and a way of getting the tools to do experiments. It is, in Thomas Kuhn's terminology, a paradigm.

Molecular electronics. This speculative idea means making electronic or computing devices out of single molecules, or small groups of molecules. Talk about switches that are made of a single protein molecule leads to computers with greater-than-human powers, which could fit in a matchbox. Fairly obviously this is speculative, but may not be as speculative as nanotechnology in general. First, conventional computer chips are made by layering materials on silicon, and it takes only a leap of imagination to suggest that layering proteins could achieve similar things. Secondly, many proteins do have charge-transfer and charge-switching properties, which could, with much greater understanding of the properties, which could, with much greater understanding of the properties of proteins in general, be harnessed to provide some aspects of the information processing capability of a semiconductor device. Thirdly, Langmiur-Blodgett films (thin films of lipids) are known to be an essential part of the electrical properties of nerve cells and can be made quite readily in the laboratory. Nerve cell proteins are inserted into the lipid film and alter the film's ability to let ions pass depending on what other ions are present or the electric field that they are exposed to. This has got to the stage of building films, putting proteins into them, and demonstrating the electrical characteristics of the protein, rather like where we were with transistors in 1930. A lot of work has focused on using bacteriorhodopsin, a pigment from Halobacterium halobium (a purple, salt-loving bacterium), which has electrical and light-absorbing properties as well as being able to self-assemble into membranes. Membranes of bacteriorhodipsin assembled in the right way can store light images when exposed to electric fields, a first step to a holographic computer memory system. Molecular computing was a trend term a few years ago, but has now been partly supplanted by nanotechnology.

Molecular evolution. Evolution viewed at a molecular level. It encompasses the evolution of molecules such as proteins and nucleic acids through the selective competition that occurs in the dynamic biochemistry of the cell as well as the molecular aspects of the evolution of species and organisms resulting from the fact that although selection is at the level of the phenotype, this is itself partially determined by the genotype. Since mechanisms that translate genotype into phenotype are themselves molecular, it is clear that evolution has a very important molecular basis.

Molecular genetics. Study of how genes function to control the activities of a cell.

Molecular graphics. This is the display of a molecule's shape, usually on a computer. It has gained a lot of publicity because of its application to rational drug design. Molecular graphics takes the description of how the atoms of a molecule are arranged in space from a database and draws a picture of what the molecule would look like if it were made of solid balls (the atoms) or thin sticks (the bonds between atoms). Some examples are shown in the figure. Usually, molecular graphics does not calculate the structure of the compound (see Computational chemistry). Because the human brain is extremely good at preceiving patterns in complex pictures, but rather poor at seeing patterns in large collections of numbers, molecular graphics is a good way of allowing people to see the similarities in structures between molecules, and also of seeing if two molecules fit together well. This in turn is useful when, as part of a rational drug design programme, a scientist wishes to find a molecule that will fit into the known structure of the active site of an enzyme, or the hormone-binding site of a receptor. Molecular graphics packages often produce extremely attractive pictures as part of their output, which is another reason why they have a high profile in the public relations material of pharmaceutical and biotechnology companies. The more sophisticated display methods can produce 3D pictures that the user can manipulate as if he were in a room full of bits of molecule that he can move around with his hands, a type of computer interaction called `virtual reality'. However, these pictures can be misleading. molecules are not `really' like this, and all the pictures here are only approximations of one aspect of reality.

Molecular hybridization. Formation of duplexes of complimentary RNA or DNA strands under controlled conditions *in vitro,* this single-stranded nucleic acids become double-stranded. The progress of the reaction is monitored using methods which can detect this change, such as spectro-photometry or Sl nuclease sensitivity.

Molecular imprinting. This is an approach to synthesizing a non-biological chemical that imitates some of the properties of an enzyme of antibody. A polymer material, often polyacrylamide, is `imprinted' with gaps that exactly fit one, and only one, species of small molecule, in the same way that the binding site of an antibody exactly fits its antigen. This is done by forming the polymer matrix in the presence of the small molecule, so that the chains fold around those molecules. When the polymer has set solid the small molecule is washed out using suitable solvents, leaving `holes' in the polymer. These can have quite a high affinity for the molecule that has been washed out, and so could be used to purify those molecules away from many others. In addition,just as antibodies that are raised against a transition state analogue can have catalytic activity (see Catalytic antibodies), so an imprinted polymer, which has gaps that are `shaped' to fit a transition state analogue, could be a catalyst. So far, molecular imprinting techniques have been used experimentally for purification, but not for catalysis. Molecular imprinted matrices have been used to separate a wide range of small and large molecules, including separating chiral isomers from each other. Their binding capacity can also be used to form the binding moiety of a sensor.

Molecular modeling. This is the use of computers to draw realistic models of what molecules look like. At one end of this range of techniques is molecular graphics, which is simply drawing 3D representations of what a molecule would look like (if the atoms were solid balls). At the other end it shades into computational chemistry (the calculation of what the physical and chemical properties of a molecule are). Usually, molecular modelling falls more into the graphics end of the spectrum. Using molecular modelling, rational drug design programmes can look at a range of different molecular structures of drugs that may fit into the active site of an enzyme, and by moving them on the computer screen it can be decided which structures actually fit the site well. Molecular modelling can add sophistication to picture drawing by calculating the hydration (degree to which individual bits of the molecule bond neighbouring water molecules) and charge distribution across the molecule. These also affect how molecules bind together.

Molecular system of heredity. The cytoplasmic system of heredity in which units of heredity, or determinants are carried by the cytoplasm (cf. Nuclear and Corpuscular systems).

Molecular weight marker. A protein. polypeptide, nucleic acid, oligoucleotide, etc. used as a reference material in analytical techniques such as gel filtration chromatography, electrophoresis and density gradient centrifugation.

Molecular Weight. The sum of the atomic weights of the constituent atoms in a molecule.

Molecule. A particle made up of two or more atoms joined by covalent bonds or ionic attractions.

Molisch test. A quantitative test for the presence of carbohydrates in a biological sample.

Mollusk. A member of the phylum mollusca such as a snail, clam, or octopus.

Moloney murine leukaemia virus. Replication competent retrovirus (*Oncovirinae*) which causes leukaemia in mice, isolated by Moloney from cell-free extracts made from a transplantable mouse sarcoma.

Moloney test. Skin test for immunity to diphtheria in which active toxin is injected into one site and toxoid into another. This is to control for pseudo-positive reactions to the toxin.

Molting. The process of shedding part or all of an outer covering, as the shedding of feathers by birds or of the entire exoskeleton by arthropods.

Monad. A single cell arising (instead of a tetrad) from a spore mother-cell as a result of the failure of meiosis.

Monera. The kingdom of procaryotic organisms with unicellular and simple colonial organization.

Moneran. A member of the kingdom monera- a bacterium.

Monoamine oxidases (MAO). Enzymes that inactivate monoamine neurotransmitters (e.g. dopamine, serotonin, noradrenaline). Inhibitors of monoamine oxidases are used clinically as antidepressant drugs.

Monocentric. Having a single centromere.

Monocional antibody. An antibody preparation which contains only a single type of antibody molecule. Monoclonal antibodies are produced naturally by myeloma cells. A myeloma is a tumour of the immune system. A clone of cells producing any single antibody type may be prepared by fusing normal lymphocyte cells with myeloma cells to produce a hybridoma.

Monocistronic. Descriptive of a length of DNA that codes for only one peptide, in contrast for several related peptides or enzymes. Monocistronic DNA is typical of eukaryotic genes.

Monoclonal antibodies. Antibodies produced in the blood are made by a large number of different lymphocytes (B cells). Each B cell makes a unique antibody, so the antibodies that recognize any particular antigen are a mixture o molecules. This mixture is called `a' polyclonal antibody. an antibody preparation that reacts with only one antigen, but which nevertheless derives from many different `clones' of B cells. While this is useful for the body (covering the antigen with antibodies) it is a problem for the biotechnologist who wants defined materials to work with. Monoclonal antibodies are a way round this. They are antibodies made from a single clone of B cells which has been isolated and immortalized for growth in vitro. The invention of the methods for producing monoclonal antibodies won Cesar Milstein a Nobel Prize and, characteristically, very little money. Monoclonal antibodies are generated as follows.

Immunization. A mouse (usually) is immunized with the target antigen. This is done by injecting the antigen, sometimes with another material (an adjuvant) to stimulate the immune response (see Immunization).

Splenectomy. The spleen (a useful concentrated source of B cells) is removed from the mouse.

Fusion. The lymphocytes are fused with an immortal cell line (myeloma cell). This immortalizes them, i.e. means that they will grow for ever in culture.

Selection of clones. The fused cells are grown in HAT (hyopxanthine/aminopterin/thymine) medium. Because of the genetic defects in the myeloma cell line and the metabolic peculiarities of normal B cells, neither myeloma nor B cell can grow in this medium, but the combined, fused cell can. Therefore a small proportion of fused cells is selected.

Cloning. The fused cells are put, at very low concentrations, into the wells of a multiwell plate. On average each well with a cell in will only have one cell in it, so that, on average, the cells that grow in each well will be a clone, i.e. derived from a single cell. This ensures that you have a pure cell line. This cell line is termed a **hybridoma**.

Selection. The clones are screened by whatever method is to hand to find the one producing a good antibody against the antigen we want. Most of the cells will not be making the antibody we want. A good antibody is one that binds tightly to the antigen (in chemical terms, has an affinity of 109 or better), does not bind significantly to anything else, and is the right class and subclass (IgG, IgM, etc.; see Antibody). If the target molecule is a very small one (like a drug molecule) then injecting it into a mouse will rarely produce an antibody response. In this case the molecule is chemically linked on to a larger molecule, usually a protein and often bovine serum albumin (BSA) or keyhole limpet haemocyanin (KLH), so that the immune system can `see' it. The small molecule is called a `hapten' in this case.

Monoclonal antibody production. Monoclonal antibodies can be produced commercially in a number of ways, depending on the scale of production. As mouse ascites fluid. Mice can be injected with the hybridoma cell line that makes the monoclonal antibody. The cells form a tumour (an ascitic tumour) which secretes the antibody into their ascitic fluid (the fluid that surrounds the lungs) and the blood plasma. Either can be collected and the antibody purified from it. This is simple, and does not require sterile culture conditions. However, it does require sterile culture conditions. However, it does require animal facilities, and only produces around 50 mg/mouse. Ascites is generally not used in the West very much, partly because of controls on the use of animals, the partly because of the expense involved.

Tissue culture methods, which are used to make the hybridoma in the first place, can be used to make antibody. the tissue culture `supernatant', i.e. what is left of the medium once you have removed the cells, is a source of antibody. However, this is rarely effective at producing more than 10 mg of antibody.

Suspended cell fermentors. The traditional biotechnology bioreactor technology has been used to grow hybridoma cells in bulk. Celltech (now Lonza), for example, has a 1000 litre airlift fermentor which can produce 100 g of antibody in a two-week fermentation with a hybridoma. This is a similar technology to medium-scale microbial fermentation, although because mammalian cells are very sensitive to chemicals, temperature changes, shear (squashing), and other environmental disruptions it can be more difficult to get it to work reliably, and also costs a lot more in expensive culture media.

Immobilized cell reactors. Several types of immobilized cell reactor have been used to make monoclonal antibodies on a scale of a few grams. The most popular is probably the hollow fibre reactor. A few grams of antibody is enough for several million tests to be used for medical diagnosis, for example, and so suffices most commercial needs. A rising new technology is to use bacteria to produce antibodies. The genes for the heavy and light chains must be spliced into one bacterium, but when this is done the bug is very much easier to grow than mammalian cells. This also makes genetic engineering of chimeric or humanized antibodies easier since the cloning technology necessary to do this is done in E. coli. Production in bacteria is preferred because it is cheaper and faster than in cell culture, and the systems are usually quicker to set up on a large scale. Against this, the antibody gene has to be engineered into a suitable bacterium. This is a minor problem if the antibody has already been created by recombinant DNA techniques such as phage display or antibody fragment engineering (see Dabs and other engineered antibodies).

Monoclonal. Derived from a single clone, for example, monoclonal antibodies which are produced a single clone and are homogenous.

Monocot (Gr. *monos.* one + *kotyledon.* a cup-shaped hollow). Any member of the angiosperm class monocotyledonae, plants in which the embryo produces but a single cotyledon (seed leaf). Leaves of most monocots have their major veins arranged parallel to each other.

Monocotyledonae. The more advanced of the two major subgroups of the angiosperms.

Monoculture. The cultivation of a single crop species over large areas without provision for diversity or use of the land in any other way.

Monocytes. The largest leukocytes (macrophages) found in the blood; they are phagocytic, amoeboid cells.

Monod kinetics. A theoretical expression of the rate of growth of microbial systems.

Monoecious (Gr. one house)**.** Organisms in which both sexes are "housed" in a single individual, which produces both eggs and sperm. (In some plants these are found in different flowers within the same plant). Examples; corn, peas, earthworms, hydras. (Contrast with dioecious, perfect flower.)

Monoecy. In plants the condition in which male and female gametes are borne by different flowers of the same individual.

Monogenesis. (1) Reproduction by asexual means. (2) Theory which suposes evolution to have taken place from a single entity.

Monogenic or Monohybrid ratio. The 3.1 ratio obtained in the F_2 of a cross between two parents of a particular strain differing in respect of only one pair of contrasting characters.

Monohybrid cross. A cross between two parents of a strain or a variety in which the parents differ in respect of only one pair of contrasting characters.

Monokaryon. A uninucleate cell or a tissue composed of such cells.

Monolayer (Greek **mono,** single)**.** Uniformly thick, often single layered cultures of cells grown in the laboratory out of individual clones (q.v.).

Monolayer (Greek mono, single). Uniformly thick, often single layered cultures of cells grown in the laboratory out of individual clones (q.v.).

Monomer. Usually referred to the simplest of chemical compounds having the same emperical formula; here indicates simplest of protein units comprising a capsomere (q.v.); normally constituted of one or two polypeptide chains.

Mononucleotide. The basic building block of nucleic acids (DNA and RNA).

Monophagous. Restricted to the use of only one plant or animal species as host or prey.

Monophyletic. Derived from one original or ancestral type (race, species, family, etc.).

Monoploid. It is a haploid having $2n=x$ chromosomes. It should not be confused with haploid because haploids of polyploids contain more than one basic set of chromosomes. For instance, haploids of hexaploid wheat contain three basic set of chromosomes $2n=3x=21$.

Monopodial growth. A type of growth in plants, characterized by a branching system in which the primary apex continues to grow and remains dominant.

Monosaccharide. A simple sugar. Oligosaccharides are made up of monosaccharides.

Monosodium glutamate. A flavour enhancer produced on a large scale by fermentation.

Monosomia. An individual in which only one chromosome is less than the diploid number ($2n$-1).

Monosomic (monosome). With a certain chromosome type only represented once. In a diploid species a monosomic individual has one chromosome less than in the normal euploid condition and may therefore be denoted by $2n-1$ or $2x-1$.

Monospermy. Fertilization by only one spermatozoon.

Monotrichous. Having a single flagellum at one end of the cell.

Monozygotic twins. Twins developed from a single fertilized ovum that gives rise to two embryos at an early developmental stage; also called Identical Twins.

Morbidity rate. The number of reported cases of a given disease present at a given time per 100,000 population.

Morbidity rate. The number of reported cases of a given disease present at a given time per 100,000 population.

Morbidity. The number of cases of a disease per unit of the population within a given time period.

Morbific factor. A genetic factor which governs the predisposition of an individual to any particular disease or which directly controls a particular morbid condition (e.g. haemophilia).

Morbilli virus. Genus of viruses (of the *Paramyxovindae*). Type species is *measles virus;* other species include canine distemper virus and the related seal virus.

Mordant. Salt of aluminium and/or iron that react with certain stains, for example haematein, to form a complex known as a `lake', carrying a positive charge. Such a complex gives better penetration and binding than an uncomplexed stain.

Morgan unit (morgan). A unit of relative distance between genes on a given chromosome; it equals a cross-over value of 100 percent.

Morph. Phenotypic variant in a population of organisms, most often applied to individuals within a conspicuously polymorphic population.

Morphactin. One of a number of structurally diverse compounds that possess plant growth activity and react synergistically or antagonistically with endogenous plant growth substances, possibly as inhibitors of auxin transport.

Morphogen. Substance that induces specific changes in morphogenesis. The word is most frequently applied to a substance that acts like a hormone, in that the target site or tissue is other than or distant from the site of synthesis. The term is probably synonymous with inductor.

Morphogenesis. (1) The process of growth and development of differentiated structures. (2) The evolution of a structure from an undifferentiated to a differentiated state.

Morphogenetic response. The effect on the developmental history of a plant or its parts exposed to a given set of growth conditions or to a change in this environment.

Morphogenetic stimulus. Stimulus exerted by one part of an embryo on another, such that the cells receiving the stimulus are induced to adopt a form or a movement characteristic of a pattern of morphogenesis.

Morphology. The scientific study of organic form including both its development and function.

Morphoplasm. Formative protoplasm. A modified type of cytoplasm which appears in the form of strands (e.g. spindle fibres) or of channels in the un-modified cytoplasm.

Morphosis. Non adaptive, usually unstable variation in Morphogenesis. Associated with certain environmental changes. These phenocopies may mimic the effects of known mutations.

Mortality. Fatality. The ratio of the number of deaths to a given population in a defined situation.

Morula. A solid group of cells derived from a fertilized animal cell that gives rise to the blastula.

Mosaic. Referring to an individual having a chimeral structure or a mixture of tissues representing different genotypes or chromosome numbers. Disease symptom of leaves with mixed green, light green, and yellow patches.

Most probable number (MPN). A statistical expression of estimating the cell number in culture.

Mother cell. A cell with a diploid nucleus which, by meiosis, gives four haploid nuclei, e.g. the spore mothercell in Bryophyta and Pteridophyta, the Microspore or Pollen mother-cell and the Mega-spore or Embryo-sac mother-cell in Flowering Plants. In animals the sperm mother-cell is known as the spermatocyte and the egg mother-cell as the oocyte.

Mother plant. A source plant or donor plant from which an explant used to initiate a culture is taken.

Motifs. Neither protein nor DNA sequences are random. If nature `wishes' to evolve a protein to do something, she starts out with proteins that already exist to do something else and, usually, shuffles bits of the relevant genes around to make the new entity. Thus certain strings of bases or amino acids crop up time and again in different genes and proteins. These are often called motifs. Usually they are significant because they denote some bit of the molecule has a particular function. Thus `zinc finger motifs' in proteins suggest that the protein has a section that binds to DNA. Similarly in DNA the `TATAA' motif is suggestive of a promoter sequence in eukaryotic cells. `Motifs' are similar to signal sequences in proteins. However, signal sequences are meant to be `read' by the cell. Motifs are only read by the biotechnologist as clues to what a particular part of a gene or protein does. Amongst the signal sequences known are the leader sequences that lead to secretion (see Secretion), another that tags the protein as headed for the endoplasmic reticulum and lysosomes, one that sends the protein to the cell nucleus. the `stop transfer' sequence which anchors a protein in the cell membrane, an so on. Being able to read signal sequences is also helpful to the biotechnologist since it gives clues as to where in the cell a particular protein is meant to end up, and hence what its function might be. Signal sequences are only of relevance to proteins (although, of course, they are coded in the DNA), whereas sequence motifs can be found in DNA protein.

Motility. The ability of an organism to move by itself.

M-ring. Innermost (motor) ring of the bacterial flagellar base, located in the outer leaflet of the plasma membrane. It is this ring which is linked to the hook region (and thus to the flagellum itself), and which rotates. Composed of 16 or 17 subunits (one more or less than the S-ring).

MS. An abbreviation for the **Murashige, T. and F. Skoog (1962)** medium formulation. MS is a defined medium formulated for tobacco callus culture and now used more than any other for the culture of a very wide range of plants species. It is characterized by higher levels of nitrogen (both nitrate and ammonium), potassium and calcium than most other plant culture media.

MS2. Bacteriophage when an RNA genome and from which an RNA replicase has been isolated.

Mu. (1) Map unit used in mapping distances between gene loci or chromosomes. (2) Virus particle that endows *Paramecium* cells containing it with mate-killing capacity.

Mucoprotein. Protein containing a substantial carbohydrate group. Glycoproteins are very similar in structure, but have less carbohydrate (less than 4 percent).

Mucor. A genus of moulds, members of which are used in the production of enzymes such as amyloglucosidase and artificial rennet.

Mulatto. The F_1 offspring of a pure Negro and a pure white.

Mullerian mimicry. Resemblance as between two unpalatable species. In this form of mimicry the two species, by sharing a similar type of aposematic colouration, share jointly the risk of mistaken attack by a predator. cf. Batesian.

Multi or polygenes. Two or more pairs of genes contributing to the expression of one and the same quantitative character such as yield, size, height intelligence. In other words, each gene has an additive effect and the character expressed is dependent on the cumulative effect of the genes present.

Multicellular (L. *multus.* much + *cella.* chamber). Consisting of more than one cell, as for example a multicellular organism (Contrast with unicellular).

Multicopy plasmids. Plasmids that are present in cells in numbers exceeding one per cellular genome copy.

Multi-copy. An adjective used to describe plasmids which replicate to produce many plasmid molecules per host genome, e.g. pBR 322 is a multi copy plasmid, there are usually 50 pBR 322 molecules (or copies) per *E. coli* genome.

Multienzyme complex. Molecular aggregate containing molecules of more than one enzyme. Such complexes are an efficient way of increasing reaction rates since reaction intermediates of one enzyme are passed directly to the second.

Multienzyme system. A group of enzyme active in the sequential steps of a metabolic pathway and in physical proximity to one another.

Multifactorial inheritance. The control of one character by two or more genes. This gives rise to continuous variation in such characters (e.g. height) and a continuous gradation is apparent from one extreme to the other. Compare Single-Factor Inheritance.

Multigene family. Groups of genes which are similar in fine structure, have overlapping functions, and are closely linked.

Multigenic inheritance. Inheritance determined by several genes with cumulative effect, e.g., ear length in corn.

Multimer. Molecule, usually a protein, made up of two or more subunits.

Multimeric (L. *multus*, many + suffix mer). Refers to protein units formed out of several monomers and themselves comprising the capsomeres.

Multinominal series. The series obtained by expanding, to any power, the sum of three or more quantities $(a+b+c.)^n$.

Multinucleate. A cell containing many nuclei.

Multiparasitism. a condition resulting from the simultaneous use of a single host individual by two or more species of primary parasitoids.

Multiple alleles. A series of three or more alternative alleles occupying the same locus and conditioning the expression of the same character but in a different form. ABO blood group of man is a typical example of multiple allelic series.

Multiple allelism. The existence of more than two different forms (alleles) of a gene.

Multiple allelomorph. The characteristic produced by a multiple allele.

Multiple chromosome. The production of fusion of two chromosomes.

Multiple factor hypothesis. A hypothesis which explains the blending inheritance often in quantitative characters on the assumption that there is a series of independent cumulative genes governing such a quantitative trait.

Multiple factor. The control of a character by a large number of factors, which may be located at different loci on different chromosomes. In F_2 segregation clear cut classification of phenotypes is not possible.

Multiple resistance factors (MRF). R factors; episomes responsible for multiple drug resistance.

Multiple-factor inheritance. The determination of a phenotypic trait by genes at more than one locus.

Multiplication or proliferation. (1) Any tissue culture undergoing rapid growth. (2) Commonly used to describe Stage II culture, in which consecutive subcultures promote an increase in the number of propagules. (3) The media employed to induce culture proliferation (multiplication media).

Multiplication rule. If an event occurs in n_1 ways, and a totally independent event can occur in n_2 ways, the number of ways both events can occur at the same time is $n_1 \times n_2$.

Multipolar Spindle. A type of spindle with several poles found in cancerous tissue.

Multivalent. An association of more than two homologous chromosomes at meiosis (from zygotene to first metaphase).

Multivoltine. Having two or more complete generations annually.

Muscle proteins. Those proteins present within muscle fiber which are involved in the contractile process. Includes actin, myosin, tropomyosin, and -actinin.

Must. Pressed grape juice used for wine-making.

Mustard gas. Chemical mutagen, used also as an antipersonal weapon in warfare.

Mutability. The propensity of an individual's genes or genotype to undergo heritable mutation.

Mutable gene. Gene with a significant rate of spontaneous mutation. Such genes can be detected in some multicellular organisms by somatic mutation leading to the development of a Mosaic.

Mutable Sites. Sites along the chromosome at which mutations can occur. Genetic experiments tell us that each mutable site can exist in several alternative forms.

Mutafacient. Of one gene or genetic element which determines, or increases, the chance of mutation of another. (*Darlington,* 1944).

Mutagen. (L. *mutare*: change + Gr. *gignomai*: causing). A chemical or physical treatment or agent capable of producing genetic mutation. Causes insertion, deletion or alteration of a base or part of the nucleic acid chain (DNA). Artificial mutations can be induced in cells using chemicals or ionizing radiation. The process is **mutagenesis**. Electromagnetic and particle radiation, such as u.v., x-rays and b particles all increase mutation frequency. So do chemical mutagens such as ethylmethanesulphonate, acridine, nitrous oxide and many others.

Mutagenic agent. Substances which increase the occurrence of the overall number of mutants.

Mutagenicity tests. There are a range of tests using biological systems to see whether compounds can cause mutations. It is argued that chemicals that cause mutations are also likely to cause cancer in humans, a correlation that has generally been found to be true. The main single-cell test systems used are. **Ames test.** Named after Bruce ames, this test exposes Salmonella typhimurium to the chemical. New mutants are detected as bacteria that can grow without being provided with histidine (`back mutations'). The test is one of the standard tests required for mutagenicity testing of products.

SOS-Chromotest. This is an alternative bacterial test which detects when an E.coli has had its DNA repair enzymes activated. Muagens activate specific enzymes that repair damage in DNA, and the test uses a side-effect of these enzymes to detect their activity. The test is not generally accepted.

Micronucleus test. This looks for a characteristic aberration of the chromosomes (formation of small fragments of genetic material outside the nucleus, which are called `micronuclei') in cultured mammalian cells, usually Chinese hamster ovary (CHO) cells.

Mutant. A gene that has mutated. There is some ambiguity here since all genes have probably mutated some-time in their history. Generally, the most common allele of the gene is known as wild type, and the rarer allele is regarded as the mutant.

Mutation breeding. To (experimentally) introduce or remove a character from a cell or organism by exposure to mutagenic agents followed by screening for the desired attribute.

Mutation change. Usage usually restricted to change in the DNA sequences of an organism, which may arise in any of a variety of different ways.

Mutation pressure. Genes in a population spontaneously mutate at a certain rate. The mutation pressure for a certain gene is said to be high, for instance, if the mutation rate from a dominant to a recessive allele is high.

Mutation rate. The frequency with which mutations take place in a given variety or species, or the frequency with which any specified mutation occurs in a given population.

Mutation spectrum. A term to indicate that on organism may produce different kinds of mutants due to the differential action of various mutagens.

Mutation Theory, That new species arise by single mutation. (De Vries, 1890).

Mutation trend. A series of slight gene mutations in the same direction resulting in gradual intensification of a character change (Pincher, 1946).

Mutation. (1) The alteration of a gene. This rare event can be made more frequent by treatment with radiation or certain chemicals; the mutation is then an induced mutation; otherwise it is known as a spontaneous mutation. (2) A change of heredity not ascribable to segregation or recombination. (De Vries 1890). Classifiable as follows.-Gene Mutation. Intragenic change, i.e. from one gene to an allelomorph. A residual category after the exclusion of identifiable structural changes.

A mutation is a change in the genes. The original gene is usually called the `wild type' or, in human medical genetics, the `normal' gene. Strictly, the `mutation' is an identifiable change in one allele of the gene, i.e. one of the two copies in a cell. However, some scientists assume that if a gene is different from its form in the majority of individuals, or if the variant of the gene does not work, then it is a mutant. Mutations come in a variety of forms, depending on exactly how the DNA sequence of the gene is altered. These are.

Point mutation. a single base is altered, usually to another base.

Deletion. a region of DNA is missing. This can be anything from a few bases to millions of bases.

Insertion. a new bit of DNA is inserted into a gene. Usually this is a result of an insertion element, a piece of DNA that can put copies of itself in new places round the genóme, such as a transposon. A variation one the deletion/insertion theme is mismatch recombination, which alters the number of copies of a sequence in a repetitive array, like a VNTR (see DNA fingerprinting).

Rearrangements. everything else is a rearrangement. These are usually rearrangements of large regions of DNA rather than of a few bases. Chromosomal rearrangements shuffle whole chromosomal regions, switching them endo-to-end (inversion) or swapping regions between chromosomes (transposition). Sometimes transpositions are straight swaps (balanced), sometimes some DNA gets lost in the process (unbalanced). These are natural mutations. The other sort are:

Artificial mutations. These can be induced mutations, where we bombard an organism with radiation or mutagenic chemicals to accelerate the normal rate of mutation-any of the above mutation types can result. Alternatively, we can manipulate the DNA itself, a process called in vitro **mutagenesis** or **site-directed mutagenesis**. Mutations in genes are detected by a wide range of DNA probe or

PCR technologies. If you do not know what the mutation is, then it can be detected by sequencing the DNA and looking for differences. There are also el electrophoresis techniques that will show if a target piece of DNA is different from one you have already cloned. The most common are denaturing gradient gel electrohoresis (DGGE) and single strand conformational polymorphism analysis (SSCP). These will rapidly tell you if a difference exists, but not what the difference is. Gene `chips' are also being developed to detect mutations.

Mutational load. Accumulation of deleterious genes resulting from Mutations. genetic disadvantage sustained by a population of organisms as a result of such mutations.

Mutator gene. A gene that increases the mutation rate of other genes in the same organism.

Muton. The smallest segment of DNA or subunit of cistron capable of undergoing mutation; probably the unit is as small as a base pair (purine or pyrimidine).

Mutual translocation. Reciprocal transfer, or crossing-over, between the terminal portions of two non-homologous chromosomes.

Mutualism. The type of symbiosis, such as that exhibited by fungi and algae or cyanobacteria in forming lichens, in which both species profit form the association.

Mycelium (Gr. *maykes*: mushroom) (pl., *Mycelia*). In the fungi, a mass of hyphae, It is a matlike organization of fungal hyphae. One of the mass of interwoven threadlike filaments forming the vegetative portion of the thallus in fungi.

Mycetismus. Mushroom poisoning.

Mycobacteria. Bacteria with unusual cell walls which are resistant to digestion, being waxy, very hydrophobic and rich in lipid, especially esterfied *mycolic acids.* Staining properties differ from those of Gram negative from Gram positive organism, being acid fast. Many are intracellular parasites, causing serious diseases such as leprosy and tuberculosis. Cell wall has strong immunostimulating (*adjuvant*) properties due to muramyl dipeptide (MDP).

Mycobacterium microti. *Mycobacterium* which causes tuberculosis-like disease in small rodents (*Microtus microlus* is the vole); will infect mice but not humans, and is therefore much used as a laboratory model. Releases large amounts of cyclic AMP which may inhibit *lysosome- phagosome* fusion.

Mycobacterium tuberculosis. Obligate anaerobic non-motile bacterium, causative agent of tuberculosis in humans. Lives intracellularly in macrophages.

Mycobiont. The fungal partner of a lichen.

Mycoherbicide. A compound produced by or isolated from a fungus that shows potent anit-plant activity.

Mycology. Systematic study of the true fungi i.e., yeasts and molds.

Mycolysis (Gr. *mykes***.** fungus + *rhiza***.** root). An association of the root of a plant with the mycelium of a fungus.

Mycoplasma. Mycoplasmas are procaryote, without cell wall and have been placed under the class Mollicutes (Latin, *mollis* = soft, pliable + *cutis* = stain) and the order Mycoplasmatales. Mycoplasmas or molicutes (soft skin) are without cell wall and are bounded by triple-layered membrane. They are smallest microorganism which have been known to cause a number of diseases in animals and humankinds. In Germany, contagious bovine pleuropneumonia, a fatal disease, first recognised in 1693 and in 1843, it reached to USA. Louis Pasteur first noticed them while observing the causative agent of pleuropenumonia in culture. He was not able to isolate them in a pure culture medium. Nocard and Roux (1898) of Pasteur's laboratory cultured the microorganisms in media containing serum and demonstrated that the pleomorphic microbes could produce the diseases in inoculated healthy cattle. Elford, with the help of special filter first recorded the size of these organisms. A number of similar microorganisms were also isolated from diseased animals between 1920-1940 and were termed pleuropneumonia like organisms (PPLO).

These organisms, now termed mycoplasmas, are the smallest free living cells known, much similar to other microorganisms like bacteria, chlamydiae, rickettsiae and viruses. In fact, mycoplasmas were not suspected to be causal agents of plant diseases. However, as early as 1896 Hugo de Varies referred to cases of virescence assuming epidemic proportion. 27 plant species were severely infected in his garden. De Varies believed that disease was spread by flying insects. An 1926, an American scientist L. O. Kunkel demonstrated that insects were capable of transmitting yellow type of diseases, which is quite common in plant. Subsequently Kunel (1951), Black (1953). Maramorosch (1953) too showed that aster yellow virus could be transmitted from infected to healthy plants with the aid of grafting material and the diseases plants recovered after having been cultivated at a high temperature and by application of antibiotics of tetracycline group, but no one could isolate and purify the causal agent (virus) suspected.

In 1967, Japanese scientists Doi, Teranaka, Yora and Asuyama surprised the plant pathologists by their publication (Ann. Phytopathol. Society, Japan, 33: 259) stating that Muberry dwarf, potato witche's broom aster yellows or Paulownia witche's broom which were suspected to be the virus diseases, were in fact caused by Mycoplasma. Simultaneously Ishiie, Doi, Yora and Asuyama (1967) pointed out the suppressive effects of antibiotics of tetracycline group on symptom development of Mulberry dwarf diseases. According to GTN de Leeuw of The Netherlands, the actual discovery of the possible role of mycoplasma as causal agent of these diseases seems to be made by Koshimizu, a microbiologist of the institute of veterinary sciences in Tokyo (Japan) who observed the electron micrograph preparations of Doi and co-workers in the study of Mulberry dwarf. The organisms were closely similar to *Mycoplasma gallisepticum*, isolated from chickens.

The publications of Japanese workers inspired other plant pathologists to find out the causal agents for yellows of diseases de Leeuw (1977) stated in 1967 that these organisms were found in four plants, where as in 1976, as many as 150 plant species were known in which mycoplasmas could be demonstrated.

Mycorrhiza. The association between the hyphae of a fungus and the roots of a higher plant. Two main types of mycorrhiza exist, ectotrophic in which the fungus forms mantle around the smaller roots, as in trees, and endotrophic in which the fungus grows around and within the cortex cells of roots, as in orchids and heathers. In ectotrophic mycorrhizae the fungus, which is usually a member of the Agaricales, benefits by obtaining carbohydrates and possibly B- group vitamins from the roots. The trees benefit in that mycorrhizal roots absorb nutrients more efficiently than uninfected roots, and it is common forestry practice to ensure that the appropriate fungus is applied when planting seedling trees. In endotrophic mycorrhizae the fungus is generally a species of Rhizoctonia and again both partners benefit nutritionally from the relationship.

Mycosis. A disease of animals caused by fungi.

Mycotoxicosis. Poisoning of humans and animals following the ingestion of foods containing toxins; produced by fungi.

Mycotoxin. A poison (toxin) produced by and associated with fungi.

Mycropyrenic. (i) Having nuclei which are considerably smaller than the average for the species (or individual, etc.). (ii) An individual of this type.

Myelin sheath. An insulating covering that surrounds the axon of a neurone. It is composed of the cell membranes of Schwann cells wound tightly in a spiral around the axon. The membranes consist of a fatty material (myelin). In between each Schwann cell is a short region of bare axon (node of Ranvier). Myelinated (medullated) axons occur in most vertebrate neurones but are less common in invertebrates. It serves to insulate the axon electrically and to increase the rate of transmission of a nervous impulse.

Myelin. A material forming a sheath around some axons. It is formed by Schwann cells that warp themselves about the axon. It serves to insulate the axon electrically and to increase the rate of transmission of a nervous impulse.

Myeloma proteins. Igs or proteins of Igs produced in large quantities by neoplastic plasma cells (plasmacytoma).

Myeloma. Cancer of antibody-producing cells characterized by proliferation of a single clone of plasma cells producing a pure immunoglobulin.

Myeloma. A lymphoma (plasmacytoma) produced from cells of the B cell lineage.

Myeloperoxidase. Peroxidase found in the lysosomal granules of *myeloid* cells, particularly macrophages and neutrophils; responsible for generating potent bacteriocidal activity by the hydrolysis of hydrogen peroxide (produced in the *metabolic burst*) in the presence of halide ions. A metalloenzyme containing iron. Deficiency of myeloperoxidase is not fatal, and it is reportedly absent entirely in chickens.

Myiasis. Infestation caused by maggots (fly larvae).

Myoblasts. Precursor cells which aggregate to form the multinucleated striated muscle cell.

Myofibrill. The bundles of myofilaments that form the contractile elements of muscle cells.

Myofilament. The individual actin (thin) and myosin (thick) filaments of the myofibril.

Myoglobin gene. Gene coding for the protein myoglobin. Unlike the globin genes, which contain two introns each, the myoglobin gene contains three, and these are of such a length that the exons together make up less than 5 percent of the total gene coding sequence.

Myoglobin. A conjugated protein found in muscles (sometimes referred to as muscle haemoglobin'). It is similar to haemoglobin in being a haemoprotein capable or binding oxygen but is structurally simpler, having only one polypeptide chain combined with the haem group. Each molecule of myoglobin can attach one atom of oxygen.

Myosin light chain kinase (MLCK). Calmodulin-regulated kinase of *myosin light chains;* molecular weight varies according to source, 130kD in non-muscle mammalian cells. May regulate activity of myosin in some cells.

Myosin. Protein molecules, each composed of two coiled subunits (MW-220,000), which can aggregate to form a thick filament which is globular at each end.

Myotonic dystrophy. Single-gene congenital disorder in the human, caused by an autosomal dominant gene on chromosome 19. The disease is characterized by progressive muscle weakness in early adult life and frequent cataract formation.

Mythogenesis. Biotechnology has been extraordinarily successful in attracting scientists and investment. This has occurred despite that fact that few biotechnology companies even today are breaking even, and there are very few truly biotechnological products on the market that were not there 10 years ago. One entirely reasonable explanation for this is that much of the new wave of biotechnology is aimed at medical problems, and these take a long time to solve, are great intellectual and social challenges, and could earn enormous profits for their conqueror. Another explanation, favoured by a few arts-trained observers, is that this is a post hoc rationalization for something much deeper, and that biotechnology is attractive because it promises the realization of very ancient dreams, in Jungian terms the physical embodiment of mythic archtypes.

Mythylmercuric hydroxide, CH_3, HgOH. A powerful denaturing agent which may be incorporated into agarose gels in order to determine the single-strand molecular weight of RNA or DNA molecules.

Myxobacteria. (slime bacteria). Saprophytic gliding bacteria that may produce brightly coloured fruiting bodies visible to the naked eye. The cells, which occur in masses of slime, are flexible unicellular rods that divide by transverse fission. Myxobacteria are common in soil, animal dung, and decaying plant matter.

The mycobacteria are rod-shaped nonsporeforming, aerobic bacteria that do not stain readily but, once stained, resist decolorization by acid or alcohol and are therefore called **acid-fast** bacilli. In addition to many saprophytic forms, the group includes pathogenic organisms (e.g. *Mycobacterium tuberculosis, Mycobacterium leprae*) that cause chronic diseases producing lesions of the infectious granuloma type.

Mycobacterium tuberculosis. In animal tissues, tubercle bacilli are thin straight rods measuring about 0.4 x 3 μm. On artificial media, coccoid and filamentous forms are seen. *Mycobacteria* cannot be classified as either gram-positive or gram-negative. Once stained by basic dyes they cannot be decolorized by alcohol, regardless of treatment with iodine. True tubercle bacilli are characterized by acid-fastness, e.g. 95% ethyl alcohol containing 3% hydrochloric acid (acid-alcohol) quickly decolorizes all bacteria except the mycobacteria. Acid-fastness depends on the integrity of the structure of the waxy envelope. The Zieho-Neelsen technique of staining is employed for identification of acid-fast bacteria. In sputum or sections of tissue, mycobacteria can be demonstrated by yellow-orange fluorescence after staining with fluorochrome stains (e.g., auramine, rhodamine).

Myxomycetes. The largest class of slime fungi, which have variously been placed in both the Myxomycophyta and the Eumycophyta. The fungal body is a plasmodium, and ingests food, but reproduction is by flagellate spores formed in sporangia.

Myxomycophyta (slime fungl). A division of simple organisms showing both plant and animal characteristics. They are multinucleate, lack cell walls, and display amoeboid movement ingesting food, leading some to group them with the Protozoa as *Mycetozoa.* However, in some genera spores with cellulose cell walls are formed in sporangia, this being a plant characteristic. The Myxomycophyta includes the classes Acrasiales and Labryinthulales; in its wider sense it also includes the Myxomycetes and the Plasmodiophorales.

Myxospore. A resting cell of a myxobacterium within a hard slime capsule. Also called a Microcyst.

Myxoviruses. Group of enveloped viruses with single-stranded RNA genome and helical capsids including those viruses responsible for influenza, mumps, measles, rubella and rabies.

N

^{15}N. A nonradioactive heavy isotope of nitrogen that is used as a marker in cell biology and in studying the biochemistry of protein synthesis and nitrogen fixation.

1-Naphathaleneacetamide. A synthetic hormone analog of the auxin type. It is sometimes used in plant tissue culture media. It dissolves in base.

2-Naphthaleneacetamide acetic acid or 2-naphthoxyacetic acid or b-naphthoxyacetic acid (NOA or b NOA, $C_{12}H_{10}O_3$ mw 202.20). A synthetic hormone analog of the auxin type. It is sometimes used in plant tissue culture media and in horticulture to promote rooting of cuttings. It dissolves in base (ca. 1M KOH or NaOH).

2-Naphtoxyacetic acid or ß-naphthoxyacetic acid (NOA or BNOA,). A synthetic hormone analog of the auxin type. It is sometimes used in plant tissue culture media and in horticulture to promote rooting of cutting. It dissolves in base.

5'-Nucleotidase. Enzyme which cleaves the 5' monoester linkage of nucleotides, and thus converts them to the corresponding nucleoside.

N and 2n. Designations for the gametic and zygotic chromosome numbers, respectively, in a species. In higher plants and animals, *2n* also indicates the somatic chromosome number.

NACS™ (pronounced nacks to rhyme with sacks). An ion-exchange resin to which nucleic acids can be quantitatively bound in low-salt buffers and then released by washing with a high-salt buffer. NACS columns are used to free DNA or RNA from contaminating low-molecular-weight compounds and are especially useful for purifying DNA which has been extracted from polyacrylamide or agarose gels.

NAD (Nicotinamide adenine *dinucleotide*): A compound found in living cells, existing in two interconvertible forms: the oxidizing agent NAD and the reducing agent NADH.

NADP (Nicotinamide adenine dinucleotide phosphate). Like NAD, but possessing another phosphate group; plays similar roles but is used by different enzymes.

NADPH. Nicotinamide adenine dinucleotide phosphate, reduced form.

Naegleria. Genus of amoebae found in the soil and best known for their transition from an amoeboid to a flagellated cell form in experimental conditions involving changes in ionic content of the solution in which the cells are suspended.

Nagarse (nagarase). Broad-specifity protease from bacteria.

Nagative control. The situation in which a regulatory macromolecule (generally a repressor) functions to turn off transcription. In the absence of a regulatory macromolecule, the structural genes are turned on.

Nagler's reaction. Standard method for identifying *Clostridium perfringens.* When grown on agar containing egg yolk, an opalescent halo is formed around colonies which produce Â-toxin (lecithinase).

Nail patella syndrome. Inherited disease in the human, due to a dominant gene on chromosome 9, and expressed in the phenotype as small or absent kneecaps and deformed fingernails.

Nalidixic acid. Drug used in molecular biology for its specific action in inhibiting DNA gyrase in *E. coli.*

Names. One of the most fiercely competitive areas for biotechnology start-ups is finding a good name. As well as the obvious ones (`Monoclonal Antibodies Inc', `Affinity Chromatography Ltd.') biotechnology companies' names are assembled from a wide range of standard units. Start with one of: **bio**-: almost an essential part. Means, to do with life. **immu**-or **immuno**-: to do with the immune system, usually to do with antibodies. **hyb**- or **hybri**-: usually to do with DNA hybridization. Can also be related to making hybrid species. Hybritech is the maverick here, being concerned with monoclonal antibodies. **trans**-: across, suggesting multidisciplinarity. `Transgenics' is a special case. **eco**-: now needs no introduction-anything that can be considered `ecological'. **agro**- or **Agri**-: agricultural. **myco**-: to do with fungi. **onco**-: to do with cancer. **cyto**-: to do with cells (usually means mammalian cells). **gen**-: to do with genes, and hence with recombinant DNA. **enz**- or enzo-: to do with enzymes. and finish with one of: -**gene** or **-gen**: anything to do with genes. **-zyme** or **-ase**: to do with enzymes. **-med** or **-medix** or -medic or -medics: implies an application in the heathcare industry. -**tech**: superfluous, and obvious. **-probe**: either something to do with DNA probes, or something to do with medical diagnostics. Ideally both. **-clone**: suggests recombinant DNA technology. The name can also have **Sciences**, **Systems**, or **Technology** added to the end. If the name has several words in it, a memorable acronym can be helpful: hence DNAX, ABC, etc.

Nanometer. Unit of measurement; one nanometer (nm) equals 0.001 μm, or 10^{-9} meter; formerly referred to as millimicron.

Nanometer: Unit of measurement; one nanometer (nm) equals 0.001 æm, or 10-9 meter; formerly referred to as millimicron.

Nanotechnology. Nanotechnology is the production of machines of nanometre size components, popularized by K. Eric Drexler. Enthusiasts say that biotechnology can provide the methods necessary to build such small machines. This is because many biological molecules assemble themselves chemically into large complexes and the nanotechnologists hope to use the same self-assembly processes to get the nanomachines to build themselves. A much quoted example is that of a tiny submarine which can be injected into a patient to clear out arteries clogged by atherosclerosis. Biology could provide some of the elements of this (for example, the world's smallest screw propeller is the flagellum of a bacterium). However, this is definitely 21st century stuff, and is unlikely to be realised in any of the forms being discussed today. Just think how Charles Babbage saw his calculating engine, and what computers actually looked like when they became real. **Micromechanics**, building engineering structures on silicon chips, works on a scale of tens of micrometres rather than the 100-nanometre scale needed of nanotechnology, and has focused on a few well-defined products such as pressure and strain gauges. The success of micromechanics in a few fields does not imply that molecular electronics or nanotechnology are realistic for the next few years.

Naphthaleneacetic acid (NAA, $C_{12}H_{10}O_2$ mw 186.20). A synthetic hormone analog of the auxin type. It is frequently used in plant tissue culture media and in horticulture to promote rooting of cuttings. It dissolves in base (ca. 1M KOH or NaOH).

Naphthaleneacetic Acid. A growth regulating substance which has been used with considerable effect, at low concentrations, in the induction of flowering.

Nascent. Describing a molecule that is in the process of synthesis and therefore still incomplete. The term is applied chiefly to the growing chains of polypeptides or polynucleotides on ribosomes or on the DNA.

Native DNA. Double-stranded DNA isolated from a cell with its hydrogen bonds between strands intact (cf. Denatured DNA).

Native immunity. A state of resistance conferred by one's genetic make-up, not acquired by exposure to infectious disease agents.

Natural acquired immunity. Resistance to disease following exposure of the individual to the virus or organism causing the disease at some earlier time.

Natural active immunity: Immunity following recovery from disease.

Natural complex. A complicated, non-synthetic, often undefined addendum in plant tissue culture media, such as orange or tomato juice or coconut milk.

Natural control. The process of dynamic equilibrium which maintains the characteristic mean density of a wild population within particular upper and lower limits, over a period of time, by a complex combination of all the additive, conditioning, and subtractive processes impinging upon that wild population.

Natural enemies. Strictly, the parasitoids, predators, and pathogenic microorganisms associated naturally with a specific wild population of plants or animals, and causing mortality or debility to the individuals thereof; often used in general sense for all parasitoids, predators, and pathogens.

Natural passive immunity. Immunity resulting from the passage of performed antibodies from the mother to the fetus or newborn.

Natural products. Natural products are any chemical or material derived from any organism, as opposed to ones that are synthesized chemically. In biotechnological terms this often means drugs or therapeutics derived from living things, usually fungi or bacteria. There are a range of examples. Microorganisms produce a wide range of chemicals which are used directly as drugs or as the starting materials for semi-synthesis of drugs, for example, penicillin and cephalosporin antibiotics. Major research groups and several specialist companies are dedicated to discovering new natural product drugs of this type of screening bacteria and fungi.

Natural selection. A theory which attempts to explain the mechanism by which evolution takes place. Like the theory of evolution; it is associated with Charles Darwin. It is an idea which deserves a great deal more attention than may seem to be called for at first glance. Briefly and superficially, it says that species will become fitter and fitter because the fittest individual of each generation will leave more children than the unfit and the children of the fit will tend to resemble their parents.

Nature and nurture. The antithesis between Genotype and Environment.

Nearest neighbour analysis. A technique used in the structural analysis of oligonucleotides. It is based on the fact that when a new nucleotide is added to the growing point of a transcript the Â-phosphate on its 5'-hydroxyl is linked to the 3'-hydroxyl of the preceding nucleotide.

Nebulizer. A device used to produce fine sprays from liquids.

Necktie Association (of chromosomes). Two attached bivalents connected by a `hump' on each of them. Such a hump is thought to be an indication that chiasmata cannot be terminalized, either by reason of lack of homology at the ends of the chromosomes or through the inhibiting effect of secondary constrictions.

Necrohormones. Products of degenerating nuclei which are thought to stimulate mitosis.

Necrosis (Gr. *necros*, dead + suffix *sis*). Refers to pathologic death of tissue.

Nef (Nephritic factor). An autoantibody to the alternative pathway C3 convertase which stabilizes the enzyme and potentiates the action of the alternative pathway amplification.

Negative assortative mating. The tendency for dissimilar individuals to mate.

Negative complementation. Situation when complementation between alleles permits a mutant subunit to inactivate the wild-type subunit with which it associates in a multimetric protein.

Negative interference. Tendency for one recombinational event and its resulting cross-over to increase the likelihood of another similar exchange in its vicinity. Compare Positive Interference.

Negative regulation. Negative feed-back in biological systems mediated by allosteric regulatory enzymes.

Negative stain. Stain which colors the background while the cells appear clear, or unstained.

Negative staining. Staining technique used in the electron microscopy of viruses and other particulate material. S suspension of the material to be examined is mixed with a solution of phosphotungstic acid and sprayed onto a carbon-coated grid by an atomizer. The fine detail of the contours of the specimen is rendered more evident by this technique.

Negative supercoiling. Supercoiling of a duplex DNA molecule upon itself in a direction opposite to the turns of the double helix strands. Compare Positive Super Coiling.

Negri bodies (After Dr. P. Negri, Italian scientist)**.** Crystalline inclusion bodies formed due to aggregation of virus crystals in cells infected by rabies virus; named after their discoverer. These inclusion bodies are found in brain cells of animals infected with rabies virus.

Neighbourhood. A population in which the individuals are neighbours in the sense that their gametes may come together.

Neisseria. Genus of Gram-negative diplococci; genus named for bacteria of gonorrhea and meningococcic meningitis.

Nekton (Gr. *nekhein* = to swim)**.** Animals, such as fish, that can swim against currents of water. (Contrast with plankton).

Nematocyst (Gr. *nema* = thread + *kystis* = cell)**.** An elaborate, threadlike structure produced by cells of jellyfish and other cnidarians, used chiefly to paralyze and capture prey.

Nematode. A member of the phylum Nematoda; a roundworm.

Neo-Darwinism. The theory that evolutionary change is due to variability of individuals from mutation and natural selection.

Neoergosterol. A steroid, of the Vitamin D group, which shows oestrogenic properties.

Neomorph. Mutant allele that produces a qualitatively new and different effect to that produced by the wild-type allele.

Neomycin resistance. The ability to survive treatment with the antibiotic neomycin carried on a plasmid (pUB110) specific to *Bacillus* species.

Neomycin. Aminoglycoside antibiotic produced by certain species of *Streptomyces* with a broad spectrum of action against bacteria. Toxic side-effects make it unsuitable for systemic use. The biochemical action of neomycin is by selective inhibition of protein synthesis on the 70S (prokaryotic) ribosome; in this respect it closely resembles streptomycin.

Neonatal mortality rate. The number of deaths reported among infants under one month of age in a calendar year per 1000 live births in the same year and place.

Neonate. A newborn infant (up to 4 weeks old).

Neoplasm. Localized cell multiplication or tumor; a collection of cells which have undergone genetic transformation. These cells differ in structure and function from the original cell type.

Neoplastic growth. Indicates tumorous growth arising out of abnormal localized multiplication of certain types of cells; malignant, if the growing cells infilterate surrounding tissue and/or are carried by blood or lymph to other locations in the body, growing there more vigorously. Otherwise benign, i.e, localized at a particular place only and relatively less harmful.

Neoplastic. A disease.

Neoteny. The retention of larval or other juvenile features beyond the normal stage in the development of an animal. It may be either temporary, because of climatic or other factors, or permanent, in which case the animal breeds in the larval stage. Neoteny is thought to be important in the evolution of some groups including man, who has certain resemblances to young stages of apes.

Nephron. The excretory unit of the vertebrate kidney, comprising a Mal-pighian body (glomerulus and Bowman's capsule) and a uriniferous tubule. Water, salts, nitrogenous wastes, etc., are filtered across the walls of the glomerulus and collected by the Bowman's capsule. As the filtrate passes trough the tubule, useful substances are selectively reabsorbed into surrounding capillaries, which join the renal vein. In mammals, the remaining waste (urine) is conducted via collecting ducts to the renal pelvis and hence to the ureter.

Neritic zone. The shallow part of the ocean, situated over the continental shelf.

Nerurofilament. Filament within a nerve cell, composed of three different intermediate filament polypeptides and thought to have a role in strengthening axons.

Net primary production. Total photosynthesis minus respiration by plants.

Neural cell adhesion molecule (NCAM). Cell surface glycoprotein of vertebrates initially defined by antiserum capable of inhibiting cell adhesion of chicken neural retina cells. Believed to be important in calcium-in-dependent intercellular adhesion of neural and other cells, including cells in early embryos.

Neural crest. A band of ectodermal cells formed on either side of the neural tube in vertebrate embryos. It gives rise to the dorsal root gangila of the spinal nerves.

Neural. Having to do with the nervous system.

Neuraminic acid. Important constituent of many phospholipid membranes. It is a glycolipid and is particularly abundant in the plasma membranes of neurones.

Neuraminidase (sialidase). Enzyme catalyzing cleavage of neuraminic acid residues from oligosaccharide chains of glycoproteins and glycolipids. Since these residues are usually terminal, neuraminidases are generally exoenzymes, although an endoneuraminidase is known. For use as laboratory reagent, common sources are from bacteria such as *Vibrio* or *Clostridium.* A neuraminidase is one of the transmembrane proteins of the envelope of influenza virus.

Neurobiotaxis. A change in the position of nerve centres in the brain during phylogenetic evolution due to a change in the nerve-fibres which any given nerve-centre habitually receives its impulses.

Neuroblast. Any of the cells of animal embryos that become or produce nerve cells, frequently by unequal mitotic divisions. Although many of the cells of the neural tube of vertebrates will produce nerve cells, the term neuroblast is usually restricted to those cells whose immediate progeny will differentiate as nerve cells.

Neuroendocrine systems. The systems involving both nervous and endocrine factors that control functions of the body. Many examples are known, particularly involving the pituitary gland, from which hormones are secreted under direct nervous stimulation.

Neurofibromatosis. Single disorder in humans caused by an autosomal dominant gene with full penetrance. It is characterized by numerous skin neurofibromatoma tumours, coffee-coloured patches (cafe-au-lait patches), and sometimes mental retardation and spinal cord compression.

Neurohormone. A hormone that is produced by specialized nervous tissue. Examples are noradrenaline, serotonin, vasopressin, and oxytocin.

Neurokinin. A biologically active peptide; one of a class of peptides known as tachykinins.

Neurological mutant. Organism displaying abnormalities of locomotion or behaviour traceable to malformation or malfunction of the central nervous system.

Neuromuscular junction. The synapse between a motor neurone and a muscle cell membrane. It is similar in structure and function to the synapse between neurones.

Neurone theory. The theory, now accepted, that the nervous system is composed of many discrete nerve cells (neurones), closely contacting each other at synapses. It replaces the reticular hypothesis that there is cytoplasmic continity between cells.

Neuropathy. General term for behavioural or neuronal diseases, some of which are presumed to have a hereditary basis.

Neuropeptides. Peptides with direct synaptic effects (peptide neurotransmitters) or indirect modulatory effects on the nervous system (peptide neuromodulators).

Neurophil. A leukocyte in which the granule do not have an apparent affinity for acid or basic dyes.

Neurosecretion. Production of neurohormones by specialized secretory cells of the nervous system.

Neurospora. A fungus belonging to the Ascomycetes, which include the molds, yeasts, and many plant parasites. It is a favorable organism for research because it grows quickly under laboratory conditions and requires no complex substances in the growth medium except biotin, and because its sexually produced spores are arranged in packets of eight, representing twin products of the four cells of a meiotic division.

Neurotensin. Tridecapeptide hormone (sequence: elyenkprrpyil) of gastrointestinal tract: has general vascular and neuroendocrine actions.

Neurotoxin. Toxin that affects the nervous system specifically. Neurotoxins are found in some snake venoms, as well as other sources, and include cobratoxin and Â-bungarotoxin which act by binding to acetylcholine receptors on motor endplates.

Neurotransmitter. A chemical that is released from neurone endings to cause either excitation or inhibition of an adjacent neurone or muscle cell. It is stored in minute vesicles near the synapse and released when a nerve impulse arrives. The most widespread neurotransmitter is acetylcholine.

Neuter. Describing individuals that lack sex organs. For example, some flowers in the family Compositae lack stamens and pistils.

Neutral alleles. Alleles that differ so slightly that the proteins for which they code function identically.

Neutral mutation. Mutation that confers neither selective advantage or disadvantage on the organism concerned. According to a theory advanced by M. Kimura, most nucleotide substitutions results in neutral or near-neutral mutations. Such mutations may spread in a population by chance since there is no selective pressure against them. However, many geneticists disagree with Kimura's interpretation and hold that neutral mutations are rare or non-existent, the argument turning on how fine an advantage or disadvantage must be to be significant or measurable.

Neutral protease. Protease which is optimally active at neutral pH.

Neutral red. Biological stain and acid/base indicator which gives a red colour in acid; and an orange-brown in basic conditions. It is normally employed as a solution in 70 percent ethanol.

Neutral substitution. Amino acid substitution in a protein that does not affect the activity of the molecule.

Neutralizing antibodies. Antibodies which render invading viruses noninfective.

Neutron. One of the three most fundamental particles of matter, with mass approximately 1 amu and no electrical charge.

Neutrophile. Micoorganism having optimum growth at neutral pH.

New diseases. Because it was a panoply of new and powerful techniques at its disposal, biotechnology is always looking for new ways of using them. One way is to identify a disease that has not been identified before, or one that is now thought to be much more serious than it was previously, and develop a treatment for it. For other diseases, of course, treatments already exist, which makes it harder for a new one to be accepted. Some of the `hot' diseases discussed as targets for biotechnological solutions are: Any viral disease (since there are so few antiviral drugs). Particularly AIDS, but also:

Hepatitis, which is degenerative liver disease. Hepatitis A, B, C, and D viruses are recognized, as well as environmental causes such as alcohol or solvent abuse.

Herpes simplex, and particularly genital herpes, which can be life-threatening to new babies if caught from their mothers. Cytomegalovirus (CMV) is another of the herpes group for which better treatment would be useful.

Lyme disease. A debilitating bacterial disease caused by the spirochete Borrelia burgdorfei first recognized in 1982 and now affecting thousands of people. A vaccine is wanted. *New* viruses, such as *Lassa, Ebola, Marburg*, and *Hanta viruses*. These generally are not natural pathogens for humans, but if humans get infected the viruses are almost lethal, usually very quickly. Despite the film

Outbreak, starring Dustin Hoffman, these viruses are rare in the West and seldom spread beyond the immediately infected individuals. *Helicobacter pylorii*, a bacterium that scientists now accept is a major cause of gastric ulcers.

New reunion (of chromosomes). Rejoining between a broken end and any other broken and except the one to which it was originally attached.

Newcommer's fluid. A fixative used in chromosome analysis composed of 6 parts isopropyl alcohol : 3 parts propionic acid : 1 part petroleum ether : 1 part acetone : 1 part dioxane (with or without ferric acetate).

NG (N-methyl-N-nitro-N-nitrosoguanidine). Potent mutagen acting only *in vivo*. its mode of action is primarily through alkylating the 0-6 of guanine bases in DNA which leads to mispairing or errorprone repair synthesis. NG has a tendency to produce clusters of mutations at the replicating fork of DNA, but is widely used to induce mutations in bacteria.

Niacin or nicotinic aid ($C_6H_5NO_2$ mw 123.11). Vitamin B_3 forming part of a respiratory coenzyme. It is a common micronutrient addition to plant tissue culture media. It is sometimes also known as vitamin PP.

Niacin test. A diagnostic test used to detect those species of *Mycobacterium* which produce-and liberate into an agar medium an easily detectable amount of niacin (i.e. nicotinic acid).

Niacin. A component of the vitamin B_2 complex.

Niacinamide or nicotinamide ($C_6H_6N_2O$ mw 122.12). An amide of niacin occasionally added to plant tissue culture media.

Niche. The biological role of an organism in its community, which includes its location and its activities.

Nick translation. A means of introducing radioactively labeled nucleotide triphosphates into native DNA *in vitro,* employing *E. coli* DNA polymerase l. Using a 5'-PO_4/3'-OH single-strand nick in a native DNA molecule, DNA polymerase l catalyzes the addition of an exogenous radioactive nucleotide to the 3'-OH side of the nick and hydrolyzes a nucleotide from the 5'-PO_4 side of the nick. This moves the nick along the native DNA molecule by one nucleotide. Subsequent polymerization and concomitant hydrolysis translates the nick along the DNA molecule while introducing more radioactive nucleotides into the DNA.

Nick. A break or cut in one of the two strands of a DNA duplex.

Nick-closing enzyme. A form of DNA polymerase that can restore base sequences. A nick is introduced adjacent to an incorrect base. The base is then removed and the correct base inserted.

Nicking. A fortunate combination of genes which produces a more positive effect than the sum total of the single gene effects (Pincher, 1946).

Nicking-closing enzyme (topoisomerase l). Enzyme involved in the normal DNA replication. The enzyme in *E. coli* is a monomer (Mr = 110 000) which uses negatively supercoiled DNA as a template. Its function is to relax supercoiling stress which would other-wise build up in the DNA during unwinding for replication.

Nidation. The embedding of a fertilized ovum in the uterine wall.

Nif. The genetic notation for the genes involved in nitrogen fixation. The *nif* operon is a complex array of seventeen genes. The proteins encoded by the *nif* genes will fix atmospheric nitrogen (N_2) into ammonia (NH_4^+) and nitrate NO_3^-). Many soil bacteria will fix nitrogen and there is much interest in manipulation of *nif* genes from bacteria to allow plants to fix nitrogen.

Nigrosin. A black composite dye which includes one or other of the indulins; it is used e.g. for negative staining in light microscopy.

NIH guidelines. Recommended procedures for the conduct of recombinant DNA experiments to which all NIH grant-holders must adhere.

Ninhydrin reaction. Biochemical test used widely to detect amino acids and proteins separated by paper chromatography. Triketohydrinadene hydrate (ninhydrin) when heated with the test chromatograph produces a blue colour in the presence of proteins or amino acids.

Ninhydrin. A reagent used to test for the presence of proteins and amino acids. An aqueous solution turns blue in the presence of alpha amino acids in solution. When dissolved in an organic solvent it is used as a developer to colour amino acids on chromatograms. If a chromatogram treated with ninhydrin is heated strongly the amino acids appear as purple spots that can be identified by measuring the Rf value. Ninhydrin is carcinogenic.

NIP (-hydroxy, 5-iodo, 3-notrophenylacetyl). A commonly used hapten.

Nipple flask. A flat bottomed, round sided flask with many side projections or nipples. Suspension cultures are agitated and aerated in these flasks as medium flows into and out of the projections while the apparatus rotates. The flasks and rotating apparatus were designed by Steward et al. ca. 1952.

Nitrate assimilation. The conversion of nitrate to ammonia. For most plants in their natural environment nitrate is the usual source of nitrogen.

Nitrate reductase. An enzyme that catalyzes the conversion of nitrate to nitrite. Eukaryotic nitrate reductase is an enzyme complex containing flavin (FAD), haem (cytochrome b557) and molybdenum. Electrons are provided by NAD(P)H derived from photo-synthesis or carbohydrate oxidation. Prokaryotic nitrate reductase is a smaller protein containing molybdenum, but no haem or cytochrome. The enzyme has a fast rate of turnover and is present at high concentrations when nitrate is available, but is repressed by the presence of ammonium ions.

Nitrate reduction. Reduction of nitrates to nitrites or ammonia.

Nitrate respiration (dissimilatory nitrate reduction). Under anaerobic or microaerophilic conditions, some bacteria can use nitrate as a terminal electron acceptor for respiratory metabolism.

Nitrification. The conversion of ammonia to nitrite, and nitrite to nitrate, carried out by certain bacteria in the soil. The chemosynthetic bacteria *Nitrosomonas* and

Nitrobacter carry out the first and second stages respectively of this conversion. The process is important in the nitrogen cycle since nitrate is the only form in which nitrogen can be used directly by plants. Compare Denitrification.

Nitrite reductase. An enzyme that converts nitrite to ammonia. Plant and photosynthetic prokaryotes possess a ferredoxin-linked nitrite reductase which accepts electrons from reduced ferredoxin, but not from NAD(P)H. The enzyme contains a tetranuclear iron-sulphur centre and a special haem (siro) similar to that found in sulphite reductase. Nitrite is reduced directly to ammonia without the formation of free intermediates.

Nitrocellulose filter. Widely used material in molecular biology, consisting of a paper-like mat of nitrocellulose fibres. Double-stranded nucleic acids adhere to it selectively, whereas single-stranded molecules pass through the filter. Filters are used in southern blotting of gels.

Nitrogen balance. Measurement of nitrogen equilibrium of the body. Positive balance occurs when new body protein is being formed (pregnancy, growth, and healing); negative balance occurs when body protein is being broken down (fever, surgery, burns).

Nitrogen cycle. A geochemical cycle in which atmospheric nitrogen gas is converted into a usable form for plants by microbial enzymatic reactions.

Nitrogen fixation. Nitrogen is an essential macronutrient (something we need a lot of in our diet) for all living things. Air comprises 80% of nitrogen gas: however, plants and animals cannot convert this into protein. They rely, instead, on other forms of nitrogen: ammonia and nitrates for plants, proteins and amino acids for animals. Only a few organisms can convert atmospheric nitrogen into these forms of nitrogen, which can be assimilated readily, a process called `fixing'. The rate at which fixed nitrogen can be supplied to growing crop plants is one of the limiting factors in their growth and yield. Nitrogen-fixing organisms are bacteria. Some live free in soil and some in symbiosis with plants. The symbiotic ones are the most interesting to the biotechnologist, although the free-living organisms like *Azobacter* and *Klebsiella* are easier to handle in the laboratory, and so most researchers prefer to use them. Symbiotic nitrogen-fixing organisms live in nodules in the roots of a few plants, and convert atmospheric nitrogen to ammonia for the plants in return for a supply of C4 acids, made by the plant from carbon dioxide. The genes that code for the enzymes that fix nitrogen (the nif genes) have been cloned and characterized in some detail. *Klebsiella pneumoniae's nif* genes have been studied the most: there are at least 17 of them, showing how complex this process is. The nod genes, which induce the plant to make nodules in which the bacteria can live, are less well characterized, but the subject of intense study. Biotechnology has tried several routes to fixing nitrogen for agriculture more efficiently. Only a few types of crop plants (legumes, clover, rice, lupins) fix nitrogen via symbiotic bacteria of the genera *Rhizobium* and *Bradyrhizobium*, which live in their root nodules. Some other non-legumes fix nitrogen, but are not widely used as crops. One route is to force the *Rhizobia* to live in other plants, by introducing the bacteria to the plants in tissue culture or by engineering the cell-surface receptors of the plant root cells so that the bacteria absorb on to these roots in the same way as they do on to beans and clover. This route has been moderately successful on a laboratory scale. Another route, much touted 10 ears ago, was to transfect the nif genes into the plants themselves so that they did not need the bacteria at all. This is now thought to be a route that is not likely to work. The bacteria provide a lot more enzymatic machinery than just the nif genes to fix nitrogen, and the rots also provide specific proteins (such as the plant haemoglobin protein, leghamoglobin) which are an essential part of the nitrogen-fixing process: the nodules are not just passive containers for the bacteria. The simplest use of biotechnology is in producing legume inoculants, to increase the soil population of rhizobia around a growing legume. Since each plant has to pick up the bacteria from the soil (there are no bacteria in the seed), nitrogen fixation can be limited by the rate of infection of the growing roots. Thus, dosing the soil with suitable bacteria can give a better rate of fixation. (It is controversial as to whether this is an effective economic measure.) A variant on this approach is to improve the efficiency of the bacteria in fixing nitrogen. Bio Technica tried an engineered *Rhizobium meliloti* in 1988, in which there were several copies of the gene for **nitrogenase** instead of the usual single copy. Nitrogenase is the enzyme that actually takes nitrogen molecules from the air and splits them open. The engineered bacterium was used to infect alfalfa, but it did not result in any increased yield from the plant and so the trial was stopped. If fixing nitrogen frees a plant from dependence on soil nitrates, why don't all plants fix their own nitrogen? The reason is that fixing nitrogen takes a great deal of metabolic energy, so if there is any other way to get nitrogen for the plant (or indeed for bacteria), then they will take it as the more energy efficient course. It is not clear, therefore, whether getting plants that do not normally fix nitrogen to do so would actually decrease crop yields because they would divert a lot of energy away from producing the edible portions of the plant and into nitrogen fixation.

Nitrogen fixer: A microorganism that fixes atmospheric nitrogen, it converts nitrogen gas to ammonia. Few examples are as follows: Name of the plants *Rhizobium* spp. (Clover - *R. trifolii*, Bean- *R. phaseoli*, Lupinus- *R. lupini*, Ornithopus- *R. lupini*, Soybean- *R. japonicum*, Melitotus, Trigonella- *R. maliloti*, Pea - *R. leguminosarum*). The species of *Agrobacterium* (*A. radiobacter, A. tumefacions*) also fix atmospheric nitrogen to ammonia.

Nitrogen. An essential element found in all amino acids and therefore is all proteins, and in various other important organic compounds, e.g. nucleic acids. Gaseous nitrogen forms about 80% of the atmosphere but is unavailable in this form except to a few nitrogen-fixing bacteria. Nitrogen is therefore usually incorporated into plants as the nitrate ion, NO_3^-, absorbed in solution from the soil by roots. In animals, the nitrogen compounds urea and uric acid form the main excretory products.

Nitrogenase. An enzyme complex that catalyzes the reduction of atmospheric nitrogen (dinitrogen) to ammonia in nitrogen-fixing organisms. The enzyme has two major components, each of which has two or four subunits. The large component contains molybdenum, non-haem iron and sulphur. The enzyme, which is strictly anaerobic and is easily inactivated by molecular oxygen, also acts as a hydrogenase. Hydrogen production is a wasteful reaction, and those species with a low level of hydrogen production are more efficient nitrogen fixers.

Nitrogenous base. An aromatic N-containing molecule having basic properties (tendency to acquire an H atom). Important nitrogenous bases in cell are the purines and pyrimidines.

Nitrous acid (HNO_2). Acid that acts on cells as a mutagen by converting the purine and pyrimidine base amino groups to hydroxyl groups.

NK (Natural killer) cell. Natural killer cells are group of lymphocytes which have the intrinsic ability to recognize and destroy some virally infected cells and some tumour cells.

Nobilization. A term used in sugar cane breeding to denote the crossing of *Saccharaum officinarum* with related species, followed by one or more back-crosses to *S. officinarum.*

Nocardia. A genus of Gram-positive, aerobic bacteria of the Nocardiaceae; the genus characteristics are those of the family. According to species, colonial appearance and texture range from glossy and butyrous to matt and leathery; the former description is characteristic of those species which tend to fragment more readily. Species which undergo early fragmentation often give rise to microcysts coccoid forms with a heat resistance greater than that of the ordinary vegetative cells.

Nodulation and fixation genes. The genes involved in the Rhizobium-legume symbiosis, which is one of the most effective nitrogen-fixing systems. Identification of some *Rhizobium nif* genes has been achieved by hybridization with the *nif*YKDH fragment of *Klebsiella pneumoniae,* with cloning and analysis of special *Rhizobium nif* fragments in *E. coli.* Fixation and nodulation genes can also be identified by transposon mutagenesis. By this technique it is possible to analyze the whole symbiotic process, and hence to identify the genes involved by cloning the genes in *E. coli.* The next step is to alter the fixation and nodulation genes by genetic engineering techniques in order to construct Rhizobium strains that are more effective in nitrogen fixation.

Nomenclature. The method of assigning names in the classification of organisms.

Nomogenesis. The theory that the evolution of organisms is the result of certain processes inherent in them and that it follows definite laws.

Non-adaptive radiation. The evolution of several closely related and morphologically divergent forms without apparent ecological diversification (Cain, 1944).

Non-coding DNA. DNA that is in the nonsense strand (negative strand) of the DNA duplex, or is out with the coding sequence of any gene. It may or may not be taken to include intron sequences. Centromeric region DNA is non-coding.

Non-competent. Cells or tissues incapable of undergoing morphogenesis. The antonym is **competent.**

Noncompetitive inhibition. A form of enzyme inhibition in which a compound binds at a site other than the substrate binding site, causing a reduction in enzyme activity. This may be a form of allosteric inactivation or it may be due to irreversible damage to the enzyme.

Non-conjunction. Failure of metaphase pairing.

Noncontaminated area. An area in the laboratory building with definite boundaries designed to be free of harmful microorganisms.

Non-disjunction. Non-separation of chromosomes or chromatids at anaphase. In non-disjunction, chromosomes constituting bivalents or chromatids or mitotic chromosomes fail to move to opposite poles at anaphase. Various kinds are:

Non-equivalence. Term used in cell determination for cells which will give rise to the same sorts of differentiated tissues but which have different positional values (e.g., cells of fore-limb and hind-limb buds).

Non-histone proteins. Proteins found in the nuclei of eukaryotic cells that are not histones. Such proteins are acidic, unlike the positively charged histones, and include HMG protein, and the greaty variety of polymerases and gene regulatory proteins necessary for the control of transcription.

Nonhomologous. Of chromosomes or chromosome segments that contain dissimilr genes and that do not pair during meiosis.

Nonionic surfactant. A detergent that is uncharged in solution. Typical commercial products are obtained by condensing ethylene oxide on to the polar end of a fatty acid, fatty alcohol or alkyl phenol to provide a polymeric ether chain that terminates in a hydroxyl group.

Non-mendelian. Describing ratios in the progeny of a cross which fail to follow Mendel's Laws, due to phenomena such as Gene Conversion.

Nonoccluded viruses. Viruses which do not occur within inclusion bodies: also called noninclusion virus: cf. Occluded Viruses.

Non-permissive cell. Originally a cell of tissue type or species which does not permit replication of particular virus. Early stages of the virus cycle may be possible in such a cell, which is the case of tumour viruses may become transformed. Now used in a more general sense, of agents and treatments other than viruses.

Non-permissive condition. Condition that prevents survivors of conditional lethal mutations.

Non-polar. Referring to chemical groups that are not water-soluble since they lack the polar group necessary for aqueous interaction.

Nonpreference. Here used in the sense of Painter (1951) as an insect pest's response to a potential host (variety) which displays various characteristics that discourage its use for oviposition, food, or shelter.

Nonrandom mating. A mating system in which the frequencies of the various kinds of matings with respect to some trait or traits are different from those expected according to chance.

Non-reciprocal contact inhibition. Collision behaviour between different cells types in which one cell shows contact inhibition of locomotion, and the other does not. An example is the interaction between sarcoma cells and fibroblasts (the former not being inhibited).

Non-recurrent parent. Donor parent: that parent from which, by back-crossing, one or more genes are transferred to the backcross parent (recurrent parent). That parent of hybrid which is not again utilized as a parent in back-crossing.

Non-repetitive DNA. DNA that shows reassociation kinetics indicative of unique DNA.

Nonsense (chain-terminating) mutation. Mutations that change a codon for an amino acid to one of the codons (UAG, UAA or UGA) that signal termination of translation. The resulting gene product is a shortened polypeptide that begins normally at the amino-terminal end and ends at the position corresponding to the altered codon.

Nonsense mutation: A change in the cell's DNA with the effect that an amino acid has been left out o an essential protein.

Nonsense suppressor. Gene that codes for a molecule of tRNA which is able to respond to one or more nonsense codons.

Nonsense. A codon that does not specify any amino acid in the genetic code.

Nonseptate. Coenocytic; lacking cross walls, thus multinucleate.

Non-sibling species. When two species that do show close morphological identify and therefore would not be likely to be seen as sibling species.

Non-specific immunity (immunol). The constitutive or acquired resistance of the body to foreign materials in general (living and non living) including those with which it has had non previous contact. Factors which may be involved in non-specific immunity include e.g. (a) Phagocytosis. (b) The presence of Lysozyme in secretions (e.g. Teats). (c) The activitis of interferons and Properdin.

Non-spiking. The ability of a neuron to convey information without generating action potentials. As passive electrical potentials are attenuated over distances greater than the space constant for a neuron (typically 1 mm), this implies that most non-spiking neurons are involved in signalling over relatively short distances. Typical examples are invertebrate stretch receptors and interneurons in the central nervous system.

Non-transcribed spacer. DNA sequence that lies as a spacer between tandom copies of a transcribed gene sequence, but is not itself transcribed. Note that the so-called non-transcribed spacer, between the copies of the large ribosomal genes in Xenopus is now known to be transcribed.

Nopaline. One of the opine compounds that are synthesized by plant cells when present in crown gall and stimulate the *Agrobacteria* causing the gall to grow rapidly.

Norientation. Descriptive of the situation in which the vector and the inserted fragment of DNA have the same orientation.

Normal curve. A bell-shaped curve which represents the distribution of individuals segregating for a quantitative character, such as, height, yield intelligence etc. Usually the curve is humped in the midway between two extremes indicating that the majority of individuals in the population are grouped in this class called mode or model class. There may be two modes. The mode may be nearer to one end of the curve in which case it is called a skewed curve.

Normal deviate. The ratio of an observed deviation to the appropriate or corresponding Standard Deviation as fixed by hypothesis.

Normal distribution (normal curve of errors). The limit which is reached either by the Binomial or the Multinomial series where the power is large and none of the summed quantities very small in relation to the power and to one another. This frequency distribution is expected from a series of observations on a variate whose magnitude is affected by a large number of agents having small independent effects.

Normal flora. Microorganisms that have become established in a given area; e.g., those organisms that have found a permanent home in some area of the human body.

Northern blotting. Procedure analogous to southern blotting, except that the nucleic acids being transferred are RNAs and DNAs. Nucleic acids, previously resolved into subfractions by agarose gel electrophoresis, are transferred to nitrocellulose filter by capillary action. They are then bound to the sheet by heating and can be probed with radioactive labelled nucleic acids followed by auto-radiography, Note that although southern blotting was so called after its inventor, E. Southern, northern blotting was so called simply as contrasting name.

Nosocomial infection. Hospital acquired infection.

Nosocomial. Pertaining to the hospital.

Notochord. A rod of tissue that lies dorsally along the length of the bodies of members of the phylum chordata. In vertebrates, the notochord is partially or totally replaced by the vertebral column.

NP (4-hydroxy, 3-nitrophenylacetyl). A hapten which partially crossreacts with NIP.

N-terminus. The amino (-NH_2) end of a peptide chain, by convention written as the left end of the structural formula.

Nu bodies. Nucleosomes which, when visualized, appear like beads on a string on interphase chromatin viewed using an electron microscope.

Nuclear budding. The division of a nucleus into daughter nuclei by constriction.

Nuclear cage (nuclear lamina, nuclear matrix). Fibrous layer that lies immediately inside the inner membrane of the nuclear envelope. This laminar structure is attached to all of the nuclear pore complexes, as well as being associated with many loops of DNA of the chromosomes. It thus serves to provide a skeletal structure to the nucleus and perhaps also ensures continuity between actively transcribing chromatin and the nuclear pores themselves.

Nuclear disc. The star-like group of chromosomes on the equatorial plate at metaphase.

Nuclear emulsion. Photographic emulsion that is used to visualize the tracks of ionizing particles during radioactive decay. Such emulsions are used in autoradiography.

Nuclear envelope. The double membrane that surround the nucleus and is continuous with the endoplasmic reticulum.

Nuclear magnetic resonance spectroscopy. (NMR spectroscopy). Technique used in chemistry and biochemistry to identify different compounds in complex mixtures and to quantify the amounts present by analysing absorption specta following changes in the resonance frequencies of the atomic nuclei.

Nuclear membrane. The membrane surrounding the nucleus, dividing the nucleoplasm from the cytoplasm.

Nuclear pore. An annular-shaped hole that occurs in the nuclear membrane; usually large numbers are present. The inner and outer membranes of the nucleus become fused together at the annulus. Nuclear pores are the site of active transport of materials into and out off the nucleus. Nuclear imports include the nucleotide precursors of DNA and RNA, histones and ribosomal proteins, whereas exports include mRNA, tRNA and ribosomal subunits.

Nuclear RNA. The nucleus contains RNA which has just been synthesised, but in addition there is some which seems not to be released, or is only released after further processing, the heterogenous nuclear RNA (hnRNA) and small RNA molecules associated with protein to form small nuclear ribonucleoproteins (snRNPs).

Nuclear sap (karyolymph). The fluid phase within the nuclear membrane is which the nuclear membrane in which the chromosomes and the nucleolus (nucleoli) are suspended.

Nuclear system of heredity. The normal, mendelian, system in which inheritance is controlled by genes located on the chromosomes, as opposed to the corpuscular (=plastid) system and the cytoplasmic (=molecular) system.

Nuclear transplantation. Experimental approach in study of nucleo-cytoplasmic interactions, in which a nucleus is transferred from one cell to the cytoplasm (which may be anucleate) of a second.

Nuclease-free reagent. A chemical used to extract or purify DNA or RNA that has been tested to ensure very low levels of ultraviolet-absorbent material and enzymes capable of denaturing or hydrolyzing nucleic acids (RNA and DNA).

Nucleases. A kind of hydrolytic enzyme (hydrolase), capable of hydrolysing nucleic acids into their constituent nucleotides, are of two types, amely, DNase (deoxyribonuclease) and RNase (ribonuclease) hydrolysing DNA and RNA respectively.

Nucleic acid sequence analysis. A technique for determining the sequence of bases in a specific nucleic acid and from which the amino acid sequence of the protein for which it codes can be determined. It is also used in determination of taxonomic relationships between viruses and other organisms, as well as determining control sequences governing initiation and termination of transcription.

Nucleic acids. Compounds of pentose sugars, phosphoric acid and nitrogen containing base (purine and pyrimidine); these three units combine to form nucleotides which in their turn join to form polynucleotide strands generally referred to as nucleic acids. These are of two types, ribonucleic acid (RNA) and deoxyribonucleic acid (DNA); the former has ribose as the sugar and is generally single stranded, the latter has deoxyribose as the sugar and is generally doublestranded. RNA is mostly found associated with cytoplasm and DNA with nucleus in a cell. Both the types are characteristic of cells of all living organisms. However, in viruses only, either of the types is observed. DNA is considered to be the universal genetical material and is the repository of the genetic code. RNA plays an important role in translating this code in terms of proteins, particularly enzymes.

Nucleocapsid. The fundamental nucleic acid of a virus, plus its surrounding *crystalline* protein coat or shell. Some nucleocapside are naked, but a mature, ultimate phase virus particle is enclosed by a limiting membrane, and is known as a virion.

Nucleo-cytoplasmic ratio. The ratio of cell nuclear to cytoplasmic volume. This ratio is elevated in meristematic cells and low in differentiated cells.

Nucleoid. The region of a prokaryotic cell in which the DNA is situated. It is analogous to the eukaryotic nucleus, but is not enclosed within a membrane at any time. Similar regions containing fibrils of nucleic acid are also found in chloroplasts and mitochondria.

Nucleolar organizer. A region on a chromosome that is associated with the formation of a new nucleolus following nuclear division. The site of the genes that code for ribosomal RNA.

Nucleolus. A body not containing deoxyribose nucleotides and secreted by a specific organizer gene or super-gene, in the resting nucleus. (2) A spherical nuclear organelle, one or sometimes several, present in the prophase stages. These are formed through the interactions of specific segments (nuclear organizers) on specific chromosomes, that as a rule, are provided with satellites, (satellite chromosomes). Nucleoli are formed anew during the interphase stage. The nucleolus is rich in RNA and is thought to be the seat of ribosomal RNA synthesis as well as its storage.

Nucleomicrosomes. Nuclear chromatin granules.

Nucleoplasm. The denser protoplasm of which nuclei are composed as distinct from the less dense cytoplasm outside the nuclear membrane.

Nucleoprotein. A protein which is combined, or combinable, with a nucleic acid.

Nucleosidase. Enzyme that hydrolyses a nucleoside to a free nitrogenous base (a purine or a pyrimidine) and a pentose (ribose or deoxyribose).

Nucleoside triphosphate (NTP). A general term used to include all the triphosphate of biological importance (i.e. ATP, CTP, GTP, ITP, TTP, and UTP).

Nucleoside. A nucleotide lacking the phosphate group is known as nucleoside. It is the combination of a nitrogenous (purine or pyrimidine) base and a pentose sugar. Thymidine is a nucleoside consisting of the nitrogenous base thymine plus deoxyribose.

Nucleoskeletal DNA. DNA which is proposed to exist mostly to maintain nuclear volume and not for coding protein.

Nucleosome phasing. The arrangement of the nucleosomes with respect to specific DNA sequences within a eukaryotic chromosome. Nucleosome phasing is thought to control the expression of some eukaryotic genes. If the DNA sequence containing a gene's promoter is buried within a nucleosome it is inaccessible to RNA polymerase and the gene is not expressed. On the other hand, if the gene's promoter is contained in an internucleosome space, then RNA polymerase may bind to the promoter and express the gene. Changes in nucleosome phasing have been correlated with developmental switches in some eukaryotic organisms.

Nucleosome. A portion of a eukaryotic chromosome, consisting of part of the DNA molecule wrapped around a group of histone molecules, and held together by another type of histone molecules. The chromosome is made up of many nucleosomes. It is the a structural arrangement on the eukaryotic chromosome in which DNA of 130 to 140 base pairs is associated with eight histone molecules.

Nucleotide. The basic chemical unit (monomer) in a nucleic acid. A nucleotide in RNA consists of one of four nitrogenous bases linked to ribose, which in turn is linked to phosphate. In DNA, deoxyribose is present instead of ribose.

Nucleus (from L. *diminutive of nux:* kernel or nut)**.** (1) The dense central portion of an atom, made up of protons and neutrons, with a positive charge. Surrounded by a cloud of negatively charged electrons. (2) Centrally located chamber of eukaryotic cells that is bounded by a double membrane and contains the chromosomes. The information center of the cell. It is the structure within cells which contains DNA, the hereditary material.

Nuclide. Atom designated according to the constitution of its nucleus in terms of proton and neutron numbers. Thus radioisotopes may be described as radionuclides.

Nude mouse. A genetically athymic mouse. It is also carries a closely linked gene producing a defect in hair production.

Null allele. Allele that produces no functional product and therefore behaves as a recessive.

Null hypothesis. The hypothesis from which the expectations are formulated for the purpose of a test of significance.

Nulliplex. Condition of a polyploid in which all the chromosomes of one homologous type carry the recessive allelomorph of a particular gene, as opposed to simplex, duplex, triplex, quadruplex, etc., in which the dominant allelomorph is represented one, two, three, four, etc., time.

Nullisomic. An individual that completely lacks chromosomes of a certain kind that otherwise normally occur in the species. In a diploid species a nullisomic may be indicated by $2n-2$ or $2X-2$. Nullisomics are, however, only viable in alloploids or certain structural heterozygotes.

Nullisomy. Condition in which both members of a pair of chromosomes are absent. If N is the haploid number, a nullisomic individual will be styled $2N-2$.

Numerical aperture. Applied to the objective lens of a light microscope; a measure of that fraction of light leaving the spectrum under examination that can be `collected' by the objective will provide a brighter image with more information about the specimen (within the resolving power of the system).

Numerical response. The derivative reproductive, migrative, and survival response of a parasitoid or predator population to changing densities of its host prey; a positive response involves increased natality, survival, and immigration with increased host-prey density, and vice versa; cf. Functional Response.

Numerical taxonomy. A technique for determining the relationships among organisms by determining the number of properties the organisms have in common.

Nutrient agar. A nutrient both solidified by the addition of agar.

Nutrient broth. A medium used in the culture of pathogenic and other bacteria that contains peptone, meat extract and sodium chloride.

Nutrient film technique (NFT). Hydroponic plant growth whereby plant roots are suspended in shallow, slowly circulating nutrient solutions delivered as a continuous film of liquid rather than in bursts of liquid.

Nutrient gradient. A diffusion gradient of nutrients and gases is set up in tissues where only a portion of the tissue is in contact with the medium. This is common in callus culture and causes differential growth rates and senescent regions of callus.

Nutrient limitation. The condition in which growth of a culture is limited by the availability of a nutritional requirement. The lack of a carbon source or any of the essential elements, vitamins or other compounds leads to a cessation of growth. Lack of other nutrients may produce changes in metabolism.

Nutrient. A nutritive substance or ingredient, as are the major and minor mineral elements (macroelements and microelements) necessary for plant growth and development as well as the organic addenda, sugars, vitamins, amino acids and others employed in plant tissue culture media. **Nutritional requirement.** The minimal nutritional need for the proper growth and maintenance of an organism or culture. It includes both macronutrients and micronutrients (trace elements, vitamins, essential amino acids, etc.).

Nutrition: The study of foods in relation to health and disease. It is the means by which an organism assimilates its food.

NZB/W. A strain of mouse which is a model for systemic lupus erythematosus. The parental NZB strain also suffers from autoimmunity.

O

O. Used to denote the absence of a sex chromosome e.g. *XX* and *XO* as compared with the *XX* and *XY* system of sex inheritance.

"O" antigen. Somatic antigen, or that associated with the cell body, in contrast to flagellar; (Ger. *ohne hauch*, "Without breath").

Oat coleoptile test. A quantitative auxin assay. This determination is based on the degree of curvature, a result of asymmetric growth, in a detipped oat coleoptile affixed to agar containing the auxin.

Obelin. Calcium-activated photoprotein in the photocyte of the colonial hydroid coelenterate, *Obelia geniculata.*

Obesity. Being overweight.

Obligate aerobe. An organism that will only grow is the presence of oxygen.

Obligate anaerobe. An organism whose growth is inhibited or that is killed in the presence of oxygen.

Obligate parasite. One that is unable to multiply apart from a living host.

Obligate parasitism. Here in reference to nematodes which must develop parasitically and cannot reproduce and complete growth away from a host. cf. Facultative Parasitism.

Obligate pathogen. A disease-causing microorganism which requires a living host in which to grow and reproduce. cf. Facultative Pathogen.

Obligate. Descriptive of an organism that requires a certain environmental condition or nutrient for growth or survival.

Obligate. Necessary, as in obligate anaerobe. (Contrast with facultative.)

Occluded viruses. Viruses characterized by enclosure of the mature replicated virions in the matrix of proteinaceous or crystal-like inclusion bodies, such as capsules or polyhedra. also called inclusion viruses. cf. Non-Occluded Viruses.

Occult virus (Latin *occultus*: hidden). Refers to a hidden virus i.e., a virus causing inapparent infection (q.v.) and thus remaining latent (q.v.).

Oceanic zone. The deeper ocean basins.

Ochre codon. The codon UAA (in mRNA) which codes for polypeptide chain termination.

Ochre mutant. In phage, a suppressor sensitive mutation that blocks transcription due to the presence of the nonsense codon UAA.

Ochre suppressor. A gene which codes for a altered transfer RNA so that its anticodon can recognize the ochre codon and thus allows the continuation of protein synthesis. A suppressor of an ochre mutation is a tRNA which is charged with the amino acid corresponding to the original codon or neutral substitute. Ochre suppressors will also suppress amber mutants.

Octopamine. A biogenic amine found in both vertebrates and invertebrates (identified first in the salivary gland of Octopus). Octopamine can have properties both of a hormone and a neurotransmitter, and acts as an adrenergic agent.

Octopine. Plant growth substance-an opine-produced by cells in crown galls in plants and active in promoting growth of the *Agrobacterium in* the crown gall cells.

Odontoblast. A cell that secretes the dentine of a tooth. Odontoblasts lie around the outside of the pulp cavity and their long fine processes extend into fine canals (canaliculi) in the dentine.

Oestradiol. (USA stradiol.follicular hormone). A hormone (MW 272) synthesised mainly in the ovary, but also in the placenta, testis, and possibly adrenal cortex. A potent oestrogen.

Oestrin-withdrawal, Theory of menstruation. The theory that the fall in the oestrin level is the essential factor in the genesis of all types of menstrual bleeding.

Oestriol. A metabolite of oestradiol and oestrone, found in the urine in high concentrations. It has only weak oestrogenic activity.

Oestrogen (USA estrogen). A type of hormone which induces oestrus ("heat") in female animals. It controls changes in the uterus that precede ovulation, and is responsible for development of secondary sexual characteristics in pubescent girls. Some tumors are sensitive to oestrogens.

Oestrone. A metabolite of oestradiol, found in the urine in high concentrations. It has only weak oestrogenic activity.

Oestroscope. An instrument for measuring the viscosity of the cervical mucus which closes the uteri. The oestroscope detects the onset of oestrus since the viscosity of the cervical mucus reaches its minimum at oestrus.

Oestrous cycle. The reproductive cycle that occurs in sexually mature, non-pregnant female mammals. It is regulated by the hypothalamus which release factors that stimulate the periodic secretion of pituitary gonadotrophins.

Offset. A short runner growing horizontally above ground away from the parent, seen in daisies and houses-leeks. It stores no food but turns up at the end and produces a new·plant from the apical bud. Like stolons and suckers, offsets are a means of vegetative propagation in angiosperms.

Ogive. The integral of a frequency distribution (Galton, 1883).

Ohno's law. Genes present on the *X* Chromosome in one species are likely to be *X*-linked in other species because of suggested unique regulatory features of the *X* chromosome; proposal advanced by S. Ohno.

Oigodendrocyte. A class of glial cell that myelinates axons in the central nervous system.

Oil immersion. A microscopic technique using special high-powered objective lenses. A drop of immersion oil (e.g. cedarwood oil) is placed on the coverslip of a microscope slide and the objective lens carefully lowered into it. The oil has the same refractive index as the lens glass and increases the resolving power obtainable by letting a wider angle of rays enter the objective lens.

Oil overlay. A thin film of oil added to a slant culture or liquid stock culture of microorganism to reduce the rate of dehydration of the culture due to water loss from the surface.

Oilgo dT. Homopolymer of deoxyribothymidylate.

Okazaki fragment. Segments of approximately 1, 000 to 2, 000 nucleotides of the daughter strand of replicating DNA. These are joined in succession by the enzyme DNA ligase to form a normal double-stranded DNA molecule.

OKT. A group of monoclonal antibodies used to identify T cell surface markers in humans.

Oleagnous. Descriptive of a cell that accumulates large amounts (over 20 percent of the dry weight) as lipids.

Oleic acid. cis-9.10-Octadecanoic acid; an 18 carbon fatty acid with one double bond. It is one of the most widely occurring long-chain fatty acids.

Oleoresin. A natural mixture of essential oils and resin.

Oleosome. Any fatty inclusion in the cytoplasm.

Olfactory epithelium. The epithelium lining the nose.

Olfactory nerve. (Cranial nerve l) the nerve that connects the olfactory cells of the nose with the olfactory bulb at the front of the vertebras. The cell bodies of its sensory fibers are located not in a ganglion but in the olfactory membrane.

Olfactory organ. An organ of smell. In vertebrates, olfactory organs are paired structures located in nasal cavities that communicate with the exterior environment through the external nares. In vertebrates, olfactory organs include structures such as the antennae of insects.

Olfactory. Having to do with the sense of smell.

Oligideoxyribonucleotide. An oligonucleotide that contains deoxyribose; a short length of DNA.

Oligo dA. Homopolymer of deoxyriboadenylate that can be made as an artificial polymer *in vitro* and used as a synthetic template for polyermases.

Oligo(dT) cellulose. Short lengths of deoxythymedilic acid (dT) cellulose is used in the purification of eukaryotic mRNA by affinity chromotography. In this process the poly(A)-tail of the mRNA molecule is immobilised by hybridization with the oligo(dT).

Oligogenes. Major genes controlling qualitative characters which show normal mendelian inheritance.

Oligomer. A compound molecule of intermediate size, made up of two to a few monomers. (Contrast with monomer, polymer.)

Oligomycin. Antibiotic active against the fungi, but not against bacteria.

Oligonucleotide synthesizer. An automated machine designed to synthesize short polynucleotide chains (oligonucleotides) similar in structure to DNA or RNA.

Oligonucleotides. Oligonucleotides are short DNA (or, rarely, RNA) molecules, usually defined as 100 bases long or less. This is the length of DNA that an automated DNA synthesis machine (a DNA synthesizer, oligonucleotide synthesizer, or `gene machine') can make in one go and still have a significant yield of product. Oligonucleotides are usually defined by their origin. if it is made chemically it is an oligonucleotide, if it is cloned then it is a gene or a gene probe. Oligonucleotides are usually named for their length. The naming follows the monomer-dimer-trimer scheme, up to decamer (10 bases). Beyond that, the name of an oligonucleotide is generally its length as a number followed by `-mer'. Thus a 17-base oligonucleotide is called a `17-mer'. Automated DNA synthesizers, often called gene machines, use a series of chemical reactions to build up the DNA chain one base at a time. Each reaction consists of four steps since the chemistry has to make sure that only one base is added each time, so building up a 50-base oligonucleotide (a `50-mer') requires 200 reaction steps. Clearly, if one of those steps is slightly inefficient, then the overall efficiency will be very poor. this is why synthesizing oligonucleotides of greater than 100 bases becomes quite difficult. Most gene machines are completely automated, so all the biotechnologist has to do is type in the DNA sequence required and collect the DNA. Oligonucleotides have become critical to biotechnology for three reasons. 1. They can be linked together to form larger lengths of DNA which can function as completely synthetic genes (see Gene synthesis). 2. They can be used as DNA probes for a variety of genetic studies. In this they are particularly useful since they can distinguish between versions (alleles) of a gene that differ by only one base. Such oligonucleotides are called allele-specific oligonucleotides (ASOs). 3. They are the primers for the widely used PCR technique.

Oligopeptide. Peptide polymer containing somewhere between two and 10 amino acids.

Oligophagous (stenophagous). Adapted to the use of only a limited number of plant or animal species (e.g. members of a genus) as hosts or prey; cf. Monophagous, Polyphagus.

Oligopyrene. Of spermatozoa. having less than the normal complement of chromosomes.

Oligoribonucleotide. An oligonucleotide that contains ribose in the backbone; a short length of RNA.

Oligosaccharide. A carbohydrate containing a few monosaccharide.

Oligotroph. A species that grows at low nutrient concentrations.

Oligotrophic. Describing lakes that are deficient in nutrients and consequently low in productivity. Compare Eutrophic.

O-micron DNA. An alternative name for the 2% plasmid of *Saccharomyces cerevisiae.*

Ommatidium (Gr. **omma.** an eye)**.** One of the units which, collected into groups of up to 20, 000 make up the compound eye of arthropods.

Omnivore (L. *omnis:* all, everything *vorare:* to devour)**.** An organism that can eat both animal and plant material. (Contrast with carnivore, detritivore, herbivore.)

ONC gene. Oncogene in a retrovirus. Such gene sequences are closely similar to proto-oncogenes which occur as normal coding sequences in vertebrate genomes.

Oncogene hypothesis. An hypothesis which postulates that most or all vertebrate cell contain, as a normal part of the genome, genetic information which specifies a C-type RNA tumour virus genes which specify the virus, and which are normally in a repressed state, are termed virogenes. Certain of the virogenes are termed oncogene(s); if expressed, the oncogene(s) specify changes which lead to cell transformation. The hypothesis postulates that, in the presence of oncogenic agents, some or all of the virogenes become derepressed; according to which of the virogenes are derepressed, the effects of the oncogenic agent may include viron synthesis and/or cell transformation. The oncogene hypothesis is consistent with a number of observation which include. (a) the `spontaneous' intracellular appearance of C-type RNA tumour viruses following the repeated subculturing of certain cells which gave no indication of previous virus infection; (b) the induction of viruses (by oncogenic agents) in cells which previously, gave no indication of containing viruses or their antigens; (*c*) the demonstration, by hybridization studies, of the occurrence of C-type viral genetic information in cells which are not synthesizing C- type virus. These observations are also compatible with the provirus hypothesis.

Oncogene theory. A theory to explain the viral etiology of cancer. It proposes that the cells of all vertebrate species contain, incorporated in their chromosomes, a potentially carcinogenic provirus, called an oncogene, and that oncogenes are kept in a repressed state in the normal cell by repressor proteins. The transformation of a normal cell to a cancer cell occurs when the repressor protein specific for the oncogene is blocked.

Oncogene. Oncogenes are genes that are believed to be necessary for cancers to develop. There are a large number of them and, as would be expected from the variety of cancer types, they act in many different ways. Most are present in normal cells are protooncogenes, i.e. versions of the gene that are benign, and indeed are essential to the body's normal development. A mutation turns them into the malign oncogene. Oncogenes are of interest to biotechnology because of the importance of cancer as a cause of morbidity and mortality in Western societies. Many biomedical R & D programmes are aimed at curing or palliating cancer, and hence are interested in directly or indirectly preventing the effect of oncogenes. The approach depends on the oncogene involved. Some oncogenes make proteins that are detectable on the outside of cells or in the blood. these proteins can be `tumour markers', i.e. markers to show where a tumour is developing. In turn they may be used to diagnose cancer o to target a biotherapeutic to the cancer cell and so destroy it specifically. Oncogenes that act only inside the cells cannot be used as tumour markers in this way. Among the more talked about oncogenes or oncogene proteins are: **erb**, a family of proteins, of which erbB-2 is associated with breast cancer. **Myc**. a protein in the cell nucleus, one of the earliest oncogenes identified (see also `myc-y mouse' below). **Fos**, a nuclear protein. **Neu**, a membrane protein that is similar to the receptor for growth factors. it is believed that the mutant from `looks' to the cell like a growth factor receptor that always has its growth factor bound, i.e. is always giving the cell a `grow' signal. **Ras**, a cell membrane protein that is associated with the protein kinase cascade, a complex series of enzymes that regulate many of the cell's functions in growth and differentiation. **tat**, a gene from HIV and many related retroviruses. Many oncogenes have prefix letters. Thus there is c-myc (the cellular gene), v-ras (a viral, cancer-forming version of ras), H-ras (the human gene, to distinguish it from a number of homologues in other species). There also exist anti-oncogenes, called tumour suppressor genes, whose normal function is to suppress some genetic activity that could promote the development of a cancer. If a tumour suppressor gene is mutated, then it `releases' the activity of another gene and so hastens the development of the disease. Examples are P53. Mutations in oncogenes usually occur at only a few places in the gene, turning the protein product into a cancer-triggering molecule. By contrast, mutations in tumour suppressor genes can be anywhere in the gene. This makes detecting them harder (in essence you have to sequence the whole gene from a patient to tell if they have a mutated tumour suppressor gene). Tumour suppressor genes can also be inactivated by speeding up their breakdown (p53) or directing them to the wrong part of the cell (BRCA1).

Oncogenic (Latin **oncos,** growth plus Latin **genic,** to lead to). Indicates agents capable of inducing cancerous growth.

Oncogenic virus. A virus that carries an oncogene which may be inserted into the genetic elements of an infected host cell or organism in abnormal development. Both RNA and DNA viruses may be oncogenic. They include adenoviruses, leukoviruses and papovaviruses. Oncogenic viruses are also capable of inducing cell transformation *in vitro.*

Oncology. The systematic study of tumors (neoplasms).

Oncornaviruses. Acronym for oncogenic RNA viruses.

Oncovirinae. The family of retroviruses (Retroviridae) which can cause tumours. They are enveloped by membrane derived from the plasma membrane of the host cell, from which they are released by budding without lysing the cell. Within each virion is a pair of single-stranded RNA molecules Replication involves a DNA intermediate made on an RNA template by the enzyme reverse transcriptase.

One gene-one polypeptide chain hypothesis. Modern restatement of an older proposition. Since many enzymes (e.g. eukaryotic RNA polymerases) are complex proteins made up of a number of different polypeptide chains, it is clear that the relationship is 1.1 for the gene and the polypeptide chain. But, although each different polypeptide chain is the product of a different gene, many genes do not code for any polypeptide chains.

On-line. Descriptive of the continual transfer of data from a recording device to a data logger, chart recorder or microprocessor.

ONPG (o-nitrophenylagalactoside). Synthetic substrate for the enzyme ß-galactosidase which is cleaved by the enzyme into two separate substances, one being the yellow-coloured o-nitrophenol. This compound is therefore used to assay for the presence of this enzyme, as for example in mutants of *E. coli* with lesions in the lac operon genetic sequence.

Ontogeny. The developmental (**ontogenetic**) life history of an organism, its development from fertilized egg (zygote) to seed-forming adult (reproductive phase), with emphasis on embryonic development.

Oocyte (primary), 2n. The egg mother cell giving rise to the egg and a polar body at the first division, both with *n* chromosomes.

Oocyte (secondary), n. The haploid cell following the first meiotic division that divides equationally to produce the egg (female gamete) and an additional polar body which disintegrates.

Oocyte (Gr. *oon.* egg + *kytos.* cell). The cell that gives rise to eggs in animals.

Oogamy. The fusion, in fertilization, gametes of unequal size typically a large female gamete and a small active male gamete. Adj. Oogamous.

Oogenesis. The process in sexual organisms, particularly in animals and in man, by which female gametes (egg) are formed.

Oogonium. A diploid primary germ cell in the female animal, usually found in the early embryo. The oogonium gives rise to primary oocytes, secondary oocytes, and eggs (the female gametes.).

Ookinesis. The mitotic phenomena in the egg-cell during maturation and fertilization.

Oomyctes. A class of fungi that includes organisms noted for causing `downy mildew of grapes' and parasitizing green algae.

Ooplasm. The cytoplasm of the oocyte or ovum.

Oosphere. A female gamete or egg-cell. In angiosperms it is enclosed in the embryo sac and protected by the integuments of the ovule. In certain fungi (e.g. Pythium) the oosphere is thin-walled but, after being fertilized, develops a thick resistant wall and becomes an oospore.

Oospore. A diploid zygote of certain algae and fungi, produced by fertilization of the female gamete, the oosphere, by the male gamete, the antherozoid. The zygote may form a thick coat and go through a resting period before germination. The term `oospore' distinguishes a zygote produced oogamously form a zygote produced isogamously or anisogamously and called a zygospore.

Ootid. The haploid egg cell after meosis. cf. Spermatid.

Opal codon. The codon UGA, one of the three that causes termination of protein synthesis.

Opal mutant. In phage, a suppressor sensitive mutant that prevents transcription due to the presence of the nonsense codon UGA.

Opal suppressor. A gene which codes for an altered transfer RNA so that its anticodon can recognize the opal codon and thus allows the continuation of protein synthesis. A suppressor of an opal mutation is a tRNA which is charged with the amino acid corresponding to the original codon or a neutral substitute. Some eukaryote cells normally synthesis opal suppressor tRNAs. The function of these is not clear and they usually do not prevent normal termination of protein synthesis at an opal codon.

Open circle. One of the three conformations which a plasmid molecule may adopt. In an open circle there has been either a single break or a number of staggered breaks made in the strands of the duplex. Such a molecule may not form supercoils and therefore is referred to as an open or relaxed circle.

Open continuous culture. A cell suspension culture with a continuous influx of fresh medium, maintained at constant volume by the efflux of cells and spent medium.

Open reading frame (ORF). Region of a DNA sequence that has the potential to be read as codons specifying amino acids (i.e. it is not interrupted by nonsense codons as stop signals). ORF sequences are frequently encountered in the sequencing of gene banks and only occasionally can they be assigned to a particular gene.

Open system. A system that exchanges matter as well as energy with its surroundings.

Operator gene. (1) A gene that in some way controls the activity of certain structural genes adjacent to it in the chromosome. (2) A region of DNA which controls the transcription of a structural gene.

Operator site. A segment of DNA in an operon that affects activity or nonactivity of associated cistrons; it may be combined with a repressor and thereby "turn off" the associated cistrons.

Operator. The term introduced by Jacob, Perrin, Sanchez and Monod (1960) for the site at one end of an operon where a repressor molecule binds to the DNA and thereby inhibits transcription. The term is appropriate also for the corresponding site when a gene does not form part of an operon.

Operon concept. It is hypothesis which explains the mode of action of a gene in both active and quiescent stage. It envisages that there are two kinds of genes, structural and controlling genes. The latter, according to this concept, are of two types, regulators and operators. The operator and structural gene together constitute an `operon', In the absence of the substrate (inducer), the regulator gene synthesizes a protein repressor which binds itself to the `operator gene' thus preventing the functioning of the structural gene. In the presence of the inducer the regulator gene forms a complex with a small part of the inducer, thereby leaving structural genes to operate and transcribe mRNA.

Operon. The term which Jacob, Perrin, Sanchez and Monod (1960) introduced for a group of closely-linked genes which appear to affect different steps in a single biosynthetic pathway and which appear to function as an integrated unit. With the discovery that the genes of an operon are transcribed in one messenger RNA molecule, an operon could be defined as an aggregate of genes with a single messenger.

Ophthalmic. Pertaining to the eye.

Opiates. Naturally occurring basic (alkaloid) molecules with a complex fused ring structure. Have high pharmacological activity.

Opine synthetase. (opine synthase). Enzyme system responsible for the synthesis of opines of which are a number of different types. The genetic information coding for opine synthetase resides in the T-DNA region of the Ti plasmid. These plasmids confer virulence to strains of Agrobacterium, which are responsible for the crown gall disease. The T- DNA is transferred to the plant cells, where it is stably integrated into the nuclear genome and opine synthetase gene is transcribed and translated.

Opine. Amino acid derivative synthesized only in plant tumors caused by *Agrobacterium tumefaciens.*

Opioid receptors. Membrane proteins, widely distributed in animal cells, but especially in the brain (enkephalin receptors) and gut. The natural ligands are the opiate peptide neurotransmitters, but the name is given because opiates are potent agonist occupy the receptors and mimic the action of the natural transmitters.

Opportunist pathogens. Organisms which exists part of the normal body microflora but which may become pathogenic under certain conditions - e.g. when the normal antimicrobial defence mechanism of the host have been impaired. Thus, e.g. Bacteroides spp-common inhabitants of the large intestine in man-may give rise to peritonitis etc. following bowel surgery.

Opportunist. A microorganism that causes infection only under especially favourable (e.g. when defense host mechanism are not fully functioning).

Opportunistic infections. A disease caused by an organism that is not usually associated with the particular host (e.g. infection of humans with dry rot fungi, of the wide range of organisms that exploit the loss of immune resistance in the disease AIDS).

Opportunity cost. The sum of the benefits an animal forfeits by not being able to perform some other behavior during the time when it is performing a given behavior.

Opsonic index. An *in vitro* assessment of the relative concentration of opsonins in a sample of blood. The opsonic index is the ratio of phagocytic activity in the patient's blood to that in the blood of a normal (non-immune) subject. It is determined by initially incubating both (citrated) blood samples for 15 minutes with a standard number of the relevant antigen particles (usually microbial cells. Blood smears are then prepared form each sample and the average number of particles engulfed by the phagocytes in each sample is determined; the two averages are expressed as a ratio (the Opsonic Index).

Opsonin. A substance (usually an antibody or complement fragment C3b) that adheres to bacteria, viruses, and antigen-antibody aggregated molecules and makes those particle more susceptible to ingestion by phagocytes.

Opsonization. A process by which phagocytosis is facilitated by the deposition of ospinins (e.g. antibody and C3b) on the antigen.

Optic chiasma. The site in the floor of the vertebrate forebrain at which fibers entering via the optic nerves cross to the opposite side of the brain, forming an *X*-shaped pattern. In man and other mammals with good stereoscopic vision, only half the fibers from each side cross over. In lower vertebrates, nearly all the fibers may cross over.

Optical activity. Ability of a molecule to rotate the plane of polarized light when it is passed through a solution of the molecule. Sugars such as monosaccharide may exist in either of two isomeric forms, as a function of their asymmetric carbon atoms; they exist in two configurations known as D-form (dextrorotary) and L-form (laevorotatory) bending light either to the right or to the left, respectively.

Optical biosensors. A type of biosensor where the effect of a chemical on a biological system is detected using light rather than electrochemistry. Several systems have been taken to commercial development in the last few years. They are based on the following principles.

Optical density (OD). Degree of absorbance of light by a solution as a function of the concentration of the solute. For solution that absorb light, the relative optical density provides information about concentration and the character of the solution. Nucleic acids absorb light maximally at a wavelength of 260 nm, whereas proteins absorb maximally at 280 nm. Thus the protein content of an impure solution of DNA can be measured. Since the optical density of a DNA solution increases when the DNA is melted (the hyperchromic effect) changes in optical density can be used to provide detailed information about DNA complexity, especially in the use of reduction of optical density after melting, as is used to construct cot plots.

Optical diffraction. A technique used to obtain information about repeating patterns. Diffraction of visible light can be used to calculate spacings in the object.

Optical isomerism. A form of isomerism in which the isomers differ in their optical activity.

Optical isomers. Molecular isomers that induce rotation of plane-polarized light when it is passed through a solution of the molecule. The rotation, which results from the asymmetric form of the molecule, may be to the right (dextro-) or to the left (laevo-) for different isomers of the same molecule. These isomers are referred to as in the case of, for example, glucose, as *d*-glucose or *l*-glucose.

Optical rotation. The angel through which the plane of polarized light is rotated when passed through an optically active substances.

Optimal proportion (optimal ratio) (immunol). In a titration involving antibodies and soluble antigens, the ratio in which antibody and antigen are present when precipitation occurs at a maximum rate. For a given antigen/antibody system two different values may be obtained for the optimal ratio-each value corresponding to a particular experimental procedure. In the Dean and Webb Titration the constant-Antibody Optimal Ratio is obtained. In this titration an optimally reacting mixture often contains antibody and antigen at Equivalence, i.e. following antigen-antibody combination no free antigen or antibody remains. The Ramon Titration gives the Constant-Antigen Optimal Ratio. In this titration an optimally-reacting mixture frequently contains an excess of free antibody following antigen-antibody combination.

Optimal yield. The largest rate of increase that a population can sustain in a given environment. There exists a particular population size, less than the carrying capacity, at which this yield is realized.

Optimum temperature range. A narrow range of temperature, which may cover only two or three degrees, in which cultures of an organism show maximum growth rate. Below the optimum growth rate, the rate in general declines only-slowly with the decrease in the temperature, whereas at temperatures only a few degree above the optimum a precipitous decline in rate to zero may occur.

Optimum. Condition most favorable.

Oraganotypic. Resembling an organ *in vitro* in three dimensional form or function or both. For example, a rudimentary organ in culture may differentiate in an organotypic manner, or a population of dispersed cells may become rearranged into an organotypic structure and may also function in an organotypic manner. This term is not meant to be used along with the word `culture' but is meant to be used as a descriptive term. Neither should the word `growth' be used along with this term since the tissue need not actually be increasing in size or number of cells in order to be organotypic.

Orbivirus. A genus of Arboviruses recognized by some authors. The Orbivirus virion is a non-enveloped icosahedral virus; 50-60 nm in diameter, which contains a fragmented linear double-stranded RNA genome-i.e. a genome consisting of 10 or more pieces; an RNA polymerase forms of the virion. Replication occurs within the cytoplasm of the host cell. Orbiviruses include the casual agents of e.g. African horse sickness and Blutongue.

Orcein. Dye used in cytology, especially when dissolved in acetic acid as aceto-orcein for the staining of squash preparations of chromosomes.

Orchil (orseille). A purple dye obtained from species of the fructicose lichen Roccella.

Order. In taxonomy, the category below the class and above the family; a group of related, similar families.

Ordered tetrad. Linear sequence of the products of meiotic division as visualized in the row of eight ascopores of a fungal ascus. crossing-over events can be studied by growing each spore separately and scoring for its expression of one allele at a particular locus.

Organ. A body part, such as the heart, liver, brain, root, or leaf, composed of different tissues integrated to perform a distinct for the body as a whole.

Organelle. Membrane-bound structure found in eucaryotic cell containing some organelles, including mitochondria and chloroplast, have DNA and can replicate autonomously.

Organic compound. A substance primarily composed of carbon, hydrogen, oxygen, and nitrogen and that in nature is produced only by various forms of life.

Organic consolvent. Compound used to dissolve some neutral organic substances; as in media preparation. These include alcohol (usually ethanol), acetone and dimethylsultoxide (DMSO).

Organic phase catalysis. This is the use of enzymes in liquids other than water. Organic phase catalysis (also called solvent catalysis, hydrophobic catalysis, non-aqueous phase catalysis).

Organic selection, Theory of. The hypothesis that an organism may become genetically adapted to particular conditions by first undergoing nonheritable modification so that any subsequent mutations which are favorable to the new conditions of life tend to be selected and fixed.

Organic. Pertaining to any aspect of living matter, e.g., to its evolution, structure, or chemistry. The term is also applied to any chemical compound that contains carbon.

Organiser. A part of an embryo whose presence causes neighboring tissue to develop in a particular way. Examples are the eye-cup of the vertebrates, which causes lens, and later cornea, to be produced; the gut of snail embryos, which organizes shell gland and mantel; the dorsal lip of the blastopore of frogs, which become notochord and organizes all the axial structures of the embryo; and the epidermal papillar of a hair, feather, or tooth, which organizes local epidermis and demis to form the follicle and appendages. The term primary organizer is restricted to the first or the most important initiator at gastrulation; for example, the dorsal lip of the amphibian blastopore or Hensen's node in mammals and birds. Determinants of major systems (e.g. notochord) are called secondary organisers; local centers of developmental activity (e.g. dermal papillae) are tertiary organisers.

Organising centers. Centers from which microtubles are organized or are grown. They include the cell centers and basal bodies.

Organism. Any living creature.

Organized growth. The in vitro development of organized explants such as meristem tips or shoot tips, floral buds or organ primordia, or their de novo formation from **unorganized** tissues.

Organized. Arranged into definite structure.

Organizer, embryonic. A region of an embryo which directs the development of nearby regions. In amphibian early gastrulas, the dorsal lip of the blastopore.

Organogenesis. The initiation (de novo) and growth of organs (roots and shoots, usually) from cells or tissues; as in organ culture. Organs may form on the surface of explants (direct organogenesis) or upon an intervening callus phase (indirect organogenesis).

Organoid. An anomalous organ-like structure formed in culture; as on leaves, roots or callus.

Organoleptic. Detected by the senses.

Organometallic compound. An organic compound that contains one or more metal atoms.

Organotheraphy. the use of extracts of organs (enzymes and hormones) for therapeutic purposes.

Organotroph. An organism that grows on organic material.

Oriental sore. Skin disease caused by the flagellate protozoan, *Leishmania tropica,* which is parasitic in macrophages of the dermis.

Orientation. The movement of centromeres so that they lie axially with respect to the spindle, either as to their halves at mitosis (auto-orientation) or as to members of a pair or higher configuration at meiosis (co-orientation). Determines Congression. Co-orientation or multiple configurations may be linear, convergent or parallel.

Origin of life. Geological evidence strongly suggests that life originated on earth about 4000 million years ago. The basic components of organic matter-water, methane, ammonia, and related compounds-were abundant in the atmosphere. Energy from the sun (cosmic rays) and lightning storms caused these to recombine into increasingly complex organic molecules. Particular combinations of such complex substances eventually showed the characteristics of living organisms. The experiments of Redi (17th century) and Pasteur (19th century) indicate that life cannot develop from inorganic matter under present day conditions.

Origin of replication (ori). A base sequence that is recognised as the position at which the replication of a molecule of DNA should begin. In bacteria, plasmids and viruses, there is usually only one such position. Lengths of DNA that contain the sequence for the origin of replication may be termed replicons.

Origin of species. The production of new types of organisms in descent from extant types. Natural selection is the process by which changes are favoured leading to the production of new types. The term was introduced by Charles Darwin.

Origin. In replication-that point in the chromosome where DNA replication commences. In conjugation-the region of a donar cell which is adjacent to the integrated sex factor and which is transferred first during conjugation.

Ornithine cycle. (urea cycle). The sequence of enzyme-controlled reactions by which urea is formed as a breakdown product of amino acids. It occurs in cells of the liver. The amino acid ornithine is combined with ammonia (from amino acids) and carbon dioxide, forming another amino acid, arginine, which is then split into urea (which is excreted) and ornithine

Ornithine decarboxylase. The enzyme which converts ornithine to putrescine (dibasic amine) by decarboxylation. Rate limiting in the synthesis of the polyamines spermidine and spermine.

Ornithine. Basic amino acid that is an important intermediate in the urea cycle.

Orphan viruses. Range of viruses that may be found in the respiratory and digestive tracts of healthy individuals; such viruses are not associated with any known pathology. Double-stranded RNA reoviruses are examples of orphan viruses.

Ortet. The original mother plant from which vegetatively propagated plants are derived.

Orthochromatin. Normal chromatin, as compared to unstable chromatin, deemed to be a reservoir of regulatory elements. The term is now little used.

Orthogenesis. Evolution in a definite direction and along a predetermined route, irrespective of natural selection. Since this theory implies that evolution proceeds independently of natural laws, it is impossible to test experimentally. Orthogenesis therefore conflicts with conventional theory such as neo-Darwinism.

Orthograde transport. Axonal transport form the cell body of the neuron towards the synaptic terminal. Opposite of retrograde transport and probably dependent on a different mechanochemical protein (almost definitely kinesin) interacting with microtubules.

Orthokinesis. Kinesis in which the speed or frequency of movement is increased or decreased.

Orthologous genes. Homologous genes that have become differentiated in different species derived from a common ancestral species (cf. Paralogous Genes).

Orthopoxviruses. Genus of double stranded DNA viruses (250-390 x 200-260nm) which preferentially infect epithelial cells. Includes variola (smallpox) and vaccinia.

Orthoselection. Natural selection acting continuously in the same direction over long periods of time. Often used in place of orthogenesis to avoid implication of vitalism.

Orthospiral. A spiralization of such a type that no secondary compensational twisting occurs in the fabric of the thread. cf. Anorthospiral.

Orthotropic. Vertically oriented; with an upright growth habit. the antonym is plagiotropic.

Oscillatoria princeps. Large cyanobacterium which exhibits gliding morements, possibly involving the activity of helically-arranged cytoplasmic fibrils of 6-9 nm diameter.

Osmiophilic. Having an affinity for osmium tetroxide.

Osmium tetroxide (O_5O_4). Compound used as a biological fixative for light and electron microscopy; sometimes wrongly termed osmic acid. It is highly toxic and readily vapourizes.

Osmium tetroxide or osmic acid (OsO_4 mw 254.20). A fixing agent used to prepare tissues for electron microscopy.

Osmolarity. The total molar concentration of the solutes affecting the **osmotic** potential of a solution or nutrient medium.

Osmometer. A device for measuring the osmotic pressure (water potential) of solutions.

Osmophole. A microorganism that requires a high solute concentration in the medium for growth.

Osmoregulation. The process by which animals maintain the correct concentration of water, salts, and other solutes in their bodies. This is carried out by nephridia, malphigian tubules or kidneys in multicellular organisms and by the contractile vacuole in protozoans.

Osmosis. The movement of water from one region to another where the water potential is more negative. This is often a region in which the concentration of dissolved molecules or ions is higher, although the effect of dissolved substances may be offset by hydrostatic pressure in cells with semi-rigid cell walls.

Osmotic potential. A property of any solution, resulting from its solute content; it may be zero or have a negative value. A negative osmotic potential tends to cause water to move into the solution; it may be offset by a positive pressure potential in the solution or by a more negative water potential in a neighboring solution. (Contrast with pressure potential.)

Osmotic pressure. The hydrostatic pressure generated by osmotic flow of a solvent (like water) through a semipermeable membrane into an aqueous phase containing a higher concentration of the solute. Plant cells function as osmometers, which affects water movement between plant cells and in and out of roots. Osmotic pressure is measured with an osmometer.

Osmotic shock. Passage of solvent into a membrane-bound structure due to osmosis, causing rupture of the membrane. A method of lysing cells or organelles.

Osmotic swelling. Increase in cell size (or in the size of other membrane bound vesicles) due to osmosis.

Osmoticum. An agent such as glucose or sucrose employed to maintain the nutrient medium osmotic potential equivalent to that of the cultured cells (isotonic). This prevents cell damage in vitro.

Osmotolerance in plants. Osmotolerance is a measure of a plant's ability to stand up to drought or to large amounts of salt in its water supply.

Ossification. Formation of bone. It is a process of replacement of pre-existing connective tissue, either in the form of cartilage (endochondral ossification) or simple connective tissue membrane (intramembranous ossification). The connective tissue is invaded by osteoblasts, which secrete an extracellular matrix of bone and finally become trapped in the matrix as osteocytes. When existing bone is being reformed in development or repair, some resorption of existing bone may occur, being accomplished by activity of osteoblasts.

Osteoblast. Any of the cells that form layers of bone in the early stages of ossification. They are at first on the outside of the embryonic cartilage or membrane, but after it has been eroded by osteoclasts they accompany the ingrowing blood vessels and form temporary trabeculae of bone. Later the osteoblasts lay down the permanent structure of bone, and those that become trapped between the lamellae are called osteocyte.

Osteocyte. Any of the cells that secrete the hard matrix of bone. They are found in small spaces (lacunae) between the concentric lamellae of bone that form the Haversian systems. Each osteocyte has many fine cytoplasmic processes that pass, in fine canaliculae, through the matrix and connect with each other and blood vessels to maintain, supplies of food and oxygen to the living thing.

Osteogenesis imperfecta. Inherited disease of the human, but with varying genetic backgrounds (some diseases are autosomal dominant, and others are autosomal recessive). It is characterized by brittle bones, discoloured teeth and deafness, but is very variable in its severity from complete lack of symptoms to perinatal death.

Osteoma. Benign tumour of bone.

Ostiole. Pore in fruit bodies (pycnidia and perithecia) of certain fungi and in conceptacles of brown algae through which, respectively, spores or gametes are discharged.

Ostium. (1) Of sponges, as opening through which water is drawn into the body. Much more numerous than oscula. (2) Opening from blood-filled space (sinus) surrounding the heart in Arthropoda.

Otosclerosis. Progressive conductive hearing loss in middle age inherited as an autosomal recessive condition.

Ouchterlony technique. (double diffusion technique). Precipitation test involving antibody-antigen interaction in gel diffusion systems. Two or more wells are cut in an agar slab, and a solution of antibody is placed in one central well. Other wells have various presumptive antigens in solution placed in them. By allowing simple diffusion, lines of precipitation are formed between the wells, each one representing a specific interaction.

Oudin technique. Test involving the formation of zones of precipitation resulting from antigen-antibody interaction; similar to the ouchterlony technique. A gel column is prepared with an even distribution of a selected antibody molecule. Following the deposition of an aqueous suspension of antigen molecules on the upper layer of the column, lines of precipitate form as the antigen molecules move into the column and interact specifically with the antibody.

Outbreeding. Mating of unrelated individuals or of individuals not closely related; the opposite of Inbreeding.

Out-cross. A cross between individuals of different genetic constitution, usually by natural cross-pollination.

Outside-out patch. A variant patchclamping technique, in which a disc of plasma membrane converts the tip of the electrode, with the outer face of the plasma membrane facing outward, to the bath.

Ovalbumin gene. Gene coding for ovalbumin; one of the first sequences in which the fragmented structure of introns and exons was demonstrated. It has been most thoroughly investigated in the chicken, where it is found to be some 7.8 kilobases in length and to consist of eight exons separated by six introns. Only a single copy of the gene exists in the chicken genome, but it is one of three genes in a small clustered multigene family. The other two genes in the family, code-named *X* and *Y*, are not as yet identified in terms of protein products.

Ovalbumin. One of the major proteins of egg white. It is the major nutritional protein of avian egg white and is a phosphoprotein of 386 amino acids.

Ovary. (1) The essential reproductive organ; the organ in which the eggs are produced. (2) In flowering plants an enlarged (usually basal) portion of the pistil containing the ovules.

Overdominance. Hybrids showing greater survival rate than both dominant and recessive homozygotes. Overdominance is not to confused with heterosis in which the hybrid shows greater vigour, greater dimension etc. Vigour does not necessarily mean higher survival rate because some vigorous F_1s, much more vigorous than either of the parents, are partially to completely sterile.

Overlap index. A measure of the extent to which a population of cells in culture form multilayers. The predicted amount of overlapping is calculated knowing the cell density, the projected area of the nucleus (usually), and assuming a Poisson distribution. The actual overlap is measured on fixed and strained preparations and the ratio of actual-predicted is derived. A value of 1 implies a random distribution with no constraint on overlapping; normal fibreblasts may have values as low as 0.05. Although a useful measure it does not unambiguously indicate the reason for the effect which may be contact inhibition of locomotion or differential adhesion of cells between substratum and other cells.

Overlapping code. The genetic code is in general non-overlapping, in that each triplet of bases codes for a distinct amino acid, and there is no sharing of common bases between adjacent triplets. Some overlapping genes do not provide examples of uses of the same sequence with a change in reading frame, which involves triplet overlap.

Overlapping deletion. Compound deletion occurring as two separate events, in which one overlaps another.

Overlapping genes. Genes whose nucleotide coding sequences have some areas of overlap. Apart from some overlapping regulatory genes in bacteria, such as the operator and promotor sequences of the tryptophan operon of *E.coli,* the only clear example known in the genes of certain phages. For example in the genome of the phage X174, one gene lies entirely within the coding sequence of another, but is translated using a different reading frame.

Overlapping inversion. Situation in which a second inversion overlaps a previous existing inverted sequence.

Overproducer. A mutant cell that produces large quantities of a chemical that is not usually accumulated. This is due to a fault in metabolic regulation. Overproducers generate large amounts of secondary metabolites or produce intermediates of central metabolism. Many industrial fermentations are carried out using mutants or strains that are overproducers.

Overstepping. This occurs when an animal, during its own ontogeny, passes through the ontogenetic stages of its ancestor but beyond the final adult stage of that ancestor; hypermorphosis. cf. Anaboly.

Oviduct. A duct in animals that conveys ova from the ovaries to the exterior. It may or may not connect directly with the ovary; in vertebrates, ova are shed into the coelom before entering the open ciliated funnel at its anterior end.

Oviparity. The production of undeveloped eggs, which are laid or spawned by the female. Fertilization may occur before their release, as in birds and some reptiles, or after, as in most invertebrates, fish, and amphibians. Large numbers of eggs are usually produced because of their poor chances of survival, due to lack of maternal protection. Each egg contains a large yolk store to nourish the developing embryo. Compare Ovoviviparity. Viviparity.

Ovotestis. Organ of some hermaphrodite animals, e.g. snail, which functions both as ovary and testis.

Ovovivipary. Situation in which the reproduction of some animals involves the development of discrete eggs, but the embryos develop within the egg inside the body of the female, although no direct contact between egg and female is established. Certain fish and reptiles are Ovoviviparous.

Ovulary culture. A culture in which the explant is an ovary containing the ovule(s). The ovules may be fertilized in culture (in vitro fertilization). This technique is used primarily when the ovary is essential to proper embryo development.

Ovulation. The release of an egg from an ovary.

Ovule (L. *ovulum:* little egg)**.** In plants, an organ that contains a gametophyte and, within the gametophyte, an egg; when it matures, an ovule becomes a seed.

Oxalic acid. Occurs in plants, and is toxic to higher animals by virtue of its calcium-binding properties; it causes the precipitation of calcium oxalate in the kidneys, prevents calcium uptake in the gut, and is not metablised.

Oxaloacetic acid (OAA). A water soluble carboxylic acid, structurally related to fumaric acid and maleic acid. Oxaloacetic acid forms part of the citric acid cycle, it is produced from L-malate in an NAD-requiring reaction and itself is a step towards the formation of citric acid in a reaction involving pyruvate ion and coenzyme A.

Oxidant or oxidizing agent. A substance that accepts electrons (is reduced) in an oxidation-reduction reaction.

Oxidase test. A test to distinguish colonies of *Neisseria* from some other bacteria; it does not differentiate *Neisseria* species. The test may be carried out by flooding the colonies on a plate with either dimethyl or tetramethylparaphenylene diamine hydrochloride and observing for color changes in the colonies; *Neisseria* are oxidase positive and the colonies first turn pink, then dark red, and finally black.

Oxidase. An enzyme that catalyzes an oxidation (e.g. glucose oxidase).

Oxidation. Relative loss of electrons in a chemical reaction; either outright removal to form an ion, or the sharing of electrons with substances having a greater affinity for them, such as oxygen. Most oxidations, including biological ones, are associated with the liberation of energy.

Oxidation/reduction reaction. Chemical reaction involving transfer of electrons from one molecule to another. The atom or molecule losing the electron is termed the reductant and is therefore oxidized, whereas the atom or molecule receiving the electrons is termed the oxidant and is reduced in the reaction.

Oxidation-reduction potential (E). The electrode of a half cell containing a mixture of oxidized and reduced forms of the reactants. In practice, E_o is used as the electrode potential when the concentrations of the oxidized and reduced forms are equal. In biological reactions, the potential is determined at pH 7.0 and is denoted as E_o.

Oxidative phosphorylatin. The generation of energy in the form of adenosine triphosphate (ATP) that results from the passage of electrons through the electron transport chain to the final electron acceptor, oxygen.

Oxidoreductases. Enzymes that catalyse oxidation/reduction reactions. Because all oxidation reactions must be coupled to a reducing reaction the enzymes catalyzing such reactions (often called dehydrogenases or reductases) are more properly called oxidoreductases. For example, the enzyme that catalyses the reversible oxidation of alcohol to acetaldehyde using NAD^+ as coenzyme may be called alcohol dehydrogenases, aldehyde reductase or alcohol. NAD^+ oxidoreductase.

Oxychromatin. That part of the nucleus which stains readily with acid dyes.

Oxygen concentration. The amount of oxygen, expressed in terms of equivalent volume under standard conditions. It is expressed as moles or weight per unit volume of a gas or liquid phase.

Oxygen cycle. The general cycle by which oxygen is exchanged and distributed with the biosphere.

Oxygen debt. A physiological state that occurs when a normally aerobic animal is forced to respire anaerobically during a temporary shortage of oxygen (anoxia), e.g. due to violent muscular exertion. Pyruvate, a product of the first stage of internal respiration, is converted anaerobically to lactic acid, which is toxic and requires oxygen for its breakdown, thereby building up an oxygen debt. The debt is repaid when oxygen is made available and allows oxidation of the lactic acid in the liver.

Oxygen electrode. An electrochemical cell in which the current generated is proportional to the activity of oxygen present in a solution.

Oxygen quotient (QO_2). The rate of oxygen consumption of an organism or tissue. It is usually expressed in. microlitres of oxygen per milligram of dry weight per hour. Small organisms tend to have higher oxygen quotients than larger ones.

Oxygen reaction. A reaction carried out in a reactor in which oxygen, rather than air, is introduced. In fermentation, such a reaction may use commercial-grade oxygen to provide the dissolved oxygen required for biochemical oxidation by the aerobic organism.

Oxygen transfer rate. A measure of the rate at which oxygen passes from the atmosphere into solution. Such rates are determined using an oxygen electrode or by measuring the rate of oxidation of a solution of sulphate in the presence of a catalyst (cobalt or copper).

Oxygen uptake rate. A measure of the actual rate of oxygen use by a culture expressed on a volumetric basis. It is dependant on the cell concentration times the average rate of oxygen consumption for that culture.

Oxygen yield. The amount of biomass and/or end products produced in an aerobic reaction expressed as a function of the amount of oxygen consumed.

Oxygen. An odorless, invisible, gaseous element occurring as O_2 and comprising 21% (volume) of the atmosphere. A component of almost all matter, it readily combines with most other elements. As a macroelement it is essential for plant and animal life. Plants produce O_2 during photosynthesis by splitting water.

Oxygenase. An enzyme enabling an organism or system to utilize atmospheric oxygen.

Oxygen-development killing. One of the most important microbicidal mechanism of mammalian phagocytes involves the production of various toxic oxygen species (hydrogen peroxide, superoxide, singlet oxygen, hydroxyl radicals) through the metabolic burst. Although oxygen independent killing is possible, the oxygen-dependant mechanism is crucial for normal resistance to infection, and a defect in this system is usually fatal within the first decade of life (chronic granulomatuous disease).

Oxygenic. The ability to use a photosynthetic process in which oxygen is a byproduct.

Oxyhaemoglobin. A bright red form of haemoglobin produced when the iron atom in haem binds with molecular oxygen. In many vertebrates, oxyhaemoglobin is used to transport oxygen in the blood.

Oxytocin. A peptide hormone, produced by the hypothalamus and posterior pituitary gland, that acts on smooth muscle. At the end of pregnancy the uterus becomes very sensitive to oxytocin, which promotes labour and also the release of milk from the mammary gland. It is used to induce labour artificially.

Ozone. A reactive form of oxygen produced by the discharge of electricity in air or oxygen in the atmosphere by the action of solar ultra violet (UV) or by UV in laboratory sterilizing lamps. It has a faint blue colour and faint chlorine-like odour. It is used to sterilize and purify water and air and in bleaching. It is injurious to plants and people at concentration greater than 2 μl/1.

ß-Oxidation. The process whereby fatty acids are degraded in steps, losing two carbon as (acetyl)-CoA. Involves CoA ester formation, desaturation, hydroxylation and oxydation before each cleavage.

P

Pacemaker. That part of the heart which undergoes most rapid spontaneous contraction, thus setting the pace for the beat of the entire heart. In mammals, the sino-atrial node. Also, an artificial device, implanted in the heart, that initiates rhythmic contraction of the organ.

PAF (platelet activating factor). A factor released by basophils which causes platelets to aggregte.

Pagophagia. The compulsive eating of ice.

Paleobiology. The attempted reconstruction-from fossil evidence, from knowledge of the properties of chemical substances, and from theories about primitive earth conditions- of the early stages in the evolution of life.

Paleontology [Gr. *palaios*. ancient, *old* + *logos*. discourse]. The scientific study of fossils and all aspects of extinct life.

Palindromic sequences. The same DNA nucleotide sequences that appear on complimentary strands but in opposite directions.

Palisade parenchyma. In leaves, one or several layers of tightly packed, columnar photosynthetic cells, frequently found just below the upper epidermis.

Pancreas. A gland, located near the stomach of vertegrates, that secretes digestive enzymes into the small intestine and releases insulin into the blood-stream.

Pandemic. An epidemic which has become very widespread, or is worldwide.

Panmyxia. The condition in which mating is random.

Panning. In a discovery context, also called biopanning. This is a technique for finding which of a library of molecules (often generated by combinatorial chemistry or by phage display technology) will bind to a target molecule. One of the partner molecules is immobilized to a surface, usually plastic but sometimes glass, and the other is swirled over it like a gold-rush `panner' looking for gold in river silt. If any of the mobile molecules bind to the surface, then you known that a match has been found. This has been used extensively in receptor binding studies, to find which of many millions of chemicals will bind to a receptor. It is also used to select from phage display libraries of antibodies to find the antibody that binds best to an antigen (see Phage display). Panning can also be used to purify cells from a population. It has been used to purify stem cells from blood. (Stem cells are cells with the capacity to proliferate into many types of blood cells, usually found only in the marrow, which can be used to regenerate the blood system after massive irradiation as part of cancer therapy.) Here the surface is coated with a ligand which binds to a receptor on the cell you are looking for, and then blood is passed over it. This is, in fact, a whole-cell version of affinity chromatography.

Papule. Small, raised area on the skin.

Parallel evolution. The evolution of organisms that were originally very similar and have both evolved in the same direction. (Contrast with convergent evolution.)

Parapatric speciation. Development of reproductive isolation among members of a continuous population in the absence of a geographical barrier. (Contrast with geographic, sympatric speciation.)

Parasite. An organism which lives on or within a living host from which it derives nourishment. An organism that survives on or at the expense of another living host.

Parasitism. The form of symbiosis in which one species lives at the expense of the other, but not ordinarily to the point of killing its host (such destruction would then be designated predation).

Parasocial route. Development of sociality by the aggregation of adults of the same generation to form nonsocial groups around clumped resources. (Contrast with familial route.)

Parasympathetic nervous system. A portion of the autoomic (involuntary) nervous system. Activity in the parasympathetic nervous system produces effects such as decreased blood pressure and decelerated heart beat. The neurotransmitter for this system is acetylcholine. (Contrast with sympathetic nervours system.)

Parathyroids. A pair of glands involved in controlling calcium balance, among other activities.

Paratope. The part of an antibody molecule which makes contact with the antigenic determinant (epitope).

Parenchyma [Gr. *para.* beside + *enchyma.* infusion]. A plant tissue composed of relatively unspecialized cells without secondary walls.

Parenteral. Injuction by a route other than oral, that is, intravenous, intramuscular, subcutaneous.

Parity. The condition of a woman with regard to the number of children she has borne.

Parthenogenesis [Gr. *parthenos.* virgin + *gignomai.* be born]. The production of an organism from an unfertilized egg.

Partial containment. An enclosure which is so constructed that contamination between its interior and the surroundings is minimized by the movement of air. A laminar flow clean bench is an example of a partial barrier in which contamination is minimized by the outward flow of air. A chemical hood is also a partial barrier in which contamination of the surroundings is minimized by the inward flow air.

Passive immunity. Immunity in which preformed immune products are acquired by an individual from an external source either naturally (via colostrum or transplacental passage) or artificially (via injection).

Passive transport. Transfer of molecules without the need for energy.

Pasteurization. Heating to destroy pathogenic microbes. Methods. heating to 62.9ø C for 30 minutes (holding method); or heating 71.6øC for not less than 15 seconds. The heating is followed by rapid cooling.

Patch test. A tuberculin skin test in which the tuberulin preparation is applied to a clean skin area on a piece of gauze on tape. A positive reaction, indicating sensitivity to the tubercle bacillus, will show an inflamed area on the skin.

Patents. Whether a biotechnological process or product can be patented, and if so how, has been one of the biggest legal stumbling blocks to the application of biotechnology since the early days of genetic engineering. The first patent for an organism altered by recombinant DNA technology was granted to Chakrabarty in 1980, for a hydrocarbon-degrading Pseudomonas strain to be used for oil clean-up. Since that patent established that recombinant DNA, as well as some more conventional biotechnology techniques and products, can be patented, legal battles have been fought generally over. (i) whether something is a patentable invention or not, and (ii) if it is, who invented it. In principle, patents are based on claims of novelty, utility and enablement. a patent must describe something new, something useful, and must describe it in a way that enables someone else to duplicate it. Perpetual motion machines fail at this last hurdle, for example, which is why most patent offices now require a working model. Living things are not usually patentable, but biological materials obtained by non-biological means (one in which `the hand of man' has had a part) are patentable. Nearly 23.5% of all patents granted in OECD (organization of Economic Cooperation and Development) countries in 1987 were granted in Japan. 30.5% in the USA, 8.8% in the FRG, and <6% in the rest for any one country. However, Japan has a tradition of patenting everything (nearly 50% of all applications are Japanese). Patents often form a type of trade barrier between countries, making it difficult for non-residents to obtain protection, and hence to use their invention in that country. For example, the US patent office claimed that the Japanese patent system puts any foreign language applicant at a disadvantage. The material that is patentable varies from country to country. In addition to patents covering things (`composition of matter' patents), patents covering processes for making or using microbes are allowed in all areas, although methods for breeding are not allowed by the EPO. Apart from the differences and ambiguities in patent law, biotechnology companies face a substantially longer time between filing their patent and getting it granted than companies in most other fields, especially in the USA. This means that they cannot defend their patent in the courts for several years after it has been made public.

Pathogen [Gr. *pathos*. suffering + *gignomai*. causing]. An organism that causes disease.

Pathogenic. Disease producing.

Pathogenicity. Ability of an microorganism to cause disease.

Pathovar. A pathological variety of a species of bacteria.

PC (phosphorylcholine). A commonly used hapten which is also found on the surface of a number of microorganisms.

PCA (passive cutaneous anaphylaxis). The technique used to detect antigen-specific IgE, in which the test animal is injected intravenously with the antigen and dye, the skin having previously been sensitized with antibody.

Pebrine. A disease of silkworms caused by a protozoan.

PEG (polythylene glycol). Polythylene glycol (PEG) is a water-soluble polymer used in many guises in biotechnology. It is the basic structure of a family of polymers that can have PEG units linked to other types of polymers. PEG polymers are typically soluble in many solvents, and are surfactants, i.e. they can promote mixing of water with other solvents. Biotechnological uses include. Large-scale purification of proteins and viruses, through two-phase systems or precipitation.

Pelagic zone [Gr. *pelagos*. the seal]. The open waters of the ocean.

Pelicle. A film due to bacterial growth on the surface of a liquid medium; also refers to the tough flexible covering of protein surrounding certain protozoa, e.g., Class Mastigophora.

Penetration. Refers to the process of injection of the bacteriophage nucleic acid into the host bacterium.

Penicillinase. Enzyme that destroys penicillin.

Penis. The mammalian male organ used in coitus (sexual intercourse).

Pentamer (Greek *pente*, five + suffix mer). Capsomeres (q.v.) formed out of five monomer (q.v.) units.

Pentose. A five-corbon sugar, as ribose.

Peplomer (Greek *peplos*, a covering or an envelop). Indicates the units of envelop that covers some virus particles and which is usually a lipid membrane.

Pepsin [Gr. *pepsis*. digestion]. An enzyme. in gastric juice, that digests proteins.

Peptide bond. Bond between two amino acids with water split off between the carboxyl and amino groups; CO-NH.

Peptide linkage. The connecting group in a protein chain, -CO-NH, formed by removal of water during the linking of amino acids, -COOH to -NH_2. Also called an amide linkage.

Peptide synthesis. Peptides are very short strings of amino acids, usually 10-20 amino acids long, but sometimes only two or three. They are made by different routes from proteins, for two reasons. First, peptides are usually broken down rapidly by bacterial cells, so it is difficult to make them by recombinant DNA methods. Secondly, because they are relatively small molecules, it is feasible to make them by chemical or enzymatic methods.

Peptides. Peptides are short protein molecules, but are usually produced using different methods from other proteins. In general, something is a peptide if it contains 20 amino acids or less, a protein if it is 50 amino acids long or more; in between, it depends on who you talk to. Peptides were very popular in the 1980s since it was discovered that a large number of hormones and neurotransmitters (the hormones that carry signals between nerve cells) are peptides. They can be produced by chemical, biochemical, or genetic means, unlike larger proteins, which are usually produced solely by genetic or cell biological methods. Chemical synthesis adds amino acids one at a time to a growing chain using a cycle of reactions (see Peptide synthesis). Peptides that have been made commercially include calcitonin (for osteoporosis), glucagon (for hypoglycaemia), and thyrotropin-releasing hormone (TRH, for thyroid disease). Aspartame, the artificial sweetener marketed under the Nutrasweet label, is a two-amino acid peptide, and is produced in amounts that dwarf the other, pharmaceutical products.

Peptidoglycan. The cell wall material of prokaryotes, consisting of a single enormous molecule that surrounds the entire cell.

Peptonization. The digestion of milk casein by proteolytic enzymes.

Perennial [L. *per.* through + *annus.* a year]. Referring to a plant that lives from year to year. (Contrast with annual.)

Pericycle [Gr. *peri.* around + *kyklos.* ring or circle]. In plant roots, tissue, just within the endodermis, but outside of the root vascular tissue. Meristematic activity of pericycle cells produces lateral root primordia.

Perinatal. Pertaining to the period of childbirth and shortly thereafter; in medical statistics, the period usually begins with the birth of a fetus 20 weeks or more of gestation and ends 7 to 28 days later.

Periplasm. Contents of a space (periplasmic) that lies between the outer membrane and cytoplasmic membrane of the gram-negative cell.

Peristalsis [Gr. *peri.* around + *stellein.* place]. Wavelike muscular contractions proceeding along a tubular organ, propelling the contents along the tube.

Peritrichous. Having flagella around the entire cell.

Permeability. Allowing the passage of molecules through cell membranes.

Permeabilization of cells. Cells are normally surrounded by a thin membrane of lipids and proteins, the plasma membrane. This is meant to keep out anything not essential for the cell's survival (and, for cells from animals and plants, they function as part of the whole). However, they can also keep out materials that biotechnologists want to get in. To get round this, cells can be `permeabilized'. This effectively makes small holes in the plasma membrane, so that we can get material into the cells but all their contents do not leak out. they remain capable of doing what we want. Permeabilization can be done by treating the cells with organic solvents (which dissolve out small patches of the lipid membranes), detergents such as bile salts, some special purpose ionophores (molecules that introduce channels of molecular size into the membrane, and which usually only let in a limited number of types of molecule), or physical treatments like freeze-drying or sonication. Many types of cell also become rather more permeable to some chemicals after they are immobilized on to solid supports.

Permeabilized cells have other advantages over intact cells for use in a bioreactor. They are rarely viable, so they do not waste metabolic energy (and hence your valuable starting materials) building more cell mass. They will also not grow inside a bioreactor, blocking it up.

Permease. A protein in membranes that specifically transports a compound or family of compounds across the membrane.

Pernicious anemia. A chronic macrocytic anemia caused by a lack of vitamin B_{12}; characterized by disturbances in the gastrointestinal, nervous, and hematologic systems.

Persistent [L. *persistere*, to stand through]. Here refers to a virus that is capable of multiplying, and thus persisting, within the body of the infection carrier or the vector (q.v.) making it an indefinite source of infection.

Pertussis. Whooping cough caused by the bacterium Bordetella pertussis.

PFA (plaque forming cell). An antibody producing cell detected in vitro by its ability to lyse antigen-sensitized erythrocytes in the presence of complement.

pH. The negative logarithm of the hydrogen ion concentration; a measure of the acidity of a solution. A solution with pH = 7 is said to be neutra; pH values higher than 7 characterize basic solutions, while acidic solutions have pH values of less than 7.

PHA (phytohaemagglutinin). A mitogen for T cells.

Phage (Greek *phagein*, to eat). Short form for bacteriophages (q.v.) in general. Other types of phages are also known, e.g., cyanophage.

Phage display. Phage display is a method of making and testing proteins, usually newly invented proteins. The gene for the protein is cloned into a bacteriophage genome in such a way that the protein appears on the surface of the phage. You can then select the phage you want to selecting the protein directly, often using panning. The protein you have selected is still bound to the phage, so you can grow the phage up to indentify the gene sequence, and hence the protein sequence, and to manipulate it further. Typically, the M_{13} phage is used for this, the protein being inserted into the Gene III. Libraries of up to 1012 phage can be made with `random' variants of a protein in them, and the half a dozen variants that we want identified by panning. Phage display has been used mainly for generating antibodies by cloning. Antibody genes are either cloned from human lymphocytes or are synthesized on a gene machine. In either case, they are then spliced into an appropriate phage DNA and put into bacteria, where they make antibody fragments (such as SCAs) on the surface of the phage. If you select the phage that bind most strongly to the antigen, not only will they have the appropriate antibody on their surface but they will also have the gene for that antibody inside. This is an alternative to monoclonal antibody technology, and one that can generate human antibodies (as monoclonal technology cannot). Phage display can also be used for a variety of protein engineering projects for other proteins, screening a large number of variants efficiently, providing that some sort of binding technique can be used to find the phage we need.

Phagocytes. Cells (especially PMNs and macrophages) that are capable of ingesting (endocytosing) substances and subsequently digesting them.

Phagocytosis. A form of endocytosis; the uptake of a solid particle by forming a pocket of plasma membrane around the particle and pinching off the pocket to form an intracellular particle bounded by membrane. (Contrast with pinocytosis and endocytosis.)

Phagosome. A cell vacuole resulting from phagocytosis of particulate materials.

Phagotroph. Organism that ingests living material as a source of food.

Pharmaceutical proteins. Pharmaceutical proteins, also often called biopharmaceuticals and sometimes also `biologics' (in regulatory contexts), are proteins made for use as drugs. The first biopharmaceutical made by recombinant DNA technology was insulin, approved for human use in 1982, followed by human growth hormone in 1985 (both proteins had been used as drugs before). The first truly new recombinant DNA biopharmaceutical was alpha-interferon, approved for treatment of Karposi's sarcoma in 1985. Usually biopharmaceuticals, which have to be human proteins to be fully effective in humans, are made from genetically engineered bacteria, as the only other source is cadavers or live human tissue. The genetic engineering of such products is covered elsewhere (see cytokine, growth factor, stem cell factor). Issues peculiar to biopharmaceuticals are usually the result of the stringent regulation that any drug must pass before it is allowed to go into general use. They include:

Demonstration of efficacy. A curiosity of the regulations is that each biopharmaceutical must be demonstrated to be effective on its own, whereas many of them are meant to be adjuncts to therapy with other drugs and not to have an effect of their own. Demonstration that the production version does what it is meant to do is usually performed by bioassay. Demonstration that the product is free of contaminants. This is particularly true of bacterial proteins and cell wall material (endotoxins), which could act as a `pyrogen', i.e. a material that could cause a feveish immune response in someone injected with it.

Demonstration of purity and stability. There can be other materials that the biopharamaceutical in the preparation-indeed, some are so powerful that a single dose is only a few milligrams, invisible to the naked eye, so something else has to go in there just to make it easier to handle. However, the `something else' must be exactly characterized. The whole preparation must also be demonstrated to be stable. This is often achieved by freeze-drying it.

Freedom from side-effects. Apart from those caused by impurities or extremely high doses, these include principally the body's ability to recognize the protein as foreign, and so mount an immune response against it. Differences as small as the removal of an N-terminal methionine from a protein can alter the body's immune response to it.

Pharmaceutical proteins include. insulin. The first major seller, now produced as a recombinant DNA product by Novo Nordisk and Eli Lilly. Insulin is used to treat diabetes.

Erythropoietin (EPO). Used as an adjunct to cancer therapy, it stimulates the bone marrow to produce more blood. It is one of Amgen's big sellers.

Human growth hormone **(HGH).** Produced by Genentech and by Pharmacia and Upjohn, it is used to treat a variety of growth disorders that result in abnormal shortness. There is also a thriving black market for HGH, in sports medicine and because it has been claimed to be a treatment for ageing.

Pharmacodynamics. Pharmacodynaics is the study of what a patient does to the drug. toxicology is the study of what the drug does to the patient.

Pharming. This is the use of transgenic animals to make pharmaceutically useful proteins.

Pharynx. The part of the gut between the mouth and the esophagus.

Phase difference. The difference in brightness, that is light and shade, of different cellular structures, as observed with the phase-contrast microscope.

Phenetic systematics. Systematics based strictly on phenotypic similarities among living organisms. (Contrast with cladistic systematics.)

Phenogram. Graphic representation of phenetic similarities.

Phenol coefficient. A test for evaluting the effectiveness of a disinfectant by compaing it with phenol under identical conditions.

Phenotype [Gr. *phaenein*. to show + *typos*. impession]. The observable properties of an individual as they have developed under the combined influences of the genetic constitution of the individual and the effects of environmental factors. (Contrast with genotype.)

Phenotypic mixing. A process during which phages (q.v.) of two different types, which are usually close mutants (q.v.) infect the same bacterial cell and induce the host to produce their respective components. Then as the multiplication cycle proceeds, during assembly, some components of one type get constituted into the other type. Thus, the progeny becomes phenotypically mixed. The process, however, is not heritable.

Pheromone [Gr. *phero*. carry + *hormon* . excite, arouse]. A chemical substance used in communication between organisms of the same species.

Pheromones. Pheromones are chemicals used to signal from one organism to another. They are more specific than scents-the target organism can detect them, in minuscule amounts that are too dilute for other organisms to notice. They are usually complicated organic molecules.

Phloem [Gr. *phloos*. bark]. In vascular plants, the food-conducting tissue. It consists of sieve cells or sieve tubes, fibers, and other specialized cells.

Phosphatase test. Test employed to determine the adequacy of the pasteurization process; the enzyme is destroyed in adequately pasteurized milk.

Phospholipids. Cellular materials that contain phosphorus and are soluble in organic solvents. An example is lecithin (phosphatidyl choline). Phospholipids are important constituents of cellular membranes.

Photoheterotroph. An organism that uses light for energy and organic molecules as a source of carbon.

Photon [Gr. *photos*. light]. A quantum of visible radiation; a "packet" of light energy.

Photoperiod. The duration of a period of light, such as the length of time in a 24-hour cycle in which daylight is present.

Photophosphorylation. Photosynthetic reactions in which light energy trapped by chlorophyll is used to produce ATP and, in noncyclic photophosphorylation, is used to reduce NADP+ to NADPH.

Photoreactivation. Recovery of cells from ultraviolet damage by exposure to visible light; an enzyme, activated by visible light, hydrolyzes the thymine dimers formed as a result of ultraviolet exposure.

Photorespiration. Light-driven uptake of oxygen and release of carbon dioxide, the carbon being derived from the early reactions of photosynthesis.

Photosynthesis. (1) Metabolic processes, carried out by green plants, by which visible light is trapped and the energy used to synthesize compounds such as ATP and glucose. (2) Photosynthesis is the process of building new chemical new chemical compounds using the energy of light. Plants build sugars using sunlight, and it is the synthesis of sugars by green plants that is referred to in biotechnology under `photosynthesis'. The rate and efficiency of photosynthesis is the major limiting factor on how fast plants can grow, which itself limits the productivity of all agriculture and food production. Photosynthesis is carried out in subcellular objects (organelles) called chloroplasts, which look like American footballs filled with green lasagne. The light-capturing part of photosynthesis happens on the membranes of the `lasagne'. There are light reactions and dark reactions in photosynthesis. The light reactions capture the energy of light. The dark reactions use that energy to capture carbon dioxide from the air and link it into growing sugar molecules. The light reactions happen in `photosystems', called I and II. The light itself is captured by an antenna complex, a large array of protein molecules that is almost 100% effective at capturing photons falling on them. The energy is then transferred to electrochemical reactions that generate ATP, or are used to split water into hydroxide (which forms oxygen gas) and hydrogen (which is used to reduce cytrochrome proteins). The two systems boost the energy of the electrons involved in this process to a high enough level in two steps-one system on its own could not push in enough energy to split the water. The captured energy accumulates as reduced NADH and ATP, which is used in the dark reactions. The dark reactions centre around the reaction of carbon dioxide with a sugar called ribulose diphosphate, making two three-carbon sugar molecules. It is catalysed by ribulose-1, 3-diphosphate carboxylase/oxygenase, always referred to as RUBISCO. A whole series of other reactions then convert the products into other sugars, and thence into plant. RUBISCO is the most common protein on earth. (The most common protein from vertebrates is collagen, which we use to make gelatin, and its total mass on earth is probably less than 1000th that of RUBISCO.) This is the basic photosynthesis process in green plants, and is called the C_3 process, because it centres around three-carbon sugars. Photosynthesis builds sugars from CO_2 and light, generating oxygen as a side product. Another process is called photorespiration, which does exactly the opposite-it uses light and oxygen to break down sugars into CO_2. Unfortunately, exactly the same enzyme is responsible for photosynthesis and photorespiration, i.e. RUBISCO. So you cannot get rid of the photorespiration. In tropical countries, photorespiration can almost equal photosynthesis, so plants adapted to these conditions use a more complex scheme called C_4. Here carbon dioxide is carried on a chemical carrier molecule to the chloroplasts, where it is concentrated before it is used in photosynthesis. Thus RUBISCO `sees' more CO_2. C_4 photosynthesis is not just a matter of different enzymes. The leaves of C_4 plants are arranged so that the CO_2-capturing enzymes are on the outside and the CO_2-releasing enzymes on the inside next to the chloroplasts.

Phototaxis. Movement toward light.

Phototroph. An organism that utilizes the energy of light.

Phototropism [Gr. *photos*. light + *trope*. a turning]. A directed plant growth response to light.

Phycology. Study of algae.

Phyletic evolution. Change through time in a lineage of organisms which is genetically continuous from generation to generation.

Phylogency. This is the study of how things are related by descent from a common ancestor. It is useful for gaining clues about what parts of living things do. For example, if we discover a new human gene of unknown function and find that it is related to a gene in yeast that is known to be concerned with controlling the cell cycle, we might guess that the human gene was also concerned with the cell cycle. If a yeast protein was a protein kinase, maybe the corresponding human protein is too. And so on.

Phylogenetic tree. Graphic representation of lines of descent among organisms.

Phylogeny [Gr. *phylon*. tribe, race + *gignomai*. be born, produce]. The evolutionary history of a particular group of organisms; also the diagram of the "family tree" that shows which species may have given rise to others. (Contrast with ontogeny.)

Phylum [Gr. *phylon*. tribe, stock]. In taxonomy, a high-level category just beneath the kingdom and above the class; a group of related, similar classes. It is a large subdivision within a kingdom which consists of a number of organisms that have certain broad characteristics in common.

Physical containment. Physical containment is keeping something inside a laboratory by putting physical barriers in the way of its escape. It is the principal way that genetically engineered organisms are kept inside a laboratory and prevented from `escaping' to the wider world. (The other route is biological containment.) There are a range of physical barriers used, many of which are similar to those used in building clean rooms. However, in a containment laboratory the idea is to keep the dirt in, not out. Technologies used include.

Air filtration. Exhaust air from air conditioning is filtered before it is vented to the outside. Often the containment laboratory is kept at a lower pressure than the outside (`negative' pressure) so that any air that leaks leak into the laboratory, not out.

Sterilization lights, usually versions of fluorescent tube lighting that give out a lot of ultraviolet light, are commonly used to sterilize the laboratory's exposed surfaces at night (when it will not give worker s sunburn).

Waste disposal. Often, all waste leaving a containment laboratory is autoclaved to sterilize it. This includes apparently innocuous waste like paper towels, as well as obviously contaminated material. An alternative is to incinerate it, but then it has to be sealed to take it to the incinerator.

Personnel protection. Personnel working in a containment laboratory must often wear protective clothing, much as they are required to do in a clean room. This, however, is so that potentially contaminated clothing can be thrown away on leaving the room and not carried into the outside world. Typical levels: **Level 0.** any laboratory. **Level 1.** `good microbiological practice'. This is equivalent to any microbiology laboratory, where normal microbiological techniques are used to ensure that relatively non-hazardous organisms are kept in the lab, and do not cross-contaminate experiments. Typically, such laboratories are used for routine gene cloning involving no expression of a gene that could be hazardous to humans. **Level 2.** The laboratory is kept at negative pressure and air filtered. Any contaminated waste is autoclaved. Initial gene cloning experiments involving high levels of protein expression may be carried out in such laboratories, as well as microbiology involving organisms that have a relatively low hazard risk. As an additional safety precaution, most work would be done inside laminar flow hoods (hoods in which the air is circulated so that any particles generated by the experiment are carried up into the hood's own filter system and not into the laboratory). **Level 3.** The laboratory is only entered through an `airlock' system, and all waste leaving it is autoclaved. Workers have to wear elementary protective clothing. Work on genetically engineered organisms that are expressing bioactive proteins, and on dangerous but relatively non-infectious organisms such as Clostridia would be done in such laboratories. **Level 4.** This is the ultimate containment level in most countries. Air is usually double-filtered on the way out, there is a double-airlock system for personnel with disinfectant bath to wash their shoes/boots in on the way out, and no one is allowed into the laboratory without substantial training (and no one who does not need to be there). Work on `live' AIDS viruses and genetic engineering of common bacteria to express highly toxic proteins such as ricin could be done in such facilities. **Level 4** facilities are very rare. usually even the most potentially hazardous biotechnology project is adequately contained by Level 3.

Physiology [Gr. *physis*. natural form + *logos*. discourse, study]. The scientific study of the functions of living organisms and the individual organs, tissues and cells of which they are composed.

Phytic acid. The ester of inositol occurring in the leaves of certain plants and cereal husks. It forms insoluble or almost insoluble compounds which pass unabsorbed through the gastrointestinal tract. It reduces the coefficient of utilization of such minerals as calcium, phosphorus, iron, and magnesium.

Phytochrome [Gr. *phyton*. plant + *chroms*. color]. a plant pigment regulating a large number of developmental and other phenomena in plants; can exist in two different forms, one of which is active and the other is not. Different wavelengths of light can drive it form one form to the other.

Phytopathogen. Microorganism pathogenic for plants.

Phytoplankton [Gr. *phyton.* plant + *planktos.* wandering] The autotrophic portion of the plankton, consisting mostly of algae.

Pica. From the Latin word for magpie, it is the craving for a food or nonfood substance in large quantities.

Pili. Fimbriae filamentous appendages, shorter, straighter, and considerably smaller than flagella they appear to make bacteria more adhesive, Sex pilus is essential for genetic transfer in bacterial conjugation.

Pilsener beer. Larger beer made according to the beer prevailing in Pilsen, Czechoslovakia.

Pinocytosis (Gr. drinking cell). A form of endocytosis; the uptake of liquids by engulfing a sample of the external medium into a pocket of the plasma membrane followed by pinching off the pocket to form an intercellular vesicle. (Contrast with phagocytosis and endocytosis.).

Pit. In botany, a small cavity in a cell wall that is not a complete perforation.

Planetary Quarantine. An area of endeavor involved in the prevention of contamination of other planetary bodies and the moon by terrestrial organisms, as well as prevention of contamination of earth by organisms from extraterrestrial bodies.

Plankton (Gr. *planktos* = wandering). The free-floating organisms of the sea and freshwater that for most part move passively with the water currents. Consisting mostly of micoorganisms and small plants and animals (contrast mekton.).

Plant cell immobilization. As well as the general methods used to immobilize growing cells in a bioreactor (see Animal cell immobilization), there are several techniques that are relatively specific for immobilizing plant cells.

Plant genetic engineering. Making `plant' products (like scents or food flavours) from plant cells in culture rather than whole plants.

Plant. A member of the kingdom Plantae. Usually multicellular, gaining its nutrition by photosynthesis.

Planula (L. *Planum* = something flat). The free-swimming, ciliated larva of the cnidarians.

Plaque. A circular clearing in a turbid layer (lawn) of bacteria growing on the surface of a infection initiated by a single bacteriophage. It also refers to the structureless debries found on the surface of teeth.

Plasma (Gr. *plassein* = to mold). The liquid portion of the blood, in which blood cells and other particulates are suspended.

Plasma cell. An antibody producing B cell which has reached the end of its differentiation pathway.

Plasma membrane. The membrane that surrounds the cell, regulating the entry and exit of molecules and ions. Every cell has plasma membrane.

Plasmid. A DNA molecule distinct from the chromosome (s); that is, an extrachromosomal element. May replicate independently of the chrormosome.

Plasmid. A plasmid is a small piece of DNA that can exist inside a cell separate from the cell's main DNA. This means that it must be able to replicate itself inside the cell, so plasmids have the correct genetic elements in them to cause the cell's enzymes to replicate them as the cell divides. Plasmids exist in most microorganisms. Those in bacteria are almost invariably circles of DNA. Some in yeast are linear DNAs, like very small chromosomes. Plasmids are sued extensively in genetic engineering as the basis for vector molecules. Because they are small, they are easy to manipulate. (By contrast the chromosome of *E. coli*, with three million bases, is a molecule 2 nn thick linked into a circle a millimetre round. A test-tube with a billion of those in is too thick to pour, and the shear forces of stirring it will break most of the molecules). Plasmids also have only a few sites for restriction enzymes in them, and so it is relatively easy to cut them open at just one place, then to splice in a piece of `foreign' DNA and join the ends up again. They can also be manipulated to be present in many copies in the cell, rather than the one copy of normal chromosomes and plasmids (see Vector). Plasmids are a specific type of episome, the generic name for any small DNA that can exist as an independent entity inside a cell free of the cell's main chromosomes. Some viruses can also be episomes, existing as DNA within a cell for a long time. (This does not include the retroviruses. These exist as DNA inside a cell, but their DNA is spliced into the chromosomes themselves.) Many plasmids can transfer themselves between cells, which is a very undesirable characteristic for a gene-cloning vector, which we want to stay in the cells we put it in. A first step to turning a plasmid into a usable vector is to delete genes such as `mob' (which mobilizes a plasmid) as `tra' (which allows it to transfer itself to another bacterium) from the plasmid.

Plasmodesma (Gr. *plasma* = formed or molded + *desma* = band). A cytoplasmic strand connecting two adjacent plant cells.

Plasmodium. In the non-cellular slime molds, a multi-nucleate mass of protoplasm surrounded by a membrane; characteristics of the vegetative feeding stage.

Plasmogamy. Fusion of protoplast (such as gametes), often followed by nuclear fusion.

Plasmolysis. Shrinking of the cytoplasm and plasma membrane away from the cell wall, resulting from the osmotic outflow of water. Occurs only in cells with rigid cell walls.

Plastid. Organelle in plants that serves for food manufacture (by photosynthesis) or food storage; bounded by a double membrane.

Platelet. Small blood components important in blood coagulation.

Pleomorphism. The state of having more than one form or shape.

PMN or PMNL. Ploymorphonuclear leukocyte of a type called a neutrophil; it is a white blood cell with a multilobed nucleus that is highly phagocytic and the predominant cell type in acute inflammation.

Pock. A pit, spot, or pustule.

Poikilotherm. An animal whose body temperature tends to vary with the surrounding environment (contrast with homotherm.).

Point mutation. A mutation that results from a small, localized alteration in the chemical structure of a gene. Such mutation can give rise to wild-type revertants as a result of reverse mutation. In genetic crosses, a point mutation behaves as if it resided at a single point on the genetic map. (contrast with deletion.).

Pokeweed mitogen. A mitogen for B cells.

Polar body. A non-functional nucleus produced by meiosis, accompanied by very little cytoplasm. The meiosis which produces the mammalian egg in addition three polar bodies.

Polar flagellation. Flagella arise at ends of cells. Types. monotrichous (single flagellum), amphitrichous (single flagellum at each end), and lophotrichous (tuft of flagella at one or both poles.).

Polar nucleus. One of two nuclei derived from each end of the angiosperm embryo sac, both of which become centrally located. They fuse with a male nucleus to from the primary triploid nucleus that will produce the endosperm tissue of the angiosperm seed.

Polluted water. Water that has undesirable appearance or taste or it may or may not contaminated.

Polycistronic. A massenger RNA in which several coded genes are part of the molecule.

Polyclonal. A term which describes the products of a number of different cell type (cf. monoclonal).

Polymer. A large molecule madeup of similar or identical subunits called monomers (contrast with monomer, oligomer.).

Polymerase. An enzyme that engages in the polymerization of monomeric units.

Polymorphism. (i) In genetics, the coexistence in the same population of two distinct hereditary based on different alleles. (ii) In social organisms such as colonial cnidarians and social insects, the coexistence of two or more functionally different castes within the same colony.

Polymorphonuclear neutrophil. A type of leukocyte in the circulating blood that contains granules which stain with neutral dyes. The nucleus is divided into large lobes. These cells function in phagocytosis and are the prevalent type of leukocyte in the circulating blood.

Polypeptide. A large molecule made up of many amino acids joined by peptide linkages. Large poly-peptides are called proteins.

Polyploid. A cell or an organism in which the number of complete sets of chomosomes is greater than two.

Polysaccharide processing. A common use of industrial enzymes in the food industry is in processing complex polysaccharides such as starch and pectins. Enzymes are involved in several processes.

Liquefaction. The dispersal of starch granules into a gelatinous suspension (i.e. essentially what happens when cornflour is boiled to thicken sauces). The starch is also hydrolysed into shorter molecules by enzymes such as pullulanase and alpha-amylase. Because liquefaction is often carried out in hot solutions, one valuable biotechnological product is heat-stable alpha-amylase and pullulanase, isolated from thermophilic bacteria, which will work at 80 or 90° C.

Saccharification. The formation of low molecular weight sugars, often mainly glucose, from liquefied starch. Acid treatment of starch will break it down into a mixture of glucose and higher sugars, called glucose syrup. The amount of hydrolysis depends on the amount of acid, the time taken, the type of starch involved, and so on. The degree of breakdown is measured as dextrose equivalents (roughly, the fractional extent to which the starch is broken down completely to glucose), and commercially usually falls in the range 40-90. There are also a variety of enzymes that will do this. amylases and pullulanases to break down the starch, invertase to break down sucrose, and glucose isomerase to convert glucose into the sweeter fructose.

Debranching. A chemical rather than a process term, this is the removal of the side branches from the long starch or pectin molecules, leaving long, straight molecules that are easier to break down further. Branched and unbranched polysaccharids also have different gelling properties, and so confer different mechanical properties on food. Enzymes such as pullulanase and isoamylase can perform debranching.

Polysaccharide. A complex carbohydrate consisting of many monosaccharide molecules. Common examples are cellulose and starch.

Polysome. A complex consisting of a threadlike molecule of massenger RNA and several (or many) ribosomes. The ribosomes move along the mRNA, synthesizing polypeptide chains as they proceed.

Population dynamics. The sum of the activities of the members of a deme.

Population. Any group of organisms capable of inter-breeding for the most part, and coexisting at the same time and in the same place.

Porin. Protein pores found in the outer membrane of the Gram-negative cell wall.

Portal of enterance. The means by which a pathogenic organism enters the body to produce disease.

Porter. A weak stout that is rich in saccharine matter and contains about 4% alcohol.

Positive control. The situation in which a regulatory macromolecule is needed to turn transcription of structural genes on. In its absence, transcription will not occur.

Posterior. Toward or pertaining to the rear.

Postmating isolating mechanism. Any intrinsic isolating mechanism (for example, hybrid sterility) that operates after mating.

Post-translational modification. This is a blanket term to cover the alterations that happen to a protein after it has been synthesized as a primary polypeptide.

Potable water. Water free from harmful and unpleasant substances; water fit for drinking.

Pour plate method. A procedure designed to obtain separate colonies on a nutrient agar plate. The process consists of inoculating a culture into cooled melted nutrient agar medium, mixing, and then pouring into a petri dish to solidify. The culture can be diluted as required, by transferring an aliquot from one melted agar tube to another before pouring the plates.

Prausnitz-Kustner Test (P-K test). A passive transfer test that is used in atopic allergy diagnosis; Serum from the allergic individual is deposited in the skin of a volunteer and the suspected allergen is injected later into the same site.

Precipitin reaction. A reaction between a soluble antigen and its antibody resulting in the formation of an antigen-antibody complex too large to stay in solution.

Precipitin. Antibody that forms precipitating lattice upon interaction with its homologous soluble antigen.

Predation. The relationship in which one animal feeds or preys on another.

Predator. An organism that kills and eats other organism that kills and eats other organisms. Predation is usually thought of as involving the consumption of animals by animals, but in broad usage of ecology it can also mean the eating of plants.

Premating isolating mechanism. Any intrinsic isolating mechanism that operates prior to the completion of mating.

Pressure potential. The actual physical (hydrostatic) pressure within a cell (contrast with osmotic potential, water potential.).

Prey (L. *praeda* = body). An organism hunted or caught as an energy source.

Pribnow box. A nucleotide sequence in the prokaryotic promoter gene that is the site of RNA polymerase binding.

Primary cell culture. Specific tissue, e.g., kidney, liver, taken directly from an broken up into individual cells, and then infected with the virus to be cultivated.

Primary culture. Tissue culture derived out of cells taken directly from the source, a plant or an animal.

Primary Non-Disjunction. The production of eggs with two or no *X*'s by an *XX* individual.

Prime. To give an initial sensitization to antigen.

Primordial germ cells (L. *primus* = first). Cells determined early in embryonic growth of animals to become gametes (or to produce gametes on their ultimate arrival at the gonads.).

Prion. Minute subviral particles associated with certain `slow' diseases such as scrapie (in sheep). It is proteinaceous infectious particle behaved to be responsible for neurological disorders of sheep and humans.

Primary structure. The specific sequence of amino acids in a protein.

Pro- (Gr. *first*, before). A prefix often used in biology to denote a developmental stage that comes first or an evolutionary form that appeared earlier than another. For example, prokaryote, prophase.

Proctitis. Inflammation of the rectum.

Prodormal period. A stage in infectious disease process in which the first symptom appear.

Producer. Organism that synthesizes organic material.

Prokaryotes (Gr. *pro* = before + *karyon* = nucleus). Organisms whose genetic material is not contained within a nucleus. The bacteria. Considered an earlier stage in the evolution of life than the eukaryotes.

Promoter region. The area on a gene or operon that binds RNA polymerase and allows transcription initiation.

Properdin. A bacterial protein component of the blood.

Prophage. Here indicates a non-infectious virion which multiplies with the growing bacteria but does not bring about lysis to the host cell; these remain associated with the bacterial chromosome and are in fact a stage in the life cycle of a temperate phage. It is the stage of a bacteriophage in which the viral genome is integrated into the bacterial genome and is duplicated each cell generation.

Prophase. The first stage of nuclear division during which chromosomes condense from diffuse, threadlike material to discrete, compact bodies.

Prophylaxis. Protection, for example, against disease.

Proplastid. A plant cell organelle which under appropriate conditions will develop into a plastid, usually the photosynthetic chloroplast. If plants are kept in the dark, proplastids may become quite large and complex.

Prostaglandins. Pharmacologically active derivatives of arachidonic acid. Different prostaglandins are capable of modulating cell mobility and immune response. It is the group of specialized lipids with hormone-like functions. It is not clear that they act at any considerable distance from the site of their production.

Prostheca. A cytoplasmic filled cell wall protrusion from a cell.

Prosthesis. Artificial component substituting for a defective body part.

Prosthetic group. Any non-protein portion of an enzyme.

Protease. An enzyme that hydrolyzes peptide bonds, thus converting proteins into polypeptides and eventually amino acids.

Proteases. Proteases are enzymes that break up proteins. There are four distinct uses for proteases in biotechnology. Their use depends partly on how cheap they are to make, and partly on how specific they are, i.e. whether they chop up all proteins indiscriminately or only a few proteins at specific points.

Protein crystallization. A key part of most ways of finding out a protein's 3D structure, and hence being able to use that structure to design drugs, is making crystals of the protein. This is difficult, since protein molecules are not as well behaved as simple salts, and the larger they are the worse behaved they are. The trick is usually to perform the crystallization very slowly and in exactly the right solutions-finding the right solutions can take a lot of expertise and time. Novel approaches to protein crystallization include crystallizing under high pressure or in free fall (i.e. in orbit).

Protein engineering. Protein engineering is the design, production, analysis, and use of altered, non-natural proteins. This can be a Herculean task if we do not use a natural protein as a starting point, so usually protein engineering involves modifying existing proteins. Protein engineering has a number of aims.

Improving protein stability. Protease enzymes that have been genetically altered for greater stability are in widespread use in detergents.

Altering antibody properties. There are a range of specialist technologies concerned with antibody engineering, and these are discussed in the entries for chimeric/humanized antibodies and Dabs and other engineered antibodies.

Altering the substrate specificity of an enzyme. Most enzymes catalyse only a very narrow range of reactions, and it would be helpful to be able to alter that range of reactions, and it would be helpful to be able to alter that range so that they acted on other, more commercially useful products. Protein engineering can seek to do this by altering the amino acids around the active site of the enzyme, the bit of the molecule that actually binds on to the substrate and catalyses the reaction. By altering the amino acids the forces holding the substrate in place are altered, and hence the molecules that the enzyme best recognizes are altered. A spectacular example of this was the conversion of malate dehydrogenase (MDH) to lactate dehydrogenase (LDH), two enzymes that catalyse similar types of reactions on different substrates. Unfortunately, neither MDH nor LDH are particularly useful enzymes, and this approach has not been successful on any commercial enzyme. Engineering of proteases such as subtilisin for use in detergents has been a more practical example, in this case broadening the substrate specificity (that is, making it less specific).

Altering pharmacological action. Much protein engineering is aimed at biopharmaceuticals (see Pharmaceutical proteins). In this field protein engineering seeks to alter the biological activity of proteins that have effects that can be harnessed as drugs, by making the effects more potent, more specific, coupling them to targeting mechanisms so that they only effect a few cells or cell types, improving their survival time in the patient, or reducing side-effects. This can include altering the way the proteins aggregate. Therapeutic insulin tends to form multimolecule complexes, which slows its rate of absorption by tissues. Replacing proline-28 in the B chain of insulin with aspartate stops it aggregating in the vial, and allows it to be absorbed much more rapidly.

Protein sequencing. Determining the sequence of amino acids in a protein is done chemically via a cycle of reactions that chop one amino acid off at a time. There are several machines that perform these quite complex sequences of reactions automatically. The number of amino acids that can be determined depends on the amount of protein available and the nature of the amino acids. None of the reactions in the cycle is 100% efficient, and the efficiency varies to some extent depending on which amino acid is being removed for analysis. Thus, after a while, the amount of each amino acid being released by the reaction cycle becomes too small to detect against the `noise' of other amino acids released from those proteins that were not broken in previous cycles. Clearly, also, the protein must be reasonably pure, otherwise the result is a mixture of amino acids at each step. The standard chemical method is called the Edman degradation. The process starts from the amino end of the protein (the N-terminus). In some proteins the N-terminal amino acid has another small chemical group attached to it-usually a methyl, acetyl, or formyl group. The presence of this group makes it impossible to start the reaction cycle. Therefore, some pre-preparation of the protein is needed before a sequence can be determined. Other methods including mass spectrometry (MS), especially fast atom bombardment (FAB) mass spectrometry, are gaining popularity.

Protein stability. Proteins are not very stable in chemical terms. they are easily denatured (i.e. converted to inactive forms). Denaturation occurs when the protein chain of amino acids, usually folded into a specific, tight-coiled conformation, unfolds. the carefully arranged 3D structure of its surface is lost, and usually whatever its function was is lost as well. This change can be brought about by heat, acids, alkali, and by some chemicals such as urea and guanidine, which are known as chaotropic agents because they are extremely effective at opening u the structure of water-soluble macromolecules into a floppy, `random coil', which has no real fixed 3D structure.

Protein structure. Proteins are highly complex structures. For convenience, and partly to reflect biological reality, their structure is divided into several different levels.

Primary structure. This is the chemical sequence of amino acids in the protein. It is unique and definitive for that protein. Implicit in it is that the primary structure is a string of the 20 `biological' amino acids joined by peptide bonds.

Secondary structure. This is how the primary structure folds into small, distinct shapes, usually consisting of a 3-10 amino acids. These units are very stable, and occur in many different proteins. Common ones are alpha helices (short rods), beta sheets (flat, twisted plants), and a variety of turns, which join them. Alpha helices are often shown in graphics of protein structure as solid cylinders, beta sheets as parallel arrays of flat bands or arrows. Anything from 30 to 90% of the amino acids in a protein can be part of a secondary structure feature.

Super-secondary structure. Some of the secondary structures assemble into groups of three or four element which again are quite stable and common amongst many proteins. The most common example is the helix bundle, a tight bundle of four alpha helices.

Tertiary structure. This is how the whole protein (secondary structure features and other amino acids) folds together in space to make a molecule. Quite often the tertiary structure is made up is structural domains, regions of the protein that are quite distinct and have a distinct function. Thus there are recognized domain `shapes' for binding nucleoside triphosphates, which turn up in many different metabolic enzymes that use ATP as a substrate. Some scientists believe this is because the enzymes we have today have been assembled out of more primitive domains that were once individual proteins. The tertiary structure of the domians is usually much more similar than the primary structure-their amino acid sequence can be very different, but they fold into similarly shaped proteins.

Quaternary structure. Many functional proteins are assembled from several amino acid chains. How the different chains (polypeptides) assemble together is their quaternary structure.

Protein. One of the most fundamental building substances of living organisms. A long chain polymer of amino acids with 20 different common side chains. Occurs with its polymer chain extended in fibrous proteins, or coiled into a compact macromolecule in enzyme and other globular proteins.

Proteolytic enzyme. An enzyme whose main catalytic function is the digestion of a protein or polypeptide chain. The digestive enzymes trypsin, pepsin, and carboxypeptidase are all proteolytic enzymes.

Proteome. The `proteome' programme, also called the `phenome', is the name given to some of the investigations that must follow the genome programme if we are to make sense of what the genes we have discovered actually do. Specially, proteome seeks to perform protein biochemistry using the same philosophy of high throughput, automated analysis that has been applied so successfully to the genome projects.

Protist. A member of the kingdom Protista, which consists of unicellular eukaryotes such as the flagellates, amoebas, ciliates, and sporozoans.

Protocooperation. A transitory association between two species in which both benefit.

Proton. One of the three most fundamental particles of matter, with mass approximately 1 amu and an electrical charge +1.

Proton-motive force. The force tending to pull protons across the cell membrane because of the electro-chemical potential generated across the membrane.

Proto-oncogene. A normal cellular gene capable of conversion to an oncogene (capable of causing cancer).

Protoplasmic streaming. The metabolically driven motion of the contents of a eukaryotic cell. This process of `mixing' allows transport of materials over distances that are too great to allow mixing by diffusion alone.

Protoplast. A cell which would normally have a cell wall, but from which the wall has been removed by enzymatic digestion or by special growth conditions.

Protoplasts. Plant, fungal, and most bacterial cells are surrounded by a tough, thick, cell wall. A protoplast is such a cell from which we have removed the cell wall, leaving the cell naked and surrounded only by its plasma membrane.

Prototroph. The nutritional wild-type, or reference form, of an organism. Any deviant form that requires growth nutrients not required by the prototrophic form is said to be a nutritional mutant, an auxotroph. For example, those bacterial strains that do not require added growth factors such as vitamins or amino acids.

Protozoa. A group of single-celled organisms classified by some biologists as a single phylum; includes the flagellates, amoebas, and ciliate. In modern classification protozoans are elevated as distinct kingdom (Protista) and each major subgroups to the rank of phylum.

Protozoology. Systematic study of protozoa.

Provirus. Inactive virus, due to repression of the viral nucleic acid, but may become active and replicate when the repressor is blocked. The virus integrated into host animal or human chromosome and transmitted from one generation to another.

Proximal. Near the point of attachment or other reference point. (Contrast with distal.).

Pseudo-. Prefix, meaning false.

Pseudoalleles. Tandem variants of gene; they do not occupy a homologous position on the chromosome.

Pseudocoelom. A body cavity not surrounded by a peritoneum, Characteristic of nematodes and rotifers.

Pseudogenes. Genes which have homologous structures to other genes but which are incapable of being expressed (e.g., Jk 3 in the mouse.). A DNA segment that is homologous to a functional gene but contains a nucleotide change that prevents its expression.

Pseudomembrane. False membrrane; in diphtheria, it consists of dead tissue cells, leukocytes, red blood cells, and bacteria which form a dull gray exudate.

Pseudoplasmodium. In the cellular slime molds such as Dictyostelium, an aggregation of single amoeboid cells. Occurs prior to formation of a fruiting structure.

Pseudopod. A temporary, soft extension of the cell body that is used in location, attachment to surface, or engulfing particles.

Pseudopodia. False feet; finger-like projections of protoplasm which provide for movement and securing food in protozoa, class Sarcondina; e.g., Amoeba. A "foot" or temporary extension of the cell membrane of ameboid cells and used for movement.

Pseudowild. The patial restoration of a function to an organism.

Psittacosis. A bird disease transmissible to humans.

Psychophile. Cold-loving microorganism, may grow at near-freezing temperatures; optimum growth temperature below 15 °C and may grow at temperatures as low as 2 °C to 3 °C.

Psychrotroph. Mesophilic organism that can grow at psychrophilic temperatures (0°C to 10 °C.).

Puerperal fever. Acute infection following child birth due to introduction of the infectious agent into the uterus; can be caused by beta-hemolytic streptococci.

Pulmonary. Pertaining to the lungs.

Pure culture. Culture containing only one species, or type of micoorganism.

Purification methods. Large scale. One of the central parts of the downstream processing of a fermentation product is purification. Large-scale purification methods are used to take a crude fermentation supernatant or cell homogenate and isolate the product from it in a fairly pure form. Industrial enzymes are often sold in this semi-pure form as a bulk product. If they need to be really pure, then they have to go through a second purification step (see Purification methods. small scale). Purification of the cells from a culture is usually called harvesting, and relies on rather different methods.

Purification methods. Small scale. Many biotechnological products have to be extremely pure, for use as drugs or in producing fine chemicals, so the relatively crude purification methods that isolate them from large-scale culture are not good enough. A further purification step is needed. There are many such methods, but since they are complicated and expensive they are usually only used on a small scale (micrograms to grams rather than grams to toons). Most are chromatography methods. Here the mixture is passed down a tube that is packed with some material to which some components in the mixture stick and others do not. It does not matter whether the product you want sticks or not, providing the contaminants do the opposite.

Purulent. Associated with the formation of pus.

Pus. A yellow white substance found in abscesses containing primarily living, dying, and dead leukocytes. It is the accumulation of dead leukocytes, along with bacteria and serum.

Putrefaction. Anaerobic decomposition of proteins by bacterial enzymes.

Pyemia. Pus in the blood, or the release of purulent material into the blood.

Pyogenic. Pus-forming.

Pyramid of energy. Graphical representation of the total energy contents at different trophic levels in an ecosystem.

Pyramide of biomass. Graphical representation of the total masses at different trophic levels in an ecosystem.

Pyrogen. An agent that induces fever.

Pyruvate. A three-carbon acid; the end product of glycolysis and the raw material for the citric acid cycle.

Q

Q banding. Chromosome banding technique so called because it involves the use of quinacrine dyes (quinacrine mustard or quinacrine dihydrochloride). The technique is most useful in demonstrating areas of heterochromatin, especially the *Y* chromosomes.

Q enzyme. Enzyme involved in the formation of (1-6) linkages at amylopectin branch points.

Q. Symbol for the frequency of a mutant allele in a population: $p + q = 1$, $q = 1\text{-}p$.

Q_{10}. Temperature coefficient; increase in the rate of a reaction for a rise of 10°C, the rate being expressed as a multiple of the initial rate.

Qb. RNA bacteriophage best known as the source of an RNA replicase.

Q-Beta. A bacteriophage which infects *E. coli.* It contains a single-stranded circular RNA genome. The viral RNA is termed the + strand because it can function as an mRNA molecule and direct synthesis of viral proteins. The + strand also acts as a template for the synthesis of a complementary (-) strand. The resulting double-stranded molecule together with the Q beta replicase directs the synthesis of a large number of + strands. This autocatalytic replication can be harness seed to produce large quantities of a desired RNA molecule using recombinant RNA technology.

QC/QA. QC is quality control, testing a product to see that it meets the required quality. QA is quality assurance, making sure that the process you are using to make the product is appropriate for making a product of the required quality. Both are general concerns in all industries and are very important to a wide range of biotechnological processes, especially ones producing pharmaceuticals or foods. Some of the QA/QC procedures that are specific to biologically based products are: **sterility** (are there any living organisms in the product?) and bioburden (how many organisms are there?), both performed using conventional microbiology. **endotoxin** (or pyrogen) testing, usually done using the Limulus amoebocyte lysate (LAL) test. Endotoxins are lipopolysaccharides from bacterial cell walls, which cause extreme immune system responses, sometimes fatal. **GLP and GMP** processes are meant to assure quality in product development and product production, respectively. In fact, they assure that the product is always the same, not that it is always of high quality. Products produced by biotechnology are not usually attacked for their technical quality, but for their deliberate performance: for example, no one disputes that BST is pure and highly effective, but people argue strenuously about whether it should be used on cattle at all. One well-publicized exception was the Showa Denko tryptophan case (see Amino acids).

QH_2-cytochrome c reductase. Membrane-bound complex in the mitochondrial inner membrane, responsible for electron transfer from reduced coenzyme Q to cytochrome c. Contains cytochromes b and c_1, and iron-sulphur proteins.

Quadrat. A square area (standard size is one metre square) taken at random, within which the composition of organisms is noted. The quadrate sampling technique is mostly used in plant ecology to study plant communities but quadrats are also used as a sampling unit to count and weight animals for an estimate of density or to discover animal distribution in a selected area. Permanenet quadrats can be established that are examined at given intervals as a means of assessing changes in species composition in an area over a period of time.

Quadriallel analysis. Analysis based on double crosses obtained by crossing *n* homozygous lines in a diallel fashion, and then crossing the F_1's so generated according to a diallel scheme with the restriction that in any double cross a homozygous line must not occur more than once as a parent.

Quadripartite structure (chromosome ring). One having four corresponding parts.

Quadriplex. Having four dominant genes-used in polyploids (e.g., *AAAA*) (cf. Nulliplex).

Quadripolar spindle. A type of spindle with four poles found in sporocytes.

Quadrivalent (tetravalent). An association of four segmentally paired homologous or partially homologous chromosomes. This association occurs from the zygotene to first metaphase in of meiosis and most clearly seen at diakinesls and first metaphase.

Quadruple Diploid. An octoploid carrying in its somatic nuclei a double set of each of four specifically distinct genoms.

Quadruplets. Four individuals arising from a single birth whether monozygotic, dizygotic, or multizygotic.

Quail. Small galliform bird. Quail embryos are often used in developmental studies because quail cells can be distinguished from chicken cells, yet the two are sufficiently closely related that it is possible to graft embryonic tissue from one to the other to obtain chimeras.

Qualitative analysis. An analysis which seeks to determine what components are present in significant amounts in a substance or mixture of substances; such analysis is designed to indicate the presence or absence of specific chemical species, and not their respective quantities.

Qualitative characters. Controlled by genes, which have ordinarily discrete effect, dominance and discontinuous variation.

Qualitative traits. A trait easily classified into discrete groups.

Quantal mitosis. A controversial concept in cellular differentiation proposed by H. Holtzer and defined by him as a mitosis "that yields daughter cells with metabolic options very different from those of the mother cell as opposed to proliferative mitoses in which the daughter cells are identical to the mother cells". Implicit in this is the idea that the changes in cell determination that occur during development take place at these special quantal mitoses.

Quantasome. A particle seen in the electron microscope to occur in large numbers in a regular arrangement over the surface of chloroplast thylakoid membranes. Latest evidence suggests at least two distinct sizes for the particles, smaller ones (8 nm diameter) being in the outside and larger ones (16 nm diameter) on the inside of the membrane. Functionally the small and large particles are thought to represent Pigment Systems (photosystems) I and II respectively. Both are usually associated with accessory pigment molecules and contain electron carriers. They may be described as photosynthetic units.

Quantitative analysis. An analysis which seeks to determine the relative amounts of significant compounds present in a substance or mixture of substances. There is no sharp demarcation between qualitative and quantitative procedures.

Quantitative characteristic. A continuously varying characteristic; a measuring instrument is often used to distinguish quantitative characteristics.

Quantitative characters. The expression of such characters is controlled by a series of independent genes, which are cumulative in effect, resulting in a continuous variation and absence of dominance.

Quantitative inheritance. Inheritance determined by several or many genes, so that no discrete classes occur in segregating generation; blending inheritance.

Quantitative trait loci (QTL)**.** A quantitative trait is a trait or characteristic of an organism that varies continuously across the population, and hence is almost always controlled by a combination of several genes and environmental factors. Human height is one such trait: we clearly inherit a general tendency to be tall or short from our parents, but many genes contribute to this, as well as nutrition. A quantitative trait locus (QTL) is a gene that contributes towards a quantitative trait. A synonym is metric trait locus. QTLs are important in agricultural breeding programmes. Traits such as plant productivity, protein quality, growth rates in plants and animals, fat content in animals, etc. are all under the control of QTLs, and so breeding an animal or plant with the right collection of alleles at their QTLs is a major aim of the breeding industry. Mapping QTLs is therefore a major concern. A critical aspect to be discovered is the variance caused by the QTL. How much of the variability of the trait is owing to genes as a whole, and to that QTL in particular? Is the variance additive with other genes (e.g. if you find two gene alleles that will add six inches to your height, will someone possessing both `tall' alleles of both genes be 12 inches taller than average)? A related concept is penetrance, which is a measure of how often in the population the gene has any effect: sometimes an allele for a trait will be present, but not show its effects because of interference from other genes or from environmental effects. (This is not the same as it being a recessive allele, when we expect it to be `hidden'.).

Quantum (L. *quantus* = how great). An indivisible unit of energy. A packet of energy; according to quantum theory, the amount of energy; according to quantum theory, the amount of energy released or absorbed in a physical process is always an integral multiple of a discrete unit of energy, known as quantum (of energy). The quantum of electromagnetic radiation is known as a **photon.**

Quantum evolution. Relatively rapid evolution involving a major adaptive shift.

Quantum number. A measure of the number of light quanta required for a given photochemical reaction. In photosynthesis, it is a measure of the number of light quanta required for the assimilation of one molecule of carbon dioxide or evolution of one molecule of oxygen. The theoretical minimum quantum requirement for photosynthesis, if the two photoreaction *Z* scheme is correct, is eight. The quantum number is determined experimentally as the reciprocal of the quantum yield.

Quantum speciation. The rapid rise of a new species, usually in small isolates, with the founder effect and random genetic drift playing important roles; also called Saltational Speciation.

Quantum yield. Number of molecules of oxygen produced or carbon dioxide fixed for each light quantum absorbed in photosynthesis. The higher the efficiency of photosynthetic chemical reactions, the higher the quantum yield.

Quantum. This is the name given to a packet of radiant energy of value hv joules or $hv/1.\ 6 \times 10^{19}$ electron volts where h is Planck's constant and v is the frequency of the radiation. This unit was introduced by Planck in 1900. An alternative name usually used when light energy is being considered is photon. The energy in a quantum of radiation having a wavelength 5600 A is about $3.\ 55 \times 10^{19}$ joules so that a 100 watt bulb which emits 5% of its energy is visible light gives approximately $1.\ 4 \times 10^{19}$ visible quanta per second.

Quarantine (L. *quaranta* = forty). Refers to the compulsory period of isolation (40 days for a ship and hence the term) to prevent spread of infection; usually suspected or actually infection carrying elements, including patients, are subjected to the procedure.

Quarantine block. An isolated block in which all plants imported into the area are first grown to obsêrve any disease or pest that may be carried by the plants in question and prevent their further spread to those grown in the locality.

Quartet. The four cells (before separation) arising from a spore mother cell. In higher plants they become grains when mature.

Quaternary ammonium compound. A cationic detergent with a central nitrogen atom to which four groups are attached.

Quaternary hybrid. A hybrid derived from four grand-parental types all of which are different.

Quaternary structure. Of aggregating proteins, the arrangement of polypeptide subunits.

Quellung reaction. Swelling of the capsule surrounding a bacterium as a result of interaction with anticapsular antibody, consequently the capsule becomes more refractile and conspicuous.

Quench correction. A method used to correct the observed counts per minute produced by a scintilation counter to give a true value in terms of disintegration per minute.

Quenching. A process that reduces the number of photons that impinge upon the window of the photomultiplier of a scintillation counter. Quenching results in a counting rate (counts per minute, *cpm*) that is lower than the true disintegration rate (disintegrations per minute, *dpm*). The difference is due to dilution effects, or effects of coloured quenchers or chemical quenchers.

Quercetin. Mutagenic flavonol pigment found in many plants. Inhibits F_1F_0-ATPases.

Quick stop. Type of bacterial mutant in which DNA and cellular replication cease abruptly at 42° C. Compare Slow Stop.

Quiescent centre. A group of cells in the centre of the apical meristem in which mitotic divisions are rare or absent. The cells may begin dividing if another part of the meristem is damaged.

Quiescent culture. Culture of cells in which active cell growth and replication are minimal, but in which cells continue to survive adequately. Confluent cell cultures may behave in this way.

Quiescent. Quiet, at rest, not necessarily dormant and having the potential for resumed activity; can apply to cells unlikely to divide, the non-meristematic cells.

Quinacrine mustard. Cytotoxic drug; alkylating agent used as an anticancer drug.

Quinacrine. Derivative of acridine used as an antimalarial drug and also in cancer chemotherapy. Quinacrine dyes (quinacrine mustard or quinacrine hydrochloride) are also used as chromosome stains and will fluoresce under ultraviolet light.

Quinate. NAD oxidoreductase. A plant enzyme converting hydroquinic acid (a derivative of the shikimate pathway) to quinic acid. The enzyme is activated by a calcium-and calmodulin-dependent phosphorylation.

Quinone reductase. Enzymes that reduce quinones to phenols usually using NADH or NADPH as a source of reductant.

Quinone. One of a class of molecules active in biological oxidation/reduction systems.

Quintuplets. Five individuals arising from a single birth whether monozygotic or multizygotic.

Qulescent centre. A group of cells in the centre of the apical meristem in which mitotic divisions are rare or absent. The cells may begin dividing if another part of the meristem is damaged.

Qulnone reductase. Enzymes that reduce quinones to phenols usually using NADH or NADPH as a source of reductant.

R

R factors. Bacterial plasmids carrying genes for drug resistance.

R loop. A region of a double-stranded DNA molecule in which one DNA strand is displaced and paired with a complementary strand of RNA.

R. That portion of the *R* plasmid carrying genes for all drug resistance, with the probable exception of tetracycline resistance.

R^+m^-. strains would commit suicide by restricting their own, unmodified, DNA).

R_1, R_2, R_3 etc. The first, second, third, etc. generations following any type of radiation to induce mutations.

R_1. Designates regenerant from *in vitro* culture; as a shoot from callus or a plantlet from a somatic embryoid. R_2, R_3 are the second and third generations from crosses between the R_1, R_2 generations respectively.

Rabbit reticulocyte system. A cell-free system prepared from lysed rabbit reticulocytes which is able faithfully to translate eukaryotic mRNAs from a wide variety of heterologous sources.

Rabi. The winter season.

Rabies. An acute disease of humans and other animals, caused by a virus. Transmitted by direct contact from infected animals to humans through bites.

RAC (pronounced rack). Recombinant DNA Advisory Committee. A committee of the United States' NIH which formulates guidelines governing the performance of genetic engineering experiments.

Race. A population or group of populations distinguishable from other such populations of the same species by the frequencies of genes, chromosomal arrangements, or hereditary phenotypic characteristics. A race that has received a taxonomic name is a sub-species.

Racemases. Anenzyme in the group of isomerases that convert L-isomers to D-isomers and vice versa.

Rad. Acronym for radiation absorbed dose. A rad is equal to 100 ergs of absorbed energy per gram of absorbing material. Adopted by Seventh International Congress of Radiation, July 1953. 1 r of X-rays or gamma rays equals 1 rad with an RBE of 1.

Radial symmetry. The condition in which two halves of a body are mirror images of each other regardless of the angle of the cut, providing the cut is made along the centre line. Thus, a cylinder the cut is made along the centre line. Thus, a cylinder cut lengthwise down its centre displays this form a symmetry (Contrast with bilateral symmetry.).

Radiation. Process of emission or diffusion of radiant energy from a given source.

Radioactive decay correction. A compensation made for the loss of radioactivity due to decay that occurs during the course of an experiment.

Radioactive decay. The breaking up of a radioactive atom with the emission of radioactive particles or gamma-rays.

Radioactive half-life. The time taken for a radioactive isotope to loose half its radioactivity.

Radioactive Isotope. An isotope with an unstable nucleus that stabilizes itself by emitting ionizing radiation.

Radioactive label. A radioactive compound or element. It is used as a means of marking a compound and following it in a biological system.

Radioactive tracer. Refers to definite quantity of radioisotope introduced into a biological system so that its path through the system and its concentration in particular areas can be determined by the measuring of radioactivity.

Radioactive. A substance that shows radioactivity.

Radiobiology. The branch of biology concerned with the effects of radiation on living organisms and the behaviour of radioactive biological materials.

Radiochemistry. The study of radioactive elements and their use in the investigation of chemical processes.

Radiograph. An image produced by the action of X-rays or other radiation from a radioactive substance on a photographic plate.

Radioimmunoassay. A method for finding the concentration of peptide hormones in a biological system. Radioactive isotopes are used to label hormones which are then introduced, with specific antibodies, into the tissues or fluids that are being analyzed. The proportion of labelled hormone that becomes bound to the antibody is recorded. This figure shows a negative correlation with the amount of unlabelled hormone in the sample since the unlabelled hormone is competitively inhibiting the binding of labelled hormone. The concentration of unlabelled hormone can be found by comparing the degree of competitive inhibition it produces with a set of standards.

Radioisotope. A radioactive isotope of an element. Examples are carbon-14 and tritium (hydrogen-3).

Radionuclide. An unstable atomic species characterized by a given mass number, atomic number and energy state, that decays emitting radiation (alpha, beta and/or gamma-rays).

Radiotherapy. The treatment of disease, usually tumours; using X-rays or radioactive substances.

Radon. A radioactive gas that is released from the soil into the atmosphere. It is also released during the combustion of fossil fuel, and for this reason there is more radon in the air over cities than in country air.

Ramet. The vegetatively propagated offspring of an ortet. An individual of a clone.

Random genetic drift. Variation in gene frequency from one generation to another due to chance fluctuations.

Random mating. The situation within a population in which, subject to known restrictions of determinable effect, such as Dioecy, Incompatibility and intrauterine copulation, each individual has an equal chance of mating with any individual, including itself.

Random sample. A sample obtained in such a way that each individual in the population, or each gene in the genome, has the same chance of being selected.

Randomization. The process of arriving at a random combination of types of event.

Randomized blocks. An experimental design in which the treatments are arranged in random order within the blocks or replicates.

Randomness. That combination of types of event in which the distribution in classes of one has no causal relation with the distribution in classes of the others, *i.e.* arrived at by chance without discrimination or choice.

Range of reaction. The range of all possible phenotypes that may develop, by interaction with various environments, from a given genotype; also called *norm* of reaction.

Raphe. A linear slit on the valves of diatoms.

Rapid lysis mutant. Type of *E. coli* mutant in which infection by T even phage leads to particularly fast cell lysis.

Rapidly reassociating DNA. Fraction of DNA that anneals or renatures rapidly after denaturation. The annealing involves the recovery of double-strandedness by specific hydrogen bonding, and its rate in solution is a function partly of the frequency of a particular sequence in the mixture and partly of the nature of the sequences involved. The DNA that reassociates most rapidly is the palindromic sequence material, which can hybridize with itself by doubling over of a single-stranded molecule. It is thus called snap back or fold back DNA. A fraction that reassociates somewhat more slowly than fold back DNA, but still much faster than bulk eukaryotic DNA, is highly repetitive, largely composed of centromeric DNA. DNA of intermediate speed of reassociation is termed moderately repetitive DNA and comprises material such as ribosomal and transfer RNA genes, as well as genes coding for histone which are present in multiple copies.

Rare genotype advantage. Hypothesis in connection with frequency-dependent selection which suggests that some genotypes would be advantageous only when rare. If it is increased in frequency due to selection the fitness of such individuals would decrease, thus stabilizing the correlation between rarity and fitness.

Rare phenotype advantage. Situation in a polymorphic population in which commoner phenotype or phenotypes are disadvantage, in some or all circumstances.

Ras gene. Oncogene from a retrovirus that causes sarcoma and leukaemia in rats.

Rate constant. Of a particular chemical reaction, a constant which, when multiplied by the concentration (s) of reaction (s), gives the rate of the reaction.

Rate-gene. A gene controlling the speed of any particular developmental process.

Rational drug design. This approach to drug discovery seeks to model the molecular structure of the target of the drug, and then design a drug molecule that will fit it. This is therefore also called structure-based drug design (SBDD). This contrasts to the alternative, which is to screen a large number of compounds for drug activity, choose the most promising and make a whole lot of variants, choose the most promising of them, and repeat until a suitable drug is found. Rational drug design (RDD) involves knowing the chemical structure of the drug's target, usually a protein. Protein structures are difficult to obtain: it is relatively easy to obtain the amino acid sequence of a protein if you can purify it, but determining how the peptide chain folds up in space is difficult. Finding out the structure usually involves cloning the genes for the proteins to which drugs bind and making them in large amounts in an expression system. The protein must then be crystallized, and the structure of the crystals deduced using X-ray analysis (see Protein crystallization). Increasingly, protein structures are extrapolated from the structures of related proteins, rather than being discovered de novo. The roche drug Saquinavir, an inhibitor of the HIV protease, approved in 1996, is a successful example of a drug `designed' almost entirely by RDD methods. However, this is unusual, and most drug discovery programmes use RDD methods to complement more traditional biology and chemical screening approaches. The disproportionate level of attention paid to RDD arises in part because it offers an alternative to the exhausting extensive screening programmes by which drugs used to be discovered, and in part because it is done on a computer and produces coloured pictures.

Ratoon. A shoot from the base of the stem, or from the root, of a plant which has been cut back.

RBE. Relative biological effectiveness. Ratio of the absorbed dose in rads of X-rays or gamma rays (of a specified energy) to that of the rads of the given radiation having the same biological effects. RBE for X-rays and gamma rays equals one.

rDNA (ribosomal DNA). The genes encoding ribosomal RNA. There are usually many copies of these genes and in eukaryotes they may be extrachromosomally located.

Reaction norm. A specific way of reacting to changes in the environment depending on the nature of the genotype.

Read through. Action of an RNA polymerase in continuing to transcribe a DNA sequence beyond a stop codon, or the action of the translational machinery in translating a messenger RNA molecule beyond the normal codon for chain termination. The latter form of read through results, for example, in transfer RNA which acts as a nonsense suppressor.

Reading frame shift. Mutation that induces the addition or deletion of a base within a coding sequence, resulting in misreading of the code during translation. This is because the DNA sequence is read in triplets (codons) from a fixed point of initiation, and any addition or deletion will lead to all triplets downstream of the mutation differing from the original.

Reading frame. Particular sequence of nucleotide triplets (codons) employed in translation. The reading frame depends on the location of the initiation codon; addition or deletion of one base (or two bases) will alter the reading frame whereas three base deletions or additions will have no effect on the reading frame.

Reading mistake. The incorrect placement of an amino acid residue in a polypeptide chain during protein synthesis.

Readthrough. In the presence of a suppressor -tRNA, an amino acid can be inserted into a growing polypeptide chain in response to a stop codon. Termination is thus prevented and a longer than usual polypeptide synthesised. This process is known as readthrough and the protein product is a readthrough protein.

Reagin. The antibodylike substance in serum and spinal fluid that reacts with the antigens used in flocculation and complement-fixation reactions for syphilis.

Realized niche. The actual niche occupied by an organism; it differs from the fundamental niche because of the presence of other species (Contrast with fundamental niche.).

Reannealing. Renaturation; specifically, the restoration of duplex DNA regions through complementary base pairing of single stranded DNA molecules.

Reassociation. Reformation of a DNA double helix by renaturation following earlier DNA melting to form single-stranded molecules.

RecA. The genetic notation for a gene which encodes a protein involved in homologous recombination in *E. coli.* The recA gene product also acts as a protease which specifically degrades certain proteins such as that encoded by the *cl* gene of phage Cells containing a mutant recA gene are often used when fragments of the *E. coli.* chromosome are cloned in *E. coli.* This prevents homologous recombination between the chromosomal gene and its cloned copy.

Recalcitrant molecule. A compound that is slowly mineralized or that is not metabolized by micro-organisms in nature.

Recall phenomenon. Anamnestic response, or secondary response to an antigen with rapid production to anti-bodies.

Recapitualation. The theory that ontogeny recapitulates phylogeny; that is, that the development of the individual passes through phases resembling the adult forms of its successive ancestors.

RecBC. The genetic notation for an *E. coli* gene whose product mediates recombination. The enzyme encoded by the recBC gene is an ATP-dependent nuclease specific for double-stranded DNA.

Receiving room contaminated. An area for holding contaminated equipment and materials until they can be sterilized and passed through double-door autoclaves or gas sterilizers that open into the non-contaminated receiving room.

Receiving room noncontaminated. A service room generally at the rear of the building that is maintained as a noncontaminated area. Supplies delivered to the building are placed in the receiving room before transfer are placed in the receiving room before transfer through a UV air lock to the contaminated receiving room.

Receptive field. Of a neuron, the area on the retina from which the activity of that neuron can be influenced.

Receptor binding screening. This is one of the biotech-based methods for discovering conventional (i.e. `chemical') drugs. The method relies on the fact that many drugs act by binding to specific proteins (receptors) on or in cells. Normally, these proteins bind to hormones or to other cells, and control the cell's behaviour, although they may be enzymes or structural elements of the cells. The drug interfers with the normal role of the protein. Finding a drug that has a particular effect on a cell or animal involves exposing the cell or animal to the drug and then looking for the often subtle effect. Receptor binding assays isolate the receptor protein, and then search for chemicals that latch on to that receptor. The ones that do may be good drugs, but the ones that do not are pretty certain not to be, so you have narrowed the field.

Receptor sites. The specific sites of the host, particularly bacterial, cell wall to which the infecting virus particles get adsorbed.

Receptor substance. A substance present at definite places in a bacterial wall that permits infection by a specific type of bacteriophage.

Receptor. A cell surface molecule which binds specifically to particular proteins or peptides in the fluid phase.

Receptor-mediated endocytosis. A phagocytic process in which specialized segments on the plasma membrane called coated pits, act as receptors for substance to be transported.

Receptors. Cells signal to each other by chemical means-only nerve cells, as far as we know, use direct electrical signals. The proteins that receive the chemical messages are generically called receptors. They are favoured targets for drug discovery since they are usually very specific to one cell type, but they are also important in fields as diverse as nitrogen fixation biology and cow rumination. Cell receptors and other cell surface molecules are responsible for cell-cell recognition-the mechanisms by which a cell `knows' which type of cell it is next to. ICAMs are examples of such cell-cell recognition molecules (see Cell adhesion molecules). Receptors generally are on the outside of cells, where they can `see' the chemical signals. (Exceptions are receptors for steroid and thyroid hormones, inter alia, which are in the cell nucleus because these fat-soluble hormones can penetrate the cell membrane.) They therefore need a `signal transduction' system to carry the signal from the outside, where the chemical binds to the part of the cell that receives the message. Two types of signal transduction systems are common. Second messengers are small molecules that act as `hormones' inside the cell, triggering other proteins to act. One common such second messenger is cyclic-AMP (cAMP). Protein kinases (serine kinases or tyrosine kinases) put phosphate groups on target proteins and alter their activity. Often, the resulting proteins then become kinases, and phosphorylate other proteins, and so on down a `kinase cascade'. Often second messenger systems and kinase systems interconnect. Many receptors are members of the G protein-coupled receptor gene family, also called seven-helix membrane spanning (SMS) receptors. These form a tube across the cell membrane and interact with a protein called G protein on the inside of the cell. When the message molecule binds to the receptor, it signals the G protein to catalyse further actions inside the cell, such as synthesis of cAMP. There are hundreds of SMSs: the family includes receptors for hormones, proteins, neurotransmitters, and (in the eye) light. There are a range of other signal transduction mechanisms as well. Several biotechnology companies, including Pharmagenics and Sugen, are working on drug discovery by specifically investigating the signal transduction mechanisms inside cells, because altering these mechanisms should allow one to alter whether a cell proliferates or differentiates or not.

Recessive gene. When the parents with contrasting characters are crossed, the character of one parent does not appear in F_1 generation even though genes for both the factors are present in the hybrid. The gene which does not appear is said to be recessive. (Contrast dominance.).

Recessive. An inherited trait only expressed in the phenotype when the allele controlling it is in the homozygous condition. Thus a recessive trait is not expressed in a hybrid.

Reciprocal Chiasmata. Chiasmata formed between two chromatids in such a way that a double crossing-over occurs in which the other two chromatids are not involved.

Reciprocal cross. A cross repeated in the same combination but with the direction of the cross reversed. The cross *A* (female) X *B* (male) is the reciprocal cross of *B* (female) X *A* (male).

Reciprocal recurrent cross. A recurrent selection breeding system in which genetically different groups are maintained, and in each selection cycle, individuals are mated from the different groups to test for combining ability.

Reciprocal translocation. The exchange of segments between two homologous chromosomes.

Reciprocating shaker. A plateform shaker with a back and forth action used for agitating culture flasks at variable speeds.

Recircularization. The ligation of the two ends of a linear DNA molecule to reform a circular plasmid.

Recognition site. The nucleotide sequence to which a restriction enzyme binds initially.

Recombinant DNA technology. This is the blanket term for the technologies that have made the recent boom in biotechnology possible. It is also called biomolecular engineering, especially in France (`ingenieur biomoleculaire'). Recombinant DNA techniques allow the biotechnologist to isolate and amplify a single gene out of all the genes in an organism, so that it can be studied, altered, and put into another organism. The technique is also known as gene cloning (because you produce a lot of genetically identical genes), and the result sometimes called a gene clone, or simply a clone. An organism manipulated using recombinant DNA techniques is called a genetically manipulated organism (GMO). Recombinant DNA technology covers the following areas:

Isolating genes. This involves splicing the gene of interest with a vector and putting the result into a suitable organism, usually a bacterium or yeast. This new DNA is made of at least two bits of DNA (the target gene and the vector) and is called a recombinant DNA. The organism then grows, multiplying the gene-vector combination as it does so, to produce a clone of cells. The DNA is then said to have been `cloned into' the vector. The piece of DNA we have inserted into the vector is called, with uncharacteristic obviousness, the `insert'.

Identifying and characterizing genes. This involves finding which clone contains the gene you want. This is done using biochemical methods of increasing power to discriminate one gene from another, culminating in DNA sequencing.

Sub-cloning is a related technique. This takes a large gene clone and breaks it up into smaller pieces, making a new clone from each. This means that what was originally a large piece of DNA is now in smaller, more convenient pieces. This is often used to take a large piece of DNA with many genes on it and separate the genes into one-gene clones.

Modifying genes. This involves replacing anything from a single base to a whole chunk of the gene with other DNA, using (inter alia) site-directed mutagenesis.

Putting the genes into another organism. In some cases this is not necessary because you are looking for information about the gene. However, for biotechnologists putting the gene to use it is usually essential, and so they put the gene into another organism using either transfection, transduction, biolistics, electroporation, or microinjection. Other key technologies mentioned in this book are the polymerase chain reaction (PCR), and homologous recombination.

Recombinant DNA. *Bits* and *Kits*. There are a number of pieces of the technology of DNA cloning that are usually referred to without further explanation. The more common ones are:

Adaptor/linker. These are short oligonucleotides that are used to join disparate DNA molecules together. To do the actual joining, DNA ligase is needed.

DNA polymerase. An enzyme that makes DNA. To do so, it must have a DNA molecule to copy (the template) and a short DNA molecule to start with (the primer). It then adds bases on to the primer, copying the template until it gets to the end.

DNA ligase. Also sometimes T4 DNA ligase. This enzyme joins two double-helical DNA molecules together to make one longer one.

Klenow enzyme. A version of DNA polymerase.

Methylation. This is a process (again carried out by specific enzymes, methylase) that puts methyl groups on specific bases on DNA. The presence of these methyl groups can stop some restriction enzymes cutting at that site, and in mammalian cells it is important in controlling genes.

Restriction enzymes. These are enzymes that cut double-stranded DNA at very specific base sequences, and nowhere else. Thus, they cut cloned DNA into a few pieces only. The place at which they cut is called a restriction site, and the map of all such sites on a clone is called a restriction map.

Reverse transcriptase. An enzyme that makes DNA, but uses an RNA template to do so, not a DNA template.

RNA polymerase. There are several of these around, notably SP4 RNA polymerase. These are used to make an RNA copy of a DNA. It needs a template but does not need a primer.

Taq polymerase. Another DNA polymerase made from Thermus aquaticus, and an enzyme that is stable when heated to 95° C, used for PCR. There are a lot of `kits' on a market, collections of reagents, enzymes, DNAs, and even organisms that have been developed as a package that works together to process the purchasers' samples. Among the more common are packaging kits (which are used in bacteriophage cloning), in vitro transcription and translation kits (which carry out transcription and translation in a `test-tube'; see Genetic code and protein synthesis), kits for site-directed mutagenesis, labelling DNA with radioactive, fluorescent or chemical labels, and so on. There is a school of though that there are so many kits around that molecular biology has been reduced to a game of putting together the right kits and writing up the result. Having done it both with and without kits, I think that kits have a lot going for them in allowing the scientist to concentrate on doing creative experiments rather than making all the reagents needed.

Recombinant type. An association of genetic markers, found among the progeny of a cross, that is different from any association of markers present in the parents (cf. *parental type*).

Recombinant. An individual, meiotic product, or single chromosome in which genetic materials originally present in two individuals end up in the same haploid complement of genes. The reshuffling of genes can be either by independent segregation, or by crossing over between homologous chromosomes. For example, a human may pass on genes from both parents in a single haploid gamete.

Recombination (prefix *re* + L. *combinare:* to join). Refers to the process of formation in offspring of gene combinations not present in either of the parents, but are formed as result of combination of both the parental genomes. In viruses, this involves exchange of fragments of replicating DNA molecules of two different virions simultineously infecting one host.

Recombination factor. The gametic output in a hybrid of new combinations of two genes expressed as a fraction or percentage of the total gametic output of new and old combinations combined. The formulae for the gametic output of the double heterozygote *AaBb* are (i) Coupling : *1/2(1-p)AB, 1/2pAb, 1/2paB, 1/2(1-p)ab; (ii)*

Repulsion : *1/2pAB, 1/2(1-p)Ab, 1/2(1-p)aB, 1/2pab,* where *p* represents the recombination fraction. Where there is no linkage, p=0.5.

Recombination Index. A measure of the mean number of independently segregating chromosome segments in the species, individual, etc. The recombination index is calculated by adding the gametic chromosome number to the average total number of chiasmata in the mother-cell.

Recon. The smallest unit of DNA segment between which recombination may occur; probably two adjacent nucleotide pairs constitute a recon.

Reconstitution. Here refers to the process of assembly of artificial disintegrated virus particles, such as TMV, into complete ones.

Reconstructed cell (recon). The viable cell reconstructed by the fusion of a karyoplast with a cytoplast.

Reculture or subculture. The aseptic transfer of a culture or a portion (division when necessary) of that culture (inoculum) to fresh nutrient medium.

Recurrence risk. The probability of any trait showing up again.

Recurrent parent. That parent of a hybrid with which it is again crossed or with which it is repeatedly crossed; the backcross parent.

Recurrent selection (cumulative selection). This method involves hybridization between plants with varying characters and from as many diverse sources as possible; isolation of desirable complexes in the F_2. The selected lines without further purification, inter-crossed in all combinations possible and carried to a new F_2 bulk. Selection for character is done and the cycle repeated again till desired standard is reached.

Red-green colour-blindness. The inability to distinguish red and green colours. One type is determined in man by a sex-linked gene located in nonhomologous part of the X chromosome.

Redifferentiation. Cell or tissue reversal in differentiation from one type to another type of cell or tissue.

Redirected activity. The direction of some behaviour, such as aggression, away from the primary target and toward another, less appropriate object.

Redox couple. Compounds that occur in both the oxidized and reduced forms and that are participants in oxidation-reduction reactions, such as NAD^+-NADH.

Redox. Oxidizing and reducing reactions involving transfer of an electron from a donor to an acceptor molecule; the donor becomes oxidized and the acceptor is reduced. Arranged in order of intensity these reactions are known as the redox series.

Reduced gamete. A sex cell with chromosome number half of the somatic number.

Reduced number of chromosomes. The gametic or haploid number of chromosomes. Some lower organisms have reduced number in somatic cells, e.g., *Neurospora*.

Reducer. In biology, a heterotrophic individual which utilizes the chemical energy of organic matter while breaking it down or reducing it to more simple substances.

Reductase test. A test employed to determine the approximate bacterial count in milk; the time required for the reduction of methylene blue to a colourless compound is proportional to the number of bacteria present.

Reductases. Enzymes that carry out a reduction.

Reduction division. A special type of cell division in which the chromosomes pair before dividing, then divide to give gametes with just half the somatic number of chromosomes. One of two divisions at meiosis is reductional.

Reduction. Gain of electrons; the reverse of oxidation. Most reductions lead to the storage of chemical energy, which can be released later by an oxidation reaction. Energy storage compounds such as sugars and fats are highly reduced compounds.

Reductional separation. Separation of homologous parts of non-sister chromatids at anaphase of one or other meiotic division, as proposed to equational separation of sister parts of chromatids.

Reduplication. The occurrence of a segment of chromosome twice in the same haploid set.

Refuse incinerator. A fuel-furnace for the combustion of organic wastes, in which all gases will have reached a minimum temperature of 1350° F before discharge.

Regenerate. 1. To form or create again; reform. **2.** To give or gain new life or to renew (heal) by a new growth of tissuer; as **regenerants** (shoots, plantlets) form from explants in plant tissue cultures. The process is **regeneration.**

Region. In biogeography, a major division of the world distinguished by its peculiar animals or plants. For example, Africa south of the Sahara is recognized as constituting the Ethiopian region.

Regression coefficient. The rate of change of the dependent variate on the independent, variate.

Regression Coefficient. the rate of change of the dependent variate on the independent, variate.

Regression line. Line superimposed on a graph that indicates to what extent an increase or decrease in one variable will affect the increase or decrease in a second variable.

Regression, filial. The tendency for the offspring of exceptional parents to revert towards the norm of the species.

Regression. (1) Galton's Law, that in any population of families the deviation of the mean of the offspring from that of the population is less than the deviation of one specified parent, adjustments being made for sex differences (Galton, 1889).

Regressive curve. Curve that expresses the change in rate of a dependent variable, with respect to an independent variable.

Regulation of organism release. Regulation about the deliberate release of organisms, and particularly genetically manipulated organisms, vary widely. The USA has a fairly consistent set of regulations controlled by the EPA. European regulations vary enormously from the extremely restrictive (Denmark) to the extremely liberal (Italy, Greece) as judged by the American yardstick. By the end of 1989 there had been 140 deliberate release experiments in the USA, about half that number in Europe.

Regulation. Biotechnology sometimes complains that it is a heavily over-regulated industry, but in practice it is no more regulated than many other industries, and especially those that rely on relatively new technology.

Regulator gene. A gene that contains the information for making a regulatory macromolecule, often a repressor protein.

Regulator site. The specific segment of DNA responsible for production of a regulatory protein that may serve as a repressor in some cases or as an activator in others.

Regulon. A system of more than one operon controlled by a single regulatory protein.

Reinert, J. (1958). Among the first to observe adventive embryogenesis in cell cultures of carrot.

Reiterated. Nucleotide sequences which occur many times within a genome are said to be reiterated.

Rejuvenation. 1. Synonymous with dedifferentiation. **2.** Treatment that leads to culture invogoration (such as subculture) or revival (dormancy breaking).

Relational coiling. The loose coiling of chromatids about each other.

Relative biological effectiveness. It is the ratio of the dose (in rads) of gamma radiation from cobalt-60 to the dose (in rads) of any type of ionizing radiation which produces the same, specified, biological effect.

Relative plating efficiency. The percentage of cells that give rise to colonies when compared to a control culture in which the plating efficiency is arbitraily established at 100%.

Relative sexuality. The ability of gametes to function as male gametes in certain unions and as female gametes in other unions.

Relaxed control. Situation in which the numbers of a particular plasmid present in a bacterial cell are not closely regulated by coupling of plasmid DNA synthesis to host cell DNA synthesis. In relaxed control upto 100 copies of plasmid per cell may be found; this may be further increased using chloramphenicol.

Relaxed mutant. Bacterial mutant that does not shut down RNA synthesis abruptly in a depleted growth medium.

Relaxed plasmid. A high copy number plasmid, generally represented by more than 20 copies chromosome.

Relaxin. A sex hormone produced in the female reproductive glands of animals (including humans), which causes the symphysis pubis to relax thereby facilitating parturition.

Release (Old French *relassaire.* to relex). Indicates the process of letting loose of mature virus particles from bacterial cells disintegrated as a result of bacteriophage infection; equally applicable to other host-viron systems.

Release factors. Specific proteins involved in the reading of genetic stop signals for protein synthesis.

Releaser. A sign stimulus used in communication. Often the term is used broadly to include any sign stimulus.

Releasing factor. One of several hypothalmic hormones that stimulate the secretion of anterior pituitary hormones.

RELP. This abbreviation stands for **restriction fragment length polymorphism** and is usually pronounced `riflip'. It means a piece of DNA that varies between two individuals: whether the DNA has a function or not, or whether the variation is important is irrelevant. The term refers only to the way of detecting the variant, which is by the use of the very specific DNA-cutting enzymes called restriction enzymes. The essence of an RELP is that one variant DNA is cut by a particular enzyme at one site, the other is not. This means that the fragments produced by that enzyme on those DNAs have different lengths. RELPs have found wide use as `marker' genes for genetic studies. Here, the RELp is used to detect when a piece of DNA has been inherited by an individual from one parent (rather than the other). If the RELP is near to a gene that we wish to track but cannot detect directly, then there is a fair chance that the target gene has been inherited along with the RELP. The RELP is termed a linked marker since it is physically and genetically linked to the gene we are actually interested in. A related term is the allele-specific oligonucleotide (ASO). This is an oligonucleotide that will hybridize to the DNA from one individual but not to that of another, because the DNAs differ by one or two bases. The variant forms of the DNA are called alleles. Both RELPs and ASOs have found substantial use in human genetics, and the animal and plant breeding programmes.

Rem. Acronym for *Roentgen equivalent man.* The quantity of radiation which produces the same biological damage in man as that resulting from absorbing *1r* or X-rays or gamma rays. The dose in rem is equal to the dose in rad multiplied by the RBE.

Renal. Relating to the kidneys.

Renaturation. The return of a protein or nucleic acid from a denatured state to its "native" configuration.

Renner effect. Competition among the four genetically different spores formed by one meiosis in regard to which shall form the embryo sac. (Renner 1921, Darlington 1932).

Rennet curd. Sweet curd; coagulation of casein in milk catalyzed by the enzyme rennin.

Rennin. An enzyme that transforms the soluble casein of milk into insoluble paracasein. It is obtained from the gastric juice of a calf.

Rep protein. Enzyme isolation from certain strains of *E. coli* that opens the strands of the DNA duplex and therefore acts as a helicase. The enzyme also hydrolyses ATP.

Rep. Acronym for *Roentgen equivalent physical.* The quantity of radiation which upon absorption in the body liberates the same amount of energy as *1r* of X-rays or gamma rays, i.e. 86 ergs per gram of tissue. Almost obsolete.

Repair enzyme. An enzyme that catalyzes the insertion of a correct base or nucleotide into a sequence where the wrong base pair has been previously inserted. Repair enzymes include polymerases, enconucleases and glycosidases.

Repair mechanism. An enzymatic mechanism for the repair of abnormal DNA.

Repair nuclease. Nuclease that recognizes a DNA strand with a missing base and cleaves out the damaged nucleotide, plus some neighbouring nucleotides, allowing later repair by insertion of an undamaged sequence by DNA polymerase plus DNA ligase.

Repair synthesis. Enzymatic excision and replacement of regions of damaged DNA. Repair of thymine dimers by UV irradiation is best understood example.

Repeat. A chromosomal duplication in which the duplicated segments are adjacent (in tandem), inverted or not; also called Tandem Duplication.

Repelcote. The brand name of a material (dimethyl dichloro-silane) used to coat glassware (soda glass); for tissue culture containers and for other purposes.

Repetitious DNA. DNA in which the base sequences are repeated many times in the genome of any cell; in highly repetitious DNA the base sequences may be represented more than 1000 times. Such DNA is concentrated in heterochromatin and in chromosomes where it is located in the region of the centromere. Moderately repetitious DNA includes the genes for rRNA and tRNA, as well as structural genes coding for the synthesis of histones.

Repetitive DNA. Repeated sequences of nucleotides that may occur in great numbers of reiterated copies in a chromosome complement.

Repetitive sequence (satellite DNA). A base sequence of nucleic acids which may be repeated many times in the genome.

Replacement control. A particular type of cultural manipulation involving the encouragement of nonalternate-host plants to succeed and competitively replace the weedy alternate hosts of pest insects, thus eliminating pest breeding areas (Piemeisal and Carsner, 1951).

Replacement site. Site within a gene sequence at which a mutation causes a change in the amino acid determined.

Replacement substitution. A change in a gene, leading to a change in the amino acid sequence of the polypeptide for which gene codes (Contrast with silent substitution.).

Replacement vector. A phage cloning vector in which a portion of the phage genome is replaced by the foreign DNA fragment to be cloned. Replacement vectors are usually based on phage.

Replacement, gene replacement. A method of substituting a cloned gene, or part of a gene, which may have been mutated *in vitrol,* for the wild-type copy of the gene within the host's chromosome.

Replica plate. This is a simple technique for reproducing and selecting bacteria. A number of bacteria are grown on a petri dish. A pad (traditionally sterilized felt) is lowered carefully on to the plate, and when it is taken off some of the bacteria stick to it. It is then lowered on to another plate, where some of the bacteria stick.

This second plate then carries a replica of the pattern of organisms on the first plate. The replica plate may now be incubated, and the bacteria on it tested quite destructively for some property. The ones that come out as having the best result are then identified, and the corresponding group of organisms on the original plate can be identified because they are in the equivalent place. A related technique is the plaque lift and colony blot. Here the `pad' is a filter membrane, and after picking off some microorganisms the filter is treated to liberate their internal DNA and proteins and treated to detect specific proteins or genes. Again, the bacteria or bacteriophage that contained those proteins or genes can be identified by their position on the original plate.

Replica plating. A technique designed to show that mutations are spontaneous and not specifically induced in bacteria. In the test, a large number of bacterial cells are streaked out on the nutrient agar plate and allowed to develop into microcolonies; then a sterile velvet pad is passed over the plate; a few bacteria from each colony adhere to the velvet so that when the velvet is pressed down on a second plate, it produces an exact copy of the original plate. The copy plate can be used to test resistance to various chemical agents without previous contact with the agent; e.g., penicillin.

Replicase. An enzyme (e.g., a polymerase) that catalyzes the replication of macromolecules such as RNA and DNA.

Replicate. To form replicas from a model or template; applies to synthesis of new DNA from preexisting DNA as part of nuclear division.

Replicating fork. Y-shaped region of chromosome that is a growing point in DNA replication.

Replicating forms (RF). The structure of a nucleic acid at the time of its replication - the term most frequently used to refer to double-helical intermediates in the replication of single-stranded DNA and RNA viruses.

Replication. The process by which viruses increase in number in a host cell. Or the conversion of one double-stranded DNA molecule into two identical double-stranded DNA molecules.

Replicative form (RF). Refers to the newly formed DNA strand, complementary to the already existing single stranded DNA synthesised during the early stages of the replication cycle of the bacteriophage Ox174. The replicative form is synthesised using the phage DNA as template. It is primarily responsible for the formation of new phage components.

Replicator. A DNA segment which contains an origin of replication and is able to promote the replication of a plasmid DNA molecule in a host cell.

Replicon. A sequentially replicating segment of a nucleic acid, controlled by a subsegment known as a replicator. A single replicator is present in the bacterial "chromosome, whereas the chromosomes of eukaryotes bear large numbers of replicons in series.

Replisome. Complex of DNA polymerase and other enzymes that initiates and extends DNA synthesis at the replicating fork. Although replisomes have only been intensively studied in prokaryotes, they are presumed to occur in eukaryotic replicons.

Reporter gene. This is a gene that we introduce into a cell to signal when some cellular event is happening. Often reporter genes are used to signal the activation of a whole lot of other genes. The reporter gene makes an enzyme that is easy to detect: often luciferase, the enzyme that catalyses luminol-based bioluminescence, or green fluorescent protein (GFP). The gene for the `marker' protein is joined with a controller gene, which responds to what we want to measure. Thus, if we want to measure the effect of steroids on a cell, we might tie the luciferase gene on to a steroid-sensitive promoter-when steroids get into the cell, the cells will make luciferase and will glow. When we succeed in blocking the effect of steroids, the glow will die. Examples of the use of reporter genes are: studying how signal mechanisms affect gene control in cells and tissues (for example, kinase cascades; see Receptors), detecting viral infection of cells, showing what effects a new gene or mutant has, and studying the different effects of hormones or protein factors like cytokines on different types of cells. Reporter genes are responsible for pictures of glowing plants and bacteria, which make good press.

Repressible enzyme. An enzyme whose synthesis can be decreased or prevented by the presence of a particular compound.

Repression. A decrease in the rate of synthesis of an enzyme specifically caused by a small molecule which is usually the end product of a biosynthetic pathway, e.g., an amino acid or nucleotide.

Repressor gene. A gene which produces a hypothetical repressor substance which prevents the formation of an enzyme. The repressor gene may or may not be closely linked to the structural gene of the enzyme or enzymes concerned.

Repressor. Usually a protein, the product of a regulator gene. This protein in the absence of the substrate binds itself to the operator site and thus inhibits the transcription of mRNA which would have been otherwise transcribed by the structural genes concerned.

Reproduction (self-propagation). The production of an organism, a cell, a chromosome, or any other cell component, by one like itself either as parent or sister.

Reproductive isolating mechanism (RIM). Any biological property of an organism that interferes with its interbreeding with organisms of other species.

Reproductive isolation. Inherent blocks to crosses between members of different breeding populations.

Reproductive stage. The stage in the life cycle of an organism in which sexual or asexual reproduction occurs.

Repulsion (trans-state). A type of linkage met with in those cases where the dominant genes studied in an experiment are located in different members of a pair of homologous chromosomes. If two pairs of alleles are involved, the parents will be *AAbb* and *aaBB* and the F_1 hybrid will be *Ab/aB*.

Repulsion phase (trans position). Location of the dominant alleles of linked genes on different homologs.

Research kit. A package that contains all the required reagents, in a suitable form, to carry out one or more specific assays which may be based on immune, chemical or enzyme-catalyzed reactions. A research kit uses a procedure or reagents that have not been validated or accepted by health authorities.

Reservoir of infection. Animate or inanimate objects that support the growth of microorganisms.

Resistance (Latin **resistere,** to stand against). It is the ability on the part of the host cell to naturally prevent the occurrence of infection from any source.

Resistance factor (R Plasmid). Type of plasmid that carries genes for antibiotic resistance and confers the resistant property on bacteria which carry such a factor.

Resistance gene. A gene which determines resistance to some inhibitor or toxin, e.g. an antibiotic or heavy metal. Resistance genes are frequently included in vector molecules since they enable the selection of organisms containing the vector.

Resistance transfer factor (RTF). A group of genes that permits the transfer of resistance factors during the conjugation process.

Resistance. The relative amount of inherited qualities which allow an organism to influence or reduce the damage done to it by its enemies (Painter, 1951).

Resistivity. The degree of resistance (in ohm-cm or ohm-m) or interference to the flow of an electrical current or the movement of particles from place to place. A measure of water purity; the purer the water, the greater its **resistivity.** The reciprocal of conductivity.

Resolvase. Enzyme that facilitates site-specific recombination between two transposons which represent direct copy repeats in a cointegrate structure.

Resolving power. Of an optical device such as a microscope, the smallest distance between two lines that allows the lines to be seen as separate from one another.

Resource management. The comprehensive decision-making process regarding the rational use and conservation of our natural resources and environment: cf. Pest Management.

Respiration (L. *spirare:* to breathe). (1) Cellular respiration; the oxidation of the end products of glycolysis with the storage of much energy in ATP. The oxidant in the respiration of eukaryotes is oxygen gas. Some bacteria can use nitrate or sulfate instead of O_2. (2) Breathing.

Respirator. A conventional device covering the nose and mouth, which provides a filter for inspired air.

Respiratory chain. The terminal reactions of cellular respiration, in which electrons are passed from NAD or FAD, through a series of intermediate carriers, to molecular oxygen, with the concomitant production of ATP.

Respiratory pigments. Coloured compounds that can combine reversibly with oxygen. Haemoglobin is the blood pigment in all vertebrates and a wide range of invertebrates. Other blood pigments such as haemoerythrin (containing iron) and haemocyanin (containing copper) are found in lower animals, and in many cases are dissolved in the plasma rather than present in cells. Their affinity for oxygen is comparable with haemoglobin, though oxygen capacity is generally lower.

Resting phase. Interphase; the phase in the nuclear cycle when it is at rest in the sense of not dividing.

Resting potential. A study-state potential difference across a biological membrane when it is not excited by a stimulus or metabolic activity associated with energy conversion or active transport.

Resting stage. Of the nucleus, that in which the linear structure of the chromosomes is not visible.

Restitution nucleus. The cell nucleus arising at meiosis owing to the chromosomes not separating into two groups at anaphase, but on account of the presence of univalents or for other reasons all being included within the same nuclear membrane. In this way the halving of the chromosome number is prevented, and the formation of restitution nuclei therefore, as a rule, leads to the origin of unreduced gametes.

Restitution. The reunion of the two broken surfaces of a fragmented chromosome, the original structure thereby being restored.

Restorer gene. A gene, usually dominant, that effectively overcomes the effect of male sterile cytoplasm on male fertility, i.e., produces functional male gametes even in the presence of the male sterile cytoplasm.

Restraint. A limitation of the random arrangement of combinations in an experimental design so that error variation, while still capable of being estimated without bias, is reduced or potentially reduced.

Restricted. Refers to the bacteriophage mutants that are capable of infecting only a few bacteria.

Restriction endonuclease. Any one of the enzymes, produced by bacteria, that break foreign DNA molecules at very specific site. Some produce "sticky ends". Extensively used in recombinant DNA technology.

Restriction enzyme. An endonuclease that recognizes specific nucleotide sequences in DNA and then makes a double-strand cleavage of the DNA molecule. The DNA of an organism possessing such as enzyme is usually modified at the recognition sites, to prevent self-cleavage, by a modification enzyme that recognizes the sites and methylates specific nucleotides at each site.

Restriction fragments. Products of the complete digestion of a chosen target DNA with a restriction endonuclease. Fragments are flanked by the specific recognition sites for that restriction endonuclease.

Restriction map (= *cleavage map*). Genetic map determined by use of fragments from restriction endonuclease digestion of chromosomes.

Restriction site mapping. A technique in which genomic DNA is cleaved with restriction endonuclease into hundred of fragments of varying size. The fragments are separated by gel electrophoresis and blot transferred to a cellulose nitrite film.

Restriction site. A sequence of base pairs in a DNA molecule which is recognized by a restriction endonuclease.

Retardation. The slowing-down of the rate of action of a gene so that the character it controls is not affected till the late adult stage and thus becomes vestigial.

Reticulocyte. Immature red blood cell active in haemoglobin synthesis.

Reticuloendothelial system. A diffuse system of phagocytic cells derived from the bone marrow stem cell which are associated with the connective tissue framework of the liver, spleen, lymph nodes and other serous cavities.

Retroregulation. Ability of a sequence downstream from a coding sequence to affect the translation of parts of a messenger RNA molecule upstream from it.

Retrovirus. A group of RNA viruses that possess reverse transcriptase for transcribing the RNA genome into a DNA intermediate. The agent of some cancers and AIDS.

Retroviruses. Retroviruses are viruses whose RNA genes are copied on to DNA as part of their life cycle. The DNA is then usually inserted into the DNA of the host cell, and it can remain there for many cell divisions as a `provirus', until some signal triggers it to be transcribed on to RNA and so be translated into viral proteins, making more virus. The only thing distinguishing the provirus from any other DNA in the cell is its base sequence. Several retroviruses are of medical importance.

Reverse genetics. Reverse genetics is the type of genetic analysis that starts with a piece of DNA and proceeds to work out what it does. By contrast, normal genetics (`forward genetics') starts with the phenotype (what the organism looks like) and proceeds to work out what the genetic structure is, ultimately decoding the DNA itself. Such feats of gene cloning as the isolation and characterization of the cystic fibrosis gene are often called reverse genetics; however, although these use an impressive panoply of recombinant DNA techniques they still start with an observed phenotype (the disease) and work via ever more detailed genetic techniques to a genetic explanation of what is going on. Reverse genetics has been used, for example, in understanding the genetic structure of a range of viruses, including the AIDS virus. Here, the DNA structure is known in detail, but what it does is not known. So mutations are found or made in the DNA, and then their effect on the phenotype is discovered. In this way the function of those bits of the gene is worked out.

Reverse mutation. A mutation of a mutant allele back to the original allele, i.e., a back mutation. In such cases it is usually a question of a mutation of a recessive allele to a dominant wild-type allele.

Reverse osmosis. A technique used to concentrate solutions. the sample to be concentrated is passed under pressure through a system bounded by a semipermeable membrane. The pressure forces water through the membrane, but larger molecules are retained, and hence their concentration increases. The membranes may be supported on a tubular matrix or on large flat plates in order to give a large surface area.

Reverse transcriptase. An enzyme coded by certain RNA viruses which is able to make complementary single-stranded DNA chains from RNA templates and then to convert these DNA chains to double-helical form.

Reversed phase biocatalysis. Some enzymes work on reactants or products that are almost completely insoluble in water. Others work using water as a substrate, and it would be useful to be able to remove the water from the reaction to make it `run backwards'. In both cases, it is useful to be able to operate an enzyme reaction in a solvent other than water.

Reversion (genetic). A mutational event that restores wildtype phenotype to a mutant.

Reversion mutation (=*back mutation*). Mutation of a mutant gene to original wild-type.

Reversion. A second mutation that restores the genetic information altered by a first mutation.

Reyniers slit sampler. The commercial version of the Fort Detrick `pot' or slit sampler. An impactor which collects airborne particles onto a 150 mm agar plate. The plate slowly rotates under the impacation slit, making it possible to obtain the microbial profile of a given air space over a period of one or two hours on a single agar plate. The sampler consists of a metal body and a spring wound clock mechanism which rotates the agar plate. The impaction slit width and its height above the agar surface are adjustable.

RF form. Replicating form. The structure of a nucleic acid at the time of replication. Refers to double stranded complexes involved in the replication of single stranded DNA or RNA.

RF value. In paper chromatography, the distance travelled by the solute divided by the distance travelled by the solvent front, the latter always being taken as one. The Rf (relative front) value is thus always between nought and one and is characteristic of a particular molecule. In this way various amino acids, chlorophylles, etc., may be identified.

R-factor. A transferrable plasmid that carries the information for resistance to antimicrobials or other agents.

RH positive (+). A person who manufactures antigen D.

Rhesus factor (RH factor). An antigen found on red blood cells. If the blood cells contain this factor, the individual is Rh-positive; those without it are Rh-negative.Antigen carried on the erythrocytes of certain humans that is responsible for the problems of rhesus incompatibility involved in the rhesus system.

Rheumatoid arthiritis. Common disease of humans affecting about one in 50 of western populations. There is a weak hereditary element and some association with specific HLA antigens.

Rhiboflavine. $C_{17}H_{20}O_6N_4$, one of the components of the vitamin B_2 complex.

Rhinitis. Inflammation of the nose.

Rhinovirus. A genes of viruses of the family picornavirdae; in size and morphology the rhinoviruses resemble the enteroviruses but they differ from them e.g. in that they are intolerant of low pH-being inactivated at pH 3-5. Rhinoviruses occur in the upper respiratory tract in man and other animals; approximately 90 serotypes have been isolated from man-in whom rhinoviruses are responsible for a proportion of cases of the common cold.

Rhizines. Short twisted hyphae which serve as anchors in an alga-fungus lichen-like relationship.

Rhizobium. Bacteria are found in special nodules on the roots of the plants and within the bacteria nitrogen fixation takes place. The plants are provided with NH_2 and the bacteria obtain carbon compounds for growth from the plant. Many, if not all, Rhizobia contain one or more large plasmids some of which give the bacterium the ability to colonise a particular species of plant.

Rhizogenesis. Root formation and growth; as in root development *de novo* from callus.

Rhizoids. Hair like extensions of cells in mosses, liverworts, and a few vascular plants that serve the same function as roots and root hairs in vascular plants. The term is also applied to branched, root-like extensions of some fungi and algae.

Rhizome. A stem that grows horizontally below ground. Rhizomes may be fleshy (e.g. iris) or wiry (e.g. couch grass) and may serve as an organ of perennation or vegetative propagation, or occasionally both. *Compare* stolon.

Rhizoplane. The root surface and its adhering soil.

Rhizosphere. The soil region subject to the influence of plant roots and characterized by a zone of increased microbiological activity.

Rho factor. The protein factor which specifies the termination of RNA molecules.

Rhodamine isothiocyanate or tetramethylrhodamine isothiocyanate ($C_{25}N_3SO_3H_{21}$ mw 536.10): A dye used to stain plant protoplasts in the identification of fusion products.

Riboflavin or lactoflavin ($C_{17}H_{20}N_4O_6$ mw 376.36). A water soluble, vitamin (B_2) essential to cellular respiration. It is important in carbohydrate metabolism and implicated in pohotooxidation of auxins and perception of phototrophic stimuli. Riboflavin is an occasional addition to plant tissue culture media. It is sometimes called vitamin G.

Ribonuclease. An enzyme that can cleave the phosphodiester bonds of RNA.

Ribonucleic acid, messenger (mRNA). A ribonucleotidal chain produced within the cell nucleus which codes for a specific protein synthesized on ribosomes in the cytoplasm.

Ribonucleic acid, transfer (tRNA). A class of RNA molecules that covalently bond to specific amino acids. These in turn hydrogen-bond to an mRNA triplet or codon for that amino acid.

Ribonucleic acid. Single stranded nucleic acid consisting of phosphateribose backbone and four nitrogenous bases, adenine, guanine, cytosine and uracil. Note the sugar here is ribose instead of deoxyribose (as in DNA) and in place of thymine there is uracil. RNA's are of three principal kinds: messenger RNA, transfer RNA and ribosomal RNA.

Ribonucleoprotein (RNP). Complex of RNA and protein. Particles of RNA occur in both nuclei and cytoplasm of eukaryotic cells, and are formed in order to protect RNA from degradation and probably also to assist in transport and localization within the cell.

Ribonucleoside. Portion of an RNA molecule composed of one ribose molecule plus either a purine or a pyrimidine.

Ribonucleotide. Nucleic acid consisting of ribose, phosphate, a purine and pyrimidine base. The pyrimidines and purines are uracil and cytosine; adenine and guanine. Unusual nucleosides in some tRNA'S are inosine, pseudoridine, dihydrouridine, methylguanosine, methylinosine.

Ribose. A sugar of chemical formula $C_5 H_{10} O_5$, one of the building blocks of ribonucleic acids.

Ribosomal biosynthesis. The assembly of ribosomal particles from RNA and protein components. This is coordinated in such a way in eukaryotes and prokaryotes that neither excess protein nor nucleic acids accumulate.

Ribosomal cistron. Gene coding for ribosomal RNA.

Ribosomal DNA (rDNA). The genes at the nucleolar organizing region that code for ribosomal RNA.

Ribosomal proteins. A group of proteins that bind to rRNA by non-covalent bonds to give the ribosome its three dimensional structure.

Ribosomal RNA (rRNA). The nucleic acid component of ribosomes, making up two-thirds of the mass of the ribosome in *E. coli,* and about one-half the mass of mammalian ribosomes. Ribosomal RNA accounts for approximately 80 percent of the RNA content of the bacterial cell.

Ribosome. A small oval body within the cell which attaches itself to an initiation site on the mRNA resulting in the synthesis of a protein molecule. Ribosomes consist of protein and RNA.

Ribosome-binding site, Shine-Dalgarno sequence, S-D sequence. A particular sequence of nucleotides in an mRNA molecule to which a ribosome will bind.

Ribozyme. Also called catalytic RNA, these are RNA molecules that catalyse chemical reactions, often the breakdown of other RNAs. Their discovery in the mid-1980s overturned the idea that only proteins could be biological catalysts. Ribozyme have potential in two areas. They are widely touted as potential pharmaceutical agents since their action against other RNAs can be extremely specific. They could, for example, attack a viral RNA without affecting the normal RNAs in a cell. Thus they have potential as antiviral agents and, through their potential ability to attack RNAs from oncogenes, anticancer agents. Ribozymes as therapeutics are still a research technology, however. Although very specific in the test-tube, like antisense RA they can have unexpected effects when introduced into cells. There is also the problem of how to get them into cells. RNA is destroyed very easily by chemical or enzymatic attack, so has to be protected by encapsulating in (for example) liposomes to get it to the cells it must affect. The other area is to use ribozymes as industrial catalysts, selecting suitable catalytic activities through Darwinian cloning.

Ribulose-1,5-bisphosphate (RuBP). The compound in chloroplasts which reacts with carbon dioxide in the first reaction of the Calvin-Benson cycle.

Rickettsiae. Small, disease-causing bacteria-like objects which are obligate intracellular parasites. They contain both DNA and RNA, as well as a protein synthesizing machinery. Q fever, typhus, and Rocky Mountain spotted fever are caused by Rickettsiae.

Rifampicin. (rifampin). Antibiotic that inhibits RNA synthesis. Rifampicin, a derivative of ifamycin SV, specifically inhibits DNA-dependent bacterial and chloroplast, but not mammalian RNA polymerase. It is also reported to inhibit the assembly of DNA and protein into mature virus particles.

Rifamycin. Antibiotic that inhibits RNA synthesis by binding to RNA polymerase and blocking the initiation of RNA chains. the activity is limited to prokaryote RNA polymerase. *Compare* Rifampicin.

Ring bivalent. One with terminal or nearly terminal chiasmata at both ends of the chromosomes; Rod bivalent, one with such a chiasma at only one end; Unequal bivalent, with the two partners of unequal size Daughter bivalent-the separating chromosomes after pairing at meiosis, as opposed to univalents and daughter chromosomes.

Ring chromosome. Association of chromosomes end to end to form a ring in meiosis.

Ringer's solution. An aqueous solution of calcium, potassium and sodium chloride, buffered with bicarbonate or phosphate, of such a pH and osmotic strength to be a suitable medium in which to suspend living cells or tissues under investigation *in vitro.*

Rings. (1) At mitosis, chromosomes with no ends. (2). At meiosis, chromosomes associated in rings, usually by terminal chiasmata, and in twos, fours, sixes, etc.

Ringworm. Fungal infection caused by a group called dermatophytes.

Ripening. The aging of a food product to enhance the chemical or microbiological development of specific flavors and textures.

Risk cost. The increased chance of being injured or killed as a result of performing a behavior, compared to resting.

r^-m^-. Shorthand genetic nomenclature for a bacterial strain deficient in both the restriction and modification of DNA.

r^-m^+. Shorthand genetic nomenclature for a bacterial strain deficient in the restriction but not the modification of the DNA. Such strains are useful hosts for the cloning of heterologous genes.

RNA (Ribonucleic acid). A polymer consisting of nucleotides formed from the bases adenine, guanine, cytosine and uracil. In living organisms, RNA is principally concerned with protein synthesis (mRNA, rRNA, tRNA). and is essentially single stranded although the strand may fold back on itself to form double-stranded regions. In viruses it sometimes forms the genetic material and can be either single stranded (ssRNA) or double stranded (dsRNA).

RNA ligase. An enzyme which can join RNA molecules together.

RNA polymerase. An enzyme that catalyzes the formation of RNA from ribonucleoside triphosphate, using DNA as a template.

RNA primase. Enzyme that synthesizes the short primer RNA of about 10 nucleotides that is elongated by DNA polymerase to form an okazaki fragment of

DNA during DNA replication. The leading strand of DNA is presumed also to be initiated by such a primer.

RNA primer. A short length of RNA that binds onto the lagging strand of the replication fork prior to ligation of Okazaki fragments.

RNA replicase. (RNA-dependant RNA polymerase). Enzyme found in certain RNA phages (e.g. MS2).

RNA sequence analysis. The determination of the base sequence in an RNA using enzymes in a method directly analogous to the Maxam-Gilbert DNA technique, except that base-specific ribonucleases are employed that cleave specifically next to *G,A,* or *A* and *U*, are specific for pyrimidines.

RNA splicing. *In vivo* phenomenon by which exon sequences are joined following intron removal. Although snRNA molecules are believed to be involved, the mechanism of RNA splicing remains nuclear.

RNA transcriptase. The enzyme responsible for transcribing the information encoded in DNA into RNA; also called Transcriptase or RNA Polymerase.

RNA virus. A virus in which the genetic information is carried in RNA. These include the picornaviruses, the aboviruses and the myxoviruses.

RNA, messenger. (mRNA). The RNA, formed in the nucleus, which enters the cytoplasm, attaches itself to a ribosome, and acts as a template for protein synthesis. Following the protein manufacture, the messenger RNA breaks down, leaving the ribosome free again.

RNA, ribosomal (rRNA). A constituent part of the ribosome that with the messenger RNA manufactures proteins at the direction of the messenger RNA.

RNA, soluble (sRNA). The RNA molecule in the cytoplasm that attaches itself to a particular amino acid and carries it to the ribosome where protein synthesis takes place. Some-times called Transfer RNA.

RNA-coding triplets. Triplets found on mRNA that code for the amino acids, the bode being recognized by the anticodon present on the molecule of tRNA charged with the specific amino acid.

RNA-dependant DNA polymerase. A group of enzyme catalyzing formation of DNA molecules from RNA templates. These occur in some viruses (e.g. those that produce tumors).

RNase. An enzyme that hydrolyzes RNA.

Robertsonian change. A chromosomal mutation due to centric fusion or centric fission.

Rodac plate. A plastic culture plate so designed that when filled with sufficient agar culture medium, a meniscus is formed which rises above the rim of the plate. The plate can then be used to assay the microbiological contamination of smooth surfaces (tabletops, floors, skin) by pressing the raised agar against the surface. A number of microorganisms are transferred to the agar. And after suitable incubation visible colonies develop. The number of colonies appearing is a reflection of the degree of contamination of the surface.

Rodamine. Red fluorescent dye often employed as a fluorescent marker when conjugated to protein.

Rodent-proof. To incorporate prescribed structural and architectural features in building to prevent access of rodent and other vermin.

Roentgen (r). The international unit of the quantity of X-rays or gamma rays. One *r* produces 1.6 x 10^{12} ion pairs. Named for William Roentgen, the discoverer of X-rays.

Rogue. 1. To critically evaluate and eliminate unwanted plants from a population; as undesirable phenotypic variants, weeds and diseased plants are destroyed in plant propagation practices. **2.** A variant plant in a population (short).

Roguing. The removal of undesirable individuals from a variety in the field.

Roller bottle apparatus. A culture system that is used for the propagation on anchorage-dependant cells or liquid cultures of aerobic organisms. The culture is placed in a sealed bottle that is placed horizontally on a set of revolving rollers which, by turning the bottle slowly round on its longitudinal axis, keeps the particles in suspension.

Rolling circle. A mechanism for the circular DNA. This model begins with the specific cut of one parental strand of the double helix, which is thus converted to a chain bearing distinct 5' and 3' ends. the 3' ends serves as a primer for DNA replication upon the uncut, circular strand (the template) while the 5' end is rolled off as an increasingly long free tail.

Root culture. Isolated root tips of apical or lateral origin may produce *in vitro* root systems with indeterminate growth habits. These were among the first kinds of plant tissue cultures (White, 1934) and remain important research tools in the study of developmental phenomena; mycorrhizal; symbiotic and plant-parasitic relationships.

Root hair. Outgrowth from an epidermal cell in the root, thereby increasing the surface area of the root. Root hairs occur in a part of the root behind the region of root extension known as the piliferous layer. Root hairs make close contact with the soil particles and are responsible for the absorption of water and solutes.

Root-nodule bacteria. Bacteria belonging to the *Rhizobium,* family *Rhizobiaceae,* which live symbiotically in the nodules of roots of leguminous plants and fix atmospheric nitrogen.

Rootstock. A vertical usually short, underground stem. It is found in many angiosperms, e.g. rhubarb and strawberry, and in certain pteridophytes, e.g. *Osmunda.* It can reproduce vegetatively.

Rose-water test. A serological test used to detect and quantify the Rheumatoid factor by its ability to agglutinate sheep erythrocytes which have been sensitized to a subagglutinable degree; unsensitized cells are used as a control.

Rotary shaker. A platform shaker with a circular motion used for agitating culture flasks at variable speeds.

Rotating culture or rotator. A wheel-like device for slowly (ca. 1 rpm) rotating and gently agitating cultures, usually in a vertical plane.

Rotator fermentar. A continuous pressure dialysis reactor in which the usual simple fixed membrane is replaced by a rapidly rotating membrane cylinder.

RP4. A promiscuous antibiotic-resistance plasmid first isolated from *Pseudomonas aeruginosa.*

rRNA, ribosomal RNA. The RNA molecules which are essential structural and functional components of ribosomes, the organelles responsible for protein synthesis.

RSF 1010. A small 8.5 kb non-conjugative, but highly mobilisable, broad host-range plasmid of the incompatibility (inc) Q group.

Rubella. German measles.

Rubeola. Red measles.

Ruffled edges (Lamellipodia). Extensive, lamellar projections which are involved in cell attachment to solid surfaces.

Rumen. The first chamber of the ruminant stomach.

Run out. To separate nucleic acid or protein molecules by gel electrophoresis.

Runaway replication. Uncontrolled replication of plasmid molecules which multiply within the host building up to several thousand copies. This usually halts cells division but is a very powerful way of amplifying a gene product. Plasmid vectors, which are temperature sensitive for a copy number control element has been, constructed as runaway vectors.

S

S (Svedberg unit). A measure of the sedimentation velocity of a molecule in a gravitational field, this being dependent on the size, density and shape of the molecule.

S gene complex. Group of genes in certain plant families (e.g., Cruciferae) that prevent self-fertilization. Thus pollen tubes are inhibited from growth by ovules carrying identical alleles of such genes.

S phase. The cell cycle phase during which DNA synthesis occurs.

S(Svedberg unit). A measure of the sedimentation velocity of a molecule in a gravitational field, this being dependent on the size, density and shape of the molecule.

S_1, S_2, S_3. S_1 : The immediate, selfbred, progeny of a plant (typically of a non-hybrid plant). S_2 : the generation produced by sowing seed of an S_1. S_3; the progeny of an S_2, etc.

Saccharification. The hydrolysis of a polysaccharide to produce a sugar solution.

Saccharolytic. Capable of splitting or degrading sugar compounds.

Saccharomyces (yeasts). A genus of unicellular ascomycete fungi that can live in both aerobic and anaerobic conditions. They are important in the brewing and baking industries respectively for the alcohol and carbon dioxide they produce by anaerobic respiration. Reproduction is generally asexual by budding although in adverse conditions sexual spores may be formed.

Saccharomyces cerevisiae. The yeast which is used commercially in bread-making and in the production of alcoholic beverages and industrial alcohol.

Sacculus. A term sometimes used to denote the single giant hollow molecule of peptidoglycan which functions as the rigid component of the cell wall in the majority of bacteria and in blue-green algae.

Sacrophyte. An organism which obtains its nutrients in the form of organic matter in solution from dead or decaying plant or animal tissue.

Safety shower. Provided in chemical and radiological laboratories for same function as in conventional, nonbiological laboratories.

Safranin. Stain used to indicate the presence of various structures in plant cell sections, especially those containing suberin. It is commonly used in alcoholic solution.

Salivary chromosomes. The chromosomes in the nuclei of the salivary glands of Diptera.

Salmonella. A genus of gram-negative, rod-shaped bacteria that are aerobic or facultative anaerobes. Many are pathogens, occurring in spoilt food and contaminated water.

Salt tolerance. The ability to tolerate or resist (withstand) a concentration of sodium (Na^+ ion) in the soil (or in culture) which is damaging or lethal to other plants. Sodium toxicity is a global problem on irrigated lands. Breeding and screening for increased tolerance in agronomic plants is of great current interest.

Saltant. Mutant of a fungus or bacterium especially in culture.

Saltation. A discontinuous variation produced by mutation; a mutation.

Saltatory replication. Increase in the number of gene copies of a specific DNA sequence by laying down numerous strands of DNA in parallel without nuclear division.

Salting out. Procedure for bringing molecules such as nucleic acids out of solution by gradually increasing the ionic concentration of one salt, often sodium chloride.

Sam. An amber mutation in the S gene of phage? The S gene is involved with damaging the host cell membrane to initiate cell lysis. Vectors often contain the Sam mutation as a means of biological containment.

Sample (random). A finite series of observations or individuals taken at random from the hypothetical infinitely large population of potential observations or individuals.

Sampling error. Deviation of a sample value from that of the population due to the small size of the sample.

Sandwich technique. An indirect immunofluorescence technique used e.g. to demonstrate the presence of specific antibodies at the surface of antibody-forming cells.

Sanger method, chain terminator technique, dideoxy technique. A method of sequencing DNA named after its main developer, F. Sanger. The technique uses a single-stranded DNA template, a short DNA primer and Klenow enzyme to synthesis a complementary DNA strand.

Sanitize. To reduce the microbial flora in materials or on articles such as eating utensils to levels judged safe by public health authorities.

Sanitizer. Agent that reduces microbial numbers to safe levels according to public health requirements.

Sap. An aqueous solution of nutrients, minerals and other substances that passes through the xylem of plants.

Saprobe. An organism that derives its nourishment from dead or decaying material.

Saprobity system (saprobiensystem). The system by which a body of polluted water (river, stream etc.) that is undergoing self purification may be divided into a number of ecologically-distinct zones-each zone being characterized e.g. by its level of pollution, content of dissolved oxygen, and type of microorganism present (indicator organisms).

Saprophyte (Gr. *sapros.* putrid + *phyton:* plant). An organism (usually a bacterium or fungus) that obtains its carbon and energy directly from dead organic matter. During the period of their obtaining food for themselves, they gradually break complex organic compounds like animal dung, rottenwood, stagnant water into simpler compounds. This process of distruction is done after secreting certain enzymes.

Saprozoic. Describing an organism that feeds on organic material in solution, rather than on solid organic material.

Sarcina. Spherical cells which divide in three planes at right angles to each other to form cubical packets.

Sarcoma. A solid tumor growing from derivatives of tissue such as embryonal connective tissue, bone, muscle and fat.

Sat-chromosome. A chromosome possessing one or more satellites, or a chromosome associated with the formation of the nucleolus; the letters SAT stand for `sine acido thymonucleinico'.

Satellite DNA. Eucaryotic DNA that bands at a different density than that of most cellular DNA upon equilibrium centrifugation. in many cases it consists of highly repetitive DNA; in other cases it arises from organelles.

Satellite. A short end segment of a chromosome that is separated from the rest of the chromosome by a secondary constriction that is often of thread-like appearance and of somewhat variable length. In certain cases the diameter of the satellite is smaller than in the other parts of the same chromosome.

Saturation density. The maximum cell number attainable under specified culture conditions in a culture vessel. This term is usually expressed as the number of cells per square centimeter in a surface substrate culture or the number of cells per cubic centimeter in a suspension culture.

Saturation. The complete absorption of the `pairing-attraction' force of one chromosome by its homologous partner during pachytene pairing, as opposed to the absence of saturation at zygotene when several homologous chromosomes can be associated together as a multivalent.

Sau 3A. A type II restriction enzyme from Staphylococcus aureus 3A which recognises the sequence and cleaves at the sites denoted by the arrows.

Sauerkraut. A fermented food product produced by the lactic acid fermentation of shredded cabbage.

Sausage fermentation. A bacterial fermentation catalyzed by species of *Lactobacillus, leuconostoc* and *Pediococcus,* as well as strains of *Micrococcus* (which reduce nitrate to nitrite) aiding preservation of these meat products.

Scab (plant pathol). Any of a variety of unrelated plant disease characterized by the formation of discrete, scab-like surface lesions-e.g. common scab potato and apple scab.

Scaffolding. A technique which enables any gene or DNA sequence to be integrated into the genome of a desired organism. It was originally described for *Bacillus subtilis* but is applicable to most organisms.

Scal-up. The development of a full-scale production facility from an initial laboratory concept or experiment.

Scanning electron microscopy. Electron microscope technique which permits three-dimensional structures rather than thin sections to be viewed.

Scarce mRNA. Refers to the many mRNA species that exist as very few copies per cell.

SCFs have been used for several model enzyme reactions. In general it helps to include a small amount of water (which also dissolves in some SCFs) to assist enzyme stabilization; it is also essential if the enzyme uses water as a substrate.

Schick test. Susceptibility test to diphtheria; diphtheria toxin in injected intracutaneously.

Schiff's reagent. A mixture of pararosaniline. HCl (an analine dye) and sodium bisulfite ($NaHSO_3$), used in staining chromosomes and nuclear material. See **Feulgen's test.**

Schizogony. Asexual reproduction by multiple fission of a trophozoite (a vegetative protozoan).

Schizont. A stage in the asexual life cycle in the malaria parasites.

Schizophyllaceae. Basidiocarp: fan-shaped or lobed; basidiospores; colourless and smooth. The organisms are characteristically lignicolous. Genera: e.g. *Schizophyllum.* Sparassidaceae. Basidiocarp: branching, flattened, monomitic lobes, the hymenium being on both sides of the erect lobes. Genera: e.g. *Sparassis.*

Schizosaccharomyces. A genus of yeasts characterized by their method of asexual reproduction, which involves fission rather than budding.

Schultz-chariton reaction. Diagnostic test for scariet fever; antitoxin to **B**-streptococcus erythrogenic toxin is injected intracutaneously and the skin rash is observed for blanching.

Schultze's solution. A solution of zinc chloride, potassium iodide, and iodine used mainly for testing for cellulose and hemicellulose. Both materials stain a blue colour with the reagent, that of hemicellulose being weaker.

Scion or cion. The portion of a bud or shoot used for grafting onto another plant (the stock).

Scion. In a plant graft, the part that is grafted on to the stock. The scion may be a bud or shoot.

Sclerenchyma. A plant tissue, characterized by thick walls. The main function of schelerenchyma is to provide mechanical support.

Scleroglucan. An uncharged microbial polysaccharide composed of glucose found in *Sclerotium glucanicum;* also known as polytran. It is highly viscous and pseudoplastic, stable to pH and temperature, and lacks sensitivity to salt. It is used in drilling muds latex paints, and to enhance oil recovery.

Scleroprotein. One of a group of proteins obtained from the exoskeletal structures of animals. They are insoluble in water, salt solutions, dilute acids, and alkalis. This group exhibits a wide range of both physical and chemical properties. Typical examples of scleroproteins are keratin (hair), elastin (elastic tissue), and collagen (connective tissue).

Screen house. A metal or plastic mesh enclosure similar in function to a greenhouse but with more exposure to the elements.

Screening test. (1) Any test, often a shortened (or qualitative) form of a more comprehensive (or quantitative) test, used for the detection of given agent (e.g. a particular type of micro-organism) in each of a relatively large number of specimens. In such tests a certain number of false-positive results may be tolerated; specimens which give a positive screening test may then be examined by more detailed or quantitative tests. (2) Any test the results of which enable a given (unknown) species to be assigned to one of a possible range of taxa.

Screening. A technique for mutant enrichment. e.g. in penicillin screening the mutagenised culture is placed under conditions in which wild-type, but not mutant, bacteria will grow. This culture is then treated with the antibiotic penicillin, which kills the growing (wild-type) cells but has no effect on the non-growing (mutant) cells. The population of viable bacteria is thus enriched for mutant cells.

Scum. A layer of floating solids formed on the surface of a mixed liquor. It may occur during fermentation due to inefficient mixing or be produced deliberately during clarification.

Scurvy. Disease caused by lack of ascorbic acid. Symptoms include superficial haemorrhages from skin and gums.

Scutellum. That part of the cotyledon in Gramineae (grasses) that absorbs food from the endosperm at germination. An explant source for plant tissue culture.

SDS-polyacrylamide gel. Modification of polyacrylamide gel electrophoresis in which the solution of proteins to be analysed is dialysed against a solution of the detergent sodium dodecyl sulphate (SDS). The proteins bind to the SDS and all become negatively charged, having similar weight-to-charge ratios. In an electric field, electrophoretic separation will occur dependent on the proteins' size and shape. Use of differing concentrations of polyacrylamide gel permits different molecular weight ranges to be analysed.

Second division. The second of two divisions of Meiosis, formerly known as the homotypic or equation division.

Second law of thermodynamics. States that in any real (irreversible) process, there is a decrease in free energy and an increase in entropy.

Second messenger. Molecule that carries the stimulus from cell surface to cytoplasm and nucleus following attachment of a polypeptide hormone to the outer surface of the plasma membrane of a cell.

Secondary association. The side by side association of bivalents in meiosis. As indication to the polyploid nature of the organism.

Secondary balance. A new balance derived by change in the proportion of genes, as in secondary polyploid, from an old balance and capable of competing with it.

Secondary cell wall. A structure of cellulose and other materials (lignins) inside of and adjacent to the primary cell wall. It functions in support.

Secondary cells. Those cells arising from proliferation of primary cells. They generally are able to divide only a finite number of times; then most die.

Secondary constriction. A permanent constriction in a chromosome arm. When the terminal segment outside the constriction is short, it is called a satellite. The secondary constriction often functions as a nucleolar organizer.

Secondary immunodeficiency. Immunodeficiency caused by such factors as malnutrition, malignancies, radiation exposure, burns, and various drugs that interfere with the functions of the lymphatic system.

Secondary metabolites. Metabolic products which appear to play no part in cell growth and which are formed maximally under conditions of restricted growth or in the absence of growth; typically, a given secondary metabolite is produced only by a limited number of organisms, i. e. one or a few species. Examples of microbial secondary metabolites include penicillins, polyenes, gibberrelic acid, etc.

Secondary nondisjunction. The phemonenon which results in gametes with extra chromosomes owing to the abnormal karyotype of the original germ cell and not to a nondisjunctional event during gamete formation.

Secondary nucleus. The nucleus resulting from the fusion of the two polar nuclei in the embryo-sac.

Secondary parastoid. An insect parasite of a primary parasitoid; cf, Hyperparasitoid, Primary Parasitoid.

Secondary polyploidy. Cases of polyploidy that for different reasons are supposed to occur in material with high basic chromosome numbers. In such cases the apparent diploids are probably derived from species with low chromosome numbers, although these are not known with certainty.

Secondary product synthesis. The often large scale utilization of plant cells for industrial production of specific plant products such as gums, resins, alkaloids, flavorings enzymes or pharmaceuticals.

Secondary protein structure. The helical or extended structure of a polypeptide chain (e.g., Â helix, ß sheet).

Secondary response. The vigorous response of the immunological system upon exposure to an antigen previously encountered. Largely characterized by IgG synthesis.

Secondary sexual characters. A character without direct reproductive function whose expression is controlled by sex hormones, A character associated with particular sexes especially in animals.

Secondary structure. Those aspects of the structure of macro-molecules that depend on hydrogen bonding. These include the helical arrangement found in DNA and the alpha-helix or beta-pleated sheet structures found in proteins.

Secondary thickening. Cambial activity giving rise to additional (secondary) vascular tissue (conduction and support tissue) resulting in increased stem or root girth.

Second-generation anaerobic digester. A sophisticated anaerobic digester with high rates of gas production at short retention times.

Secrete. To transfer a protein molecule through a membrane. Proteins may be secreted into (i) an intracellular compartment, e.g. a vacuole or mitochondrion, (ii) the periplasmic space, or (iii) the culture medium. Secretion out of the cell and into the culture medium is often described as excretion.

Secretin. A polypeptide hormone secreted by the mucosa of the duodenum and jejunum when the stomach empties its contents into the intestine. It stimulates alkaline pancreatic secretions to neutralize the acidic chyme from the stomach. Its secretion is also stimulated by the release of bile.

Secretion. In the kidney, addition of a material to the urine after the glomerular filtration step.

Secretor gene. Dominant autosomal gene in the human that allows secretion of A and B type blood group antigen into saliva and other body fluids.

Secretory IgA (sIgA). Composed of two (dimer) or three (trimer) basic IgA molecules, a secretory piece, and a J chain; provides topical or local immunity via SIgA-containing secretions (mucus secretions, colostrum, saliva).

Section. A thin slice, usually for microscopy, as a tangential section or a transverse section.

Sectioning. Technique in cell biology and histology in which thin sections are cut from tissues or organisms prior to microscopy.

Sectorial chimera. A chimera in which the tissues of different genotype appear in distinct sectors (cf. Periclinal Chimera).

Sediment survival. The volume of sedimentary rocks preserved and recognized per million years of geologic time.

Sedimentation (Latin **sedimentum,** to sit). It is the process of setting down of suspended particles due to the forces of gravity.

Sedimentation coefficient. The rate of sedimentation of a solute in an appropriate solvent under centrifugation. An *s* value of 1×10^{-13} second is one Svedberg unit.

Seed certification. Seed certification consists of field inspection and seed tests to ensure genetic and physical purity and good germination of seed lots, and issuance of a certificate to that effect.

Seed Èlite. Seed used for carrying on a special selected line from which stock seed will subsequently normally be produced. Extra pure seed.

Seed parent. The female (pistillate) parent of plant hybrids.

Seed plants. The spermatophyta; plants that form true seeds. These are divided into two groups; the Gymnospermae and the Angiospermae.

Seed test. A series of tests on purity, moisture content and germination of a seed sample to determine its quality.

Seed. A fertilized, ripened ovule of a gymnosperm or angiosperm. Consists of the embryo, nutritive tissue, and a seed coat.

Seed:Stock. Seed which is used by the seedsman to sow for the immediate production of commercial seed.

Seeding (attachment efficiency). The percentage of cells planted which attach to the surface of the culture vessel within a specified period of time. The conditions under which such a determination is made should always be stated. This term is often incorrectly referred to as Planting Efficiency.

Seedling. A young plant that has grown from a seed (rather than by grafting or by other means). It is differentiated into radicle, hypocotyl and green cotyledon (s).

Segment allopolyploid. An allopolyploid in which some chromosome segments from the parent species are still homologous.

Segment. A portion of a chromosome considered as a unit for a given purpose.

Segmental Interchange. Mutual translocation; an exchange of non-homologous segments as between two chromosomes.

Segmentation gene. Gene in insects and other segmented animals that regulates the number or polarity of the body segments.

Segmentation nucleus. The fused gametic nuclei in a zygote.

Segregating generations. The F_2 and the subsequent generations obtained through continued selfing of a hybrid between two or more strains.

Segregation (genetic). The separation of alleles, or of homologous chromosomes, from one another during meiosis so that each of the haploid daughter nuclei produced by meiosis contains one or the other member of the pair found in the diploid mother cell, but never both.

Segregation. Separation at meiosis of the chromosomes, or parts of chromosomes such as genes, of paternal and maternal origin (Bateson, 1902).

Selection coefficient. The measure of reduced fitness of a given genotype; it equals one minus the adaptive value (which see) of that genotype.

Selection culture. Utilizes difference (s) in environmental conditions or more usually in culture medium composition, such that preferred variant cells or cell lines (presumptive or putative mutants) are favored over other variants or the wild-type.

Selection differential. In artificial selection, the difference in mean phenotypic value between the individuals selected as parents of the following generation and the whole population.

Selection gain. In artificial selection, the difference in mean phenotypic value between the progeny of the selected parents and the parental generation.

Selection pressure. The measure of the action of selection in tending to alter the frequency of a gene in a given population (c.f. Mutation Pressure).

Selection unit. Single cells or small clusters, units of optimum size for isolating and regenerating variants or mutants; the minimum number of cells effective in the screening process.

Selection. (1) (statistical) discrimination in sampling or in arrangement. Opposed to randomness. (2) (biological) any non-random process which will lead to individuals of different genotypes being represented unequally by their progeny in later generations of a population of self-propagating units. Can be Natural or Artificial (Darwin, 1858). Natural Selection is responsible for the Survival of the Fittest (Herbert Spencer, 1861).

Selective advantage. That genotypic condition of a cell or individual or genetic class of individuals which increases its chances, relative to others, of representation in later generations of cells or individuals.

Selective agent. An environmental or chemical agent that imposes a lethal or sublethal stress on growing plants, or portion thereof in culture, enabling selection of resistant or tolerant individuals.

Selective mating. Nonrandom mating. Phenotypic differences influence the selection of a mate.

Selective medium. A growth medium containing a component that will inhibit the growth of certain microorganisms and either enhance or not affect the growth of other organisms.

Selective plating. Technique for the isolation of recombinant colonies of bacteria by placing two auxotrophic mutants in mixture together on agar plate of minimal medium.

Self. (1) In an immunological sense, the cells and molecules specific to one unique individual, recognized by the immune mechanisms as self rather than non-self. (2) To undergo or to use a procedure involving self-fertilization or self-pollination, and in this sense often used as a verb.

Self-annealing. The process in which complementary single-stranded sequences at either end of a DNA duplex fuse through base pairing to produce a molecule of circular DNA.

Self-antigen. A compound produced by an organisms that stimulates its own autoimmune system, resulting in damage to its body.

Self-assembly. Spontaneous aggregation of performed macromolecules into supra-molecular structures such as multienzyme complexes, protein filaments, ribosomes, membranes and viruses in which the components are non-covalently linked.

Self-differentiating. Said of behaviour developing without the experience provided by the normal objects toward which it is usually directed and without any practice.

Self-fertile. Organisms in which male gametes are able to fertilize female gametes in the same individual.

Self-fertilization (autogamy). Seed set after spontaneous self-pollination occurs in several species of higher plants (wheat, oats, barely, tomatoes, peas, beans, etc.) and may be enforced in inbreeding experiments with cross-fertilizing plants.

Self-ligation. The joining of the two ends of a restriction fragment to each other using DNA ligase. The reaction is carried out at low DNA concentrations to reduce the likelihood of two molecules being ligated together. The procedure is often used in sub-cloning.

Self-pollination. The fertilization of a plant by its own pollen. (Contrast with cross-pollination).

Self-propagation. Reproduction.

Self-replication. Property possessed by most cells given a suitable environment and molecules such as DNA and, in rare cases, RNA.

Self-sterility gene. Gene that operates against self-fertilization in monoecious plants, usually by control of the rate of pollen tube growth downs the style.

Self-sterility. Self-incompatibility. Sometimes incorrectly deduced from mere sterility and vice versa.

Self-thinning rule. This states that in plant populations the number of individuals decreases under competition, but not as fast as the mean weight of the surviving individuals increases; so, the weight of the population as a whole increases.

Semelparous organism. An organism that reproduces only once in its lifetime. (Contrast with iteroparous.)

Semen. The thick, whitish liquid produced by the male reproductive organ in mammals, containing the sperm.

Semiconservative replication. The common way in which DNA is synthesized. Each of the two partner strands in a double helix acts as a template for a new partner strand. Hence, after replication, each double helix consists of one old and one new strand.

Semicontinuous culture. The maintenance of cells in a culture vessel in an actively dividing state by periodically draining the medium and adding fresh medium.

Semi-heterotypic division. A first division of meiosis which gives rise to a Restitution Nucleus following defective pairing (Rosenberg, 1927).

Semilethal (sublethal). With a lethal effect that is not immediately fatal but that instead usually lowers viability markedly.

Semine breeding. Breeding especially in cattle with the aid of artificial insemination.

Semipermeable membrane. A porous membrane that permits the passage of small molecules such as water and inorganic ions, but does not permit the passage of larger molecule such as sugars and proteins. Most biological membranes are semipermeable.

Semisolid. Gelled but not firmly so; as are media made with small amounts of a gelling agent.

Semi-sterile. Of gene heterozygotes or structural hybrids in which approximately half the male and female gametes are inviable (Belling, 1914).

Semisterility. Plants that are heretozygous for a segmental interchange (a reciprocal translocation) are usually semisterile. In such cases about half of the gametes are nonfunctional.

Semisynthetic. Descriptive of a compound that has been derived by the chemical modification of a biological product. Examples include some penicillins and steroids.

Senescence (L. *senescere.* to grow old). Aging; deteriorative changes with aging.

Sense codon. A codon specifying a particular amino acid in protein synthesis.

Sense strand. The strand of duplex DNA which is transcribed into a complementary mRNA (or other functional RNA) molecule.

Sensitive Period. The period of development during which the action of a gene is sensitive to the influence of external conditions.

Sensitive volume. That hypothetical volume in which the occurrence of an ionization must produce a given mutation. (Timofeeff-Ressovsky et al., 1939).

Sensitization. The increase in the reaction of an organism or cell to an antigen to which it has been previously exposed. It may occur naturally or be artificially induced, e.g. following vaccination.

Sensory neuron. A neuron leading from a sensory cell to the central nervous system. (Contrast with motor neuron.)

Sepal. One of the outermost structures of the flower, usually protective in function and enclosing the rest of the flower in the bud stage.

Sephadex. Trade name for a group of absorbent resins used in chromatography.

Sepharose. A proprietary form of agarose supplied in bead form for use in gel filtration and other types of liquid chromatography. Sepharose is a registered trademark of Pharmacia Biotechnology AB.

Sepsis. A localized or general bacterial infection, especially such infections that are associated with a rise in body temperature (fever).

Septate hypha. Hypha with cross walls, or septa, which divide each filament into individual cells with individual nuclei.

Septate. Septations or cross walls separate the individual cells; characteristic of fungi in Classes Ascomycetes, Basidiomycetes, and usually Deuteromycetes.

Septic tank. A watertight sedimentation tank for organic wastes in which the sludge setting on the bottom is allowed to digest and liquefy by anaerobic bacterial action. The tank may be emptied at intervals by a special vehicle, or the effluent may pass into a seepage tank of system or agricultural pipes for final absorption into suitable ground. If an all-wastes septic tank is combined with a sand filter and a chlorination unit, it resembles a miniature sewage works capable, with good maintenance, of handling the wastes from an individual premise or small group of premises.

Septic. Infected; usually descriptive of a pus-forming bacterial infection.

Septicemia. A systemic disease caused by the invasion and multiplication of pathogenic microorganisms in the blood stream.

Septum. A dividing or cross wall characteristic of the filaments of molds with septate hyphae. The septa divide each hypha into many cells with individual nuclei.

Sequelae. Conditions following or resulting from a disease.

Sequence analysis. Determination and study of the sequence of bases in DNA or of amino acids in proteins.

Sequence map. A pictorial representation of the sequence of amino acids in a peptide or of bases in a oligonucleotide.

Sequencing gel. A long polyacrylamide slab gel which has sufficient resolving power to separate single-stranded fragments of DNA or RNA which differ in length by only a single nucleotide. Electrophoresis is carried out at high voltage and with the gel in a vertical position.

Sequencing. The determination of the order of nucleotides in a DNA or RNA molecule or that of amino acids in polypeptide chain.

Sequestering agent. An agent that will hold metallic ions in solution usually be inclusion of the ions in an appropriate coordination complex.

Sequestration. the process of complexing metal ions, usually to chelating agents, to render them ineffective.

Sequestrene 330 Fe. The brand name for an iron chelate used in some tissue culture media.

Serial float culture. A technique of floating anthers on liquid medium and subculturing them to new medium at several day intervals as anther dehiscence. pollen release and development occur, increasing anther productivity.

Serial section. Series of consecutive tissue sections, usually made with a microtome.

Serine proteases. Family of enzymes including the digestive enzymes chymotrypsin, trypsin, elastase and proteases such as thrombin.

Serological test. A test in which antibodies are used in the detection of an antigen. The serum may be complexed to a fluorescent or radiolabelled marker to enable the extent of the reaction to be followed.

Serology. The study through antigen-antibody reactions of the nature and specificity of antigenic materials from different sources.

Serotype. Subtype of a species that is based on antigenic differences. It is an isolate or strain which gives a specific immunological reaction. A single category of antibodies will only react with one serotype. A species of bacteria may consist of many serotypes, reflecting the genetic diversity within the group.

Serum protein. Protein found in serum (cell-free) component of blood. Includes immunoglobulins, albumin, clotting factors, and enzymes.

Serum sickness. A reaction that sometimes follows the injection of a foreign serum. The sometimes follows the injection of a foreign serum. The patient forms antibodies of an antigen in the foreign serum and the antibodies react with residual circulating antigen; the resulting antigen-antibody complexes are toxic and may cause fever and join pains.

Serum. The clear portion of the blood minus the factors necessary for clotting.

Servomechanism. A mechanism within a barrier system controlled by a mechanism external to the system.

Sessile. Permanently attached; not moving.

Sett (seed canes). Stem cuttings used for vegetative propagation of sugar-cane.

Sewage farm. A place where sewage is treated. More specifically it is a process of aerobic treatment of sewage solids in which they are treated by being stacked and turned over periodically by mechanical means.

Sewage farming. Sewage farming, an alternative method of sewage treatment is in effect the use of sewage for irrigation. It is a method which may be employed in areas of low rainfall. This method of treatment may be seen in Melbourne, Australia, where the sewage farm at Werribee, about thirty kilometres from the city centre. The Werribee farm has served nearly two million people; 520,000 cubic metres per day of sewage has been disposed of the 108 square kilometres of land. The precise procedure used (land filtration, grass filtration or lagooning) has depended on the season. While the nutrient load has been reduced by about 50 percent, the run-off passes into Port Phillip Bay contributing a significant nutrient load. The existing sewage treatment facilities have been overloaded, but there will be adequate alternative and additional treatment facilities during the latter part of the decade when the Southeastern Purification Plant at Carrum comes into operation.

Sewage fungus. Unsightly slimes resulting sometimes from the pollution of rivers by discharges or organic matter.

Sewage treatment. A process designed to purify sewage, to remove the BOD and pathogenic bacteria, and to produce clean water. It may include filtration, chemical treatment, aerobic biological treatment and/or anaerobic biological treatment.

Sewage. Liquid or solid refuse (domestic and industrial wastes) carried off in sewers.

Sewal Wright Effect. Non-adaptive differentiation into different types due to "drift" followed by random fixation in small isolated populations.

Sewerage system. The system that collects and carries sewage from source to point of treatment and disposal.

Sex (sexual differentiation). The production by an individual or group of individuals, of gametes of two types, differing in size and mobility (Eggs and Sperm) such that the one can fuse only with the other. The individual or organ producing eggs is female, and that producing sperm is male.

Sex chromatin. A strongly stained body in the cell nucleus, occurring in females but not in males of certain animal species with *X* and *Y* chromosomes. The number of sex chromatin bodies is always one less than the number of *X* chromosomes. Supernumerary *X* chromosomes in the female are heterochromatic and hence appear as sex chromatin.

Sex chromosome. In organisms with a chromosomal mechanism of sex determination one of the chromosomes involved in sex determination one of the chromosomes involved in sex determination. One sex chromosome the X chromosome, is present in two copies in one sex and only one copy in the other sex. The autosomes, as opposed to the sex chromosomes, are present in two copies in both sexes. In many organisms, there is a second sex chromosome, the Y chromosome, that is found in only one sex the sex having only one copy of the X.

Sex Determination. The process by which a spore (haploid) or egg (haploid or diploid) comes to develop the properties of one or other sex.

Sex factor (F) (= sex particle). An episome of bacteria which, when integrated into the chromosome, causes chromosome breakage at conjugation and facilitates genetic exchange; also exists autonomously in F^+ cells.

Sex factor. Genetic factor in the cytoplasm of some bacteria. During conjugation it is transferred to recipient cells. The sex factor, or F factor, is found in male strains, and after conjugation, receipient cells are converted to F+.

Sex hormones. A group of hormones, belonging to the steroids, which affect or control the sexual organs and secondary sexual characters and manifestations.

Sex linkage. The pattern of inheritance characteristic of genes located on the sex chromosomes of organisms having a chromosomal mechanism of sex determination. The sex that is diploid with respect to sex chromosomes can assume three genotypes; homozygous wild type, homozygous mutant, or heteozygous carrier. The other sex, haploid for sex chromosomes, is either hemizygous wild-type or hemizygous mutant.

Sex ratio. The ratio of numbers of the two sexes at conception, at birth, at a given age or in the total population, usually expressed in terms of the females as 100.

Sex reversal. A change in the sexual character of an individual from male to female or vice versa.

Sex-chromosome. Specific chromosomes which either singly or in combination with autosomes condition the expression of sex. Sex chromosomes may be homogametic, in which two sex chromosomes are identical in morphology or may be heterogametic in which these are of different morphology.

Sex-conditioned character. Phenotypic character that, although determined by an autosomal gene, is conditioned by the sex of the individual.

Sexduction. Transmittance of bacterial genes together with the sometimes mobile sex factor F.

Sex-index. In insects (particularly Drosophila), the number of *X*-chromosomes per set of autosomes, e.g. a Drosophila male has a sex index of 0. 5, a female is 1. 0, a super-male 0. 33 and a super-female 1. 5.

Sex-influenced factors. Factors whose dominance is dependent on sex so that in one sex the heterozygote shows the character concerned whilst in the other sex the character is recessive.

Sex-influenced traits. The difference in hormonal systems between male and female individuals may cause differences in gene expression (e.g., the presence of beards in male humans and the absence of this trait in females).

Sex-influenced. Autosomal traits expressed predominantly in one or the other sex owing to sex hormones.

Sex-limited characters. Characters determined by an autosomal genetic difference whose expression is modified by the sex difference.

Sex-limited genes. A gene which is expressed in only one sex. The gene may be on any chromosome, but its expression is under control.

Sex-limited traits. A female individual may carry genes that can only be expressed in the opposite sex and vice versa (e.g., milking ability in cows, presence of horn in male individuals of other animals).

Sex-limited. Traits which are governed by autosomal genes but expressed in one or the other sex owing to the influence of sex hormones.

Sex-linked gene. A gene or locus located in the *X* chromosome in the species having *XY* type of sex mechanism with Y chromosome being almost inert. The gene in *X* whether dominant or recessive finds expression in the phenotype since there is no corresponding gene in Y to modify the action of the gene in *X*.

Sex-promoting alleles (Habrobracon). Certain alleles xa, sb, sc, etc., concerned with sex determination. Females are always heterozygous, males homozygous or hemizygous.

Sexual isolation. Reproductive isolation due to a tendency toward homogamic mating.

Sexual reproduction. Reproduction by the production of male and female gametes, followed by fertilization.

Sexual selection. Selection based on male competition or female choice and responsible for sexual dimorphism.

Sexual spores. Spores produced by the fusion of two cells.

Sexual. A mode of reproduction normally involving fusion of gametes and genetic recombination.

Sexuality. The ability, by any of a multitude of mechanisms, to bring together in one individual genes that were originally carried by two different individuals. It is the capacity of genetic recombination.

Shake culture. An agitated suspension culture. Usually a flask (commonly an Erlenmeyer flask) containing the culture is attached to a horizontal or platform shaker, or agitated with a magnetic stirrer, to provide adequate aeration for cells in the liquid medium.

Shakeror platform shaker. A platform fitted with clips for grasping flasks (usually Erlenmeyer flasks), or with surfaces suitable for attaching flasks, with set or variable speed control. It is desirable to adjust the **shaking** speed for gentle, even agitation of suspension cultures.

Shear. (1) It is the movement of one layer relative to a parallel adjacent layer. (2) To fragment DNA molecules into smaller pieces. DNA, as a very long and fairly stiff molecule, is very susceptible to hydrodynamic shear forces. Forcing a DNA solution through a hypodermic needle will fragment it into smaller pieces. The size of all the fragment obtained is inversely proportional to the diameter of the needle's bore. The actual sites at which the shear force breaks a DNA molecule are approximately random. Therefore DNA fragments may be generated by random shear and then cloned (by either tailing their ends or using likers) so as to create a complete gene library of an organism.

Sheath. A tubular structure surrounding certain microbial cells.

Shift. The phenomenon by which the extracted types from F_2 are not exactly the same as parents in respect of any quantitative character. They are slightly different from the parents.

Shingles. A viral disease of adults resulting from activation of latent measles virus.

Shoot apical meristem. Undifferentiated tissue, located within the shoot tip, generally appearing as a shiny dome-like structure distal to the youngest leaf primordium and measuring less than 0. 1 mm in length when excised.

Shoot tip (apex) culture. A structure consisting of the shoot apical meristem plus one to several primordial leaves, usually measuring from 0. 1-1. 0 mm in length; in instances where more mature leaves are included the structure can measure up to several centimeters in length.

Shoot tip graft or micrograft. The grafting of a very small shoot tip or meristem tip onto a prepared seedling or mcropropagated rootstock in culture. Meristem tip grafting is used for in vitro virus elimination with Citrus and for other plants as an alternative to grafting in the greenhouse.

Shoot. The aerial photosynthetic portion of a plant that generally consists of a stem upon which leaves, buds, and flowers are borne.

Short chain peptides. Short peptide chains consisting of two to several amino acids, usually without enzymatic function.

Shotgun experiment. An experiment in which the entire genome of an organism is exposed to restriction endonuclease fragmentation, then random ligation to vector for cloning into host cells.

Shot-gun. A shot-gun experiment is one in which random fragments of an entire genome are cloned into a vector. A particular gene may then be selected for, or a gene bank established which is subsequently screened for sequences of interest. The word is also used as a verb, e.g. `We shot-gunned **Xenopus** DNA into pBR322'.

Shunt. An alternate pathway, a bypass.

Shuttle vector, bifunctional vector. A vector molecule which is able to replicate in two different host organisms and can therefore be used to `shuttle' genes from one to the other.

Sialic acid. Derivatives of neuraminic acid in which the free amino group is acetylated.

Sib mating. The mating of sisters with brothers; hence the name.

Sibling species. Species that are so similar that they are difficult to distinguish from one another.

Sibling. A brother or sister.

Siblings (sibs). Two or more offspring from the same cross. In animals, brothers and sisters are siblings. In plants the products of a self-pollination are termed sibs.

Sibship. A generation of children in a single family.

Sickle cell anemia. A condition produced by an abnormal hemoglobin molecule in the homozygous condition, h^8/h^8. Red blood cells become sickleshaped under reduced oxygen tension.

Sickle cell trait. Trait shown by individuals characterized as heterozygote for the sickle cell gene h^8/h^8 All of red blood cells will become sickle-shaped if oxygen is reduced sufficiently. More reduction is tolerated than by an h^8/h^8 individual.

Sickle haemoglobin. An abnormal type of haemoglobin found in carriers of the genetic defect that results in sickel cell anaemia.

Siderophore. Metal-binding molecule.

Sieve elements. Elongated cells that, placed end to end, make up the sieve tubes in angiosperms. Most of the organelles, including the nucleus, break down during the development of the sieve elements so all that remains of the cell contents is the cytoplasm, which runs from cell to cell through the pores in the perforated end walls or sieve plates.

Sieve tube. A column of cells formed from sieve elements, in which food is translocated in plants.

Sigma factor. A subunit of RNA polymerase, which recognises specific sites on DNA for the initiation of RNA synthesis.

Signal compound. A compound used in an affinity assay as a means of detecting the presence of a particular group. For instance a fluorochrome or an enzyme linked to an antibody can function as a signal compound.

Signal sequence, signal peptide. A short, 15-30 amino acid, segment at the N-terminus of a secreted or exported protein. This signal sequence is recognized by some part of the cell's protein processing machinery and the protein is then secreted through the membrane of the cell or one of its organelles. The signal sequence is usually removed at some point in the secretion process by a specific protease and is therefore not present in the mature protein.

Significance. In statistical treatments, probability values > 0. 05 are termed *not significant,* those < 0. 05 but > 0. 01 *are significant,* those < 0. 01 but > 0. 001 are *highly significant,* and those < 0. 001 are *very highly significant.* Significant probability values indicate that the results, although possible, deviate too greatly from the expectancy to be acceptable when chance alone is operating.

Silage. An animal feed prepared by the controlled fermentation of certain green crops, e.g. grasses, maize etc. (Young grasses with a high carbohydrate content are ideal. The vegetation is packed tightly into a closed container (a silo) and molasses may be added to encourage fermentation. Fermentation is carried out by the microflora of the vegetation; the material becomes anaerobic and the pH falls are fermentation proceeds.

Silent mutation. A mutation (change in DNA base), the effect of which is not observable. For example if a base change in DNA alters a codon *XYC* to *XYU*, change in the amino acid in the protein may not take place.

Silent substitution. A change on a gene that does not lead to a change in the amino acid sequence of the polypeptide product (Contrast with replacement substitution.).

Silica gel. A high gelatinous form of silica used as a drying agent and catalyst support.

Silicone. A compound made by substituting silicon for carbon in a organic substance.

Silk. Fibre produced by both insects and spiders, but best known as the product of the larva of the silk moth. *Bombyx mori,* being used to form the cocoon in which the larva pupates. Fibroin is the chief protein constituent of silk, the other being sericin.

Silkless maize. A male type producing pollen but no silks, due to a recessive gene *sk/sk.*

Simian virus 40. A papovavirus which normally infects monkey (simian) cells. SV 40 has a small, circular double-stranded DNA genome with a molecular weight of 3×10^6. SV40 DNA and its derivatives have proved to be useful vector molecules in mammalian cloning system.

Simple correlation coefficient. A measure of the relationship between two variables.

Simple staining. A procedure using only one dye.

Simplex. Having one dominant gene-used in polyploids (e.g. *Aaaa* of an autotetraploid).

Single burst experiment. A method for studying the results of phage-mediated lysis of a single bacterial cell. The single burst experiment can be used e.g. to determine the burst size of an individual cell under particular conditions.

Single cell isolation. A technique used in microbiology to obtain a pure culture. This entails the physical isolation of a single cell from a colony or liquid medium using a micromanipulation, or the use of serial dilutions of culture until samples containing only a few cells are available for inoculation of fresh medium. This second method is suitable for use with fungal spores where the germinating spore may be recognized and isolated from an agar plate.

Single cell line or cell strain. A culture started from a single plant cell, usually from suspension cultures of single cells and small aggregates plated on solidified medium. The latter may incorporate a selective agent, from which tolerant or resistant individual cell lines or cell clones can be selected.

Single cell protein (algal). Algae are of interest as a source of SCP since they can be grown in large quantities in open ponds using sunlight as an energy source. Productivity, and thus the economics of the process, depends on the intensity and duration of illumination. Hence successful systems have been associated with subtropical or tropical regions, especially desert area. Algae of the genera

Chlorella and *Scenedenmus* have been used as food in Japan, and Spirulina is being produced in Israel, USA, Mexico Thailand and elsewhere. Part of the commercial production is being sold as a `health food'. Algal SCP production is of potential where it is associated with the production of shellfish, shrimps, fin fish, etc.

Single cell protein (bacterial). Bacteria are of interest as a source of SCP due to their high growth rates and the amino acid spectrum of their protein. The large-scale propagation of bacteria as a source of animal feed protein is a fairly recent technology, having been developed in the last decade.

Single cell protein (fungal). A wide variety of fungal mycelia have been grown as SCP in aerated, agitated deep-tank cultures, in tower fermenters or using solid substrate fermentation. A variety of fungi, mushrooms in particular, have traditionally been used as food. Some SCP is derived from mycelia of organisms such as *Agaricus campestris* or *Morchella* species. Other fungi, including species of *Mucor, Aspergillus, Trichoderma, Gliocladiu, Verticillium* and *Fusarium* have been used, grown on wastes, molasses or artificial media based on glucose syrups.

Single cell protein (SCP). The product obtained from a wide variety of processes in which yeasts, bacteria, fungi or algae are grown for their protein content. SCP is used as food or animal feed as it also contains carbohydrates, fats, vitamins and minerals.

Single cell protein (yeast). Yeasts have been used in baked and fermented foods for generations. Dried brewer's yeast, a byproduct of the brewing industry, has an established use as an animal feed. These include the growth of *Saccharomycopsis lipolytica* on n-alkanes, *Kluyveromyces fragilis* on cheese, whey and *Candida utilis* on starchy wastes. The main yeast used on n-alkanes is *Candida lipotica,* otherwise known as *Endomycopsis lipotica.*

Single copy. A gene sequence which occurs only once per (haploid) genome. Most structural genes, those encoding proteins, are single-copy genes.

Single cross. The hybrid of two pure lines (primarily in corn breeding).

Single seed descent method. A method of advancing the generations; one seed from each F_2 plant of an F_2 and the subsequent generations is bulked to raise the next generation.

Single variation. Darwin's designation for a mutation.

Single-cell proteins. Microorganisms cultivated on industrial wastes or by-products as nutrients to yield a large cell crop rich in protein.

Single-copy plasmid. Plasmid present in bacterial cells at a frequency of only one copy per cell.

Single-factor inheritance. The control of one character by one gene. This gives rise to discontinuous variation in such characters and intermediates between the dominant and recessive forms of the gene do not usually occur.

Single-stranded DNA. A form of DNA that is not base paired with a second strand of DNA or RNA, and thus can hybridize with other suitable polynucleotide.

Single-stranded. A term used to describe nucleic acid molecules consisting of only one polynucleotide chain. The genomes of certain male-specific phages. e.g. M13, are single-stranded DNA molecules. rRNA, mRNA and tRNA are all single stranded nucleic acids, but they all contain double-stranded regions formed by the intra-strand base-pairing of self-complementary sequences.

Sink. A region receiving translocated sugars and other substances synthesized or stored elsewhere in the plant.

Sinus. Any cavity having a relatively narrow opening.

Sister chromatids. The two identical component strands of duplicated chromosomes during cell division.

Sitala (Sanskrit. **sital,** cool). Hindu mythological goddess, thought to be responsible for causing smallpox. Origin obscure.

Site. The position occupied by a mutation within the gene (Cf. Locus).

Site-specific drug. A therapeutic agent, the action of which depends on the specific interaction with a recognition site on the target enzyme or organism.

Site-specific mutagenesis, site-directed mutagenesis, directed mutagenesis. A technique in which a cloned gene is specifically mutated *in vitro* and then used to replace the wild-type copy of the gene in the donor organism. The technique has two advantages over conventional mutagenesis procedures; (i) The proportion of desired mutants obtained is high, up to 50% of the viable cells are mutant in some methods. (ii) No other genes in the organism are mutated and therefore all the desirable characteristics of the strain are retained.

Site-specific recombination. Recombination of short, specific nucleotide sequences. It does not depend on DNA homology. *Compare* General Recombinations.

Site-specific. A term used to describe any process or enzyme which acts at a defined sequence within a DNA or RNA molecule.

Six base-pair culture. A type ll restriction enzyme which recognizes a six base-pair DNA sequence and makes a double-stranded cut within it.

Skewness. Asymmetry in a frequency distribution.

Skoog, F. and C. Tsui (1948). Established that shoot and root formation were chemically (hormonally) regulated in tobacco callus cultures.

Sl system. System International d' Unites. An international system of units for the measurements of length, volume, and mass. It is based on the metric system.

Slant culture. A method used for maintaining stock cultures of a microorganism. The cells are placed on a jar contained in a sealed bottle. The bottle or tube is held at an angle while the gel sets.

Slant tube fermenter. A modification of the tower fermenter giving a higher flow rate capacity. A 3-centimeter diameter tube 14 metre long is mounted at a 45° to the horizontal, and sugar solution is pumped through the tube from the base. Fermentation takes place progressively up the tube. As the cell settling depth is very shallow, high flow rates are possible. Cells settle on the lower tube wall and roll down, while the carbon dioxide evolved rises to the upper tube wall and bubbles rapidly up out of the fermenter. There is thus a three phase flow. A bank of many slant fermenter tubes operated in parallel from a single pump could in theory achieve industrial-scale production, but a special flow distributor system would be required to maintain an equal flow in all the tubes.

SLE (systemic lupus erythematosus). A autoimmune disease of humans usually involving anti-nuclear anti-bodies.

Sliding filament model. A mechanism which explains muscle contraction on the basis of the making and breaking of cross bridges between adjacent thick (myosin) and thin (actin) filaments.

Sliding growth. Growth of plant cells in which new cells slide over previously formed cells without disturbing the underlying organization.

Slime layer. a gelatinous covering of the cell wall; sometimes used synonymously with capsule.

Slime molds. A group of higher protists which resemble both protozoa and the true fungi; they are like protozoa in their ameboid plasmodial stage and similar to the true fungi in spore formation.

Slow stop. Mutants of bacterial cells that are able to complete a round of cell division after being raised to 42°C.

Slow-cooling. Incremental temperature decrease used in cryopreservation of plant cells.

Sludge. The semisolid part of sewage that has been sedimented or acted upon by bacteria.

Sluge cake. The residue left on the filter after the filtration of a sludge.

Sluge digestion. A final biochemical reduction stage in sewage treatment.

Sma I (pronounced smar one). A type II restriction enzyme from the bacterium *Serratia macesccens* which recognizes the DNA sequence shown below and cuts in at the sites indicated by the arrows:

Small lymphocytes. White blood cells found in large numbers in higher vertebrates, which are characterized by a seemingly inert nucleus surrounded by only traces of an endoplasmic reticulum-poor cytoplasm. They fall into two groups, *B* and *T* cells, which are indistinguishable under the light microscope.

Smear. A thin layer of material, e.g. bacterial culture, spread on a glass slide for microscopic examination. Also referred to as a *film.*

Smooth endoplasmic reticulum. Extensive inner membranous sacs (endoplasmic reticulum) which are free of ribosomes. The smooth ER appears to be a major site for attachment of sugar residues to nascent proteins to form glycoproteins.

Smooth muscle. Involuntary, non-striated muscle cells which are found in the lining of internal organs (e.g., gizzard) and blood vessels.

Smuts. Basidiomycete fungi of the order Ustilaginales. Many are important parasites of cereals and form a mass of sooty black spores in place of the grain. Examples of smuts are *Tilletia* and *Ustilago.*

Snake venom. Source of important biochemical compounds. The toxins of venomous snakes operate in quite diverse ways.

Social insect. One of the kinds of insect that form colonies with reproductive castes and worker castes; in particular, the termites, ants, social bees, and social wasps.

Society. A group of individuals belonging to the same species and organized in a cooperative manner; in the broadest sense, includes parents and their offspring.

Sodium azide. An antimicrobial agent. It is used at 0. 2 percent to prevent infection of chromatography columns and gel suspensions.

Sodium dodecyl sulphate (SDS). An ionic detergent used for denaturation of protein and electrolysis of protein sub-units.

Sodium hypochlorite (NaOH mw 74. 44). A frequently used plant tissue sterilant at 0. 5-2% w/v. The usual source is dilute (10-21% v/v) commercial laundry bleach, sometimes with surfactant is common. Tissue damage may occur if contact with tissue is prolonged. Thorough washing with sterile water generally follows treatment.

Sodium lauryl sarcosine. An ionic detergent with properties similar to sodium sulphate, but soluble in high salt concentrations, thus enabling its use in caesium - chloride gradient centrifugation.

Sodium pump. An active transport system within a nerve cell membrane, by which sodium ions entering the cell are transported back into the extracellular medium against a concentration gradient. It maintains the differential concentration of ions either side of the membrane (the outside being kept positive) necessary to produce a resting potential. A carrier molecule, probably a protein, is thought to be involved. Energy required for the process is derived from ATP.

Soft rot. A plant disease caused by pathogenic coliform bacteria often the genus *Erwinia.* The bacteria produce pectinase which dissolves the intracellular cement of plant tissues, allowing the spread of the bacteria through the organism.

Soil free. Both tissue culture and hydroponics utilize soilless nutrient solutions for plant growth in contrast to traditional soil use.

Sol. A liquefied gel.

Solenocyte. A long tubular cell with a flagellum at the base of the tube that extends into the tube and forms the closed end of a protonephridial tubule.

Solid or solidified medium. A medium solidified with agar, a synthetic starch polymer or some other gelling agent. **Solid media** are widely used in plant tissue culture. For suspension cultures and for many research purposes liquid media are preferred.

Solid substrate fermentation. The growing of microorganisms, often fungi, on a solid material rather than in a liquid medium. Such techniques can be divided into two basic types; thin layer and deep bed.

Solid waste management. A planned system of effectively controlling the production, storage, collection, transportation, processing and disposal or utilization of solid wastes in a sanitary, aesthetically acceptable, and economical manner. It includes all administrative, financial, legal, and planning functions, as well as the physical aspects of solid waste handling.

Solidified medium. A medium solidified with agar, synthetic starch polymer or some other gelling agent. Solid media are widely used in plant tissue culture. For suspension cultures and for many research purposes liquid media are preferred.

Solid-phase hybridization. A technique in which one of the hybridizing species is complexed with an insoluble support.

Solid-phase sequenter. A device in amino acid sequencing of proteins in which the sample is covalently attached to a solid-phase support of derivatized glass or polystyrene beads and packed into a microcolumn, prior to initiating the Edman degradation.

Solid-state synthesis. A technique used for the chemical synthesis of polypeptides or polynucleotides.

Solitary parasitoid. An insect parasite which normally develops at a rate of one individual per arthropod host, cf. Gregarious Parasitoid.

Solute. A substance that is dissolved in a liquid (solvent).

Solution. A liquid (solvent) and its dissolved solutes.

Solvent. A liquid that has dissolved or can dissolve one or more solutes.

Soma (*adj*. somatic). The body, cells of which in mammals and flowering plants normally have two sets of chromosomes, one derived from each parent.

Somaclone. A plant regenerated from a tissue culture originating from somatic tissue.

Somatic cell embryogenesis. The production of embryos from somatic cells of explants (direct embryogenisis) or by induction on callus formed by explants (indirect embrygenesis). These two processes may not be materially different in results.

Somatic cell hybrid. The cell or plant resulting from the fusion of animal cells or plant protoplasts, respectively, derived from somatic cells which differ genetically.

Somatic cell hybridization. The *in vitro* fusion of animal cells or plant protoplasts derived from somatic cells which differ genetically.

Somatic cell variant or embryoid. An organized embryonic structure morphologically similar to a zygotic embryo but initiated from somatic (non-zygotic) cells. These develop into plantlets in vitro through developmental processes that are similar to those of zygotic embryos.

Somatic cell variant. A somatic cell with unique features not shared by the others; as selected for in a screening trial that may follow a mutational event.

Somatic cells. All body cells except the gametes and the cells from which these develop.

Somatic embryo or embryoid. An organized embryonic structure morphologically similar to a zygotic embryo but initiated from somatic (non-zygotic) cells. These develop into plantlets *in vitro* through developmental processes that are similar to those of zygotic embryos.

Somatic hybrid. A cell or plant product of somatic cell fusion; as the result of cell or protoplast fusion and implying genomic integration. The process is **somatic hybridization.**

Somatic mutation. A mutation occurring in any cell that is not destined to become a germ cell.

Somatic organogenesis. The production of shoots, roots or other organs on somatic tissues of explants (direct organogenesis) or by induction on callus formed by explant (indirect organogenesis).

Somatic pairing. A phenomenon especially occurring in the insect order Diptera that involves a pronounced tendency for homologous chromosomes to lie near each other in the metaphase plate of mitosis.

Somatic segregation. The formation by mitosis of cells differing from one another, either through mutation or somatic crossover in the nucleus or through an unequal assortment of cytoplasmic determinants such as is not required by normal differentiation, vide sport.

Somatic. With reference to the body cells (from Gr. *soma*=body).

Somatogenesis. The study of the emergence under favourable surroundings of bodily form out of hereditary sources.

Somatostatin. A peptide hormone, consisting of 14 amino acids, produced by the hypothalamus. It stimulates the release of insulin and growth hormone.

Somite (Gr. **soma:** body). One of the segments into which an embryo becomes divided longitudinally, leading to the eventual segmentation of the animal as illustrated by the spinal column, ribs, and associated muscles.

Somotoplastic sterility. Sterility caused by hyperplasia of nucellus or inner integument of ovules after fertilization.

Sonar. The mode of orientation used by bats, porpoises, and a few other animals in which the positions of objects are estimated by emitting sounds and listening for the echoes that bounce back from them.

Sonication. Technique in which ultrasound is used to homogenize tissue.

Sorbitol ($C_6H_{14}O_6$ mw 182. 17). A sugar alcohol which is the main translocatable carbohydrate in some plants. Occasionally it is added to plant tissue culture media.

Sordaria fimicola. Ascomycete fungus used widely for genetic studies.

Soredia. Reproductive bodies in lichens which are knots of hyphae containing a few alga cells.

Sorghum. An Old World tropical grass similar to Indian corn in habit.

SOS repair. Process of DNA repair involving a number of different enzymes. It is highly important because it is error-prone and thus generates mutations.

Sour cream. A fermented milk product manufactured by ripening pasteurized cream of 18 percent fat content with lactic acid-and aroma-producing bacteria.

Source clone. A source that originates from a single plant or explant within a clone; as from a specific virus tested (SVT) or specific pathogen tested (SPT) individual explant or plant.

Source of infection. Animate or inanimate objects that serve as a contact between host and microorganism.

Source plant. A mother plant or donor plant from which an explant used to initiate a culture is taken.

Souring. A drop of pH associated with the production of an organic acid die to microbial growth. In anaerobic digestion, it involves a switch from methanogenesis to acid fermentation.

Southern blot, Southern transfer. The technique is named after its inventor. E. M. Southern.

Soybean meal. A residue obtained as a result of processing soybeans or extracting oil from soybeans. The meal is rich in protein and is widely used as a animal feed supplement. Alternative sources of animal protein, such as single cell protein, may be compared with soybean meal used as a standard in terms of its amino acid composition and nutritive value.

Soybean segregation. A chlorophyll mutant with no dominance which produces one lethal golden seedling, *g/g/,* two light green *G/g,* and one normal green *G/G.*

Soy sauce. A fermented product manufactured by fermentation of a mixture of soybeans and cereal, usually wheat, with salt. The process is complicated and uses a sequence of fungi and bacteria.

Spacer DNA. DNA that is located between the regions containing the information used to determine the actual amino acid sequence in proteins. These regions may or may not be transcribed by RNA polymerase.

Spawning. The direct release of sex cells into the water.

Spaying. The castration of a female by removal or destruction of the ovaries.

Speaking diaphragm. Plastic sheet installed in wall, door, or window to permit voice communication through barrier between areas of different levels of contamination.

Specialization. The tendency of organisms to evolve and become adapted to a specific environment. Extremes of specialization are found in parasitic organisms (endoparasites in particular) where the organism is only able to survive in a specific host and requires a specific sequence of alternatives hosts in order to reproduce.

Specialized (restricted) transduction. The transduction of specific genes.

Specialized. Cells or organisms adapted anatomically or physiologically for particular functions, or habitats. The act process of adjustment is **specialization.**

Speciation. The process by which new species are formed. In the restricted sense the splitting of one species into a number of different contemporaneous species.

Species barriers. Different species are usually separated from each other by barriers of incompatibility and sterility. Spontaneous or experimental hybridization may be difficult or impossible; and if hybrids are obtained, they are generally more or less sterile.

Species diversity. A weighted representation of the species of organisms living in a region; large and common species are given greater weight than are small and rare ones. (Contrast with species richness.)

Species equilibrium. A condition in which the number of species going extinct in a given area per unit time equals the rate at which new species are arriving; thus the number of species in the area remains constant.

Species richness. The number of species of organisms living in a region. (Contrast with species diversity.)

Species. (1). A biological unit that may be characterized in the following way: (i) the individual within a species have certain external characteristics in common with the aid of which the species can be recognized; (ii) the species has a specific ecological requirements and generally a specific distributional area; (iii) as a rule, individuals of a species can intercross spontaneously without difficulty and produce viable hybrids with normal fertilities; (iv) the individuals of a species generally have the same karyotype and thus are identical in regard to the number and structure of the chromosomes.

Specific activity. (1) In radioactive work, the ratio of radioactive atoms to stable atoms in a given compound. It may be expressed in terms of bacquerels (the previously used microcurie or curie is now obsolete) per gram or per mole. (2) In enzymology, a measure of the ratio of enzymically active protein to total protein. It may be expressed in terms of millimoles or substrate converted per second (or minute) per milligram of protein.

Specific combining ability. Deviation in performance of a cross combination from that predicted on the basis of the general combining abilities of the parents involved in the cross.

Specific enzyme activity. The activity of an enzyme expressed in term of the amount or number of moles of substrate converted per unit time by unit weight of protein.

Specific growth rate constant (μ). A proportionality constant for the growth of unicellular microorganisms undergoing binary fission. In a well mixed batch culture system, growth is proportional to the biomass concentration.

Specific immunity. Resistance which is specifically directed against a particular microorganism, or its products.

Specific ionization. Number of pairs of ions per unit mass of tissue, or per unit path length of the ionizing radiation medium.

Specific modifiers. Genes which modify the effect produced by other genes. They may be without effect themselves. (Ford, 1945).

Specific pathogen free. Synonymous with Pathogen Free.

Specific pathogen tested (SPT). A culture or plant that is apparently healthy and free from infection during propagation and maintenance. Refers to freedom from the specific pathogens for which a plant is being tested (such as bacteria or virus) and is not necessarily a guarantee or good health.

Specific production rate. The rate of formation an end product from a fermenter expressed in terms of the amount formed per unit of time per unit of another variable (cell volume, wet cell weight, number of organisms, volume of reactor, etc.).

Specific rate of reaction. The rate of product formation, or substrate utilization, expressed in terms of unit mass of the microorganism involved, or per unit weight of biomass or enzyme.

Specific search image. An impression left on a predator by several chance encounters with a kind of potential prey, which eventually leads to recognition of and intensive search for the acceptable prey species (Timbergen, 1960).

Specific virus tested (SVT). A culture or plant that is apparently healthy and free form viral infection during propagation and maintenance. Refers to freedom only from the specific viruses for which the plant was tested and does not necessarily guarantee the absence of virus.

Specification. Accepted value in the manufacture of a product.

Specificity (of an antibody). Refers to the range of antigens with which a given antibody may combine; thus in addition to the homologous antigen, the antibody may function as a cross reacting antibody to any of a range of `stereochemically-similar antigens.

Specificity factors. Proteins which reversibly associate with the core component of RNA Polymerase and determine which promoters the polymerase will bind to.

Spectinomycin. An inhibitor of protein synthesis. It is used for plasmid amplification in organisms that show chloramphenical resistance.

Spectrophotometer. An instrument that measures transmission of visible light permitting accurate comparison of luminous intensities of two sources of specific wave lengths.

Spectrophotometric protein determination. A technique used to measure the amount of protein in solution. The extinction of a diluted protein solution is measured at 260 and 280 nm, and the ratio of the extinctions calculated. This is used to correct for the presence of nucleic acid contamination using standard tables.

Speltoid. Mutants occurring in cultures of common wheat *T. vulgare,* and possessing characters of *T. spelta,* spelt, wheat.

Spematogonium. A primordial germ call that gives rise, by mitosis, to *spermatocyte,* from which the spermatozoa develop by meiosis.

Spent medium. Medium discarded when a culture is subcultured. The implication is that the medium has been depleted of nutrients, dehydrated or accumulated toxic metabolic products.

Speratogenesis. The process in animals and in man by which sperms (male gametes) are formed.

Sperm (Gr. *sperma*: seed). A male reproductive cell.

Sperm bank. Store of sperm, normally in semen with additives such as glycerol, derived from many individuals, and maintained in liquid nitrogen at - 196°C for many years. Such sperm may be used in artificial insemination. Sperm banks permit storage of gene pools which might otherwise be lost for ever.

Sperm nucleus. The nucleus of a sperm cell (male gamete) in a pollen grain.

Sperm. (1.) (**Spermatozoon**, *pl.* spermatozoa). The mature male germ cell of an animal. (2.) The male gamete produced in the pollen grain or pollen tube in plants.

Spermatid. A product of meiosis on the male side in animals and in man. Meiosis takes place in *spermatocyte,* or mother cells, each of which gives rise to four spermatids. The first meiotic division occurs in the *primary spermatocyte;* the second division, in the *secondary spermatocyte.* The resulting spermatids are transformed later into sperms.

Spermatocyte. Primary: a sperm mother cell giving rise by the first meiotic division to two secondary spermatoccytes. The secondary spermatocyte at the seconds meiotic division give rise to four haploid spermatids.

Spermatogenesis (Gr. *sperma:* seed + *gignomia:* to be born). Male gametogenesis, leading to the production of sperm.

Spermatozoon (pl., spermatozoa.). A mature or differentiated male germ cell.

Spermiogenesis. (1) Synonymous with spermatogenesis (2) The development of spermatids into sperms.

Spheroplast. A gram-negative cell which lacks peptidoglycan components but possesses outer and inner membrane structures.

Spherules. Thick-walled structures within which endospores develop in tissues; characteristic of **Coccidiodes immitis.**

Sphincter (Gr. *sphinkter:* that which binds tight). A ring of muscle that can close an orifice, for example at the anus.

Spike culture. The aseptic culture of a grain inflorescence for micropore development.

Spindle apparatus. An array of microtubules stretching from pole to pole of a dividing nucleus and playing a role in the movement of chromosomes at nuclear division. Named for its shape.

Spindle attachment. That point of a chromatid to which the spindle `fibre' appears to be attached at metaphase and anaphase and which, at anaphase, starts to move towards the pole before the rest of the chromatid; centromere; insertion region; kinetochore.

Spindle fibres. Fibres which appears to connect the centromere to the poles of the spindle during cell division These are possibly artefacts of fixation which correspond positionally with some latent `structure' of the spindle cytoplasm but which do not exist as definite until coagulated by chemical reagents.

Spindle. Cellular structure, largely composed of microtubules, involved in eucaryotic chromosomal segregation.

Spinning cup sequnter. A device for the liquid-phase sequencing of peptides of proteins.

Spiral cleavage. A form of cell division which the fertilized ovum undergoes in certain Annelida, Mollusca, Nemertina and Platyhelmia in which the daughter cells are formed from the old cells in certain regular positions which follow a definite rule.

Spiral. A coil of the chromosome, chromatid or chromosome thread at mitosis or meiosis. Classified by Darlington 1935 as follows:

Spiralization, the assumption of an internal (but not a relational) spiral within the chromatids in mitosis and jointly for two chromatids in meiosis (Darlington, 1932).

Spirillum. Corkscrew-shaped cells, rigid.

Spirochaetes. Long spirally twisted unicellular bacteria surrounded by a flexible wall. An axial filament, with a similar structure to a bacterial flagellum, is spirally wound about the protoplast inside the cell wall. They swim actively by flexing the cell. They are found in mud and water and can withstand low oxygen concentrations. Many spirochetes are pathogens causing yaws, syphillis, and relapsing fever.

Spirochete. A flexible spiral or corkscrew-shaped bacterium.

Spleen. Lymphoid organ in the upper left abdomen which functions as a storage site for red blood cells and, due to the large number of resident macrophages, as a blood filter.

Splenomegaly. Increased spleen size, this is seen in a number of parasitic infections and is also used as an assay for graft versus host reactions.

Splicing enzyme. A ligase associated with joining of length of DNA or RNA *in vivo* or *in vitro.*

Splicing. (1). The enzyme-catalyzed joining of fragments of DNA in gene manipulation. (2) The joining of conserved sections of mRNA following excision of introns.

Split gene. A eukaryotic gene in which the coding sequence is divided up by a number of non-coding regions called introns.

Spongy parenchyma. In leaves, a layer of loosely packed photosynthetic cells with extensive intercellular spaces for gas diffusion. Frequently found between the palisade parenchyma and the lower epidermis.

Spontaneous fusion. Uninduced protoplast fusion which may occur between freshly isolated protoplasts or following adhesion of adjacent cells during enzymatic cell wall degradation.

Spontaneous generation. The idea that life is generated continually from non-living matter. Usually distinguished from the current idea that life evolved from nonliving matter under primordial conditions at an early stage in the history of Earth.

Spontaneous mutation. A mutation that arises without the use of any known mutagen, whose cause is unknown.

Spontaneous process. It is the process accompanied by a decrease in free energy.

Spontaneous reaction. A chemical reaction which will proceed on its own, without any outside influence. A spontaneous reaction need not be rapid.

Spontaneous variation. The variation in plant derived from tissue cultures not exposed to mutagens but occurring as a result of the culture conditions.

Sporadic. Only occasional occurrence, when used in reference to a disease.

Sporangiophore. A specialized mycelial branch bearing a sporangium.

Sporangium (Gr. **spora:** seed + **angeion:** vessel or reservoir). In plants and fungi, any specialized structure within which one or more spores are formed.

Spore mother cell. The cell that after meiosis gives rise to four, or in certain cases, eight spores.

Spore. A body which some species of bacteria form within their cells which is considerably more resistant to unfavorable conditions (heat, germicides, ultraviolet light) than the vegetative cell. When favorable conditions return, the spore usually germinates, or transforms back to the vegetative form.

Sporicide. An agent that will kill spores.

Sporocyte. The spore mother cell of a plant. Microsporocytes produce pollen, megasporocytes produce the embryo sac following meiosis.

Sporogenesis. The formation of microspores (pollen) and megaspores (embryo sac) in plants.

Sporogony. The sexual phase of sporozoans such as *Plasmodium* spp, which cause malaria.

Sporophore. Specialized mycelial branch upon which spores are produced.

Sporophyll (Gr. *spora:* seed + *phyllon:* leaf). Any leaf or leaflike structure that bears sporangia.

Sporophyte (Gr. *spora:* seed + *phyton:* plant). In plants with an alternation of generations the diploid phase that produces the spores. (Contrast with gametophyte.)

Sporophytic budding. The production of an embryo from the tissue surrounding the embryo sac.

Sporophytic polyembryony. Polyembryony brought about by the formation of Adventitious embryos by budding from the nucellus or integument.

Sporozite. An elongated sickle-shaped body formed in an oocyst as the result of the sexual reproduction phase of sporozoans such as *Plasmodium* spp.

Sport. A dissimilar form of an individual, or part of an individual, arising directly or indirectly from gene, plastid, chromosome or nuclear (polyploid) mutation. The indirect origins are due to segregation of a recessive gene, cellular sorting-out of plastids, and tissue sorting-out in chimaeras. The name is usually confined to new and unforeseen forms.

Sporulation. The formation from vegetative cells of bacteria and other organisms of dry, metabolically inactive cells with thick surface coats (spores), which can resist extreme environmental conditions.

Spreading position effect. Phenomenon in which the effect of an inversion or translocation in inactivating adjacent gene sequences affects more than one such sequence.

Springwood. Early part of the yearly growth ring.

Spropel. The anaerobic, putrefying matter- rich in hydrogen sulphide-which forms the mud of stagnant waters.

Sputum. Mucopurulent matter expectorated form air passages.

Square, Latine square. An experimental design involving two restraints.

Sreptococci. Cocci that divide in such a way that chains of cells are formed.

SsDNA. Single-stranded DNA.

Sst I. A type II restriction enzyme produced by the bacterium *Streptomyces stanford* which recognises the DNA sequence shown below and cuts at the sites indicated by the arrows:

Stab culture. Culture of bacteria made by stabbing a long needle, carrying some of the bacteria from another culture, deep into an elongated test-tube containing a solid nutrient agar. It represents an easy way of culturing relatively non-fastidious anaerobic bacteria, since the oxygen concentration deep in the agar is low.

Stabilizing selection. Selection against the extreme phenotypes in a population, so that the intermediate types are favored. (Contrast with disruptive selection).

Stable age distribution. That distribution of individuals among different age groups that results when particular, unchanging birth and death rates occur in a population.

Stable. This has its usual English meaning. Stable plasmids are ones which are lost only infrequently from their-host cells even under non-selective conditions. In bacteria, plasmid stability appears to depend on two factors, copy number and the plasmid locus par which ensures the equitable partition of plasmid molecules into daughter cells at division.

Stacking gel. Technique employed in gel electrophoresis in which a gel of differing concentration is placed above the normal gel in order to help concentrate the molecules into distinct bands.

Stage I. A step *in vitro* propagation characterized by the establishment of an aseptic tissue culture of a plant.

Stage II. A step in *in-vitro* plant propagation characterized by preparation of propagule for successful transfer to soil, a process involving rooting of shoot cuttings, hardening of plants and initiating the change from the heterotrophic to the autotrophic state.

Stage III. A step in *in-vitro* propagation characterized by the rapid numerical increase of organs or other structures.

Stage IV. A step in *in-vitro* plant propagation characterized by the establishment in soil of a tissue culture derived plant, either after undergoing a Stage II pretransplant treatment, or, in certain species, after the direct transfer of plants from Stage II into soil.

Stages of culture (I-IV): Stage I: Aseptic explantation or establishment of the explant in culture. **Stage II:** Multiplication of the propagules. **Stage III:** Rooting of the progagules and preparation for transplant to soil. **Soil IV:** Establishment of Stage II or III propagules ex vitro in soil or potting mix.

Stained bacteria. Bacteria that have been colored with a chemical stain to make them easier to see and study.

Staining. A procedure designed to heighten contrast between different structures. Normally biological material is very lacking in contrast, protoplasm being transparent, and therefore staining is essential for an understanding of structure at the microscopic level. Vital stains are used to stain and examine living material. Staining is done after fixation and either during or after dehydration. *Double staining* involves the use of two stains; the second is called the *counterstain. Acidic stains* have a coloured anion, *basic stains* have a coloured cation. Some stains are neutral. Materials can be described as *acidophilic* or *basophilic* depending on whether they are stained by acidic or basic dyes respectively. Basic stains are suitable for nuclei, staining DNA. Stains for light microscopy are coloured dyes; those for electron microscopy contain heavy metals, e.g. uranyl acetate, lead citrate, and osmium tetroxide.

Stamen. A male (pollen-producing) unit of a flower usually composed of an anther, which bears the pollen, and a filament, which is a stalk supporting the anther.

Staminate. Producing stamens only (no pistils); a male plant.

Standard deviation (SD). The square root of the sum of the deviations from the mean squared and divided by one less than the number of observations. A measure of the variability of a population of individuals.

Standard error of difference in means. Measure of the significance of the difference in two sample means. Symbolized by S_d.

Standard error of mean. This is a statistical measure by which the extent of error occurring between the means of different samples is determined.

Standard error of sample mean. An estimate of the standard deviation of a series of hypothetical sample means, serving as a measure of the closeness with which a given sample mean approximates the population mean. Symbolized by Sx.

Standard error of standard deviation. A statistical measure that tells us as to the extent by which standard deviations determined from different samples, may vary.

Standard error. The standard deviation divided by the square root of the number of observations. A measure of the variation of a population of means.

Standard. A law stating accepted value (s) for a manufactured product.

Standing crop. The nutritional portion of the biomass in an area at a given moment.

Staphylococci. Spherical bacteria (cocci) occurring in irregular, grapelike clusters.

Staphylococcus aureus. A Gram-positive spherical bacterium which commonly inhibits the skin of humans. Several small antibiotic resistant plasmids have been isolated from *S. aureus* and have formed the basis of many cloning vectors for *Bacillus* species particularly *Bacillus subtilis.*

Starch (O. E. *stearc.* stiff, from Gr. *stereos:* a solid). A linked polymer of glucose; used by plants as a means of storing energy and carbon atoms.

Starch gel electrophoresis. Zone electrophoresis method in which the support medium is a starch gel. As components of the mixture migrate through the gel under the influence of the electric current. The pore structure of the gel acts as a molecular sieve. As a result, proteins of the same charge to mass ratio separate if their molecular sizes are sufficiently different.

Starch. A polysaccharide that occurs exclusively in plants. Starches are extracted commercially from maize, wheat, barley, rice, potatoes, and sorghum. They exist in the plant cells as granules dispersed in the cytoplasm. The starches are storage reservoirs for plants; they can be broken down by enzymes to simple sugars and then metabolized to supply energy needs. Starch is a dietary component of animals. In man it is digested by salivary and pancreatic amylase then further degraded by maltase to yield glucose, which may be stored as glycogen (animal starch). Excess starch, i. e. above the maximum liver and muscle storage capacity, is converted to lipids and stored as fat. Starch is not a single molecule but a mixture of amylose (water - soluble, blue colour with iodine) and amylopectin (not water - soluble, violet colour with iodine). The composition is amylose 10-20%, amylopectin 80-90%.

Start codon, Initiator codon. The trinucleotide in an mRNA molecule with which the ribosome starts the process of translation. The start codon sets the reading-frame for translation. The most commonly used start codon is AUG which is decoded as methionine in eukaryotes and as N-formyl-methionine in prokaryotes. AUG appears to be the only start codon used by eukaryotes, while in bacteria GUG (valine) may sometimes be employed.

Starter culture. A pure or mixed culture of microorganisms used to initiate a desired fermentation process, as in cheese manufacture.

Start-up. The first stage of operation of a continuous fermentation or anaerobic digestion process, during which the process is brought to steady running conditions.

Stasis. Suffix, refers to inhibition.

State-of-the-art. An iodiomatic phrase used to imply that the most recent or up-to-date technology has been used or is being described.

Static. Growth-inhibiting.

Stationary culture. A non-agitated culture. The antonym is shake culture.

Stationary phase. The growth phase of a culture when the rate of reproduction equals the death rate and the number of viable bacteria remains constant.

Statistic. The estimate of a parameter arrived at from an observed sample. Bears the same relation to the sample as the parameter does to the population.

Statolith. One of a number of large starch grains found in the statocytes, plant cells that are thought to be gravity sensitive. They sediment through the cytoplasm to the lower-most cell surface, enabling the plant to detect the direction of gravity.

Steady state. A nonequilibrium state of an open system through which matter is flowing and in which all components remain in constant concentration.

Stearic acid (Octadecoic acid). A saturated carboxylic acid, which is widely distributed in nature as the glyceride ester. It is present in most fats and oils of animal and vegetable origin, particularly the so-called hard fats. i. e. those of higher melting point.

Steatorrhea. Fatty stools.

Stele (Gr. *stele*: pillar). The central cylinder of vascular tissue in a plant stem.

Stele. The primary vascular tissue in plants. It consists of the xylem and phloem, as well as the pericycle and endodermis.

Stem body (Stemm-Korper). The part of a spindle between two groups of chromosomes separating at anaphase (Belar, 1928).

Stem cell. A cell from which other cells stem or arise by differentiation.

Stem cutting. A section of stem or shoot divided for rooting purposes. A commonly employed propagation procedure.

Stemline. The genetically uniform cell type that characterizes and mainly propagates a normal tissue, a tumor, or a tissue-culture cell line. Besides the stemline, other cell types may be present in the cell populations mentioned as secondary stemlines or more temporary lines, potentially capable of developing into new stemlines. The stemline concept, being of mainly theoretical significance, is for practical purposes often replaced by the stemling chromosome number (the numerical stemline) or s, which is the modal number of chromosomal elements without regard to frequency or structure of chromosome types.

Stenothermophile. An organism which grows at 60°C or higher; true or obligate thermophile.

Step allelomorphism. The occurrence of a series of multiple allelomorphs with overlapping effects, which can supposedly be related to a linear order in the distribution of units of change within the gene. Deduced from the scute gene in *Drosophila melanogaster* (*Serebrovsky,* 1930).

Stepwise synthesis. The production of metabolic products in discrete steps, gene-controlled enzymes initiating the reactions.

Stereaceae. Basidiocarp: appressed, effused-reflexed or stipitate, typically dimitic; basidiospores: colourless and smooth. The organisms are characteristically lignicolous. Genera: e.g. *Amylostereum, Chondrostereum, Podoscypha, Sterum.*

Stereoisomers. Molecules that have the same structural formula but different spatial arrangement of dissimilar groups bonded to a common atom. Many of the physical and chemical properties of stereoisomers are the same, but there are differences in their crystal structures, in the direction in which they rotate polarized light, and, very importantly for biological systems, in their ability to be used in an enzyme-catalyzed reaction.

Steric (Stereochemical). Pertaining to the arrangement in space of the atoms in molecules.

Sterile operation. A procedure or series of procedures that use materials free from infection or contamination with microorganisms in conditions that prevent such infection or contamination.

Sterile transfer. The process of aseptically explanting or subculturing; moving plant material from one location or culture to another under aseptic conditions.

Sterile. Free of living organisms.

Sterile-insect technique. A genetic method of pest suppression involving the mass release of compatible but sterile insects into a wild fertile pest population to overwhelm and suppress its reproductive capacity, often eventually to the point of extinction.

Sterility. Any failure (partial or complete) of an individual under given environmental conditions to produce effective gametic or viable zygotic progeny. May be due to environmental or genotypic defects or to segregation in a hybrid or polyploid. Genotypic defects in the gametes are due to segregation in a hybrid parent, and in zygotes to unbalance in the hybrid individual concerned.

Sterilization. The process of making things **sterile** through **1.** Rendering plants non-reproductive. **2.** Killing or excluding microorganisms or their spores with heat, filters, chemicals or other **sterilants.** Dry heat sterilization is useful for metal instruments and glassware. Foil-wrapped items are subjected to 150°C for a minimum of 3 hours in a hot air oven. Steam sterilization or autoclaving is useful for nutrient media, distilled water, paper products and glassware. Solutions contained in glass flasks, plugged with cotton and capped with foil are subjected to 1. 05 kg/cm^2 (121°C) for 10 to 20 minutes (depending on the total volume and how it is distributed). Foil wrapped capped glassware are autoclaved in the same way. Filter sterilization through a "Millipore-type" membrane is useful for thermolabile solutions such as those containing vitamins and urea. Chemical sterilization (most commonly sodium hypochlorite) is useful for plant materials in preparation for excision (surface sterilization) and for working surfaces. Some medium constituents that are thermolabile or insoluble in water may be sterilized through dissolving in organic solvents, such as chloroform or alcohol. They are then dispensed to sterile filter paper for solvent evaporation and the filter paper with the residue is added to or below the sterile medium. Metal instruments are flame sterilized by immersion in 70% ethanol until required, then flaming.

Sterilizer. An autoclave or similar heated container used to sterilize growth media, biochemical or health care reagents and solutions, small apparatus or instruments, etc.

Steriloid. Mutant types occurring in cultivated oats and resembling *Avena sterilis.*

Steroid bioconversion. A process in which one type of steroid, or a precursor, is converted to a related molecule using enzymes or cells as catalysts. Such transformations include the hydroxylation of progesterone using *Rhizopus arrhizus,* which introduces an oxygen atom at C-11. Microorganisms are also used in the conversion of raw materials, such as diosgenin from soybeans, by cleavage of their side chain to produce the aglycone.

Steroid hormones. Hormones, such as oestrogens and androgen, which are steroids.

Steroid. A complex macromolecule, containing carbon atoms arranged in four interlocking rings, three of which contain six carbon atoms each and the fourth of which contains five.

Sterols. Organic compounds consisting of several aromatic rings and at least one alcohol group, e.g., cholesterol. Probably present in all eukaryotic membranes, but absent in prokaryotic membranes.

Sticky ends. On a piece of two-stranded DNA, short, complementary, one-stranded regions produced by the action of a restriction endonuclease. Sticky ends allow the joining of segments of DNA from different sources.

Stigma (L. *stigma:* mark, brand). The part of the pistil at the apex of the style, which is receptive to pollen, and on which pollen germinates.

Stilboestrol. A synthetic hormone; the parent substance of a group of oestrogenic agents.

Stillbirth. The birth of a dead child.

Stimulus. A change in the external or internal environment of an organism that elicits a response in the organism. The stimulus does not provide the energy for the response.

Stirred tank fermenter. The most widely used type of fermentation vessel. It consists of a tank fitted with cooling coils, or cooling jackets, and a mechanically driven impeller, as well as various parts for input and output of liquid and gases. The degree of sophistication varies and includes fully automated and instrumented systems.

Stock cultures. Known species of microorganisms maintained in the laboratory for various tests and studies.

Stock plant. The source plant from which cuttings or explants are made. These are usually maintained carefully in an optimum state for (sometimes prolonged) explant use. Preferably they are certified, pathogen-free plants.

Stock solution. A solution, usually concentrated (10 to 100 times the final medium concentration), of select medium constituents that are grouped for compatibility to avoid precipitation and prepared before hand to save time during medium preparation. Usually they are frozen or stored in the refrigerator and portions are utilized as media are prepared.

Stock. (1) In breeding, the race or strain of plants or animals from which a particular group of offspring has been obtained (breeding stock, pedigree stock). (2) In horticulture, the stem, tree or plant that furnishes slips or cuttings or a stem in which a graft is inserted and that is its support.

Stoichiometric. Descriptive of a reaction mixture in which the ingredients are present in the exact amounts required for the reaction to go to completion without any of the reactants being left.

Stoichiometry. Pertaining to the relationship and proportions in which substances react together.

Stolon. A horizontal stem of a plant that grows along the surface of the ground and takes root at the nodes where a new shoot arises. It is an organ of vegetative propagation.

Stoma (Gr. **stoma:** mouth). Small opening in the plant epidermis that permits gas exchange; bounded by a pair of guard cells whose osmotic status regulates the size of the opening. Stout a heavy-bodied brew that is darker and sweeter than porter and is made with roasted malt and a relatively high percentage of hops.

Stomatal complex. Includes the stoma together with the guard cells and, when present, the subsidiary cells.

Stomatal index. The number of stomata divided by the number of stomata plus the number of epidermal cells, all multiplied by 100, calculated per unit area (usually per mm^2). This value has been found to the reasonably constant for a particular species and is useful in comparing leaves of different sizes. During leaf development this value is affected by relative humidity and perhaps by light intensity.

Strain stability. A measure of the tendency of change to occur in the genetic make-up of succeeding generations of the descendants of a single cell or variety of breeding stock.

Strain. (1) A pure culture of microorganisms composed of the descendants of a single isolation. (2) A natural or artificial mating group uniform in some particular. Biological Strain. one detected by difference in parasitic adaptation. Sexual Strain (incorrect), in Heterothallic fungi, one differing from others in its incompatibility allelomorphs.

Strand displacement. Method of replication in some viral genomes in which the newly synthesized strand physically displaces one of the previous homologous strand of the DNA duplex.

Strata. Layers of the earth's crust.

Streak plating. Technique for isolating single colonies of bacteria (and other microorganisms) and thereby obtaining pure cultures (clones). A small amount of the bacterial specimen is drawn over the surface of a solid medium (e.g., nutrient medium solidified with agar) in a petri dish using a sterile wire. The agar surface is thoroughly covered in a cris-cross fashion so that the bacteria are well distributed and will produce isolated colonies after a period of incubation.

Streak-plate method. A culture is streaked on the surface of a nutrient agar medium in a manner that separates the organisms so that isolated colonies develop.

Streptococci (*sing., Streptococcus*). Spherical cells in chains.

Streptokinase. Substance that activates a proteolytic enzyme (plasminogen) normally present that host's plasma; this causes dissolution of blood clots and may allow for the spread of the organism.

Streptolysins. Hemolysins produced by **B**-streptococci; streptolysin S, stable in the presence of atmospheric oxygen, causes hemolysis on a blood medium; streptolysin O, inactivated in the presence of oxygen, stimulates antibody production, which provides a useful serological test for a recent streptococcal infection.

Streptomycetes. A group of Gram positive sporulating bacteria which grow in a mycelial (filamentous) manner. They are very common soil organisms and are involved in the breakdown and recycling of many polymeric carbon compounds, e.g., cellulose, protein and starch. Streptomycetes produce over 60% of the known antibiotics and because of this much attention has been devoted to these organisms by the pharmaceutical industry.

Streptomycin dependence. The need to include streptomycin in a growth medium that is found in some mutant organisms.

Streptomycin dependent *E. coli*. A strain of *E. coli* that must have small amount of streptomycin (an antibiotic) to grow.

Streptomycin. An antibiotic ($C_{21}H_{39}N_7$-O_{12}) from *Streptomyces griseus* (a soil bacterium) which binds specifically to bacterial 30S ribosomal subunits, thereby blocking protein biosynthesis.

Stress fibers. Bundles of microfilaments aligned parallel just under the plasma membrane. These fibers contain myosin, tropomyosin, actin, and α-actinin, and seem to be involved in cell motility.

Striated muscle. Muscle cells which possess myofibrils that are regularly arranged so as to create visible striations.

Stringent control. Descriptive of replication of chromosomal DNA that only occurs in the presence of protein synthesis.

Stringent mutant. Bacterial mutant that shuts down RNA synthesis abruptly in a depleted growth medium. *Compare* Relaxed Mutant.

Stringent plasmid. A plasmid whose copy number is under strict control and therefore has only one or two copies of chromosome, i.e. a low copy number plasmid, e.g. the F plasmid.

Strobilus (Gr. *strobilos:* a cone). The cone, or characteristic multiple fruit, of the pine and other gymnosperms. Also, a cone-shaped mass of sporophylls found in club mosses.

Stroma. The loosely packed membrane regions within the chloroplast where the dark reaction occurs.

Stromatolite. A composite, flat to domed structure composed of successive mineral layers. Some are known to be produced by the action of bacteria in salt or fresh water and some ancient ones are considered to be evidence for early life on Earth.

Strong promoter. A promoter for which RNA polymerase has a high affinity and which direct the synthesis of large amount of mRNA.

Strophariaceae. Basidiocarp: stipitate (stipe often elongated), pileus e.g. buff, yellow, ochre, or (in *Stropharia aeruginosa*) greenish; basidiospores: typically brown to purplish-brown. Genera include *Hypholoma, Panaeolus, Pholiota, Psilocybe, Stropharia.*

Structural gene. A gene that encodes the primary structure of a protein.

Stuffer fragment. A non-essential fragment of a phage cloning vector which is replaced by the DNA fragment to be cloned.

Style (Gr. *stylos:* pillar or column). In flowering plants, a column of tissue extending from the tip of the ovary, and bearing the stigma or receptive surface for pollen at its apex.

Stylet. Stiff, slender, hollow feeding organ of plant parasitic nematodes.

Styragel. A polystyrene support material used in gel filtration chromatography with lipophilic solvent.

Sub. A prefix often used to designae a structure that lies beneath another or is less than another. For example, subcutaneous, subspecies.

Suballeles. Correspond to groups of mutants separated by intragenic cross-overs. Very large populations are required for their detection.

Subclinical infection. A state in which the individual does not experience all of the characteristic symptoms of a particular disease, or such effects are less severe.

Subclinical. Without clinical manifestations of the disease.

Sub-clone. A method in which smaller DNA fragments are cloned from a large insert which has already been cloned in a vector.

Subculture Harvesting cells from one culture and using them to initiate a new culture. The term is synonymous with passage.

Subculture interval The time between sequential subculturings of a cell line. Synonymous with passage interval.

Subculture or passage. A culture derived from another culture or the aseptic division and transfer of a culture or a portion of that culture (inculum) to fresh nutrient medium. Subculturing is usually done at set time intervals, the length of which is called the subculture interval or passage time.

Suberin. A waxy material serving as a waterproofing agent in cork and in the Casparian strips of the endodermis in plants.

Subinert. A chromosome with a slight genetic effect is said to be subinert.

Sublethal gene. A lethal gene with delayed effect. The gene in proper combination kills its possessor in infancy, childhood, or adulthood.

Subline. A cell line regenerated from a unique cell line of a hybrid callus colony.

Submerged culture. A technique for culturing molds and other microorganisms in which the microbes are distributed through the culture fluid by vigorous mixing. It is in contrast to surface culture where the organism grows on the upper surface of the culture fluid.

Submetacentric chromosome. A chromosome with a centromere some-what nearer to one end than to the other.

Subspecies. A population or group of populations distinguishable from other such populations of the same species by the frequencies of genes, chromosomal arrangements, or hereditary phenotypic characteristics. Subspecies sometimes exhibit incipient reproductive isolation, although not sufficiently to make them different species.

Substitution line. A line in which a pair of chromosome has been replaced by a pair from another variety of the same species. Used in aneuploid analysis.

Substrain. A substrain can be derived from a strain by isolating a single cell or groups of cells having properties or markers not shared by all cells of the parent strain.

Substrate conversion efficiency. The ratio of the amount of end product recovered from a biological reaction to the amount of substrate used or introduced into a fermenter.

Substrate yield. The amount of biomass and/or end products produced in a reaction expressed as a function of the amount of substrate used.

Substrate. The molecule or molecules on which an enzyme exerts catalytic action.

Substratum. Surface on which an organism moves, or by which it is supported.

Subterminal. Situated between the center and the end of the cell, as subterminal endospores.

Subunit. One of a number of particles that associate to produce a whole unit or structure. Many enzymes are composed of several polypeptide subunits.

Subvital. A gene or genotype that, when expressed, reduces the viability of its bearers significantly below that of the "normal" or standard type but has a viability at least half as great.

Succession, Law of. The hypothesis that every species has come into existence coincident, both in time and space, with a pre-existing and closely allied species.

Succession. In ecology, the gradual and predictable series of changes in species composition of a community, from the time of colonization of an empty space to the maturing of the final, stable community.

Succinate dehydrogenase. A constituent enzyme of the TCA cycle found in mitochondria of all aerobic cells. It contains FAD as a prosthetic group and requires no separate coenzyme.

Succinic acid. A dicarboxylic acid formed by fermentation of sugars, It occurs in algae, lichens, sugars, beets, and other plant substances. Succinate ion plays on important role in the Krebs cycle.

Sucrose (Cane sugar). A sugar that occurs in many plants. It is extracted commercially from sugar cane and sugar beet. Sucrose is a disaccharide formed from a glucose unit and a fructose unit. It is hydrolysed to a mixture of fructose and glucose by the enzyme invertase. Since this mixture has a different optical rotation (laevorotatory) than the original sucrose, the mixture is called invert sugar.

Sucrose gradient. A density gradient which is performed by mixing sucrose solutions of different concentrations, e.g. 10 and 40%, so that there is a linear increase in concentration from the top to the bottom of the tube. Sucrose gradients are commonly used to separate molecules or particles on the basis of their sedimentation coefficients, which depend on their size and shape rather than their density.

Sudge digestion. Anaerobic treatment of sewage sludge.

Sufficient Statistic, one which uses all the information available in the data, even as small samples. Not always available.

Sufu. Oriental food made by fermentation of soybean.

Sugar alcohol. An alcohol derived from a monosaccharide by reduction of its carbonyl group (CO) so that each carbon atom of the sugar has an alcohol group (OH). For example, glucose yields sorbitol, common in fruits, and mannose yields mannitol.

Sugar. (1) A synonym for sucrose. (2) Any monosaccharide or disaccharide.

Sulfur cycle. A geochemical cycle by which sulfur is utilized and distributed through the biosphere.

Sulphonamide. An antibacterial drug made by chemical synthesis from the dye sulphanilamide. Sulphonamides are competitive inhibitors of the synthesis of folic acid from *P*-aminobenzoic acid.

Sulphur bacteria. Filamentous autotrophic chemosynthetic bacteria of the order Beggiatoales. They derive energy by oxidation of sulphides, and build up carbohydrates from carbon dioxide. An example is *Beggiatoa.*

Sum of Squares, the sum of squared deviations of observations from their mean.

Super female. An abnormal type of Drosophila, almost completely sterile, with an *X* chromosome/autosome ratio greater than *1*, e.g., *3X/2A* = 1. 5.

Supercoil. A double-stranded DNA molecule containing extra twists in the helix that cause the helix to coil upon itself. In order for these extra twists to be maintained, it is necessary that the ends of the double-stranded helix be constrained from rotating freely e.g., in a covalently closed circular DNA molecule.

Supercoiling (N. *Supercoil*, = *Super-helix*). Twisting of the double helicle DNA molecule to form secondary cells (superhelices).

Supercoiling. Coiling-on-coiling, as in DNA during prophase.

Supercritical fluid enzymology. All materials have a critical temperature (Tc) above which their gases cannot be turned into a liquid by compressing them. At this critical temperature, liquid and gas can coexist if the pressure is at the critical pressure (Pc). For carbon dioxide, for example, the critical temperature is 31°C. Thus at room temperature, if you compress carbon dioxide enough (as in a gas cylinder) it will turn into a liquid. Above 31°C, no matter how hard you compress it will not liquefy-the gas will just become more dense.

Supercritical fluids are also used to purify materials, using supercritical fluid extraction (SFE). SCFs can have very specific abilities to dissolve chemicals, which can be 'tuned' by altering pressure and temperature. Thus SFE is used commercially to extract caffeine from coffee, leaving virtually all the other chemical components of the coffee behind. The solvent used (supercritical CO_2) is removed by reducing the pressure, whereupon it all turns back into normal gas.

Supergene. A DNA segment that contains a number of closely linked genes affecting a single that or an array of interrelated traits.

Superinfecting phage. A bacteriophage which infects a cell which already harbours the same type of bacteriophage.

Supermale. An abnormal type of Drosophila, almost completely sterile, with the *X* chromosome/autosome ratio less than 0. 5, e.g. 1X/3A = 0. 33.

Supernate. The liquid over a precipitate or sediment; the fluid remaining after removal of suspended matter.

Supernormal stimulus. Any stimulus, or any intensity of a variable stimulus, that is preferred by animals over the natural sign stimulus.

Supernumerary or B-chromosomes. Chromosomes as are unessential to development. (cf. Accessory Chromosome).

Superoxide dismutase. An enzyme that catalyzes the reduction of the superoxide radical to hydrogen peroxide.

Superparasitism. A condition resulting from the use of a single host individual by more individual parasitoids of the same species than it can successfully sustain to maturity because of nutritional limitations; cf. Multiparasitism.

Superspecies. A monopyletic group of geographically representative (allopatric) species which are morphologically too distinct to be included in one species.

Supervital. A gene or genotype that, when expressed, is significantly more viable than the normal or standard type.

Supplementary factors. Factors which by themselves have no effect but which qualitatively alter the effect of another factor; modifying factors.

Supported growth system. A culture method in which growth of cells depends on the presence of a surface to which the cells become attached.

supports: whether the organism is to be on solid supports or in suspension.

Suppression. (1) In immunology, a state of immune non-responsiveness that is specially induced. (2) Recovery or restoration of normal genetic function due to the silencing of some aberrant or deleterious gene activity.

Suppressive. Of a plasmagene which suppresses the expression of an alternative condition in a particular respect in a hybrid individual and its descendants (*Darlington,* 1944).

Suppressor cell. A subpopulation of T cells which act to reduce the immune responses of other T cells or B cells. Suppression may be antigen-specific, idiotype specific, or non-specific in different circumstances.

Suppressor gene. A gene that can reverse the phenotypic effect of a variety of mutations in other genes.

Suppressor mutation. A mutation that totally or partially restores a function lost by a primary mutation and is located at a genetic site different from the primary mutation.

Suppressor T cell (Ts). A subset of regulatory T cells that suppresses the activity of B cells and other T cells.

Suppressor. A gene preventing homozygosity for another recessive gene to express itself. In this way an apparent reverse mutation to a dominant condition will occur.

Suppressor-Sensitive mutation. A mutation whose phenotype is suppressed in a genotype that also carries an intergenic suppressor of that mutation, e.g., *amber, ochre, or opal.* .

Suppuration. The formation of pus.

Suppurative. Pus-forming.

Supraoptimum. An amount(level) greater than required; as an inhibitory concentration of an exogeneous growth factor.

Surface sterilization. The removal of plant surface microflora prior to aseptic excision of explants. Surface sterilization is accomplished by immersion of tissue in one of many sterilants, such as calcium hypochlorite or sodium hypochlorite, hydrogen peroxide, mercuric chloride, silver nitrate or bromine water for an empirically determined period of time. Sometimes a preceding ethanol dip or spray is useful, as is the addition of a surfactant to the sterility. gitation or vacuum may improve the efficiency of sterilization. Thorough washing in sterile distilled water usually follows, to remove the sterility, although in some cases this is not necessary (hydrogen peroxide).

Surface tension. The force acting on the surface of a liquid tending to minimize the area of the surface.

Surface/volume ratio. For any cell, organism, or geo-metrical solid, the ratio of surface area to volume; this is an important factor in setting an upper limit on the size a cell or organism can attain.

Surfactant. A surface active agent or wetting agent; as is Tween 20 or Tween 80, Teepol, Lissapol F, Alconox, etc. These agents act by lowering the surface tension so are common addenda to solutions used to surface sterilize plant materials prior to aseptic excision of explants.

Survival of fittest. The mechanism of natural selection. By fittest is meant able to leave most progeny.

Survivorship curve. A plot of the logarithm of the fraction of individuals still alive, as a function of time.

Susceptibility (Latin *susceptum*, to take up). Refers to the inability on the part of the host cell to prevent infection (q. v.).

Suspension culture. Cells and group of cells (aggregates) dispersed in an aerated, usually agitated, liquid culture medium. These are obtained by adding friable callus to the medium. The plant species, explant type and treatment; the composition of the nutrient medium; and many other features determine the size and nature of the cell aggregates, which may appear different during each culture growth phase. These cultures are used to study cell division, differentiation and metabolism, in secondary product synthesis or form the basis for single cell lines, callus cultures, somatic cell embryogenesis and many other culture purposes.

SV40 T antigen. Gene regulatory protein coded on the genome of SV40 virus. It represses transcription of specific viral genes and is essential for DNA replication to be initiated at a nearby site. Thus this protein shuts down gene transcription in order to permit DNA replication.

SV40 vector. The SV40 virus is an attractive cloning vehicle or vector for several reasons. (i) The genome consists of a single, covalently closed circular DNA molecule that has been fully sequenced. (ii) The regions of the genome responsible for various functions have been accurately located in respect to the physical map. (iii) The viral genome can multiply either vegetatively or as an integral part of a cellular chromosome.

Svedberg (S). A unit of measure for the rate at which a particle sediments in a centrifugal field; $1s = 10^{-13}$ sec. In general, the greater the mass of a particle, the greater the observed sedimentation rate.

Swing-out rotor. A centrifuge rotor having buckets which are free to swing out into a horizontal position when the rotor is in motion. The long dimension of the centrifuge tube is thus held parallel to the lines of centrifugal force. and a density gradient may be formed along the entire length of the tube. This gradient will therefore cover a greater density range than in an angle or vertical rotor. Swing-out rotors are particularly useful when there is uncertainty about the buoyant densities of the molecules to be resolved.

Symbiont. A helpful organism in reference to the normal flora of the animal body. In a symbiotic relationship, the microbe and the host receive benefit from each other.

Symbiosis. Strictly, any association between two species including **parasitism, mutalism** and **commensalism** (de Bary). Until the last decade, many British biologists used the term synonymously with mutualism. The situation is now changing and most now use the term in its original broad sense.

Sympathetic nervous system. A division of the autonomic (involuntary) nervous system. Its activities include increasing blood pressure and acceleration of the heartbeat. The neurotransmitter at the sympathetic terminals is epinephrine or norepinephrine. (Contrast with parasympathetic nervous system.)

Sympathetic neurons. Cells of the autonomic nervous system which function to depress secretion, and to decrease muscle tone and blood vessel contraction.

Sympatric (Gr. *syn*: together + *patria*: homeland). Referring to populations whose geographic regions overlap at least in part.

Sympatric speciation, Instantaneous. The production of a single individual (or the offspring of a single mating) which is reproductively isolated from the species to which the parental stock belongs.

Sympatric speciation. Formation of new species even though members of the daughter species overlap in their distribution during the speciation process. (Contrast with geographic, parapatric speciation.)

Symplast. The living system of interconnected protoplasts extending through a plant body. Cytoplasmic connections between cells are made possibly by the plasmodesmata. The symplast pathway is an important transport route through the plant. Compare Apoplast.

Symptom (Greek *symptoma*, associated with or the root of an adverse condition). Phenotypic (q.v.) and/or physiological manifestations of a successful invasion of a host by a pathogen.

Symptomatic treatment. Treatment of existing or expected symptoms.

Symptomatology. The study of the symptoms and signs of disease.

Synapse (Gr. *syn*: together + *haptein*: to fasten). The narrow gap between the terminal bouton of one neuron and the dendrite or cell body of another.

Synapsis. The highly specific parallel alignment (pairing) of homologous chromosomes during the first division of meiosis.

Synaptic vesicle. A membrane-bounded vesicle, containing neurotransmitter, which is produced in and discharged by the presynaptic neuron.

Synaptinemal complex. An organelle, present during meiosis, that mediates close pairing between homologous regions of chromatids.

Synchronized cells. Synchronized mitosis in a group of cells in culture by natural or artificial means.

Synchronous culture. A culture of cells in which all the individuals are at approximately the same point in the cell cycle. Cells can be synchronized by a variety of means, e.g. temperature, shock, or drugs. Such cultures are of tremendous value in physiological and biochemical investigations.

Synchronous growth. A pattern of growth when all cells of a culture are at the same stage of division. This may be accomplished by several techniques, e.g., a chiling period followed by a favorable growth temperature.

Synchronous mitosis. The occurrence of a number of cell divisions which take place at exactly the same time in a group of neighbouring cells.

Syncitia Multinucleated cells.

Syndiploidy. Doubling, the fusion of sister nuclei to give a doubled chromosome number, particularly in the divisions immediately preceding meiosis. (*Strasbruger,* 1907).

Syndrome. A group of symptoms characteristic of the same infection or abnormal genetic condition, but not necessarily all appearing together.

Synecology. The branch of ecology which deals with the interrelationships of living communities of organisms to each other and to their environment. *cf.* Autecology.

Synergids. Two nuclei found close to the egg nucleus in the angiosperm embryo sac; they disappear shortly after fertilization.

Synergism. The ability of two or more organisms to bring about changes (usually chemical) that neither can accomplish alone.

Synergistic effect. The greater effect produced by two chemotherapeutic agents given simultaneously than would result from either alone; also, an effect produced by two species together that neither could produce alone.

Syngamy. Fertilization; e.g., union of the male and female gametes to form the zygote.

Syngen. Group of identical organisms carrying the same surface antigens. There term is widely used in relation to the ciliate *Paramecium aurelia* which contains a number of distinct syngens, each of which is genetically isolated and biochemically unique.

Syngeneic. Of the same genetic constitution. Synonomous with isogenetic and used to describe tissue transplants in animals. Compare Allogeneic.

Syngenetic relations. Blood relations; individuals which have direct or common descent from the same ancestors.

Synkaryon. A hybrid cell which results from the fusion of the nuclei it carries.

Synoecy. The condition in which male and female gametes are produced by the same haploid or diploid individual. Includes Monoecy and Hermaphroditism (*Correns,* 1928).

Synovial membrane. The membrane that forms the capsule surrounding a joint. It consists of tough connective tissue with a high proportion of white collagent fibres and it secretes the viscous lubricating *synovial fluid.*

Synteny. The occurrence of two or more genetic loci on the same chromosome. Depending on intergene distance (s) involved, they may or may not exhibit nonrandom assortment at meiosis.

Syntetic genes. Genes that are believed to be linked (i. e. reside on the same chromosome) because of their concurrent loss alongwith a specific marker gene known to be located on a particular chromosome.

Synthesize. To build up.

Synthetase (Synthase). Enzyme that catalyses molecular synthesis by the coupled breakdown of a nucleotide triphosphate (e.g. ATP).

Synthetic allopolyploid. Allopolyploid produced experimentary by man.

Synthetic medium. A medium in which the chemical formula for each ingredient can be written; it may be very complex›and may vary widely for different species.

Synthetic plasmid. A circular DNA molecule that is constructed from a replican and one or more gene DNA sequences. The synthetic plasmids have a genetic make-up that differs from wild type-plasmids. They are used as cloning vehicles.

Synthetic polyribonucleotides. RNA made *in vitro* without a nucleic acid template, either by enzymatic or by chemical synthesis.

Synthetic variety. A term used particularly with cross-pollinated plants to refer to a variety produced by the combination of selected lines or plant and subsequent normal pollination. (2) advanced generations of open pollinated seed mixtures of a group or strains, clones, or inbred, or of hybrids among them.

Syphilis. A venereal disease caused by the spirochete, **Treponema pallidum.**

Syrinx. A specialized structure at the junction of the trachea and the primary bronchi leading to the lungs. The vocal organ of birds.

System. An isolated collection of matter and energy. All other matter and energy in the universe apart from the system is said to be outside the system or its surroundings.

Systematic mutation. Mutations of major effect presumed to give rise to new species or higher categories at a single step.

Systematics. It is a term and defined as *the scientific study of organisms with the ultimate object of characterizing and arranging them in an orderly manner*. When the knowledge gained in orderly manner and is used in taxonomy, becomes the part of systematics. It includes disciplines like morphology, ecology, epidemiology, biochemistry, molecular biology, and physiology. Though new **molecular techniques** are being used in classifying microorganisms which have generated new advances and excitement in microbial taxonomy, however, more traditional approaches still have value and are being extensively used. **Systemic circulation.** The part of the circulatory system serving those parts of the body other than the lungs or gills.

Systemic. Describing a process that affects the whole organism or of a factor that is distributed throughout a whole organism.

T

2,3,5-triiodobenzoate (TIBA mw 499.81): An inhibitor of auxin movement or transport (antiauxin), sometimes included in plant tissue culture media for its growth promoting effects.

2-Thiouracil. An anti viral agent sometimes included in initiation medium for virus elimination from infected explants. Of uncertain value, its action is unclear, but may involve inhibition of viral RNA synthesis.

T- antigen. Antigen found in nuclei of cells infected or transformed by certain tumor viruses (e.g., polyoma and SV40). Many be an early viral specific protein.

Target organ. Indicates the location of the initial infection by viruses, particularly animal viruses. After getting established, the infecting particles may subsequently spread to other regions in the body.

T cell. A type of lymphocyte, involved in the cellular immune response. The final stages of its development occur in the thymus gland. (Contrast with B cell).

T lymphocytes. Small lymphocytes which possess the surface 0 antigen and direct cell mediated immune responses.

T phages. Group of bacteriophages of some what diverse form, but all having double stranded DNA genomes. An icosahedral head, and a hollow tail section.

T test. Statistical method used to determine the significance of differences between means.

T type lymphocyte. A thymus derived lymphocyte that is responsible for cell mediated immunity (CMI).

T. Symbol for thymine or thymindine in molecular biology.

T15. An idiotype associated with anti-phosphorycholine antibodies, named after the TEPC 15 myeloma proto-type sequence.

Tac promoter. A hybrid promoter constructed *in vitro* which contains the-35 sequence of the trp promoter and the-10 sequence of the lacz promoter. A pair of such promoters, tac l and tac ll were constructed with minor variations between them; both are repressed by the lac repressor and induced by lPTG. Tac l is three times, and tac ll twice, as efficient as a depressed trp promoter.

Tachykinins. Peptides with biological activity the have the N terminal sequence-Phe-X-Gly-Leu-Met NH_2. They bring about contraction of smooth muscle.

Tachyphylaxis. The process by which subsequent doses of a pharmacological reagent produce decreasing effects.

Tail. Refers to the helical component, a virus with binal (q.v.). symmetry.

Tailing. Those portions of washed out that are regarded as too poor to be treated further; also called debris.

Tandem array. The contiguous arrangement of two or more identical sequences within a DNA molecule.

Tandem duplicates. Adjacent copies of related genes linked together on a chromosome.

Tandem duplication. Situation in which a segment of a chromosome is duplicated and the extra copy or copies inserted in line in the chromosome. Thus the genes for large ribosomal RNA are tandemly duplicated in eukaryotes.

Tandem genes. Genes that are arranged tandemly in the chromosome, especially multiple copy genes where the copies are tandemly duplicated.

Tandem satellites. Two short segments of chromosome separated from each by a constriction and from the main body of the chromosome by a second constriction.

Tank Bioreactors. Bioreactors, also called fermentors, are the vessels in which fermentation takes place. Tank bioreactors are vessels in which the microorganism is grown in a large volume of liquid. This contrasts to fibre/membrane bioreactors (see Hollow Fibre) and immobilized cell bioreactors. The large majority of bioreactors used in biotechnology are tank bioreactors, and most tank bioreactors are stirred tank bioreactors, because stirring helps to distribute gas and nutrient to the growing organisms effectively. The bioreactor must provide a mechanism for introducing reagents and the microorganism into the reactor vessel, providing substrate (i.e. food) to the microorganism (including oxygen for aerobic fermentation), stirring it, and keeping it at the right temperature, pH, etc. Temperature control is especially critical for large volume fermentations since metabolizing microorganisms produce a great deal of heat. Variations in layout include different sizing and spacing of the baffles (which ensure that the volume is mixed thoroughly by the stirring) and different types of impellers (stirrers). These can be wide range of shapes and sizes: disk (Rushton) turbine, open turbine, marine impeller (i.e. like a ship's screw). The other main variation between reactors is the gas injection mechanism. These are almost always via a sparger or a diffuser (a pipe or plate with holes in it, which shoots bubbles into the base of the reactor over a wide area at the base of the tank), but a wide range of shapes of pipes or plates have been used. The shapes, which include ring, crosses (`spiders') or dead-end pipes, must be selected for the particular shape and size of the reactor, and the amount of gas that has to be injected. There is a great deal of expertise in designing a suitable bioreactor for culturing an organism or cell type. As a consequence, there are more companies specializing in bioreactor design, control, and engineering than in recombinant DNA techniques and reagents, despite the much higher profile of `gene cloning'.

Tannin. Any of several soluble astringent complex phenolic substances of plant origin.

Tantigen (pronounced large tee atigent). An enzyme encoded by SV40 which is essential for viral DNA replication. It is also essential for cellular transformation and was named T for transforming antigen.

Tape. Synonymous with "messenger RNA" and going out of common usage. It was presumably coined to suggest an analogy with the magnetic tape which stores musical information for use in tape recorders.

Tapetal cell. In an anther the pollen mother cells are surrounded by tapetal cells, which provide nourishment to the pollen mother cells and secrete the cutin on the maturating pollen grains. These cells often have deviating or polyploid chromosome numbers.

Target organ. Indicates the location of the initial infection by viruses, particularly animal viruses. After getting established, the infecting particles may subsequently spread to other regions in the body.

Target site. Restriction endonucleases bind to DNA at a specific sequence known as their recognition site and then cut the DNA within a sequence known as the target site. For type II restriction endonucleases, the kind most commonly used in genetic engineering experiments, the recognition site and the target site are one and the same thing.

Target theory. Hypothesis developed to account for the mutagenic effects of ionizing radiation on cells. It is postulated that only a small target exist in the cell and that one or more ionizing events or `hits' are necessary within the target area for radiation to produce effectively a mutation.

Target tissue. Tissue that responds specifically to a molecule or factor such as a hormone, antibody or specific dye.

Targeted drug delivery. This is using any method to deliver a drug to the site in the body where it is needed, rather than allowing it to diffuse into many sites. There are three approaches to such targeted drug delivery. The first is to ecapsulate the drug in something, usually a lipid coat (i.e. a liposome). The outside of the coat is itself coated with a material that binds to the target cells, either an antibody specific for those cells, or a glycoprotein, or a receptor molecule or ligand. The liposome travels round the blood until it finds its target, where it sticks to the cell and delivers the drug. In the case of liposomes, the lipid membrane fuses with that of the cell, releasing its contents into the cell. The second approach links the targeting mechanism directly to the drug. Here the drug must either operate outside the cell, or be able to get itself into the cell on its own. A much talked about application is linking toxin proteins to antibodies: the proteins can get inside cells and there destroy the cellular machinery, but only when carried near enough to the cell by the antibody. Such conjugates are called immunotoxins. The application here, clearly, is to destroy cancer cells or, conceivably, cells infected with long-term viruses such as AIDS or HPV. The problem with both these approaches is how to get the drug-carrier complex from the bloodstream into the target tissues; unless the target is the endothelial cells of the blood vessels or a few cell types in liver, lung, or kidney, nothing as large as a liposome is going to be able to escape from the blood vessels to get to them. The third approach is to make the drug as a `prodrug' which goes to all tissues of the body, but which is metabolized to the active drug only by one tissue because that tissue has a high level of an enzyme that cuts the prodrug into inert `carrier' and active drug. This is easier to do for tissues such as liver and kidney, which have a battery of rather specific enzymes. Several drugs, such as the antiviral drug gancyclovir, are actually prodrugs that are activated by their target cells but remain less active in other cells.

TATA box (Hogeness box). Sequences of nucleotides (TATA) found in the promoter region of most, but not all protein coding genes of eukaryotes and many viral genes. TATA usually occurs about 35 base pairs upstream of the start point of RNA synthesis and is analogous to the pribnow box of pro TATA usually occurs about 35 base pairs upstream of the start point of RNA synthesis and **TATA us** TATA usually occurs about 35 TATA usually occurs about 35 base pairs upstream of the start point of RNA synthesis and is analogous to the pribnow box **TATA TA** TATA usually occurs about 35 base pairs upstream of the start point of RNA synthesis and is analogous to the pribnow box of TATA usually occurs about 35 base pairs TATA usually occurs about 35 base pairs upstream of the start point onges its chemical properties.

Taxis (Gr. *taxis:* arrange, put in order). The movement of an organism in a particular direction with reference to a stimulus. A taxis usually involves the employment of one sense and a movement directly toward or away from the stimulus, or else the maintenance of a constant angle to it. Thus a positive phototaxis is movement toward a light source, negative geotaxis is movement upward (away from gravity), and so on.

Taxon (*pl.* taxa). A unit of taxonomic classification for example, variety species, family, class, etc.

Taxonomic key. A guide to the identification of an unknown specimen. The most useful keys are dichotomous.

Taxonomy (Greek *taxis*, arrangement or order, and *nomos*, law, or *nemein*, to distribute or govern). It is the science consists of three separate but interrelated parts:

Classification: It is the arrangement of organisms into groups or *taxa* (singular, *taxon*) based on mutual similarity or evolutionary relatedness.

Nomenclature: It is the branch of taxonomy concerned with the assignment of names of toxonomic concerned with the assignment of names to taxonomic groups in agreement with published rules.

Identification: It is the practical approach to the taxonomy, and is the process of determining a particular isolate which belongs to a recognized taxon. Its practical importance can hardly be overemphasized. It is more useful for clinical microbiology.

Taxonomic Ranks. In preparing a classification scheme, one places the microorganism within a small, homogeneous group that is itself a member of larger groups in a non-overlapping hierarchical arrangement. A category in any rank unites groups in the level below it based on shared properties. The most commonly used levels or ranks (in ascending order) are species, genera, families, orders, classes, phyla or divisions, and kingdoms. Microbial groups at each level or rank names with ending or suffixes characteristic of the level. Informal names often used in place of formal hierarchical ones. Typical examples of such names are purple bacteria, spirochetes, methane-oxidizing bacteria, sulfate-reducing bacteria, and lactic acid bacteria.

Taxonomic ranks and names

Rank	Examples
Kingdom	*Procaryotae*
Division (Phylum)	*Tenericutes*
Class	*Mollicutes*
Order	*Mycoplasmatales*
Family	*Mycoplasmatales*
Genus	*Mycoplasma*
Species	*M. pneumoniae*

The basic taxonomic in microbial taxonomy is the **species**. Species of higher organisms is define differently than microbial species.

Species of Higher Organism: Species of higher organisms are groups of interbreeding or potentially interbreeding natural populations that are reproductively isolated from other groups. This is a satisfactory definition for organisms capable of sexual reproduction, but fails with many microorganisms because they do not reproduce sexually.

Species of Bacteria: Bacterial species are characterized by phenotypic and genotypic differences. It is a collection of strains that share many stable properties and differ significantly from other groups of strains.

Strain: It is a population of organisms that descends from a single organism or pure culture isolate. Strains within a species may differ slightly from one another in many ways.

Biovars: They are variant bacterial strains and are characterized by biochemical or physiological differences.

Morphovars: They differ morphologically

Serovars: They have distinctive antigenic properties.

Type strain: It is one strain of a species. It is usually one of the first strains studied and often is more fully characterized than other strains. It does not have to be the most representative member.

Genus: Every species is assigned to genus and is the next rank in the taxonomic hierarchy. It is a well-defined group of one or more species and is clearly separate from other genera.

Taxonomy Numerical. The development of computers has made possible the quantitative approach known as numerical taxonomy. Peter H. A. Sneath and Robert Sokal defined numerical taxonomy as *the grouping by numerical methods of taxonomic units into taxa on the basis of their character states.* Numerical taxonomy has proved to be a powerful tool in microbial taxonomy.

Computer is often used for numerical analysis and the information about the properties of organisms. The resulting classification is based on general similarity as judged by comparison of many characteristics, each given equal weight. This approach becomes feasible after advancement in computers because large number of calculations are involved. The important aspects of numerical taxonomy are as follows:

1. The process begins with a determination of the presence or absence of selected characters in the group of organisms under study.
2. A character usually is defined as an attribute about which a single statement can be made.
3. Many character, at least 50 and preferably several hundred, are compared for an accurate and reliable classification.
4. Many different kinds of data like morphological, biochemical, and physiological are included.
5. After character analysis, an association coefficient, a function that measures the agreement between character possessed by two organisms, is calculated for each pair of organisms in the group.

The simple matching coefficient (S*SM*), the most commonly used coefficient in bacteriology, is the proportion of characters that match regardless of whether the attribute is present or absent. Sometimes the **Jaccard coefficient (S*J*)** is calculated by ignoring any characters that both organisms lack. Both coefficients increase linearly in value from 0.0 (no matches) to 1.0 (100% matches).

The simple matching coefficient, or other association coefficients, are then arranged to form a **similarity matrix**. This is a matrix in which the rows and columns represent organisms and each value is an association coefficient measuring the similarity of two different organisms so that each organism is compared to every other one in table.

Organisms with great similarity are grouped together and separated from dissimilar organisms; such groups of organisms are called **phenons** (sometimes called phenoms).

The results of numerical taxonomic analysis are often summarized with tree like diagram called a **dendrogram**.

TDL sampler. An impaction sampler similar in principle to the Casella silt sampler. Rotation rates for the agar plate can be selected from one revolution per 2/3 of a minute to one revolution per 180 minutes, depending on model selected.

T-DNA, transferred DNA. The segment of DNA from a Ti or Ri plasmid which is transferred from *Agrobacterium* to the genome of its plant host and causes tumour formation or root induction.

TE buffer. A shorthand nomenclature for a Tris-HCL buffer solution, pH 7.5, containing the cheating agent ethane diamine tetraacetic acid (EDTA). It is commonly used to dissolve nucleic acids, particularly DNA.

Teichoic acids: They are mainly presented in Gram-positive bacteria and are substituted polysaccharides and consist of ribotol or glycerol residues joined through diphosphoster linkages.

Telegony. An effect of a previous sire on later progeny of the same dam by another sire. Stockbreeder's myth.

Telocentric. A chromosome or a chromatid having a terminal centromere. A condition arising by Misdivision or by X-ray breakage.

Telophase (Gr. *telos:* end). The final phase of mitosis or meiosis during which chromosomes become diffuse, nuclear envelopes reform, and nucleioli begin to reap-pear in the daughter nuclei.

Telosynapsis. The association of chromosome end to end. (The pairing is now taken to be parasynaptic).

Tempeh. Oriental food made by fermentation of soybean or coconut.

Temperate bacteriophage. A bacteriophage capable of replicating in step with its host bacterium, being thus transmitted through cell divisions without necessarily causing lysis.

Temperate phage (Latin *temperare,* moderate or restrained). Refers to the bacteriophages (q.v). which on entering the host bacterial cell, do not necessarily lyse the latter; usually it remains associated with the host bacterial chromosome. This association, though indefinite, is not permanent, and the phage DNA, once detached, may again behave as a virulent (q.v). type.

Temperate sensitive daphnia. A mutant that dies at normal temperature, but is adapted to much higher temperatures.

Temperate virus. Lysogenic virus in which genome is integrated or associated with bacterial genome.

Temperature quotient (Q_{10}). The ratio of the rate of reaction at a given temperature to that of the same reaction carried out at a temperature that is 10 degrees centigrade lower.

Temperature sensitive gene. A gene which is not expressed when the temperature to which it is exposed falls outside certain specific limits (e.g., 25-30°C). Temperature sensitive genes have been used in gene engineering to improve the production of unusual secondary products that may impair the growth of the host cells.

Temperature sensitive mutation. Mutation yielding protein that is functional at low (high) temperature, but that is inactivated by temperature elevation (lowering).

Template, templet strand. (1) In general, a pattern or mould against which a complex product is made. (2) In molecular biology the nucleic acid strand (DNA or RNA) which is copied during replication or transcription.

Template. In biochemistry, a molecule or surface upon which another molecule is synthesized in complementary fashion, as in the replication of DNA. In the brain a pattern that responds to a normal input but not to incorrect inputs.

Tendon. A collagen-containing band that connects a muscle with a bone.

Tent: The enclosure or process of including plants, such as exvitro plantlets, under plastic or glass to maintain elevated relative humidity.

Tepal. In an angiosperm flower, a sterile modified leaf. This term is used to refer to such flower parts when one is unable to distinguish between petals and sepals.

Teratogenic: Refers to an agent which induces or increases the incidence of gross structural abnormalities **(teratomata)** in an individual or a population; as do x-rays.

Terminal affinity. The property by which chromosomes are held together end to end from Diplotene till first Metaphase of Meiosis (by a Terminal Chiasma) or are brought together in this way at Metaphase.

Terminal CCA. At the 3'end of all tRNA there is a sequence CCA, to the adenosine moiety of which is attached the amino acid.

Terminal chiasmata. Association of homologous chromosomes at meiosis, involving only the chromosome ends. This may be the result of a crossing over event at the chromosomal extremity, or of terminalization of chiasmata which earlier were positioned some distance away from the chromosome terminus.

Terminal deletion. Deletion involving the end of a chromosome.

Terminal disinfection. Disinfection procedures carried out at the end of an infectious period. This involves cleaning the entire area which may have been contaminated by the patient.

Terminal gene. A gene situated at the extreme end of chromosome telomere.

Terminal granules. Granular bodies which occur at the ends of some of the chromosomes in certain plant species. These granules should not be confused with satellites.

Terminal infection. An infection with pathogenic microorganisms occurring during the course of a disease that terminates in death.

Terminal redundancy. Repetition of the same DNA sequence at either end of a long DNA molecule. Some bacteriophages have DNA that is terminally redundant and can be circularized by removal of a few bases from each end.

Terminal transferase. An enzyme capable of adding nucleotide residues of a single kind to the 3' end of DNA chains.

Terminalization coefficient. The number of terminal chiasmata divided by the total number of chiasmata.

Terminalization. The process in which the chiasmata move (terminalize) towards the chromosome ends, the chiasma number thereby being reduced. This phenomenon occurs during the latter part of the prophase of meiosis.

Termination codons. The three codons UAA, UAG, and UGA, which signal termination of a polypeptide chain.

Termination sequence. A DNA sequence that appears at the end of transcriptional unit and signals the end of transcription.

Terminator. A nucleotide sequence in DNA that causes RNA transcriptase to cease transcription.

Terpene. Organic hydrocarbon or hydrocarbon derivative constructed from recurring isoperene units.

Terpenoid. Any of a large number of compounds with varying numbers of carbon atoms, but which are clearly derived from five carbon precursors. Terpenoid quinones and chromonols include phylloquinone (vitamin K), plastoquinones, ubiquinones and tocopherols. Many are important intermediates in electron transport systems.

Terramycine. A synonym for the antibiotic now generally known as oxyeracycline.

Terrestrial (L. *terra:* earth). Pertaining to the land. (Contrast with aquatic, marine).

Territory. A fixed area from which an animal or group of animals excludes other members of the same species by aggressive behavior or display.

Tertiary split. Longitudinal doubleness of the chromonema of each chromatid which may become visible during the first or second division of meiosis.

Tertiary structure. In reference to a protein, the relative locations in three-dimensional space of all the atoms in the molecule. The overall shape of a protein. (Contrast with primary, secondary, and quaternary structures).

Test cross. A cross of a dominant-phenotype individual (which may be either heterozygous or homozygous) with a homozygous-recessive individual.

Test tube baby. Production of a child following *in vitro* fertilization, after which the embryo is transplanted into the uterus of its biological mother or a surrogate mother to permit implantation and normal gestation.

Test tube fertilization. Pollination, followed by fertilization occurring *in vitro.*

Test. Shell or capsule secreted by a protozoan.

Testcross. It is cross in which an F_1 individual showing dominant phenotype is hybridised with the recessive parent in order to test whether the dominant phenotype is homozygous or heterozygous and also to test the degree of linkage, in case there is any.

Tester strain. A strain whose genetic make up is such that its contribution to the inheritance of the offspring will not hide the inheritance being received from the parent being progeny tested.

Testicular ferminization. A sex limited abnormality in which an XY male has a female appearance because the testes secrete estrogen rather then androgen.

Testis. The male gonad; that is, the organ that produces the male sex cells.

Testosterone. A primary sex hormone produced by the testes.

Tetnaus. A disease caused by the exotoxin produced by **Clostridium tetani.**

Tetra allelic. Situation in which four distinct alleles occur at a particular gene locus in polyploid organization.

Tetracycline resistance Tc^r, tetracycline sensitive Tc^s. Resistance or sensitivity to the lethal effects of the antibiotic tetracycline. Tetracycline is bacteriostatic. It halts growth by preventing the binding of charged tRNAs to the ribosome.

Tetracycline resistance. The ability to survive treatment with the antibiotic tetracycline.

Tetracyclines. Group of broad spectrum antibiotics obtained from various species of the mould streptomyces. They bind to the 30S subunit of the bacterial ribosome, preventing attachment of the aminoacyl tRNA.

Tetrad (Gr. *tetras:* foursome). Paired homologous chromosomes during the first prophase and metaphase of meiosis. Each chromosome at these stages is a double structure, consisting of two chromatids joined at a not-yet-divided centromere.

Tetrad analysis. Analysis of the four haplonts that may be formed from a single tetrad. Such analyses in mosses and fungi have given valuable information concerning the separation of genes and chromosomes and their recombination during meiosis.

Tetrad divisions. The first and second meiotic divisions which the spore mother cell undergoes to from four spores.

Tetrad. A group of four cells that in plant are formed as the result of meiosis in a spore mother cell. In flowering plants the tetrads on the male side consist of four microspores, which later separate from each other and develop into pollen grains. In the zoological literature the tetrad is sometimes used to designate a bivalent that is composed of four chromatids.

Tetrahydrofolic acid. The reduced active coenzyme form of the vitamin folic acid.

Tetramer. Structure resulting form association of four subunits.

Tetraploid. A polyploid cell, tissue, or organism having four sets of chromosomes (*4n*).

Tetrasaccharide. A polysaccharide comprising four linked monosaccharide units.

Tetrasomic. An individual showing two extra chromosome (all being homologous to a particular pair) in addition to the normal chromosome complement expressed by *2n+2*.

Tetratype. A tetrad in which the four meiotic products are different, e.g., *AB, aB, Ab* and *ab.* Crossing over has occurred in such a tetrad.

Tetrazolium or triphenyl (2,4,5) tetrazolium chloride (TCC, $C_{19}H_{15}NCl$ mw 334.81): Used in a quantitative assay of cell viability. It becomes reduced to a red, water-insoluble form, formazan, in the presence of oxidative metabolism. It is then dissolved in alcohol and measured spectrophtometrically.

Tetrazolium salts. Complex organic molecules that changes colour on mild reduction with the formation of insoluble formazans. All such compounds are characterized by a heterocyclic ring structure containing one carbon and four nitrogen atoms, one of which is quaternary. Two types of salts exist diterazoles with two such rings. Tetrazolium salts are used as qualitative markers of oxidative enzyme activity, and in tissue sections, this enables enzyme activity to be localized.

Thalamus. A region of the vertebrate forebrain; involved in integration of sensory input.

Thalassemia. An inherited anemia that shows codominance an individual heterozygous for the trait has thalassemia minor, while the homozygote exhibits the severe form, thalassemia major.

Thalidomide. A tranquilizer taken by pregnant mothers during the 1960's that was found to induce phocomelia.

Thallophyte. A plant having no true stem, roots, or leaves includes the algae and fungi.

Thallospore. A spore that develops by budding of hyhal or vegetative cells.

Thallus. A plant or microbial body, without special tissue systems or organs, which may vary from a single cell to a complex, branching multicellular structure.

Thelyotoky. A type of parthenogenetic reproduction in which only female progeny are produced cf, Arrhenotoky Deuterotoky.

Theoretical plate. In chromatography, the bed length that brings about the same relative change in concentration in a solute as would be obtained on equilibrating with fresh stationary phase. The number of theoretical plates (*N*), obtained by dividing the column length by the height of one theoretical plate, determines the efficiency of chromatographic separation.

Therapeutic. The treating or curing of a disease.

Therapy. Treatment of disease.

Thermal death point. The temperature required to sterilize a given substance in 10 minutes.

Thermal death time. The time required for sterilization at a given temperature.

Thermal denaturation. Destruction or partial destruction of the structural conformation of a molecule by heating.

Thermal sensors. Thermal sensors, sensors that detect tiny changes in heat or temperature, are well known in many applications. Such sensors are often used in gas chromatography systems to detect molecules emerging from the GC column. There have been some attempts to harness thermal sensors as biosensors. Here, a probe detects the heat given out when an enzymatic reaction occurs. This could be much more flexible than enzyme electrodes since, while relatively few enzyme reactions involve the transfer of electrons that could be picked up by an electrode, nearly all result in the release of heat. The problem is that, for small samples of dilute material, the amount of heat released is tiny, so very small and very sensitive heat sensors are needed.

Thermic shock. Exposed to a reduced temperature for several days, an allegedly beneficial pretreatment for immature flower buds or anthers prior to anther culture or pollen culture.

Thermocline. Zone, in a body of water, where temperatures change abruptly to about 4°C.

Thermoduric. The capacity to endure high temperatures which are not conducive to growth.

Thermoinducible lysogen. A strain of bacterium that contains a prophage whose repressor is temperature sensitive. Raising the temperature of the culture leads to induction.

Thermolabile or heat labile: Destroyed or altered during heating; as some viruses are killed during heat therapy (thermotherapy) of infected plants, or as some medium components which destroyed by autoclaving and must be filter sterilized.

Thermolabile. Destroyed or altered during heating as some viruses are killed during heat therapy (thermotherapy) of infected plants, or as some medium components which are destroyed by autoclaving and must be filter sterilized.

Thermophile. A thermophile is an organism that grows at a higher temperature than most other organisms. Generally, since a wide range of bacteria, fungi, and simple plants and animals can grow at temperatures up to 50° C, `theromphiles' are considered to be organism that can grow above 50øC. They can be classified fairly arbitrarily depending on their optimal growth temperature into `slight thermophiles' (50-65° C), `thermophiles' (65-85° C), and `extreme thermophiles' (> 85° C). Thermophiles and extreme thermophiles are usually found growing in very hot places: hot springs and geysers, and domestic hot water pipes, for example. Deep sea exploration submarines have brought back microorganisms from the `black smoker' volcanic vents on the mid-ocean ridges which can apparently grow at over 100øC, under pressure. Thermophiles are the interest to biotechnology because of the economics of fermentation and biotransformation. Many industrial processes could be catalysed by enzymes, but the enzymes are too slow. The reactions may be speeded up by heating, but this rapidly destroys the enzyme. If you can use an enzyme that works at high temperatures, then you get the benefit of heat and catalysis. Heating up the reaction is also desirable because it reduces viscosity and increased the rate of diffusion of reagents, and so reduces the amount of stirring and pumping energy needed, and the heat prevents other enzymes from working or (usually) contaminating organisms from growing in your reactor. Enzymes from thermophiles can, of necessity, withstand such high temperatures. They also frequently exhibit increased stability to organic solvents. Thus there is substantial interest in isolating these enzymes and using them in industrial processes. Because the bacteria themselves are usually tricky to grow (and must be grown at high temperatures), once a suitable enzymes is identified it is common to seek to `clone' its gene into a bacterium that grows at a more moderate temperature. This also means that the enzyme can be purified from all the other proteins in the bacterial cell simply by heating it up: all the other, non-thermostable proteins will precipitate, leaving a reasonably pure preparation of the target enzyme. A range of thermostable enzymes are used in industrial processes.

As with all research into isolating enzymes from bacteria, one key feature is to have a large number of diverse sources of candidate organisms to screen. This is why Iceland, with one of he world's densest concentration of different types of hot spring and geyser, has been the source of the majority of the publicly available thermophilic organisms in use.

Thermophile. An organism that grows best at high temperatures, e.g., 45° to 75° + C its optimum growth temperature is about 55°C.

Thermostable. Relatively resistant to heat. (Resistant to temperatures of 100°C).

Thermotherapy. Exposure of plants to elevated temperatures as a treatment for virus or mycoplasma elimination. thermotherapy is used alone or in conjunction with meristem culture or meristem tip culture for virus elimination.

Thiamin, thiamine. Aneurine vitamin B_1 (C_{12} H_{18} N_4 $SOCl_2$), the anti neuritic vitamin.

Thiamine hydrochloride or aneurine hydrochloride ($C_{12}H_{17}ON_4SCl$. HCl mw 337.28): A thiamine (vitamin B_1) salt commonly included in plant tissue culture media.

Thiamine or aneurine ($C_{12}H_{17}ON_4SCl$ mw 300.82): Vitamin B_1, a coenzyme involved in carbohydrate metabolism. A common constituent of nutrient media used for plant tissue culture. It is usually added as **thiamine hydrochloride.**

Thiamine pyrophosphate. A coenzyme involved in reactions concerned with carbohydrate and amino acid metabolism.

Thin layer chromatography. A chromatographic method in which a glass plate is covered with a thin layer of inertabsorbent material (e.g. cellulose or silica gel) and the materials to be analysed are spotted near the lower edge of the plate. The base of the plate is then placed in a solvent, which rises up the plate by capillary action, separating the constituents of the mixtures. The principles involved are similar to those of paper chromatography and, like paper chromatography, two dimensional methods can also be employed.

Thin layer gel filtration. A technique in which molecules are separated on a flat plate (typically 20 × 20 cm) on which a layer of xerogel 0.5-1.0 mm thick is spread.

Thioester. An ester of carboxylic acid with a thiol or mercaptan.

Thiourea (CH_4N_2S mw 76.12): An occasional additive in plant tissue culture media as a source of reduced nitrogen.

Third generation pesticides. A term proposed by Williams (1967) for juvenoids and other natural or synthetic chemicals potentially useful for pest suppression by virtue of their effective interference with hormonally directed natural processes in insects.

Thixotropy. The property of liquefying when agitated and of returning to the `gel' condition on standing. This property is shown by some colloids.

Thorax. In an insect, the middle region of the body, between the head and abdomen. In mammals, the part of the body between the neck and the diaphragm.

Three breed crossing. A system of breeding in which breed *A* is crossed with *B* and the F_1 with C, this hybrid is then crossed with *B* and the product is crossed with *A* to produce a hybrid which is crossed with *C* the next cross is with *B* and then A and so on.

Three factor crosses. Mating experiments involving three distinguishable genetic marks (e.g., $a^+b^+c^+$ X abc).

Three point cross. A trihybrid testcross (e.g. *ABC/abc* X *abc/abc,* etc.). used primarily in chromosome mapping.

Three point linkage. A testcross in which three pairs of contrasting characters are taken into consideration for the purpose of determining the crossover value between three genes in chromosome mapping.

Three way cross. A cross between a single cross an inbred line in corn, etc.

Thremmatology. Plant and animal breeding for domestic uses.

Threshold. A term used in studying effects of radiation, Below a certain dose, if there is a threshold, there is no measurable effect. No threshold exists for most genetic effects.

Thrombin. An enzyme that converts the soluble protein fibrinogen into the fibrous fibrin during blood clotting. It is formed from prothrombin under the influence of thromboplastin, calcium ions, and other factors, which are activated when blood is removed from the circulation, usually by injury.

Throwing back. The reappearance of character after a lapse of one or more generations due to the character being recessive or being dependent on complementary factors, or as a result of back mutation atavism. Such an aberrant individual is called a throw back.

Thrush. A condition affecting the oral mucous membranes caused by the fungus *Candida.*

THY. A cell surface antigen of mouse T cells which has allotypic variants.

Thylakoid. A flattened sac within a chloroplast. The membranes of the numerous thylakoids contain all of the chlorophyll in a plant, in addition to the electron carries of photophosphorylation. Thylakoids stack to form grana.

Thymidine kinase. Enzyme that catalyses the conversion of thymidine to theymidine monophosphate by phosphorylation.

Thymidine. The deoxyribonucleoside that contains the pyrimidine thymine.

Thymidylic acid. The deoxyribonucleotide that contains the pyrimidine thymine.

Thymine dimer. A covalently joined dimer of two adjacent thymine residues in DNA induced by absorption of ultraviolet light.

Thymine. One of the four nucleotide bases of which DNA is composed. Thymine is a pyrimidine.

Thymus. A ductless, glandular portion of the lymphoid system, involved in development of the immune system of vertebrates.

Thyroid (Gr. *thyreos:* door-shaped). A two-lobed gland in vertebrates. Produces the hormone thyroxin.

Thyrotropic hormone. A hormone that is produced in the pituitary gland of amphibia such as frogs, transported in the bloodstream to the thyroid gland, inducing the thyroid gland to produce the thyroid hormone that regulates metamorphosis from tadpole to adult frog.

Ti plasmid: A portion of the genome of the bacterium *Agrobacterium tumefaciens*, the agent involved in crown gall disease. This plasmid is a useful (experimental) vector for the transfer of genetic information into plant cells.

Ticking in dogs. Flecks or "ticks" of colour in white areas of the coat caused by dominant gene T.

Time limit. Limitation of time available for zygotene pairing which causes localization (Darlington, 1935).

Tinea. A name applied to fungal infections of the skin.

Tissue culture. The growth of cells, tissue, or organs in suitable media *in vitro.* Such media must normally be sterile, correctly pH balanced, and contain all the necessary micro and macronutrients, carbohydrates, vitamins, and hormones for growth. Studies of such cultures have shed light on physiological processes that would be difficult to follow in the living organism. The cytokinins were discovered through work on tobacco pith tissue culture.

Tissue explant: An excised plant portion of tissue used to initiate a culture.

Tissue type plasminogen activator. A plasminogen activator produced by the host organism, in contrast to artificial plasminogen activators, such as streptokinase and urokinase.

Tissue typing. Use of serological tests to determine the major histocompatibility complex antigens of potential transplant donors or recipients.

Tissue. A group of cells that is specialized for a particular function. Examples are connective tissue, muscular tissue, and nervous tissue. Several different tissues are often incorporated in the structure of each organ of the body.

Titer. (French *titre*, a standard). The potency of a biological reactant expressed in units per milliliter. The level of immunoglobulins is frequently given in this way.

Tk^-thymidine kinase deficient lk. Mutants are unable to phosphorylate the DNA precursor, thymidine tk^-cell lines are often used in mammalian somatic cell genetics since they are resistant to the thymindine analogue, bromodeoxyuridine (BUdR).

T_m. Melting temperature for a sample of double stranded nucleic acid. The more accurate figure is $T_{1/2m}$ being the temperature at which half the sample has become dissociated or melted into single strands.

Tn 10. A bacterial transposon which carries the genes for tetracycline resistance as well as genes involved in its own transposition.

Tn 3. A bacterial transposon which carries the gene for ampliclin resistance as well as genes for its own transposition. Tn 3 was the first transposon to have its DNA sequence completely determined.

Tn5. A bacterial transposon which carries a gene encoding resistance to the antibiotics kanamycin and neomycin, as well as genes involved in transposition.

Tocopherol. A group of alcohols which show vitamin E activity, Â-Tocopherol $C_{29}H_{50}O_2$, and ß tocopgerol, $C_{28}H_{48}O_2$ have been isolated from wheat germ oil and tocopherol occurs in cotton seed oil.

Tocopheryl acetate ($C_{31}H_{52}O_3$ mw 472.73): A form of vitamin E occasionally used in plant tissue culture media.

Tolerance. Here used in the sense of painter (1951) as a basis of resistance in which a host shows an ability to grow and reproduce itself or to repair injury despite supporting a pest population equal to that damaging a more susceptible host.

Topclines. Geolines particularly where distances involved are very lest.

Topoisomerase. Enzyme that converts DNA into different topological forms.

Topological isomers. Molecules of DNA that are identical to one another except for their linking numbers (i.e. they vary in the degree of coiling or super coiling of the DNA molecule).

Tornaria (L. *tornus:* lathe). The free-swimming ciliated larva of certain echinoderms and hemichordates; its existence indicates the evolutionary relationship of these two groups.

Torula (*pl*., torulae). False yeast, o yeast in which a sexual stage of spore-formation does not occur or has not been recognized.

Total organic carbon (TOC). A measure of the carbon content of a sample. It is used as an indication of the amount of organic pollution in a water sample.

Total oxygen demand. The amount of oxygen consumed in the combustion of all the organic carbon in a sample at about 900°C.

Total pollution loading. A concept which regards the aggregate mass of a specific pollutant, discharged to the environment in a given time period, as more important than the concentration of a pollutant as a basis of environment protection legislation. The total loading on any one segment of the environment becomes the basis of control.

Total solids. The amount of material obtained by drying a representative sample to constant weight at 105°C.

Totipotency. A cell characteristic in which the potential for forming all the cell types in the adult organism is retained.

Toxemia. The presence of toxins in the blood of a patient.

Toxic: Poisonous; as are some chemicals **(toxicants)** or any substances present to excess and detrimental to normal plant function or growth.

Toxin-antitoxin. A mixture of toxin and antitoxin containing slightly more toxin than antitoxin. Formerly used to produce an active immunity.

Toxins. Living things make some of the most dangerous compounds known, like ricin (castor bean toxin) and pertussis (whooping cough) toxin. One molecule of Botulinus toxin, delivered to the inside of a nerve cell a billion times more massive than it is, will kill the cell, or less than a tenth of a microgram will kill an adult human. By comparison, the lethal dose of plutonium is guessed to be half a microgram. Such powerful poisons have potential uses, and biotechnology has the potential to make them relatively safely. Toxins can be used on their own as therapeutics. Botulinus toxin is being developed as a way of blocking unwanted muscle spansm. Clearly it cannot be injected generally, as other drugs would be (it would kill the patient). However, if minuscule doses are injected into the muscle they can paralyse that muscle. the amount of protein is so small that it escapes the notice of the immune system, and so the body does no make antibodies that could neutralize future doses. Allergan and Porton International are producing commercial version of Botulinus toxin for this application. Toxins can also be coupled to other thins to give them a lethal `sting'. **Immunoconjugates** are probably the best example. Making such toxins is difficult, even with all the panoply of microbiological containment available (see Physical containment). People have tried cloning the genes for these protein toxins into bacteria to express them more efficiently (they are normally present in extremely small amounts). Such scientists tend to find themselves in the middle of empty rooms whenever they talk about their ambitions at conferences.

Toxoid. A toxin that has been treated to destroy its toxic property without affecting its antigenic properties. An inactivated toxin that has lost its toxigenicity but retains its immunologenicity and thus can be used for active immunization.

Tra. The genetic notation for the genes involved in the transfer of a plasmid from one cell to another.

Trace element. A chemical element required by an organism in only a trace amount.

Tracer. A marked compound introduced into culture or used as substrate for any reaction carried out *in vitro* as a means of following the course of growth or metabolism of the compound. Radioactive or heavy isotopes of the normal elements found in living organisms are often used as tracers.

Trachea (Gr. *trakhoia:* a small rough artery). A tube that carries air to the bronchi of the lungs of vertebrates, or to the cells of arthropods.

Tracheary element. Xylem cell that are heavily lignified, and including tracheids vessels with thickened walls, and xylem fibres.

Tracheid. A distinctive conducting and supporting cell found in the xylem of nearly all vascular plants, characterized by tapering ends and wall that are pitted but not perforated.

Trageted mutagenesis. Use of mutagen, such as ultraviolet light or nitrogen mustard, which is specific for particular traget within the cellular DNA.

Trailer. Any sequence that is downstream of the termination signal of a messenger RNA molecule and is therefore untranslated. It does not include the poly *A* tail.

Trait. A loose synonym of "character".

Trans acting locus. Gene locus that affects another gene locus on another chromosome, generally through the production of regulatory protein Compare Cis Acting Locus.

Trans arrangement. Linkage of the dominant allele of one pair and the recessive of another on the same chromosome.

Trans position. The position in which the two defected (mutant) sites occur one each in the two homologous DNA molecules. If the two homologous DNA molecules. If the two defective sites belong to the same locus, complementation does not occur. In `cis' position the two defective sites occur in one DNA molecule and none in the other resulting in complementation.

Trans. The term proposed by Haldane (1941), by analogy with chemical isomerism, to describe a double heterozygote in repulsion phase, that is, with the dominant factors derived from opposite parents. The term is used chiefly for closely linked mutants, to which it was first applied by Pontecorvo (1950). The trans configuration for two recessive mutants, 1 and 2, is +2/1+, where + stands for the wild type, and the line separates the contributions from the two parents. (Cf. Cis).

Transaldolase. An enzyme involved in carbohydrate metabolism that catalyzes the transfer of a dihydroxyacetone residue from a keto to an aldo sugar.

Transaminases. Enzymes catalyzing transfer of amino group from Â amino to Â keto acids; also called aminotransferases.

Transcribed spacer. Spacer sequence lying between the tandem repeats of the rRNA genes of certain eukaryotes, which is transcribed but of which the transcript is subsequently discarded.

Transcript. The RNA product of a gene. The primary transcript produced by RNA polymerase must often be processed or modified in order to form the mature, functional mRNA, rRNA, or tRNA species.

Transcriptase. An enzyme that catalyzes the flow of genetic information from DNA to RNA.

Transcription factor. Protein important in effecting transcription of a specific gene sequence.

Transcription or genetic transcription: DNA dependent RNA synthesis. The process by which cellular RNA molecules are synthesized as determined by homologous DNA sections where their sequence recurs in complementary form (the template or transcription strand).

Transcription unit. A cluster of genes that determines functionally related RNAs or proteins.

Transcription. The process whereby the message carried in the DNA is transcribed to the RNA through complementary base pairing of the RNA nucleotides with the template DNA. The resulting RNA carries the message to make the specific protein coded in the DNA segment, and it is called messenger DNA (mRNA)

Transcriptional control. The regulation of protein synthesis by regulation of the formation of its mRNA.

Transdetermination. Change in the state of determination of a cell or tissue.

Transdifferentiation. Change in the state of differentiation of a cell or tissue. It is phenomenon of relatively infrequent occurrence, but is known to occur in *Hydra,* and at least in experimental situations in plants.

Transduced element. Fragment of DNA transferred from one cell to another in the process of transduction.

Transduction. (Latin *transducere*, to lead of to pass across). A recipient bacterium's incorporation into its chromosome of genetic material obtained from a donor bacterium. The transfer of such material is performed by a bacterial virus.

Transfection, transduction, transformation. These terms are all used to mean ways of getting DNA into cells, usually animal or bacterial cells.

Transfection. Strictly, this means carrying a piece of DNA into a cell as part of a virus particle. For mammalian and plant cells, it is used more generally to mean almost any way of getting DNA into a cell.

Transduction. A relatively specialist genetic technique, this means transferring a piece of DNA from one organism to another via natural DNA exchange processes. It occurs almost exclusively in bacteria, and is a method of genetically engineering large pieces of DNA such as the *Agrobacterium tumefaciens* Ti plasmid.

Transformation. For bacteria, this means getting the bacterium to take up DNA that the experimenter has added to its medium. The bacteria that are able to do this are called `competent'. Demonstrating transformation was one of the key proofs that DNA was the genetic material. For mammalian cells, transformation means changing the cell from one whose growth is limited by its neighbouring cells into one whose growth is limited only by the media available to it. Transformation is a step in the development of cancer cells, and is also a crucial part of generating an `immortalized' cell line. Because these two meanings of `transformation' grew up beside each other, genetic engineers who are manipulating mammalian cells often say that they **transfect** the cells with DNA, rather than `transform' them, even if what they actually do is add naked DNA to the cells. There are several common methods used to place `naked' DNA (i.e. DNA that is not encapsulated into a virus particle, a liposome, or some other carrier system) into cells. Bacterial cells. Bacterial cells that are **competent** (i.e. in a suitable physiological state, which may be achieved by growing them in the right way and suspending them in he right buffer) will spontaneously take up DNA from the solution around them. The common factor involved in usually he need for magnesium salts in their medium. Bacterial protoplasts can also be transformed by fusing them together in the presence of DNA. This can be done using polyethylene glycol (PEG). The cells membranes join up in the presence of PEG forming many-cell masses, ad some of he external solution, containing the DNA, gets trapped inside the cell in the process.

Transfection. The infection of host cells with naked viral DNA. The transfer, to cells in culture, of a gene (s) from another cell.

Transfer factors. Two proteins required for peptide bond formation on ribosomes. Transfer factor 1 participates in binding of charged tRNAs to ribosomes. Transfer factor II moves peptidyl tRNAs from the "A" to the "P" ribosome.

Transfer RNA (tRNA). A category of relatively small RNA molecules (about 75 nucleotides). Each kind of transfer RNA is able to accept a particular activated amino acid from its specific activating enzyme, after which the amino acid is added to a growing polypeptide chain.

Any of at least twenty structurally similar species of RNA, all of which have a MW-25,000. Each species of tRNA molecule is able to combine covalently with a specific amino acid and to hydrogen bond with at least one mRNA nucletode triplet. Also called Adaptor RNA.

Transfer room: A small room, sometimes sterilized internally by a bactericidal lamp (u.v. irradiation) and provided with clean (filtered) air; employed for sterile transfer work.

Transfer. (1) Culture initiation; the first placement of an explant in culture. (2) Subculture; relocation of cultures to fresh medium. (3) Chamber laminar air flow cabinet or hood (transfer hood) or room (transfer room) in which this is accomplished. (4) The act or process of transferring.

Transferable. A term used to describe plasmids which are able to promote their own transfer from bacterium to bacterum.

Transferases. A very large group of enzymes that catalyze the transfer of chemical groups from one substrate to another; e.g., glucose kinase transfers a phosphate group from ATP to glucose.

Transferrin. A beta-1 globulin of blood serum concerned with binding and transportation of iron; also called siderophilin, siderophyllin.

Transformant. Cell or organism that has been transformed by exposure to foreign DNA.

Transformation. The genetic modification induced by the incorporation into a cell of DNA purified from cells or viruses. Mechanism for transfer of genetic information in bacteria in which pure DNA extracted from bacteria of one genotype is taken in through the cell surface of bacteria of a different genotype and incorporated into the chromosome of the recipient cell. By extension, the terms has come to be applied to phenomena in other organism in which specific genetic alterations have been produced by treatment with purified DNA from genetically marked donors.

Transformer gene tra in Drosophila. A recessive gene which when homozygous *tra/tra* causes a female fly to be phenotypically male. Such "males" are sterile.

Transgenation. A heritable change taking place at a single gene locus, presumably caused by chemical alteration of the individual gene; a mutation proper, point-mutation or genovariation.

Transgenic disease models. One application of transgenic animals is to model human diseases. Having an animal model of a disease can be very valuable when human patients for a disease are rare, when it is impossible to find them before the disease is well established and so the early stages cannot be studied, or when it is not ethical or practical to do studies on humans. However, quite a few human diseases are not mimicked accurately by any animal model. Transgenic technology seeks to create animals, especially ice, that get diseases that are in some specific way characteristic of a human disease. These can then be used for screening for potential new therapies or drugs. Among the models used are:

1. ***Humanized*** (humanoid) mice for AIDS research. True transgenic mice with the human CD4 gene can be infected with the AIDS virus. Another model (SCID-Hu mice) have no functional immune system of their own, but have human immune cells inserted into them to make an immune system that is

susceptible to AIDS. (These are properly called chimeric animals, because they are a mixture of cells or tissues from several different animals.) SCID mice can be made in several ways which knock out their immune system, including by massive whole-body irradiation and genetic engineering of them to include a toxin gene that is expressed at high levels in their lymphocytes, so destroying them.

2. ***Models for diabetes***, and several other diseases in which specific cells are absent or do not work properly. A toxin gene is placed in the animals, spliced on to a promoter sequence that only expresses that toxin gene in one particular tissue (in the case of diabetes, in the beta cells of the pancreas). The toxin then kills those cells off, leaving the rest of the animal healthy. Such genetic constructs are called toxigenics.
3. ***Models for cancer.*** Models for cancer usually have an oncogene inserted into them, so that they develop a specific cancer at an anomalously high rate (see Oncomouse.).
4. ***Models of immune function.*** A key aspect of the healthy immune system is its ability to tell the normal constituents of the body from other, potentially invading materials. A wide range of diseases arise from a failure of this mechanism. Transgenics are being used to find out how the immune system `learns' to discriminate self from non-self, both by introducing foreign protein genes into mice and by creating toxigenics that lack certain sets of lymphocytes. These studies have implications for many diseases, such ad diabetes (which has a autoimmune component), arthritis, allergies, and multiple sclerosis.

Transgenic. A transgenic organism is one that has been altered to contain a gene from another organism, usually from another species. While this would suggest that any genetically engineered organism could be called `transgenic', the term is usually only applied to animals. Bacteria and yeasts are always called `genetically engineered', and with plants it is an even choice which term to use. Creating transgenic plants is a relatively mature science, and is discussed under plant genetic engineering.

Transgression. The origin in the F_2 or later generations of segregation products that are more extreme than either of the parents.

Transgressive segregation. The appearance in a segregating generation (F_2 backcross, etc.). of an individual or individuals falling outside the limits of variation set by the parents and F_1 of the cross in one or more characters. Cf. Heterosis.

Transgressive variation. (Trangressive segregation). The variation appearing in F_2 segregation of a cross where the progenies show more extremes of variation than in present in the parents. This is due to complementary and cumulative effect of favourable genes originally distributed between the parents.

Transgressive. It is a phenomenon in which some individuals in the progeny show characters exceeding in value than those shown by either of the parents. The transgressive inheritance is usually seen in crosses involving quantitative characters.

Trans-heterozygote. A heterozygote of two linked genes with a mutant and wild type allele linked, *a+/+b*. This is a more appropriate term than "repulsion phase" of linkage, since like alleles do not repel.

Transient polymorphism. Temporary polyorphism observed while one adaptive type is replacing another.

Transition mutation. A base substitution mutation in which a purine (A or G) replaces a purine, or a pyrimidine (T or C) replaces a pyrimidine. (Contrast with transversion mutation).

Transition state. An activated form of a molecule, in which the molecule is capable of undergoing a chemical reaction.

Transition. A mutation caused by substitution of one purine for another or one pyrimidine for another.

Transitional. (1) A term applied to organs or plantlets undergoing phenotypic change; as during acclimatization after *ex vitro* transplant. Persistent organs (from culture) do not change substantially; new organs produced after transplant range from culture type to control type in anatomy and physiological performance, and so are termed the transitional organs. (2) An area between the stem and root of plants of intermediate tissue arrangement, termed the transitional region.

Transketolase. An enzyme of carbohydrate metabolism that catalyzes the transfer of a two carbon ketol moiety from a keto sugar to an aldo sugar. The products of the reaction are an aldosederived form but two carbon atoms shorter than the donor ketose, and a ketose derived form, but two carbon atoms longer than the acceptor aldose. Transketolase requires thiamine pyrophosphate and magnesium. The reaction is freely reversible.

Translation. The process by which the message contained in the mRNA directs the synthesis of a specific protein.

Translational amplification. Use of the same mRNA molecule many times, allowing the synthesis of large amounts of polypeptide from a single mRNA and therefore from a singe gene product.

Translational control. Regulation of protein synthesis by the rate of its translation on the ribosome.

Translocase. A membrane transport system. The term is also applied to an enzyme causing a movement, such as the movement of a ribosome along an mRNA.

Translocation. (1) In genetics, a rare mutational event that moves a portion of a chromosome to a new location, generally on a nonhomologous chromosome. (2) In vascular plants, movement of solutes in the pholem.

Transmination. Transfer of an amino group from an amino acid to a keto group of metabolite and the formation of a new amino acid.

Transmission (Latin *transmittere*, to send). Refers to the process of sending a substance from one location to another; particularly pathogens.

Transmission genetics. Aspect of genetics that is concerned with transmission of genes from parents to progeny.

Transmutation. Change of one element into a different element due to loss of some subatomic structure in radioactive decay.

Transomics. This is an injection technique, but instead of injecting pure DNA its practitioners dissect out sections of chromosomes under the microscope and inject them. As chromosomes are only a thousandth of a millimetre or so long, (and much thinner) this is not a profession for myopics or those with shaky hands.

Transovarial. Refers to the transmission of pathogenic microorganisms form mother to offspring while they are still in the eggs within her ovary. cf. Transovum.

Transovum. Refers to the transmission of pathogenic microorganisms from one generation to the next by way of external contamination on or about the egg. cf. Transovarial.

Transpiration (L. *spirare:* to breathe). The evaporation of water from plant leaves and stem, driven by heat from the sun, and providing the motive force to raise water (plusions) from the roots.

Transplacement (gene transplacement, replacement, gene replacement). A technique, originally developed in yeast molecular biology for exchanging a DNA sequence on a recombinant plasmid for the equivalent sequence in one of the host's chromosomes. Thus a wild type copy of a gene can be replaced by a mutant copy or vice versa. The host organism is transformed with an integrative plasmid. The recombinant plasmid integrates itself into the host's chromosome by homologous recombination, thereby creating a duplication of the sequence of interest. A subsequent intrachromosomal recombination event loops out the integrated plasmid together with the copy *(x)* of the gene originally contained in the chromosome. The copy (*X*) of the gene originally contained in the recombinant plasmid is left behind and is said to have transplaced *(x).*

Transplant shock. Refers to the stress involved when Stage II or Stage III cultures are transplanted to soil (Stage IV) many or all of the regenerated shoots or plantlets may die if suitable care is not taken to acclimatize them gradually to the soil environment particularly prevention of water stress.

Transplant. A part of a tissue or organ transferred from one organism to another in grafting or organ transplantation. Transplantation is putting a new organ, tissue, or (more recently) cell into a patient. Traditionally the new organ or tissue has come from another human being. Biotechnological approaches are allowing animal sources and synthetic tissues to be used.

Transplantation antigen. An antigen that initiates the immune response that leads to the rejection of a tissue or organ transplant.

Transplantation. The transfer of tissue or organ from one part of an animal to another part or from one individual to another.

Transport (across bacterial membranes). Only small, non ionic molecules (especially lipophilic ones) can diffuse passively across the hydrophobic barrier of the cell membrane; such molecules includes oxygen, carbon dioxide, ethanol, and certain amino acids. Most substances therefore enter or leave the cell as a result of specific transport mechanisms which may or may not require metabolic energy, and which are thought to involve specific protein `carriers located within the cell membrane.

Transport medium. Any (liquid) medium used for the transportation and/or storage of material from which the isolation of particular organism (s) is subsequently to be attempted. The purpose of such a medium is to maintain the viability and/or infectivity of the organism (s) during the delay between collection and culture of the specimen.

Transposable element (transposon). A segment of DNA that can be moved from one position in the genome to another.

Transposition pseudoallele. Same as trans heterozygote. Pseudoalleles are very closely linked.

Transposition. The shift of a gene or chromosome segment to a new position of the chromosome.

Transposon. A transposon is a genetic element that can move around the genome. Most genes stay where they are with respect to other genes, unless mutation rearranges the genome in which they reside. Transposons break this rule. They are able to create copies of themselves elsewhere in the genome, or even in other genomes if they are present in the same cell. Thus, for example, a transposon might copy itself out of a bacterial genome and into the genome of a bacteriophage as the phage infests the bacterium. Some transposons splice themselves out of their original site to do this, but most simply copy themselves, so the end result of the copying process is two copies of the transposon where one existed before. The process of a transposon moving is called transposition. It has been harnessed in a variety of ways by geneticists and genetic engineers to move genes around in bacteria and, to a lesser degree, in plants. Many transpososn carry useful genes as well as being `selfish' DNA that propagates itself around the genome. Most antibiotic resistance genes are carried on transposons in some bacteria, as are genes for things like heavy metal resistance. The way that many transposons move is reminiscent of the way that retroviruses reproduce, in that the transposon is transcribed on to an RNA, which is then copied back into the genome as a DNA. Because of this similarity, such transposons and retroviruses and sometimes grouped together and called retrotransposons.

Transsexual. A person who has adopted the appearance of the opposite sex with the aid of surgical procedures and hormone therapy.

Transversion mutation. A base substitution mutation in which a purine (A or G) replaces a pyrimidine (T or C) or vice versa. (Contrast with transition mutation).

Transversion. The term proposed by Freese (1959) for a mutation caused by the substitution of a purine for a pyrimidine, and vice versa, in DNA or RNA. (cf. Transition).

Trap crop. A small planting of a susceptible and highly attractive host, planted early in the season, or removed in space from the main crop, in order to divert attack and infestation by pests and allow for their easy destruction.

Traumatin. $C_{11}H_{17}O_4N$; a plant wound hormone which stimulates cell division. cf. Allantoin.

Trechalase. An enzyme that catalyzes the hydrolysis of trehalose to two molecules of glucose.

Treponema pallidum immobilization (TPI). Test for syphilis, using patient's blood serum with live spirochetes; in a positive test, the spirochetes lose motility due to specific antibodies in the blood serum.

Treponema pallidum: Spirochete, causative agent of syphilis.

Treponematoses. All of the disease caused by species of **Treponema.**

Triacylglycerol. An ester of glycerol with three molecules of fatty acid also called a neutral fat.

Tribe. A combination of one or more families, the foundresses of which trace in the female line to a common ancestress which is the foundress of the tribe.

Tricarboxylic acid cycle. See citric acid cycle.

Trichlogyne. Specialized hypha that extends form the fruiting body of some fungi and forms a receptive organ for the male gamete.

Trichloracetic acid (TCA). Acid widely used for protein precipitation, especially during acid preparation.

Trichocyst (Gr. *trichos:* hair + *kystis:* cell). A thread-like organelle ejected from the surface of ciliates, used both as a weapon and as an anchoring device.

Trichome (hair). A uni or multicellular outgrowth arising solely from the epidermis. Root and leaf hairs and glandular hairs are types of trichome.

Trickling filter. A secondary treatment process where sewage is trickled over a bed of rocks so that bacteria can break down organic wastes.

Trigeneric hybrid. An individual produced by crossing two distinct genera and crossing the F_1 with a third genus.

Triglyceride. A simple lipid in which three fatty acids are combined with one molecule of glycerol.

Trihybrid cross. A cross between individuals that have different alleles at three gene loci.

Trihybrid. An individual which is heterozygous for three pairs of genes.

Triparental recombinant. Recombinant Bacteriophage (phage) that can be recovered from phage injected bacteria which contain marker genes derived from three different original infecting phages.

Triple DNA. Most introductory textbooks will tell you that RNA is `single stranded' and DNA is `double stranded', i.e. that DNA consists of ahelix of two strands wrapped round each other. However, it has been known for a while that RNA can be triple stranded, and recently triple-stranded DNA has been demonstrated too (also sometimes called triplex DNA). It has several potential applications. The `third strand' of triple DNA binds to the other two by specific base pairing, and so it can be used as a reagent that recognizes a specific DNA sequence. (The base pairing `rules' are rather different from those in normal double-stranded DNA.)

This can be used to detect a specific sequence in a double-stranded target (rather like a conventional DNA probe detects a sequence in a single-stranded target DNA) or to block gene activity. In this second role, a short oligonucleotide is made that binds to a gene, preventing the RNA polymerase from opening it and copying it on to RNA. This has the same effect as an antisense agent made against that gene product, but acts at the DNA level. So far, this has worked to a limited extent in the `test-tube' only. This is an application of aptamer technology. Triplex DNAs can also be used to build very specific DNA-cutting reagents. If one is linked to a molecule that cuts DNA, the third strand can be shown to act as a sequence-specific nuclease, i.e. reagent that will cut the DNA at (or just next to) a very specific site. Several `artificial nucleases' of this sort have been made. DNA can also be forced into a left-handed helical form, called Z-DNA. This structure occurs naturally in gene promoters, and so making it appear (or disappear) may be another route to blocking gene function.

Triple fusion. The fusion between the two polar nuclei and a male gamete, which takes place in the embryo sac and which gives rise to the endosperm.

Triple vaccine. A vaccine containing diphtheria and tetanus toxoids and killed cell of *Bordetella pertussis* (pertussis vaccine).

Triple X. Trisomy for the *X* chromosome.

Triplet code. Three nucleotide code specifying for one amino acid.

Triplet. A group of three successive nucleotides in RNA (or DAN) that, in the genetic code (which see), specifies a particular amino acid in the synthesis of polypeptide chains.

Triplex. It is the condition in an autotetraploidy in which three genes are dominant and one recessive.

Triplicate genes. Three genes, any one of which produces the same phenotypic effect, e.g., seed coat colour in wheat may be produced by R_1, R_2, R_3.

Triplo polyploid. A relatively triploid derivative of an allopolyploid. (Janaki Ammal 1941).

Triplo. Prefix denoting trisomic individual where the identity of the extra chromosome is known, e.g., triplo IV Drosophila, or triplo 21 human beings.

Triploblastic. Descriptive of an animal that develops from three embryonic cell layers; the ectoderm, endoderm and mesoderm. Most metazoa are triploblastio.

Triploid. A polyploid with three haploid sets (*3n*) of chromosomes.

Tris borate buffer. A buffer consisting of Tris base and boric acid. It is one of the common buffers for the electrophoresis of DNA and RNA through agarose and polyacrylamide gels. It consists of 89 mM Tris base, 89 mM boric acid, and 2mM EDTA (pH 8.0).

Tris. (Tris (hydroxymethyl) Amino methane). Compound widely used as a buffering agent in biological media and solution.

Trisomic inheritance. Abnormal inheritance of a gene located in the extra chromosome of a trisomic.

Trisomic. An individual showing an extra chromosome in addition to the normal chromosome set expressed by $2n+1$. If the gametic chromosome number (n) in an organism is 12, then 12 different types of trisomics may occur each differing from the other in respect of the chromosome present in triplicate. This type is known as primary trisomics. The study of trisomic ratios provides an useful tool in assigning a gene to its proper linkage group. The trisomics may be secondary in which the extra chromosome is an iso chromosome and tertiary in which the extra chromosome is a translocated one.

Trisomy. The state of having one, or more, chromosomes present three times in the somatic cells, the other chromosomes being present only twice.

Trition detergent. Uncharged (non ionic) detergent widely used in cell biology for solubilizing the proteins of cell membranes. The commonest is Triton X 100.

Tritium (^{3}H). A radioactive isotope of hydrogen of atomic mass 3 that undergoes radioactive decay with the emission of a betaparticle and the formation of a stable isotope of helium.

Trivalent. A group of homologous chromosomes paired with each other along certain chromosome segments. Trivalents may occur from zygotene to first metaphase of meiosis and are best studied at diakinesis and first metaphase.

tRNA gene. Gene that codes for transfer RNA. The total number of these genes in the genome varies from 360 in yeast to over 1000 in humans. Each of the 20 common amino acids have specific tRNA genes coding for molecules with specificity of binding to their coding triplet on the message and to their own structure.

tRNA isoacceptors. Differing species of tRNA that bind the same amino acid but, because of coding degeneracy, have differing Anticodon triplets.

tRNA precursor. A form of tRNA that is about 20 percent larger than the mature molecules and contains extra sequences at both the 5' and the 3' ends. In prokaryotes, these are removed by specific enzymes, RNase *P* and RNase *Q*, which cleave at the 5' and 3' ends, respectively.

Trochophore (Gr. **trochos:** wheel + **phoreus:** bearer). The free-swimming larva of some annelids and mollusks distinguished by a wheel-like band of cilia around the middle, and indicating an evolutionary relationship between these two groups.

Trochophore (Gr. *trochos:* wheel + *phoreus:* bearer). The free-swimming larva of some annelids and mollusks distinguished by a wheel-like band of cilia around the middle, and indicating an evolutionary relationship between these two groups.

Trophic level. The position of a species in a food chain indicates which species it consumes and which consume it.

Trophic nucleus. The larger of two nuclei in a cell; vegetative nucleus; macronucleus; trophonucleus.

Trophoblast. The cells of the outer wall of the mammalian blastocyst that is attached to the wall of the uterus and it forms the part of the early placenta that is in closest contact with the maternal tissues.

Trophochromatin. Chromatin which governs the nutritive functions of the cell.

Trophoplasm. `Vegetative' or nutritive protoplasm. Unmodified cytoplasm as distinct form kinoplasm.

Trophozoite. The vegetative cell of a protozoan, i.e., active feeding stage.

Tropic hormone (tropin). A peptide hormone that stimulates its target gland to secret its hormone e.g., thyrotropin of the pituitary stimulates secretion of thyroxine by the thyroid.

Tropomyosin. A muscle protein which associates with acting to from long thin fibers and which plays a role in the regulation of muscle contraction.

Troposim. In plants, growth toward or away from a stimulus such as light (phototropism) or gravity (geotropism). It is a growth response of an organism following an external stimulus.

Troposphere. The atmospheric zone reaching upward approximately 17 km in the tropics and substropics but only to about 10 km at higher latitudes. The zone in which virtually all the water vapor in the atmosphere is located.

Trp gene. Gene concerned with Tryptophan biosynthesis in *E. coli.*

Trp promoter. Promoter of the tryptophan Operon in *E. coli.*

Trub. The coagulate protein that settles out of freshly boiled worn tubercle a nodule.

True breeding. Production of progeny which like their parents, express a particular Genotype in their Phenotype.

True to type: Applied to a plant or propagation source this term denotes correct cultivar identification and lack of variation in productivity or performance. Verification is determined visually by an expert or through biochemical, serological or other means.

Trypan blue ($C_{34}H_{24}N_6Na_4O_{14}S_4$ mw 960.83): A dye used in a quantitative method to assay cell viability. The method is based on the ability of live protoplasts to exclude the dye while dead protoplasts cannot.

Trypan red. An acidic, red diazo dye, it is used in the early days of chemotherapy for the treatment of trypanosomiasis.

Trypsin. A proteolytic enzyme (MW-23,800), secreted by the pancreas that cleaves peptide chains where the basic amino acids arginine and lysine appear.

Trypsinogen. The inactive from of trypsin, found in pancreatic tissues and pancreatic juices. Trypsinogen is converted into the active form by action of the enzyme enteropeptidase (enterokinase).

Tryptamine or 3-(2-aminoethyl) indole ($C_{10}H_{13}N_2$ mw 161.23): An occasional addition to plant tissue culture media as a source of reduced nitrogen.

Tryptophan (Trp, $C_{11}H_{12}O_2N_2$ mw 204.23): An amino acid precursor of indole compounds such as IAA. Occasionally added to plant tissue culture media.

T-system. A set of transverse tubes that penetrates skeletal muscle fibers and terminates in the sarcoplasmic reticulum. The T-system transmits impulses to the sacs, which then release Ca^{2+} to initiate muscle contraction.

TTP. Thymidine triphosphate. The corresponding ribo derivative does not exist in nature. Labeled TTP is, therefore, used to specifically label DNA.

Tube dilution technique. A quantitative technique in which the chemical agent in question is diluted in the growth medium in a series of two-fold dilutions. All tubes are then inoculated with the organism to be studied.

Tube foot. In echinoderms, a part of the water vascular system. It grasps the substratum, prey, or other solid objects

Tube nucleus. The nucleus which occurs in any growing pollen tube and which does not take part in fertilization but is concerned with regulating the growth and behavior of the tube.

Tuber (L. *tuber:* swelling). A short, fleshy underground stem, usually much enlarged, and serving a storage function, as in the case of the potato.

Tubercle. A nodule, a specific lesion in certain diseases such as tuberculosis.

Tuberculin test. A test for detecting delayed hypersensitivity to the tubercle bacillus. The various forms of the test include the Moro test (in which Tuberculin is smeared over a small area of the skin) and the Mantoux test (in which tuberculin is injected intracutaneously). A positive test is indicated (within a few days) by a characteristic local reaction. A positive test is not proof of active infection by *Mycobacterium tuberculosis.*

Tuberculin. An extract of the tuberculosis bacilli capable of eliciting an inflammatory reaction in the animal body which has been sensitized by the presence of living or dead tubercle bacilli. Used in a skin test for tuberculosis.

Tubulin. The protein subunit of microtubles. Globular protein subunits (MW's 55,000 and 57,000) whose regular helical packing forms the hollow, cylindrical microtubules.

Tumble tube. A glass tube closed at both ends with a side neck opening which is commonly attached to a slowly revolving platform. Suspension cultures are agitated and aerated as medium flows form one end of the tube to the other as it is inverted.

Tumor inducing principle (TIP): The plasmid carried by *Agrobacterium tumefaciens*, the crown gall organism. Through incorporation into the host genome the host tissue is transformed into tumor tissue.

Tumor virus. A virus that induces the formation of a tumor.

Tumor. Mass of tissue resulting from uncontrolled growth.

Tumour marker. A tumour marker is nay molecule that shows that a cancer is present. Usually each type of marker is produced by only a few types of cancer, so as well as showing that a cancer is present it helps to identify the type of cancer, and hence the most suitable herapy. Tumour markers can be used for diagnosis or, potentially, as targets for biopharmaceutical drugs such as immunotoxins. Tumour markers fall into two categories. The first type are the products of oncogenes, and hence their presence represents part of the reason that the cell is a cancer cell to start with. The others are apparently incidental to the cause of the cancer, but are always found associated with a particular type of cancer. Such proteins are usually made in only a few cell types in the healthy body, but cancer cells make them in much larger amounts (such as the hCG, human chorionic gonadotrophin, usually only made by the placenta in early pregnancy but also produced by some cancers). Quite a few markers, like CEA (Carcino-Embryonic Antigen), are proteins normally found only in the fetus (sometimes called oncofetal proteins). In addition there are a wide range of antigens (i.e. proteins to which antibodies bind) that have been identified by a monoclonal antibody as being associated with a type of cancer, but whose normal function is obscure. A number of them are glycoproteins or carbohydrates: cancer cells often add sugar units to proteins in a slightly different order from normal cells, so creating different glycoforms of those proteins. It is these differences between the glycoforms that is detected as a `marker' by the antibody. Polymorphic endothelial mucin (PEM) is an example- the protein is covered in sugar residues in normal tissue, but exposes some of its polypepide chain in cancers. Although tumour markers can be used as diagnostic for cancer, they are not infallible. What doctors want to know is whether a cancer is likely to be life threatening. What the tumour marker tells them is that a particular cell type is present. These two things may not be the same: 50% of American men aged 50 have PSA-positive `tumour' cells in their prostate, but in most cases the cancer poses no threat; only 3% of American men die of prostate cancer. Thirty per cent of those with positive PSA tests apparently have no cancer cells at all. So tumour markers have to be used carefully, especially for large-scale screening programmes. Tumour markers are also targets for tumour vaccines (also called cancer vaccines). If the marker is linked to another protein or injected with an adjuvant, it could stimulate the immune system to make antibodies that recognize the marker on the cancer cells, and so destroy the cancer. This neat idea has not been very successful, because the body becomes tolerant of the cancer during its development. If it did not, it would have destroyed it before. A compromise approach is to use a monoclonal antibody against the marker as an immunotherapy.

Tunica corpus theory. A theory of apical organization and development that distinguishes two separate tissue zones, the tunica and the corpus, in the apex of a flowering plant. The tunica is made up of one or more peripheral layers in which division is mostly anticlinal. The corpus is the inner area of tissue in which cell arrangement and division is irregular. It is suggested that the epidermis originated that the epidermis originates in the tunica region and that the other stem tissues may arise form either the tunica or the corpus, depending on the species.

Turbid. Cloudy.

Turbidimetry. A method of estimating bacterial growth or populations by measurement of the degree of opacity (or turbidity) of the suspension.

Turbidostat: An open continuous culture system wherein the inflow of fresh medium is controlled by the **turbidity** of the culture, a function of the amount of cell growth. Balancing the fresh medium inflow is a regulated outflow of cells and spent medium, restoring the original turbidity level.

Turgor. The state, in a plant or prokaryote cell, in which the protoplast is exerting a pressure on the cell wall owing to the intake of water by osmosis. The cell wall, being slightly elastic, bulges but is rigid enough to prevent water entering to the point of bursting. The cell is then said to be turgid. Turgidity is the main means of support of herbaceous plants.

Turner's syndrome. An abnormality in human being; individuals are phenotypically females, but have rudimentary sexual organs and mammary glands. Such individuals have but one *X* chromosome and *Y*, with a total of 45 chromosomes instead of the normal 46.

Turnover number (of an enzyme). The number of molecules of a substrate transformed per minute by a single enzyme molecule, when the enzyme is working at its maximum rate.

Tween 20. The brand name for a wetting agent or surfactant which breaks the surface tension of tissues. Commonly added to disinfecting solutions to make them more effective.

Tween 80. (polyoxyethylenesorbitan monooleate). Non ionic detergent. It is sometimes used as an emulsifying agent in the preparation of microbiological tissue culture (such as thioglycolate broth).

Twins. (1) Two embryos developed simultaneously within one uterus in mammals, within one vitelline membrane in other vertebrates or within one test in phanerogams. Similarly Triplets, Quadruplets, etc., also Dizygotic, Monozygotic, Polymbryony. (2) Two individuals that are born at the same time. Monozygous (identical) twins arise from the same fertilized egg (zygote) and have therefore the same genotype. Dizygous (nonidentical, or fraternal) twins arise from separate fertilized eggs (zygotes) and have different genotypes.

Twisting number. Number of base pairs in a DNA duplex sequence, divided by the number of base pairs for every helical turn.

Two dimensional electrophoresis. Method for enhancing the separation of proteins and other macromolecules that are of similar size and charge and not resolved by Electrophoresis in one dimension.

Two factor cross. A genetic recombination experiment involving two markers (e.g., a + b^+ X ab).

Two point cross. Genetic cross in which either one parent is a double mutant (--) and the other is wild type (+ +) with respect to two markers being followed in the cross (an `in coupling' cross), or each parent carries one of the two markers (+-X-+) being followed in the cross (an `in repulsion' cross).

Two stage digester. An anaerobic digester in which the acidogenic (acid forming) reactions are from the methanogenic reactions.

Two step ligation. A gene cloning protocol in which linearised vector molecules and insert DNA fragments are mixed and ligated at high DNA concentration. This favours the joining of vector and insert DNA. The DNA solution is then diluted for the second step of the ligation. This favours the circulirization of recombinant molecules.

Ty element. Transposable Element of yeast. About 30 transposable elements exist in each haploid genome.

Tylose. An in growth from a parenchyma cell into an adjacent tracheid or vessel through a paired pit. Tyloses are often found in injured tissue, older wood, and below an abscission layer, and can completely block the conducting vessel.

Tympanic membrane (Gr. *tympanum:* drum). The eardrum.

Tyndallization. Fractional sterilization; exposure to live steam for 30 minutes each day for three successive days to kill vegetative cells.

Type culture. A particular species considered representative of the characteristics of the species and used as a reference culture.

Tyrosinase. Enzyme involved in the production of Melanin from Tyrosine. Tyrosine is first converted to DOPA, then DOPA is oxidized to form DOPA quinone, and these molecules polymerize spontaneously to form melanin.

U

U. Symbol for uridine and uracil in molecular biology.

U orientation. Descriptive of the situation in which the vector and the inserted fragment of DNA have the opposite orientations.

U protein. Hypothetical protein thought to regulate the transition of cells from *G0* to *G1* phase of the cell cycle, and thus inevitably into *S* phase. The idea would be that the concentration of this unstable (*U*) protein would have to exceed a threshold level for triggering progression through the cycle, and that this would only happen if the cell had adequate access to growth factors or to nutrients. Also known as Trigger Protein.

U1 RNA. U1 RNA thought to act as an external guide sequence to facilitate the correct splicing of introns from mRNA precursor molecules. It contains sequences at its 5' end which are complementary to the conserved intron sequences near the intron/exon junction.

UAA. A codon that does not code for any amino acid. It is termed a nonsense codon and its function is to stop translation.

UAS, upstream activator sequence. A part of a eukaryotic promoter some distance away from both the transcriptional start site and the tata box. The UAS affects the extent of transcription from a given promoter without affecting the site of transcriptional initiation.

Ubiquinone (coenzyme Q). Quinone with a long isoprenoid tail; the number of isoprene units depends on the species in which the quinone occurs. In mammals there are 10 such units (hence uniquinone-10 or coenzyme Q 10). Ubiquinone is a membrane-soluble component of electron transport chains which mediate hydrogen transfer between membrane-bound dehydrogenases which contain flavoprotein and iron-sulphur proteins, and membrane-bound cytochrome complexes containing b-type cytochrome. The coenzyme alternates between the quinone and quinol forms during hydrogen transfer and the free radical semiquinone is probably an intermediate in the reaction.

Ubiquitin. Acidic protein widely, and perhaps universally, distributed among eukaryotes. It binds strongly to chromatin and may have a structural role in chromatin conformation. Bindings specifically to histone H_2A by a covalent linkage, the complex has come to be known as protein A_{24}.

Ubiquitous. Occurs everywhere, as do bacteria in the environment.

UDP. UDPG is also an intermediate in the metabolism of galactose.

UI RNA. Species of small nuclear RNA (snRNA), of 165 bases in length, that is complementary to the donor and acceptor splice junctions and is therefore believed to be involved in the excision of introns and splicing of exons. The Ul RNA is associated with a specific protein which is involved in intron excision, and together they form the Ul ribonucleoprotein particle called either the Ul particle or snRNP.

Ulcer. 1. A rounded or irregularly shaped area of inflammatory tissue destruction in the epithelial lining of a surface. 2. Circumscribed area of inflammation found in epithelial lincontiguous vessels.

Ulceration. The process of ulcer formation.

Ultra-centrifugation (Prefix ultra plus Latin *centrum*, centre and Latin *fugere*, to flee away from). Refers here to the process of sedimentation (q.v.) of suspended particles subjected to very high centrifugal force in a centrifuge (q.v.). Since different kinds of particles sediment at different rates, the process offers a convenient means for ascertaining the homogeneity of a suspension. Molecular weights and sizes of the sedimenting particles can also be determined from the rates of sedimentations.

Ultracentrifuge. A high-speed centrifuge that can attain speeds up to 60,000 *rpm* and centrifugal fields up to 500,000 times gravity and thus is capable of rapidly sedimenting macromolecules.

Ultrafiltration fermenter. A continuous fermentation system run in conjunction with continuous ultrafiltration, such that the low molecular weight product of the reaction is removed, thus reducing the effect of feedback inhibition.

Ultrafiltration. A method for the removal of all but the very smallest particles, e.g., viruses, from a fluid medium.

Ultrahigh-efficiency filter. Having a minimum efficiency of 99.97% when tested with 0.3 micron "DOP" particles; generally employs pleated glass or asbestos paper. Also commonly referred to as "Absolute filter, HEPA filter," or "superinterception filter".

Ultramicroscopic. Visible with the electron microscope; too small to be seen with the conventional light microscope.

Ultrasonic cell disintegration. A technique used to disrupt cells by bombarding them with ultrasonic waves.

Ultrasonic cleaner. A device to include high frequency vibration of materials, removing adhering substances from surfaces by mechanical action. This device is useful for cleaning glassware and for disinfecting plant material.

Ultrasonic waves. High frequency sound waves, used for destruction of bacterial cells.

Ultrasonic. Descriptive of a sound wave, or regular periodic disturbance in a medium, that is above the audible limit around 20,000 hertz.

Ultrasonography. Visualization of a foetus *in utero* by ultrasound in order to check on possible gross malformation or difficult presentation.

Ultra-structure. (Latin *prefix:* ultra, extreme plus Latin *structura,* to build by assembling components). Refers to the extremely detailed analyses of the structure of a cell, its organelles and other components. Involves sophisticated techniques, such as X-ray crystallography and electron microscopy.

Ultraviolet (UV). Continuous spectrum beyond the violet end of the visible spectrum (wavelength less than 400 nm), and above the X-ray wavelengths (greater than 5 nm), Glass absorbs UV, so optical systems at these wavelengths have to be of quartz. Nucleic acids absorb UV most strongly at around 260 nm, and this is the wavelength most likely to cause mutational damage (by the formation of thymine dimers). It is the UV component of sunlight that cause actinic deratoses to form in skin, but which is also required for vitamin *D* synthesis.

Ultraviolet absorption curve. Plot that shows the relationship between differing wavelengths of light in the ultraviolet range and the percentage absorbance by a solution of biological molecules. Such a curve helps to distinguish nucleic acids from proteins; the former absorbs maximally at about 260 nm and the latter at about 280 nm.

Ultraviolet disinfection. The destruction of microorganisms in close proximity to a ultraviolet light source of 253 nm. The technique is used in contaminated environments and clean rooms.

Ultraviolet fluorescence microscope. Ultraviolet microscope in which the ultraviolet light is directed on biological material which fluoresces in a particular colour when so irradiated. This may be because of natural fluorescence or because of staining with fluorescent dyes such as fluoresces in or association with fluorescently tagged antibody.

Ultraviolet irradiation. Irradiation with ultraviolet light. which may induce melanin formation in low doses, but is also a mutagen and, in high doses, lethal to living cells. It is widely used for the maintenance of sterility in cabinets used for tissue culture, but will not penetrate glass or plastic efficiently, Its mutagenic effect results from its induction of pyrimidine dimers in DNA.

Ultraviolet light (UV). Radiation with wavelength (100-400 nm) at the violet end of the visible spectrum. Generated by mercury vapor lamps, it is sometimes used in tissue culture for its mutagenic properties or to reduce ambient contaminants in work areas due to its bactericidal properties. Exposure is harmful to the eyes and skin and is to be avoided by workers in plant tissue culture facilities.

Ultraviolet microscope. Microscope in which ultraviolet light is generated from the light source, usually a high-pressure mercury vapour lamp, and which is provided with excitation and barrier filters so that ultraviolet light can excite fluorescent molecules in the specimen, harmful ultraviolet light can be filtered out, and fluorescent light only allowed to pass through the eyepiece.

Ultraviolet mutation. The production of a mutation in a microorganism using ultraviolet light with a wavelength of 200-300 nm. Such light is generated by a low pressure mercury vapour lamp. These effects can be reversed by photoreactivation upon exposure to longer wavelengths of ultraviolet light and the visible spectrum.

Ultraviolet radiation (UV). Denoting the actinic or chemical rays beyond the violet end of the spectrum. In this book UV refers to the germicidal line at 2537 A° produced by mercury are lamps.

Ultraviolet visualization. A technique used to detect the position of proteins, nucleic acids and other compounds that absorb ultraviolet light, or fluoresce in the presence of ultraviolet light, on paper, gel or thin layer chromatograms, or on electrophoretic gels. In some cases, the matrix is sprayed with a suitable reagent prior to inspection to enable visualization of the compounds.

Unbalanced translocation. Chromosome translocation in which some loss of chromosomal material has resulted or in which a deleterious genetic effect is evident following the translocation.

Uncoiling of chromosomeres. Situation most readily seen in the puffing of polytene chromosomes and the looping out of chromatin loops in lampbrush chromosomes.

Uncondensed chromatin. Chromatin that is not highly compacted and is therefore more likely to be transcriptionally active; there is a strong correlation between gene activity and chromatin decondensation. Both the puffs and interbands of polytene chromosomes are uncondensed and are sites of intense and moderate transcriptional activity, respectively.

Unconditioned reflex. Hereditarily determined reflex actions which appear in all normal members of the species; and which; do not result from individual experience or habit. cf. Conditioned Reflex.

Uncoupling agent. A substance that uncouples phosphorylation of ADP from electron transport; e.g. 2,4-dinitrophenol.

Undefined. A medium or substance added to medium in which not all of the constituents or their concentrations are chemically defined; as media containing coconut milk, malt extract, casein hydrolysate, fish emulsion or other complex addenda. The aversion to undefined materials or media is based on a lack of control implied by their use and a concern with their variability from batch to batch.

Underlapping. Possible outcome of collision between two cells in culture, particularly head-side collision: one cell crawls underneath the other, retaining contact with the substratum, and obtaining traction from contact with the rigid substratum (unlike overlapping, where reaction must be gained on the dorsal surface of the other cell).

Underwinding. Negative supercoiling; coiling of normal B-form DNA in a lefthanded direction, and therefore opposite to the direction of coiling of the double helix itself.

Undulant fever. Brucellosis; disease caused by *Brucella* species.

Undulating. Exhibiting a wave-like motion.

Unequal Bivalent. A bivalent in which one of the constituent chromosomes is longer than the other, so that it has an unpaired region at one end.

Unequal crossing-over. Lmprecise pairing of duplicated segments prior to crossing-over, leading to chromatids with unequal copies or numbers of copies of the segments concerned.

Unfixable. Ever-sporting from seed.

Unicellular (L. *unus:* one + *cella:* chamber). Consisting of a single cell; as for example a unicellular organism. (Contrast with multicellular.)

Unidirectional replication. Movement of a single replicating fork from a replication origin.

Unineme hypothesis. Concept that an unduplicated chromosome, that is one prior to the S phase of DNA replication has a single molecule of DNA duplex running from end to end. The concept is now generally accepted, although it is clear that the molecule is highly twisted upon itself both by reason of packing around nucleosomes, packing of nucleosomes in the 30-nm fibres of chromatin, and further packing of these fibres relative to the chromosome axis. Compare Polyneme Hypothesis.

Uniovular Twins. Monozygotic twins; twins arising from a single egg.

Uniparous. Producing only one offspring at a birth.

Unipolar spindle. A type of spindle with only one pole which occurs during meiosis in some insects; a `half-spindle'.

Uniport. Protein responsible for ion transport from the one side of a cell plasma membrane to the other. Compare Antiport Symport.

Unique DNA. Sequence of nucleotides present only once per haploid genome. The great majority of prokaryotic DNA and almost majority of prokaryotic DNA and almost all protein-coding genes in eukaryotes are unique in this respect.

Unisexual. Bearing but one kind of gamete, male or female. Such organisms are sexually dimorphic.

Unisexualism. The state of having the two sexes separated by reason of their being confined to separate individuals.

Unit character. (1) A character which behaves as a unit in inheritance and is governed by a single pair of allelomorphs. (2) One whose difference from an alternative is transmitted as a unit in heredity. Cf. Character. Factor, Gene.

Unit membrane. Single biological membrane, resolvable at the ultrastructural level into the characteristic two electron-dense leaflets enclosing an electron-lucent layer.

Univalent. A body at the first meiotic division corresponding with a single chromosome in the mitotic complement, especially when unpaired bivalent, trivalent, quadrivalent, quinquevalent, sexivalent, septivalent, octavalent, etc. are associations of chromosomes held together between diplotene and metaphase of the first division of meiosis usually by chiasmata. Trivalents and higher association are collectively termed multivalents.

Universal donor. A person with group O blood, whose erythrocytes therefore bear neither A nor B antigens, and whose blood can be donated to members of groups O, A, B, and AB if necessary.

Universal recipient. A person of blood group AB who can receive blood from members of group AB, A, B, or O if necessary.

Unmatched S Gene. Any S gene which is represented either in the style or pollen but not in both, in given pollination (An S gene is any gene controlling the specificity of incompatibility reactions in plants).

Unorganized growth. In vitro formation of tissues with few differentiated cell types and lacking recognizable structure; as with many calli.

unreduced apogamy. Euapogamy; the development of a sporophyte direct from a gametophyte without fertilization having taken place so that no zygote is formed.

Unreduced. Said of a gamete having the somatic ($2n$) chromosome number instead of half that number, i.e., the haploid (n) number.

Un-restricted (Prefix un plus Latin *restrictum*, to limit). Refers here to bacteriophage mutants (q.v.) capable of infecting a variety of bacterial strains; cf. restricted (q.v.) types which can be induced to become unrestricted.

Unsaturated fatty acid. A fatty acid containing one or more double bonds.

Unsaturated. Presence of double or triple covalent bonds in a molecule.

Unscheduled DNA synthesis. Synthesis that occurs at times other than the S phase of the cell cycle. Most of such synthesis is DNA repair, although synthesis of mitochondrial DNA in the cytoplasm is also continuous.

Unstable gene. A gene which mutates frequently.

Untranscribed spacer. DNA sequence that lies between two coding sequences and is not transcribed by RNA polymerase. Although such sequences do exist, some identified in the past (e.g., the spacers between the multiple copy of rRNA genes) are now known to be transcribed.

Untranslated sequence. A region of mRNA that is not used in the synthesis of an amino acid sequence of a given peptide or protein. They usually appear at either end of the sequence that codes for the amino acid sequence.

Untwisting enzyme. An enzyme that preserves or restores the helical structure of double-stranded DNA during replication of circular DNA molecules. These enzyme produce a nick, allowing free rotation, and then reseal the nicked DNA strand once the supercoils have been removed.

Unwinding protein. Protein that untwists the DNA double helix before the replicating fork reaches that region during DNA synthesis. Such a protein, discovered in phage T4, has been termed the gene 32 protein.

Up. Describing a mutation in the promoter that increases the frequency of transcriptional initiation. Compare Down.

Upflow sludge blanket. An advanced anaerobic digester in which the cells form flocs which are kept in suspension in the lower part of the reaction tank in the form of a bed. The system is used with a liquid feed for waste material in solution. The feed stream is introduced at the base of the fermenter so that it passes slowly up through the bed.

Up-promoter mutant. A strain carrying a mutation in a promoter sequence which increases transcription from that promoter.

Upstream. In the direction opposite to the direction of DNA transcription, and therefore going from 5' to 3' on the coding strand, or 3' to 5' on the mRNA. Compare Down-Stream.

Upwelling. The upward movement of nutrient-rich, cooler water from deeper layers of the ocean. The rise of water from a deeper to a shallower depth in an ocean; the rise brings nutrients to the surface region.

Uracil. One of the two pyrimidine bases (one-ring nitrogenous base) of the RNA molecule. It is thymine without the methyl group at position 5.

Uranyl acetate. Electron-opaque compound used to stain ultra-thin sections of tissue for viewing by transmission electron microscopy.

Urea cycle. A metabolic pathway in the liver; promotes the synthesis of urea from amino groups and CO_2.

Urea. A soluble nitrogenous compound, $H_2N\text{-}CO\text{-}NH_2$, found in the urine of humans and other mammals. A compound serving as the main excreted form of nitrogen in many animals, including mammals.

Urease. An enzyme that catalyzes the hydrolysis of urea.

Ureotelic. Describes an organism in which the final product of the breakdown of nitrogen-containing compounds (primarily proteins) is urea. (Contrast with ammonotelic, uricotelic.)

Ureter (Gr. *ourein:* to urinate). A long duct leading from the vertebrate kidney to the urinary bladder or the cloaca.

Urethra (Gr. *ourein:* to urinate). In most mammals, the canal through which urine is discharged from the bladder and which serves as the genital duct in males.

Urethritis. Inflammation of the urethra.

Urethritis. Inflammation of the urethra.

URF, Unidentified reading-frame. An open reading-frame recognised from a DNA sequence for which no genetic function is known.

Uric acid. A compound that serves as the main excreted form of nitrogen in some animals, particularly those which must conserve water, such as birds, insects, and reptiles.

Uricotelic. Describes an organism in which the final product of the breakdown of nitrogen-containing compounds (primarily proteins) is uric acid. (Contrast with ammonotelic, uricotelic.)

Uridine diphosphate galactose. (UDP-Gal) Nucleoside diphosphate sugar, found as an intermediate in the metabolism of galactose, and formed in a reaction between uridine triphosphate (UTP) and galactose-1-phosphate catalysed by the enzyme UDP-Gal pyrophosphorylase: UTP + galactose-1-phosphate - UDP-Gal + PP_1. UDPG epimerase then catalyses an inversion of configuration converting UDP-Gal to uridine diphosphate glucose.

Uridine. The ribonucleoside that contains the pyrimidine uracil.

Uridylic acid. The ribonucleotide that contains the pyrimidine uracil.

Urokinase. An enzyme that catalyzes the formation of plasmin from plasminogen.

Uronic acid. A member of the most important group of sugar acids, the common constituents of polysaccharides. In uronic acids the carbon atom carrying the primary hydroxyl group (the CH_2OH end) is oxidized to a carboxyl group (COOH); as glucuronic and galacturonic acids.

Urticaria. Inflammatory reaction characterised by eruptions that itch.

Uterus. The womb, a specialized portion of the female reproductive tract in certain mammals. It receives the fertilized egg and nurtures the embryo in its early development.

UTP. Uridine triphosphate. The corresponding deoxy derivative does not exist. Hence labeled UTP is used to specifically label RNA.

UV air lock. An air lock located between areas of different levels of contamination and air pressure. It provides a dead air space for the transfer of personnel a dead air space for the transfer of personnel and/or equipment without air flow. The interior is irradiated with UV and painted with aluminum paint to give good UV reflectance.

UV clothing discard rack. A rack that holds a standard laundry bag and is protected at the top with a curtain of UV. Clothing worn in the contaminated laboratory is discarded into this laundry bag in the contaminated change room.

UV endonuclease. An endonuclease capable of cleaving a DNA chain at the 5' side of a thymine dimer.

UV Radiation. Electromagnetic radiation with wavelength shorter than that of visible light (3900-2000 A). Causes DNA base-pair mutations and chromosome breaks.

UV. Abbreviation for ultraviolet (light).

V

V - chromosomes. Chromosomes with two arms, i.e. mediocentric chromosomes.

V factor. Nicotinamide adenine dinucleotide:required for the growth of certain *Haemophilus* spp.

V region. Variable region of immunoglobin light chain and heavy chain and containing the antigen binding sites.

V/V. May indicate simple proportion; as 3:1 (v/v). May indicate percent volume in volume; as the number of cm^3 (mls) of constituent in 100 cm^3 (mls) solution.

Vaccination. Inoculation with a biologic preparation (a vaccine) to produce immunity.

Vaccines. (Latin *vacca:* cow). Vaccines are preparations that, when given to a patient, elicit an immune response that subsequently protects the patient from a disease. Usually the vaccine consists of the organism that causes the disease (suitably attenuated or killed) or some part of it. Attenuation of a virus (or bacterium) is manipulating it so that is does not loose its ability to grow in culture, but losses some or all of its ability to cause disease in animals. Usually bacteria, and to a degree viruses, slowly loose their ability to colonize living things (and hence to cause disease) as they are cultured outside the body. There is a range of biotechnological approaches to producing vaccines.

Viral vaccines. These are vaccines that consist of genetically altered viruses.

Enhanced bacterial vaccines. Bacteria have been genetically engineered to enhance their value as vaccines (when dead). Intervet markets a genetically engineered *E. coli* as a vaccine for furunculosis in salmon by deleting some genes that are critical to causing the disease from the bacterium. The genetic engineering aims to build a bacterium that has the antigenic properties of the original (i.e.

stimulates the immune system), but is not a pathogen. Recombinant DNA techniques can also be used to generate attenuated strains, by deleting pathogenesis-causing genes, or by engineering the protective epitope from a pathogen into a safe bacterium. Thus vaccination with the safe bacterium causes the body to make protective antibodies against the pathogen. This is a research programme so far.

Biopharmaceutical vaccines. Proteins, or sections of proteins, that are the same as the proteins in a virus a bacterium's wall, can be made by recombinant DNA methods as vaccines. This is a standard biotechnological route, and has the advantage that there is no possibility that the resulting vaccine will contain any live virus particles. Peptide vaccines are often incorporated by genetic engineering into a larger carrier protein to improve their immunogenicity (i.e. how well they cause the body to become immune), or their stability.

***Multiple antigen peptides* (MAPs).** Developed by J.T. Tam, these are peptide vaccines that are chemically stitched together (usually on to a polylysine `backbone'). This means that several vaccines can be delivered in one shot.

Polyprotein vaccines. This is a similar idea to MAPs, but here a single protein is made by genetic engineering in which the different peptides from part of a continuous polypeptide chain.

Vaccines have also been produced by a range of non-standard methods. The conventional method is to produce the vaccine from an attenuated version of the disease-making organism, grown in cultured cells or hens eggs. Alternatives include: *Producing single protein vaccines in recombinant microorganisms. Producing vaccines in transgenic plants.* Plants will not have any immune response to a human virus, and will not get ill from them. They can produce lots of the protein and could, in principle, be used as food to provide an oral vaccine. Some preliminary experiments suggest that this could work.

Vaccinia virus. Cowpox virus, which is used for vaccination against smallpox. It is a poxvirus with a double-stranded DNA genome.

Vaccinia virus. Vaccinia viruses are DNA virus from the same family as cowpox and smallpox. However, they are safe viruses to work with, and so have been used for several biotechnological applications. Vaccinia has been used as the basis of an expression vector system. It can infect a wide range of cells, and has a lot of DNA, quite a bit of which can be removed using suitable genetics. Thus quite large foreign genes can be spliced into it, and then the recombinant virus can be used to infect a wide range of cells, allowing the biotechnologist to choose the cell most suitable for the process. Vaccinia virus vectors have been used quite widely in research since they can be used to express proteins in mammalian cells. Because they can hold quite a lot of DNA, they can be used to make more than one protein at once a cells, which can be useful for making proteins with more than one polypeptide chain (multi-subunit proteins). Vaccinia has also been used as the basis for `live virus' vaccines. It does not itself cause severe disease, and because it can infect a wide range of species it can be used to produce a range of animal vaccines, which are the first targets of this sort of technology. Provisional approval for a field trial of a live vaccinia virus vaccine was granted in the USA in 1990. Foreign genes are usually introduced into vaccinia by recombination, rather than by isolating the vaccinia DNA and manipulating it in vitro. This is because the vaccinia virus is too large to be manipulated conveniently. Cowpox and raccoon pox viruses, which share some of the useful characteristics of vaccinia, are being looked at as alternative vector systems.

Vacuole. Space or cavity in the cell usually filled with some substance.

Vacuum breaker. A device that is installed in a line or tank, where the breaker is not subjected to a down-stream back-pressure, to prevent reversal of flow in case of accidental occurrence of an upstream suction.

Vacuum dryer. An apparatus used to remove moisture from solids. The rate of evaporation is increased by reducing the pressure within a sealed container. The material to be dried is spread on trays or tumbled during drying.

Vacuum fermentation. A fermentation system developed for the continuous production of ethanol and other volatile products. The reactor is run under reduced pressure so that the volatile product leaves the fermenter continuously in the exhaust gas. The method, which is of interest for thermophilic fermentations, is still largely at the experimental stage.

Vacuum pan. An apparatus used to concentrate a solution by removing the liquid. The rate of evaporation is increased by lowering the pressure within a sealed container. Vacuum pans are used for controlled crystallization of solids from their mother liquor.

Vagina [L. *vagina:* sheath]. In female mammals, the passage leading from the external genital orifice to the uterus; receive the copulatory organ of the male in mating.

Vagina [L. **vagina:** sheath]. The canal leading from the uterus to the external genital opening in female mammals. In female mammals, the passage leading from the external genital orifice to the uterus; receive the copulatory organ of the male in mating.

Vagina, artificial. A double walled rubber tube to which is fitted at one end, a glass receptacle and which is used in artificial insemination in collecting semen from the male. The space between the double walls is usually filled with warm water.

Val. An abbreviation used to denote the amino acid valine in protein sequences and elsewhere.

Valans (Oenothera). A gene complex characterized by genes for red striped buds, punctuate stems, narrow leaves, white nerves, and no red flecks on rosette leaves.

Valine (Val, $C_5H_{11}NO_2$ mw 117.1). An amino acid found in seeds and proteins. Occasionally added to plant tissue culture media.

Van der Waals force. A weak attractive force, acting over only very short distances, resulting from attraction of induced dipoles.

Vanadate. Ion containing vanadium and oxygen, used in cell biology to block cross-bridge formation of dynein, and to block ATPase activity.

Variability. The capacity of an individual or group of individuals to produce gametes having genotypic variation, Mather 1941; a priori the distinction can be made (Fisher 1930, Mather 1943) between Free Variability, that which is expressed phenotypically, and Potential Variability, that which is not expressed in the phenotypes, owing to dominance, etc. Flow of Variability, the movement of variability in a population from Free to Potential and vice versa.

Variable antigen. Term usually applied to the surface antigens of those parasitic or pathogenic organisms which can alter their antigenic character to evade host immune responses.

Variable region. Portion of an immunoglobulin molecule that interacts with antigen.

Variance. Mean square deviation of a variant from its mean. Estimated as the Standard Deviation. Though strictly a parameter the name is commonly applied to a statistic (i.e. as synonymous with mean square). Analysis of Variance, a technique for the isolation of particular components of variation for assessment by comparison with Error Variance. Especially applicable to the analysis of Factorial Experiments. Variance Ratio (V. R.), the ratio of two estimated variances. Twice the natural antilog of z. Sometimes denoted by F.

Variant. A culture exhibiting a stable phenotypic change whether genetic or epigentic in origin.

Variate. A variable quantity whose measurements or frequencies from all or part of the data for analysis.

Variation. (1) The occurrence of differences in the permanent (i.e. self-propagating) structures of cells (genotypic and heritable). (Darlington, 1929). (2) The occurrence of differences between individuals due to differences between the permanent structures of their cells (phenotypic and heritable) or to differences in external conditions (phenotypic and non-heritable). Discontinuous Variation, where gradations of difference are perceptible in the phenotype. Continues Variation, where gradation of difference are imperceptible in the phenotype. Continuous Variation, where gradation of difference are imperceptible in the phenotype. Continuous variation is not to be taken to imply continuous differences in the genotype, since statistical analysis may reveal discontinuities in the determinants.

Varicella. A mild, highly infectious viral disease; also known as chicken pox.

Variegation. Incomplete and spot wise interrupted colouration, as a rule, concerning the chlorophyll development in plants. Variegation may be due to different causes; for instance, a special genotype, a chimeral constitution, often occurring mutations, or a virus infection.

Variety. A sub-division of a species owing its uniformity either to Genetic isolation in nature, or to artificial propagation in cultivation (where a variety is often a Clone).

Variola. An acute, infectious viral disease; also known as smallpox.

Vasa deferntia. A continuation of the epididymi lying between these and the urethra. In some animals each vasa deferens has an enlargement (the ampulla of Henle) which functions as a storage place for spermatozoa.

Vascular bundle. A strand of conducting tissue found in vascular plants. In plants showing secondary thickening (e.g. dicotyledons) the vascular bundles contain meristematic cambium tissue between the xylem and phloem, but this is generally lacking in plants without secondary thickening (e.g. monocotyledons).

Vascular ray. In vascular plants, radially oriented sheets of cells produced by the vascular cambium, carrying materials laterally between the wood and the phloem.

Vascular. Containing specialized vessels for the conduction of fluids: blood and lymph in animals, sap and water in plants.

Vasectomy. A form of sterilization accomplished by the removal (or ligature) of a small section of each spermatic duct.

Vasoactive amines. Products such as histamine and 5-hydroxytryptamine released by basophils, mast cells and platelets which act on the endothelium and smooth muscle of the local vasculature.

Vasoactive intestinal peptide (VIP). Peptide of 28 amino acids, originally isolated from porcine nervous system where it acts as a neuropeptide, and is released by specific interneurons. May also affect behaviour of cells of the immune system.

Vasoactive. Affecting vessels (especially blood vessels).

Vasopressin (antidiuretic hormone). A peptide hormone produced by the hypothalamus and the posterior pituitary gland. It stimulates contraction of muscles around the capillaries and arterioles, raising the blood pressure. It increases peristalsis and has some effect on the uterus. It stimulates water absorption in the kidney tubules, leading to concentration of the urine. Compare Oxytocin.

Vasotocin. Peptide hormone secreted by the posterior pituitary gland of lower invertebrates and by the pineal body in mammals. It increases skin permeability in amphibians and regulates bladder function and is structurally similar to Vasopressin.

Vector. (Latin *vector:* carrier). Indicates an organism, usually an insect, that carries and/or transmit disease-causing organism. It is a carrier of pathogenic microorganisms, in gene engineering a plasmid or virus genome used to carry genes. A vector in biotechnology is usually a DNA segment that allows another piece of DNA to be `cloned' using recombinant DNA techniques. DNA does not replicate itself: it needs a battery of enzymes to replicate inside a cell. The enzymes coordinate DNA synthesis with cell growth by only starting of a DNA molecule at a specific time in the cell's growth cycle. To allow this to happen the DNA must contain a `start here' signal, called the origin to replication. Thus any DNA that is to be cloned must contain an origin. A unit of DNA that has an origin of replication (and a signal to stop replication at the other end, if that is necessary) is called a replicon. Since most bits of DNA do not contain an origin-containing bit of DNA, called a vector. Vectors can be thought of as little replicons into which we can add other DNA. This is the basic function of vectors. To make them convenient to use, they have a range of other properties. Most cloning vectors are **episomes**, i.e. they are genetic elements that can be replicated separately from the host cell's chromosome (i.e. the rest of its DNA). Episomes can be plasmids (small loops of DNA with no function that is deleterious to the cell) or persistent viruses (bits of DNA with the potential of coding for virus particles). `Traditional' vectors, such as the pBR series and the 2æm vectors for yeast, are plasmids, while the lambda series of DNA-sequencing vectors are based on a bacteriophage (a bacterium-eating virus). Other viruses, such as T7, are also used, and bits of them have been used in the construction of more exotic beasts such as **cosmids**. (Cosmids, much used in large-scale gene cloning, are plasmids that can be packaged up into lambda virus particles, but only when you have put 40000 bases of foreign DNA into them. Thus the packaging process is an excellent way of ensuring that you have a plasmid with a large stretch of DNA inserted into it.) A few vector types integrate their DNA into the chromosomes of their targets. Examples are YIps (yeast-integrating plasmid), a plasmid that inserts itself into the chromosomes of yeast, and retrovirus vectors for mammalian cells. [Other yeast-cloning vectors, such as the YEps (yeast episomal plasmids) based on the `2-μm plasmid' remain separate from the chromosomes like a bacterial plasmid does. The DNA replication parts of a vector are specific to the organism it is intended for; a bacterial plasmid will not (usually) replicate in mammalian cells. You can construct a vector such that it has the DNA replication origins for two organisms, for example, for *E. coli* and mammalian cells, so that the vector can be manipulated in *E. coli* (which is easy to grow) and then transferred directly to mammalian cells (where we want to do the experiment). Such two-host vectors are called binary vectors or shuttle vectors. Vectors contain a range of genetic elements to make cloning with them easier. These can include: **Selectable genes.** These code for something that allows the cell to survive an otherwise hostile environment. A common one is a gene for resistance to an antibiotic: growing the engineered organism in the presence of antibiotic will select those organisms that contain the vector (and hence whatever genes we have spliced into the vector).

Polylinker. This is a piece of DNA made to contain many restriction enzymes sites, so the vector can be cut at that one place for splicing in other genes.

Specialist origins. Other variations on origins of replication are: *High copy number plasmids*, which exist as many copies inside a cell, not the usual one or two. *Runaway replication plasmids*, where, on some trigger signal (usually a change in temperature) the normal controls on how much plasmid DNA there is in a cell break down and the cell fills up with the plasmid. *Promoters, enhancers, leader peptides.* These are elements to help you express a gene that is cloned in the vector. Because there are so many possible vectors that can be assembled out of these components, some vector systems are made not as complete vectors but as cassette systems, where different selectable genes, origins, etc. can be slotted together to make the vector of your choice. A critical feature of a vector is the amount of foreign DNA that is can hold. The practical limits (in thousands of base pairs, kilobases) are:

`Conventional plasmid vectors'	15
Lambda phage vectors	20
Cosmids	45
P1 bacteriophage vectors	85
BACs (bacterial artificial chromosomes)	300
YACs (yeast artificial chromosomes)	2000
MACs (mammalian artificial chromosomes)	at least 1000

Vectors for gene therapy. A major barrier to getting gene therapy into the clinic is the lack of effective ways of delivering genes to cells in living patients. Several viruses have been engineered to act as gene therapy vectors, with varying success. The most commonly discussed are:

Retroviruses. Retroviruses efficiently transport their RNA into cells, copy their RNA into DNA, and then insert that DNA into the cell's chromosome. In principle this ability could be harnessed to carry other DNAs into a patient's cells, and the viruses' ability to grow removed (producing a replication-deficient retroviral vector, RDRV). However, such vectors are hard to engineer to be safe. Manufacturing gene therapeutics based on retroviruses is also difficult since the genetics of the cells you grow the virus on can readily generate replication competent viruses (RCVs), i.e. variant viruses that can grow in their target cell like a virus, and so start an infection. A practical difficulty is that retroviruses cannot infect non-dividing cells, like muscle or nerves.

Adenoviruses. Adenoviruses can incorporate their DNA into the cell's nucleus without inserting it into the chromosomes. They can infect cells that are not dividing, which is a major advantage over retroviruses. However, the genes that are engineered into them are not expressed permanently in the receiving cells.

***Adeno-associated virus* (AAV).** These are vectors based around the virus particle from adenoviruses, but with no viral genes. The foreign DNA has the ends of the adenovirus DNA spliced on to its ends, and is placed in mammalian cells that also contain some adenoviral genes in their chromosomes. The viral genes make adenovirus coats, which are filled with the foreign DNA. AAV virus-based therapeutics for cystic fibrosis, which take the `correct' CF gene directly to the lungs of affected patients in a spray, are in the clinic.

Alphaviruses. Viruses such as *semliki forest virus* (SMV) and *sindbis* have been research subjects for decades, and have been developed into powerful expression vectors in recent years. They could be converted into gene therapy vectors by engineering replication-defective versions of the vectors, which would then express large amounts of a therapeutic protein inside the target cells. However, this is still at research stage.

Herpesviruses. The herpesviruses carry DNA into specific types of cell, where they lie low for months or years. Engineered versions have been made to carry foreign DNA into cells either in replication-deficient viruses or in otherwise `empty' virus coats. Again, this is largely a research-stage idea.

The alternative approach to viral vectors is a chemical system, such as wrapping the DNA into liposomes or complexing it with lipids, such as lipofectin, that speed the DNA into cells. This gets round the safety problems of viruses, but is very much less efficient - less than 0.1% o cells affected, compared with over 10% for some adenvirus systems.

Vegetal pole (vegetative pole). The end of a spherical animal egg that is opposite the animal pole. The vegetal pole contains most of the yolk and is furthest from the nucleus. Compare Animal Pole.

Vegetative cell. A bacterial cell capable of multiplication, as opposed to the spore form, which cannot multiply. The vegetative form is usually considerably less resistant to adverse conditions, such as heat or germicides, than the spore form.

Vegetative nucleus. (1) The larger of two nuclei in a cell; macronucleus; trophic nucleus; trophonucleus. (2) Pollen tube-nucleus.

Vegetative petite. Petite mutant of yeast cells in which the genetic lesion is in the mitochondrial genome-When such a mutant results from a nuclear lesion, it forms a segregational petition,

Vegetative phase. Active growth period.

Vegetative Pole. The lower portion of a telolecithal ovum in which the yolk accumulates.

Vegetative propagation (Reproduction). Reproduction of the higher plants otherwise than by seed, either naturally (runners, bulbs, viviparous bulbils, etc.) or artificially (cuttings, grafts, etc.).

Vegetative segregation. The production of two genetically dissimilar cells at a somatic cell division, caused by some sudden qualitative or quantitative change in the nuclear constituents, or by the segregation of two dissimilar cytoplasmic constituents at the somatic cell divisions.

Vegetative. Nonreproductive, or nonflowering, or asexual. The concerned with the growing stage of a microorganism as opposed to the spore state.

Vehicle. The host organism used for the replication or expression of a cloned gene or other sequence. (Compare with, vector, the DNA molecule which contains the cloned gene.) The term is little used and is often confused with vector.

Vein [L. *vena*: channel]. A blood vessel that returns blood to the heart. (Contrast with artery.)

Vena cava. One of a pair of large veins that carry blood from the systemic circulatory system into the heart.

Vena cava. One of a pair of large veins that carry blood from the systemic circulatory system into the heart.

Venereal disease. A disease, such as gonorrhea and syphilis, that is spread by sexual contact and involve parts of the genitournary system.

Venereal. Transmitted by sexual contact.

Ventilated hood. Hood covering entire head, pressurized with conditioned air by hose system serving ventilated suits.

Ventilated suit. Pressurized outer garment, including head, hands, and feed, supplied by hose with conditioned air, for work in areas of high risk from infectious aerosols such as some animal rooms.

Ventral (*venter:* belly, womb). Toward or pertaining to the belly or lower side. (Contrast with dorsal.)

Verapamil. A calcium-channel blocking drug, used as a coronary vasodilator and anti-arhythmic.

Veratrine sulphate. Veratrine, a poisonous alkaloid, is obtained from dried ripe from rhizomes of *Veratrum album* and *V. viride.* The sulphate is used in inducing polyploidy.

Vermiculite. A very light mica-based substance used as a support material for plant growth in some experiments where soil would be unsuitable. It can, for instance, be sterilized and mineral solutions of known composition be fed to the plants.

Vermiculture. The culture of earthworms. it is used as a method of treating farm and household waste, with the production of compost, or as means of producing feed protein.

Vernalin. An hypothetical hormone-like substance found in plant meristematic regions, produced by vernalization. This substance is apparently graft transmissible, but has not yet been identified. Different cold requiring species may form different substances during vernalization.

Vernalization (L. *vernalis:* belonging to spring). Events occurring during a required chilling period, leading eventually to flowering. Vernalization may require many weeks of below-freezing temperatures.

Veronal buffer. Buffer prepared from sodium barbitone and used to maintain the pH of fixative such as glutaraldehyde used in the preparation of tissues for transmission or scanning electron microscopy.

Versatile anthers. Those having the filament attached near the middle, so that the anther can readily turn in any direction.

Vertebrate. An animal whose nerve cord is enclosed in a backbone of bony segments, called vertebrae. The principal groups of vertebrate animals are the fishes, amphibians, reptiles, birds, and mammals.

Vertical integration. A must term for management consultants, this means company that can perform all the parts of development, production, and sale of something. In the pharmaceutical industry, a vertically integrated company would be one that researched, developed, manufactured, marketed, and sold a drug.

Vertical resistance. Resistance, usually immunity, towards a virulent race/pathotype; the specific virulent race/pathotype produces susceptible response. Generally, controlled by oligogenes.

Vertical spread. Transfer of infectious agent from parent to offspring via ovum or sperm or by contact between placenta and offspring.

Vertical transmission. Transmission of genetic material (or integrated viruses) via conventional hereditary processes. Compare Horizontal Transmission.

Vesicular arbuscular (VA) mycorrhizae. A form of endomycorrhizae that occurs on many crop plants and temperate trees. The fungi are present in most soils and are not host - specific.

Vesicular stomatitis virus. Rhabdovirus having a single-stranded RNA genome, and bullet-shaped virion covered in an envelope with many protroding micro spikes.

Vessel (L. *vasculum:* a small vessel). In botany, a tubeshaped portion of the xylem consisting of hollow cells (vessel elements) placed end to end and connected by perforations. Together with tracheids, vessel elements, conduct water and minerals in the plant.

Vessel element. One of the cells that makes up a xylem vessel. In contrast to tracheids, the most advanced vessel elements are often more broad than long and have horizontal rather than slanting perforation plates.

Vestigial. Degenerate, not fully developed.

V_H and V_L genes/domains. V_H and V_L genes defines in part the sequence of the variable heavy and light regions of immunoglobulin molecules. V_H and V_L domains are the regions of amino acid sequence so defined. J genes and, in the case of the heavy chain, a D gene (D = diversity) also define these regions. Gene rearrangement plays a role in determining the sequences in which the genes are joined as the DNA of the immunoglobulin-producing cell matures.

Vi antigens. Polysaccharide microcapsular antigens found on the cells of certain strains of *Salmonella typhi, S. Paratyphi* C and *Citrobacter,* and certain atypical strains of *Escherichia coli.* Vi antiges may play soe part in the virulence of pathogenic species in relation to certain host animals.

Viability test. Assay of the number or percent of living cells or plants in a population that has been given a specific treatment; as do tests for cell viability after cryopreservation treatment.

Viability. Measure of, the number of individuals surviving in one class relative to another, standard, class.

Viable count. Proportion of cells in a sample of tissue culture of bacterial cells that are able to grow and replicate.

Viable particle. A discrete unit which is associated with one or more viable microorganisms. A viable particle may be an inert particle (dust, lint) on which is adhered one or more viable microorganisms, it may consist of an aggregate of two or more viable cells, or it may be simply a single cells, or it may by simply a single viable cell. Most aerosol samplers samples especially the liquid impingers, because of the severe stresses they impose on viable particles during collection, tend to break up viable particles. The results of most viable air sampling, therefore, indicate a contamination level some-where between the number of viable particles in the air and the actual number of viable microorganisms in the air.

Viable. Literally. *Capable of life*, and generally refers to the ability of microbial cells to grow and multiply as evidenced by, for example, the formation of colonies on an agar culture medium. Frequently organisms may be viable under one set of culture conditions and not under another set, making it extremely important to define precisely the conditions used for determining viability.

Vibrio. A bacterium that is curved with a twist; has less than one complete turn or twist (in contrast to a helical bacterium). Vibrio is a genus of Gramnegative bacteria. Curved organism resembling a comma; characteristic of Vibrio cholerae, the causative agent of cholera.

Vicinism, Unexpected outcrossing.

Vicinist. A plant resulting from vicinism; a natural hybrid.

Victorin. A host specific toxin produced by Helminthosporium victoriae utilizable for screening oat cells for resistance to the pathogen.

Viewing panel. Fixed window suitably sealed into an interior wall or door between two areas of different contamination level.

Vigorous biotype. A biotype of an insect pest which attacks both resistant ad susceptible varieties of the host more successfully than the other biotypes. (cf. Resistance Breaking Biotype).

Villus. (L. *villus:* shaggy hair). A hair or a hairlike projection from a membrane; for example, from many gut walls. One of the microscopic finger-like projections in the lining of the small intestine. Millions of these villus give the appearance of velvet to the lining and enormously increase the surface area for absorption. Each villus is covered by a single layer of columnar cells through which the soluble products of digestion can readily pass into the blood or lymph. Each also contains a network of blood capillaries and lacteal. Strands of muscle contract rhythmically, to shorten the villus and empty the lacteal and capillaries, so keeping up the diffusion gradient. The surface area is further increased by microvilli, visible only through an electron microscope, which cover the surface of each cell of the columnar epithelium.

Vincent's angina. An ulcerative condition of the tonsils and gums caused by a spirochete and a fusiform bacillus.

Vinegar. A condiment prepared by the microbial oxidation of ethanol to acetic acid. Ethanol is obtained by the Fermentation of malted barley or grape juice.

Viral antigens. Those antigens specified by the viral genome (often coat proteins) which can be detected by a specific immunological response. Often of diagnostic importance.

Viral genetics. The genetics of viruses and bacteriophages.

Viral specific enzyme. An enzyme produced in the host cell after viral infection from viral genetic information.

Viral vaccines. Also called live virus vaccines, these are vaccines that consist of live viruses rather than dead ones or separated parts of viruses. Clearly, the virus itself cannot be used, as that would simply give you the disease, so, instead, one of two genetic engineering methods is used to produce a virus that will elicit an immune response to a viral pathogen but will not cause the disease itself. The first method is to engineer the disease virus genetically so it is harmless, but can still replicate (albeit inefficiently, sometimes) in cultured animal cells. This has the same effect as producing an `attenuated' virus, i.e. one that has been grown in the laboratory until it loses its ability to cause disease. However, the genetic engineering route seeks to make sure that the attenuated virus has no chance of mutating back to a `wild type', pathogenic virus, by deleting whole genes or replacing key regions of genes with completely different genetic material. The other approach is to clone the gene for a protein from the pathogenic virus into another, harmless virus, so that the result `looks' like the pathogenic virus but causes no disease. The foreign protein must not itself be dangerous, of course. Vaccinia and adenoviruses have been used in this way, notably to make rabies vaccines for distribution with meat bait: a trial of such a vaccine was carried out in summer 1990 in the USA. A similar vaccinia virus engineered to produce the haemagglutinin and fusion proteins of rindepest virus (a disease of cattle) has been developed successfully by the University of California at Davis.

Viremia. The presence of viruses in the blood.

Virescent seedling. A seedling that emerges white, or pale green, and turns green.

Viricide. Agent that destryos viruses.

Virion. The virus particle, the minimum unit capable of infecting a cell. It is a complete virus particle consisting of a core of nucleic acid and a protein capsid. The extracellular inert phase of a virus. A virion consists of a protein coat surrounding one or more stands of DNA or RNA. Virions may be polyhedral or helical and vary greatly in size.

Virogene. The complete viral genome.

Viroid. A short self-replicating transmissible fragment of nucleic acid which induces phenotypic changes in a host plant. Unlike viruses, they are not surrounded by a protein coat. Little is known about them. They may be equivalent to bacterial plasmids.

Virology. Systematic study of viruses.

Viropexis. The engulfment of whole virions by cells in a phagocytic process.

Virucide. An agent that kills viruses.

Virulence. The relative pathogencity of an organism. Virulence depends on the invasiveness and toxicity of the pathogen. Virulence may very between strains of the same organism.

Virulent phage. (1) A virus that kills the host bacterium. The contrasting type is temperate phage. (2) A bacteriophage whose infection invariably kills the host cell rather than conferring lysogeny (cf. Temperate Phage).

Virus (Latin *virus*, a poisonous or slimy fluid). These may be defined as obligatory parasitic, self replicating, non-cellular organisms, composed essentially of a protein covering surrounding a central nuclic acid molecule that is either DNA or RNA; often cause disease like, smallpox or influenza.

Virus elimination. Thermotherapy, chemotherapy and meristem or meristem tip culture, used alone or in combination have been employed for the elimination of systemic viruses from plants. Plants must repeatedly test negatively for the virus in question for assurance that it has been eliminated.

Virus receptor. Specific protein, glycoprotein or polysaccharide molecule on the surfaces of eukaryotic or bacterial cells that is the specific site to which viruses absorb.

Virus-free. Free from specified viruses based on tests designed to detect the presence of the organisms in question.

Virus-tested or virus-free. A plant that appears healthy and repeatedly tests negatively for the presence of one or more identifiable viruses. Such a plant may then be used as a stock or donor plant [explant source] for propagation purposes, and may be certified as virus tested [certified virus tested]. The term virus-free is incorrect in most cases, as such a plant may contain one or more viruses which have not been assayed.

Viscidity. Stickiness; a characteristic produced by a bacterial sediment in broth that rises in a coherent swirl upon shaking.

Viscous-mechanical coupling. Method by which adjacent cillia are synchronised in a field. Coupling is through the transmission of mechanical forces, rather than of a synchronising signal.

Visicle. Blister: sac surrounded by unit membrane.

Vital stain. A dye solution of which the dye is taken up only by living cells and used for indicator purposes. Florescent dyes with specific molecular affinities are often used for the tagging of discrete cellular components (membranes, macromolecules, organelles antigens, etc.).

Vitalism. The concept that living organisms are animated by vital principle or force distinct from physical forces.

Vitamer. One of two or more substances which have a similar capacity for curing any specific deficiency syndrome.

Vitamin (L. *vita:* life). Any one of several structurally unrelated organic compounds that an organism cannot synthesize itself, but nevertheless requires in small quantity for normal growth and metabolism.

Vitamin A. The fat-soluble vitamin controlling night-blindness and the disease resisting power of mucous membranes. Vitamin A is manufactured by the liver from carotene.

Vitamin B complex. A group of ten or more water - soluble vitamins, which tend to occur together. They can be obtained from whole grains of cereals and from meat and liver. Since the B vitamins are present in most unprocessed food, deficiency diseases only occur in populations living on restricted diets. Many of the B vitamins enter into coenzymes involved in the normal oxidation of carbohydrates during respiration.

Vitamin C (ascorbic acid). A water soluble vitamin, which is widely required in metabolism. The major sources of vitamin C are fresh fruit and vegetables and severe ascorbic deficiency results in scurvy.

Vitamin D. Fat-soluble vitamin found in fish-liver oil, butter, milk, cheese, egg-yolk, and liver. Its main principal action is to increase the absorption of calcium and phosphorus from the intestine. The vitamin also has direct effect on the calcification process in bone. Deficiency results in inadequate deposition of calcium in the bones, causing rickets in young children and osteomalacia in adults. The term, vitamin D, refers in fact to a group of compounds, all sterols, of very similar properties. The most important are vitamin D_2 (calcifero) and vitamin D_3. Precursors of these are converted to the vitamins in the body by the action of ultraviolet radiation.

Vitamin D-resistant rickets. Sex-linked dominant hereditary disease in humans involving defective renal transport of phosphorus. The disease resembles rickets due to vitamin D deficiency, but persists in the presence of adequate dietary vitamin D.

Vitamin E. The reproductive, fat-soluble vitamin. It is resistant to heat light and oxidation.

Vitamin G. Vitamin B_2.

Vitamin K. A fat-soluble quione involved in photosynthetic electron transfer in plants and blood clotting in animals.

Vitamin P. A flavone derivative concerned with permeability; citrin.

Vitellogenin genes. Genes coding for vitellogenin. Three such genes exist in Drosophila, each with only one or two introns, whereas in Xenopus laevis at least four distinct genes exist. Two of the Xenopus genes -A1 and A2-have been studied in some detail. The are 21 and 16 kilobases in length, respectively, both having 33 introns interrupting the coding sequence and both yielding messenger RNA about 6.3 kilobases long.

Vitellogenin. Lipophosphoprotein: precursor of the two major proteins-lipovitellin and phosphovitin-in the yolk of oocytes. In amphibians vitellogenin (Mr-2000000) is synthesized in the liver of vitellogenic females, secreted into the serum, transported to the ovaries and is taken into the growing occytes by selective pinocytosis into small intracellular vesicles coalesce, vitellogenin is converted into lipovitellin and phosphovitin, and the yolk proteins are arranged to from yolk platelets.

Vivipary. The release of offspring by the mother free from her membranes (as in egg or seed). In animals due to delayed egg-laying, in plants due to precocious germination and (usually) apomixis.

V_{max}. Expression used to define maximal diffusion rates achieved by saturation of carrier protein binding sites for the solute substrate.

Vodka. A neutral spirit distilled from fermented grain or potato starch and treated with activated charcoal.

Voges-proskauer reaction. Test for acetylmethylcarbinol (also called "acetoin") production, used to differentiate **Escherichia** and **Enteobacter** of the coliform group.

Voltage clamp. Situation in which microelectrodes are placed in a cell and adjusted so that the membrane potential is held steady at a set voltage.

Volumeric rate of reaction. The rate of substrate utilization or product formation expressed in terms of unit volume of the bioreactor or fermenter, or unit volume of culture medium.

Volumetric loading rate. The rate of addition of raw materials to a fermenter or anaerobic digester, expressed in terms of weight of material added per unit volume per unit time. (e.g., kilograms per cubic metre per hour).

Volumetric oxygen demand. A measure of the maximum potential oxygen requirement by a fermenter broth. It is equivalent to the product of the maximum cell growth rate, or the number of cells in a steady state system, and the specific rate of oxygen consumption per cell, expressed in terms of unit volume of bioreactor or culture medium.

Volumetric oxygen transfer coefficient ($K_l a$). A measure of the capacity of an aeration system to transfer oxygen from air into the liquid phase of a fermenter. It is expressed in terms of k_l (the oxygen transfer coefficient and a (the interfacial area between air bubbles and liquid) per unit volume of liquid.

Volutin. A basophilic substance, thought to be a phosphate polymer, occurring in granules in the cytoplasm and vacuoles of algae and other microorganisms.

Volvox. Simple multicellular alga in which the component cells are held in a hollow sphere without apparent cell specialization.

Von Baer's laws. (i) In development from the egg the general characters appear before the special characters. (ii) From the more general characters the less general and finally the special characters are developed. (iii) During its development an animal departs more and more from the form of other animals. (iv) The young stages in the development of an animal are not like the adult stages of other animals lower down on the scale, but are like the young stages of those animals.

Von Gierke disease. Autosomal recessive disease in humans, presenting as defective glycogen storage due to mal-formation of glucose-6-phosphatase. It has an incidence of one in 100000.

Von Hippel-Lindau disease. Inherited disease caused by an autosomal dominant gene, characterized by retinal haemangioma and renal carcinoma.

Von pirquet test. Tuberculin skin test which makes use of small amount of the tuberculin preparation that is rubbed into a scratch on the skin. An inflamed area provides a positive reaction and indicates sensitivity to the tubercle bacillus. This test is less sensitive than the Mantous test.

Von Willebrand factor. Plasma factor involved in platelet adhesion through an interaction with Factor VIII.

VW antigens. Protein or lipoprotein antigens found in virulent strains of *yersinia pestis.* They are believed to be associated with the organism's ability either to resist phagocytosis or to resist digestion within the phagocytes.

W

W chromosome. With reference to the sex chromosome that only occurs in females in cases of female heterogamety.

W,Z chromosomes. Specification of sex chromosomes in some organisms (e.g., silk moth) in which the female is the heterogametic sex. Thus the male is *ZZ* and the female *WZ,* and the *W* chromosome is female-determining.

W/m^2 or watt per square metre. A common unit of light measurement.

W/v. Weight in volume; as the number of grams of constituent in 100 cm^3 [mls] solution.

Waagen, Mutations of. Discontinuous changes of small or moderate extent; particularly used of different forms of fossils of the same lineage but belonging to different horizons.

Waardenburg syndrome. Inherited human disease with a frequency of about one in 42000. It is inherited as an autosomal dominant and is characterized by severe deafness and other symptoms.

Waggle dance. The running movement of a working honey bee on the hive, during which the worker traces out a repeated figure eight. The dance contains elements that transmit to other bees the location of food.

Wallace effect. Hypothesis, advanced by Alfred Russell Wallace, that populations of organisms tend to become reproductively isolated at a species level.

Wallace's line. A hypothetical boundary in Indonesia between the islands of Bali and Lambok. It separates the Australasian and Oriental fauna and was drawn by A. R. Wallace, cofounder with Dar1win of the theory of evolution.

Walling-off. Separation of diseased from healthy tissue by barrier tissues produced by diseased plant.

Waltzer. Mutant strain of laboratory mice that carries a neurological defect. Homozygous animals tend to walk in circles and have a head-shaking behaviour.

Warburg respirometer. An apparatus used study enzyme reactions in which an exchange of gases takes place.

Warm antibodies. Most IgG antibodies react better at 37°C than at lower temperatures, especially against red cell antigens. These are the warm antibodies as contrasted with cold agglutinins, especially IgM, which agglutinate below 28°C.

Warning appearance. A conspicuous trait, such as bright coloration, serving to warm potential predators that the animal is poisonous or otherwise dangerous.

Warning colouration. Conspicuous colouration (especially red) of obnoxious organisms, particularly insects.

Wassermann antibodies. Antibodies against cardolipin, such as occur in syphilitic patients; also called reagins.

Wassermann test. A nonspecific complement fixation test used as a diagnostic aid in syphilis.

Wassermann test. A nonspecific complement fixation test used as a diagnostic aid in syphilis.

Waste collection treatment unit. A unit for collecting and treating liquid waste, generally serving one building consisting of a tank in which the contaminated liquid waste is collected and sterilized, either continuously or batch-wise.

Waste management. It is the comprehensive, integrated, and rational systems approach towards the achievement and maintenance of acceptable environmental quality.

Waste. Any matter, whether liquid, solid, gaseous, or radioactive, which is discharged, emitted, or deposited in the environment in such volume, constituency or manner as to cause an alteration of the environment. The concept of a waste enmraces all unwanted and economically unusable by-products at any given place and time, and any other matter which may be discharged, accidentally or otherwise, to the environment.

Water activity. The ratio of the water pressure of a food substance to the vapour pressure of pure water at the same temperature.

Water of hydration: The amount of water chemically bound to a substance. This amount may be variable and must be taken into account when solutions of salts are prepared; as in medium preparation.

Water potential. A measure of energy for water in a plant. Water potential is a fundamental thermodynamic function such that water moves from a point of high water potential to one of lower water potential. In general, the water potential will depend on a number of factors; there are contributions due to osmosis, turgor pressure of cells, surface tension (matrix components), etc.

In osmosis, the tendency for a system (a cell or solution) to take up water from pure water, through a differentially permeable membrane. Water flows toward the system with a more negative water potential. (Contrast with osmotic potential, pressure potential.)

Watson-Crick model. Double-helical, right-handed model for DNA structure, proposed by J. Watson and F. Crick in 1953. The model proposed that two polynucleotide chains were associated by specific hydrogen bonds between complementary base pairs, the sugar phosphate backbones forming a double helix and thus providing a rigid framework of the bases. Although alternative structures for DNA (e.g., left-handed Z-form) have been discovered subsequently, the model has proved correct and has provided enormous stimulus and insight in the field of molecular genetics.

Weak Interactions (also called Secondary Bonds). The forces between atoms that are less strong than the forces involved in a covalent bond (includes ionic bonds, hydrogen bonds, and van dar Waals forces.).

Weber, Organ of. The uterus masculinus; a small structure lying between the ejaculatory ducts of the male and homologous with the uterus of the female.

Weight. A differential value assigned to an estimate of a quantity, relative to other estimates of the same quantity, for purposes of combining the estimates. Usually its Invariance or Amount of Information is taken as the weight for each estimate of a series.

Weil-felix test. Agglutination test used as a diagnostic aid in certain rickettisial diseases; certain Proteus strains are used as the antigen with the patient's blood serum.

Weil-felix test. Agglutination test used as a diagnostic aid in certain rickettisial diseases; certain **Proteus** strains are used as the antigen with the patient's blood serum.

Weissmannism. Essentially a theory of complete continuity of germplasm from generation to generation, this remaining entirely distinct from somatoplasm.

Western blotting. Method for identifying proteins in complex mixtures. The proteins are separated by electrophoresis through a slab gel and then blot transferred onto a sheet of nitrocellulose by electrophoresis at 90° to the gel surface. The proteins bound on the nitrocellulose membrane can then be identified by reaction with specific reagents such as radioactive or fluorescent-labeled antibodies. Compare Northern Blotting Southern Blotting.

Wheal and flare. A circumscribed, reddened elevation of the skin.

Wheat germ lectin. Substance that specifically binds to N-acetylglucosamine and is used in cell biology to release these molecules from membranes.

Whey fermentation. A process that uses whey, or whey permeate, as the substrate for the production of food yeasts, single cell protein, ethanol, etc.

Whey permeate. A liquid containing about 35-50 grams per litre of lactose produced as a byproduct of whey protein recovery.

Whey. A byproduct of cheese making that arises following the separation of curds (the solidified casein ad butter fat).

Whirlpool separatory. A hydroclone used for the separation of cells from the effluent of fermentations. The overflow from a continuous cell recycle reactor is pumped tangentially into a vertical cylindrical vessel. The cells are deposited in a central cone, where they accumulate before they are recycled. The energy requirement for the whirlpool separator is low.

White corpuscles. Leukocytes of the blood.

Whole cell patch. A variant of patch clamp technique, in which the patch electrode seals the cell, with direct communication between the interior of the electrode and the cytoplasm.

Whole mount. An entire organism or part of an organism, prepared by fixation, staining, dehydration, and clearing (but not cut into sections by microtome), and placed on a slide, usually in Canada balsam under a coverslip, for microscopical examination.

Wick: A filter paper, chromatography paper or a strip of some other material used to support plant tissue above a liquid medium and to transport the water and nutrients to the tissue.

Widal test. Agglutination test using killed **Sal. typhi** organisms and the patients blood serum; a diagnostic aid in typhoid fever.

Wild type. An individual as it occurs in nature. The wild type is usually denoted by `+' sign with a superscript (consisting of first letter/letters of the character) to the right of the `+' sign. The wild type allele is usually dominant but it may be recessive also e.g. the wild allele of bar mutant in Drosophila.

Wild-Type Gene. The form of gene (allele) commonly found in nature.

Wild-type. Geneticists' term for standard or reference type. Deviants from this standard, even if the deviants are found in the wild, are said to be mutant.

Wilson's disease. Autosomal recessive inherited disorder in humans in which copper excretion is defective, and liver and brain damage results from copper accumulation.

Wilt. Collapse of foliage from water deficiency; may be temporary or permanent.

Wings. The lateral petals of a papilionaceous flower.

Wiskott-Aldrich syndrome. Inherited disease of immunodeficiency resulting from an X-linked recessive gene.

Witches'-broom. Broomlike proliferation of woody stems originating from closely spread nodes.

Wobble hypothesis. A hypothesis that the first two bases of the codon pair firmly with the anticodon, whereas the third base-pairing is weak, allowing pairing with different antocodons.

Wobble. Ability of third base in tRNA anticodon (5' end) to hydrogen bond with any of two or three bases at 3' end of codon. Thus, a single tRNA species can recognize several different codons.

Wolman's disease. Inherited human disorder determined by a recessive autosomal gene on chromosome 4, which involved deficiency of lipase in the lysosomes.

Woody plant medium [WPM]: A modified MS [1962] medium developed for woody plant species by G. Lloyd and B. McCown (1980). It has less nitrogen [both ammonium and nitrate], sodium, potassium and chloride than does MS [1962] medium. It has been increasingly used for the commercial propagation of ornamental trees and shrubs.

World collection. (Live herbarium) A collection of varieties of a species from the different parts of the world, both wild and cultivated, grown in isolated plots for study and to serve as a germplasm pool for use in breeding.

Wort. The raw material in the production of beer. It is derived mainly from barley malt and hops. The more important fermentable components are maltose, glucose and maltotriose. Non-fermentable carbohydrates accounts for about 20 percent of the total wort solids; these are dextrins, pentose sugars and other polysaccharides. The wort also contains 3-6 per cent nitrogenous compounds (amino acids, peptides, proteins, nucleic acids). The polyphenols or wort tannins are derived mainly from the malt and include hydrolyzable and nonhydrolyzable tannins. Specific polyphenols are also derived from the hops in the form of an essential oil, which contains mainly terpenes, and resins which produce the bitter flavour. The most abundant vitamins present are those of the B complex. The major inorganic ions are calcium, sodium, chloride, and sulphate, which are mainly derived from the water used.

Wright, Sewall, effect. Non-adaptive differentiation into different types due to "drift" followed by random fixation in small isolated populations.

Wright's inbreeding coefficient (F). Proportion of loci at which an individual organism is homozygous.

Writhing number. Number of times a super-coiled molecule of DNA is crossed or folded upon itself.

X

X. Designation or symbol for the basic number. Consequently, x=9 means that the basic number is nine, and 3x=27 indicates that it is a question of a triploid with 3 x 9 chromosomes.

X_1, X_2, X_3, etc. The first, second, third, etc., generations following X-irradiation.

X chromosome. (1) One of the sex chromosomes. A female is *XX* in Drosophila and in many mammals, including man. A male is *XY.* In poultry, the male is *XX* and the female is *XO.* (2) with diploid sex differentiation, the sex chromosome in regard to which one sex is homozygous (the Homogametic Sex.) With haploid sex differentiation (Bryophyta, etc.), the sex chromosome of the female. Attached x Chromosomes, where the two X's of a female are fused at their terminal centromeres and so act as one mechanical unit in inheritance (in Drosophila melanogaster).

X factor. Heme; required for the growth of certain *Haemophilus* spp.

X inactivation. Repression by condensation of the transcriptional activity of one of the two copies of the *X* chromosome in all somatic cells of the female mammal. One of the two *X* chromosomes is randomly inactivated in early embryonic life, as a mechanism of dosage compensation. The condensed chromosome is visible in somatic cell nuclei as the barr body, although some genes at the distal end of the short arm of the *X* chromosomes is inactivated, the pigment alleles being carried on this chromosome.

X linkage. Usual form of sex linkage in which a particulars gene is located on the *X* chromosome.

Xanthine oxidases. Dehydrogenases involved in conversion of hypoxanthine to xanthine, and xanthine to uric acid, as the final catabolism of purines. Deficient in the human disease xanthinuria.

Xanthinuria. Human disease in which the purine xanthine persists in the urine rather than being completely converted to uric acid.

Xanthophyll [Gr. *xanthos*: yellowish-brown + *phyllon:* leaf]. A yellow or orange pigment commonly found as an accessory pigment in photosynthesis, but found else-where as well. An oxygen-containing carotenoid.

Xenia. The effect of more distantly related, as contrasted with more closely related, pollen on the maternal tissue of a fruit. Originally applied (Focke 1881) to the effect on the endosperm, now known to be genetic and direct. Metaxenia has been used for true xenia in contrast to the xenia of Focke.

Xenobiotic. A chemical compound that is not normally produced or metabolized by, or associated with living organisms; a compound synthesized by man.

Xenobiotics. A xenobiotic is a chemical that would not normally be found in a given environment, and usually means a toxic chemical that is entirely artificial, such as a chlorinated aromatic compound or an organomercury compound. Biotechnology brushes with xenobiotics in three areas. First, in determining their toxicity and effects on living systems. Secondly, biotechnology has developed methods for removing them through bioremediation or enzyme-based degradation. Lastly, a range of biotechnological products aim to replace the compounds, which, if they get out of their target site, are classified as xenobiotics. Among these are the chemical herbicides and pesticides that biocontrol agents and biopesticides hope to replace. Nearly all drugs are xenobiotics, and so much of healthcare biotechnology is concerned with them as drugs rather than as chemicals. Proteins are not xenobiotics, of course. A substantial element of how xenobiotics interact with mammals is their metabolism, particularly that mediated by a group of enzymes called cytochrome P450s, which catalyse the first stage of attack on a wide range of different xenobiotics.

Xenogamy. Cross-pollination within a species.

Xenogeneic transplantation. Grafting of tissue between organisms of different species, or the transfer of organelles from a cell of one species to that of another.

Xenogeneic. Literally, of genetic stock; usually applied to tissue or cells from another species, as in xenogenic transplantation.

Xenograft. A graft between individuals of different species. A xenograft is a grafted organ or tissue that comes from another species. Because our immune system would violently reject such a graft (especially because of the alien major histocompatability complex antigens on them, discussed in the entry on transplants), strategies have to be developed to prevent immune rejection. These include: 1. Creating transgenic animals that show the human transplantation antigens, not the animal ones. 2. Creating animals that show no transplantation antigens at all, by gene knock-out. 3. Encapsulating the organ or tissue so that the human immune system does not `see' it. 4. Inducing tolerance (also called `anergy') in the human, so that they come to accept the foreign tissue as their own. So far, none of these have been successful.

The most developed practical human xenografts come from pigs, although most of the work is done in other species for convenience. For organs such as the heart, the donating animal has to be similar in size to humans, and with a similar physiology:

pigs and baboons are favorites, and since pigs are much easier to breed they are the focus of much of the recombinant DNA work.

There is substantial ethical debate about xenografts, which offend many religious groups. There are also general practical objections centering around the concern that the xenografted tissue will carry new infections, particularly new viruses, into the human population. The historical example behind this case is probably AIDS, which some scientists believe spread from being a relatively contained disease in monkeys to a pandemic in humans. It is ironic that one of the groups pushing for early clinical trials of xenografting are AIDS activists, who want to try transplanting baboom Imphocytes into AIDS patients.

Xenoplastic graft. A heteroplastic graft (i.e. an interspecific or inter-generic graft) which takes successfully.

Xenotropic virus. A virus which can be grown on cells of a species foreign to the normal host species.

Xeroderma pigmentosum. Human metabolic disorder with a genetic basis, in which defects exist in the DNA repair mechanisms. One consequence is a high incidence of skin cancer due to the inability to repair damage caused by the ultraviolet irradiation from sun-light.

Xerogel. A solution of linear molecules whose movement, relative to one another, is restricted by cross-linking or physical interaction, producing a three-dimensional network of solvated polymer chains. A given polymer matrix only forms xerogels in a limited range of solvents in which the individual polymer chains are inherently soluble.

Xerophyte. Any plant adapted to growing in dry conditions, or in a physiologically dry habitat such as an acid bog or a salt marsh, by storing available water, reducing water loss, or possessing deep root systems. Succulents such as cacti and agaves have thick fleshy stems or leaves to store water. Features associated with reducing water loss include: shedding or dieback of leaves; waxy leaf coatings coupled with closure or plugging of stomata, sunken or protected stomata; folding or repositioning of leaves to reduce sunlight absorption; and the development of a dense hairy leaf covering.

X-gal. A shorthand nomenclature for the compound 5-chloro-4-bromo-3-indoly-ß-D-galactoside. It is a colourless compound which, when hydrolysed by a ß-galactosidase, releases a bright blue non-diffusible indigo dye. It is used as a colour indicator for bacteria able to use lactose and in the M13mp phage cloning system

X-linked diseases. Any inherited disease whose controlling gene is carried on an *X* chromosome, e.g. haemophilia. Most known conditions are recessive and thus since males have only one *X* chromosome they will express any such recessive character. Female expression of such diseases is uncommon because of the relative rarity of homozygotes.

X-linked inheritance. The inheritance conditioned by genes in an X chromosome.

X-linked. Located on the X chromosome.

Xma l. (Sometimes pronounced exmar one). A type ll restriction enzyme from the bacterium *Xanthomonas malvacearum* which recognises the DNA sequence shown below and cuts it at the sites indicated by the arrows:

Xo condition. Having an X chromosome, but no Y chromosome, as the female in poultry and some other birds.

XO syndrome. Turner's syndrome, 45 chromosomes, chromatin negative.

X-radiation. Radiation of X-rays, which are produced when metal surfaces are bombarded with high-speed electrons. X-Rays are ionizing radiations and have wavelengths intermediate between ultraviolet and gamma-radiation.

X-ray breakage. The following terms are used in describing the results of X-ray breakage. They have also been applied to cases of spontaneous breakage: Chromosome Break, one which occurs in the chromosome before its division into the two chromatids observable at the following mitosis, when the break is recorded. As opposed to- **Chromatid Break**. One which follows this division, and therefore breaks only one of the chromatids at any one point. (Mather and Stone, 1933). **Isochromatid Break**. chromosome break in a cell otherwise apparently characterized by chromatid breaks. (Lea and Catcheside, 1942). Reunion, the joining of broken ends of chromosomes or chromatids. Classified by Darlington and Upcott (1941) into: **Restitution.** where the pre-breakage structure is restored. **Chromosome Reunion,** of the ends of chromosomes following chromosome breakage. **Chromatid Reunion**, of the ends of chromatids following either chromosome or chromatid breakage.

X-ray crystallography. The use of diffraction patterns produced by X-ray scattering from crystals to determine the 3-D structure of molecules.

X-ray diffraction. Basis of powerful technique for determining the three-dimensional structure of molecules, including complex biological macromolecules such as proteins and nucleic acids, which form crystals or regular fibres. Low-angel *X*-ray diffraction is also used to investigate higher levels of ordered structure, as found in muscle fibres.

X-ray or Roentgen ray (r). Electromagnetic radiation of short wave length produced by high speed electrons impacting on a metal object under vacuum.

X-ray therapy. The treatment of disease, such as cancer, using controlled levels of X-rays.

XXY syndrome. Klinefelter's syndrome, 47 chromosomes, chromatin positive.

Xylan. A polymer of xylose; a component of hemicellulose in plants. Xylants are used in the production of furfural, which is used in the manufacture of plastics.

Xylem [Gr. *xylon*: wood]. In vascular plants, the woody tissue that conducts water and minerals; xylem consists, in various plants, of tracheids, vessel elements fibers and other highly specilized cells.

Xylem parenchyma. Living parencyma cells associated with the xylem.

Xylem parenchyma. Living parencyma cells associated with the xylem.

Xylem. The water-conducting tissue in vascular plants. It consists of dead hollow cells (the tracheids and vessels), which are the conducting elements. It also contains additional supporting tissue in the from of fibres and sclereids and some living parenchyma.

Xylose [$C_5H_{10}O_5$ mw 150.1]: An aldopentose sugar [wood sugar] present in the woody tissues of many plants as polymeric xylan. It is occasionally used as a carbohydrate source in plant tissue culture media.

Xylose isomerase. An enzyme that catalyzes the inter-conversion of the pentoses xylose and xylose. The enzyme occurs in a large number of microorganisms and has been developed commercially for use as a glucose isomerase.

Y

Y cells (Immunology). Immunologically competent lymphocytes which have experienced the initial contact with specific antigen-but which are not effector cells. Y cells are also celled memory cells or primed cells.

Y chromosome. With reference to the sex chromosome that only occurs in male individuals in cases of male heterogamety or of sex determination in the haplophase.

Y linkage. Sex linkage, in which genes are located on the *Y* chromosome, and are therefore expressed only in males (holandric inheritance). In mammals the *Y* chromosome carries few, if any genes, but in some organisms (e.g., medaka fish) the *Y* chromosome does carry genes.

YACs. YACs are yeast artificial chromosomes, and are cloning vectors that are gaining a lot of use in the human genome project. They consist of those pieces of DNA that define the ends (telomeres) and the middle (centromere) of a chromosome in yeast. Both these elements are needed to allow a chromosome to be replicated in yeast cells: if there s do telomere then the ends of the chromosome are liable to be broken off, or to join on to other chromosomes. If there is no centromere, then newly made chromosomes will not be pulled into the new cells during cell division. In addition, there is an origin of replication so that the DNA is replicated. These elements are placed in a single DNA fragment, which can be used as a vector to clone foreign DNA into yeast. The advantage of a YAC is that there is no effective limit to how big the piece of DNA can be. Thus, while conventional bacterial cloning using bacteriophage or plasmids is usually limited to cloning foreign DNA fragments of a few tens of thousands of bases long, YACs can and have cloned fragments millions of bases long. This makes mapping whole genomes of DNA much easier since the whole genome map has to be assembled from far fewer YAC maps, and also makes cloning very large genes such as the gene for muscular dystrophy (which is at least two million bases long) more strength for ward. YACSA are technically difficult to use, and a common problem is that DNA from two different regions of the original genome can end up in one YAC, giving the erroneous appearance that they were originally linked together, a result called a chimeric clone. BACS (bacterial artificial chromosomes) and MACS (mammalian artificial chromosomes) use the same concept with the genetic elements from their respective species. Other vectors capable of cloning large pieces of DNA are the bacterial vectors based on cosmids and P1.

Ycp. Yeast centrometric plasmid. A *S. cerevisiae* cloning vector which contains a cloned centromere sequence. The centromere confers stability on the plasmid but also reduces its number to one or two per cell. At meiosis, a YCp segregates as a mini-chromosome.

Yeast autolyzate. An autolyzate produced using yeast as the starting material. The autolyzates are the result of the action of intercellular enzymes, primarily proteases, on polymeric substances in the cells.

Yeast classification. Yeasts which show a sexual stage (forming ascospores or basidiospores) are classified in the appropriate group. Asporogenous yeasts are classified with the fungi imperfect.

Yeast extract. A complex, undefined, B vitamin-containing addendum to some plant tissue culture media.

Yeast vectors. Yeast cells used as carriers of plasmids, in which novel genes may be replicated, or as transformed cells used to produce the products of integrated novel genes.

Yeast. A type of unicellular fungus that characteristically does not form typical mycelia. Yeasts. Microscopic fungi which exist as simple independent cells; e.g., *Saccharomyces cerevisiae*.

Yellow fat in rabbits. An interesting character conditioned by a recessive gene *y/y* causing the lack of an enzyme to break down the yellow pigment derived from xanthophylls.

Yellow fever (Yellow jack). An acute, systemic disease which, in nature, affects man and other primates the causal agent is togavirus.

YEp, yeast episomal plasmid. A *S. cerevisiae* cloning vector which relies, for its replication, on the 2U plasmid origin. YEps also contain a selectable yeast gene and, often, an *E. coli* origin, pJDB219 is a typical YEp.

Yield coefficient (Y). A number that relates the amount of product generated for a given input.

Yield test: Productivity assessment.

Y-linked inheritance. Inheritance conditioned by genes in a *Y* chromosome. Properties conditioned in this way are exclusively transmitted from males to males.

YLp, Yeast integrative plasmid. A *S. cerevisiae* cloning vector which consists of a selectable vector which consists of a selectable yeast gene inserted into an *E.coli* plasmid, e.g., pBR322. A YLp does not contain a yeast replicator and thus can only be maintained if integrated into a chromosome by, usually, homologous recombination.

Yoghurt. A fermented milk product that is produced from a standard mix of whole, partially defatted milk, condensed skimmed milk, cream and nonfat dried milk. Alternatively milk may be partly concentrated by removal of 15-20 percent water in a vacuum pan. The culture organisms are a mixture of *Lactobacillus bulgaricus* and *Streptococcus thermophilus.*

Yolk nucleus. Vitelline body; a cytoplasmic body in the young oocyte, which serves as a centre for the yolk-formation in many forms. In many cases it is traceable to the idiozome and its associated structures.

Yolk sac. The sac, connected to the gut of the embryo and developed from it, that contains the yolk in reptiles, sharks, and birds. When birds hatch from the egg, the yolk sac is drawn into the abdomen of the newly-hatched chick. The yolk gives the chick food for the first few days, until it becomes able to feed itself.

Yolk. Inert, nutrient constituent of the ovum; consisting mainly of protein and fat. Ova of viviparous animals tend to have very little yolk.

YRP, Yeast replicative plasmid. A *S. cerevisiae* cloning vector which uses a chromosomal replicator (ARS sequence) to replicate in yeast. YRps will also contain a selectable yeast gene and an *E. coli* origin.

Y-suppressed lethal gene. Sex-linked recessive lethal gene that allows survival of *XY* males, but causes the death of *XO* individuals in *Drosophila melanogaster.*

Yuk factor. A flippant term for the very real observation that the public, and indeed many scientists, judge the ethical acceptability of experimental procedures and biological manipulations in accordance with a scale of personal distate. Thus, the creation of the first cloned carrot in the 1960s was greeted with amusement in the press, while the creation of the first cloned frog in the early 1970s was treated with interest and some caution, and the creation of Dolly the cloned sheep in 1997 resulted in widespread alarm. Similarly, tests that rely on newts have less negative impact on public relations than those on rats, and rats are considered more acceptable than rabbits or dogs. In general this reflects a concern for animals that look or behave more like human beings. The ultimate public condemnation is therefore reserved for the potential scientific interference with human fetuses or children. This is a very real scale of values, and one that many scientists do not take seriously enough (hence their calling it a `yuk factor' rather than a `value scale'). In public debate the yuk factor is sometimes a deciding one: much of the opposition to Monsanto's promotion of BST as a biopharmaceutical to boost the milk production of dairy cattle was based not on arguments about farm economics but on the feeling that it must be horrible for the cow to be turned into a milk-producing machine.

Z

Z Chromosome. Sometimes used for the *Y* chromosome where the female is the Heterogametic Sex.

Z DNA. Form of DNA in which the helix is screwed in a left-handed rather than a right-handed direction. Z DNA occurs in short stretches of DNA in nature. Compare B DNA.

Z value. The degrees Fahrenheit required to reduced the thermal death time by 90% or one log value.

Z value. The degrees Fahrenheit required to reduced the thermal death time by 90% or one log value.

Z. The natural logarithm of the ratio of the two estimated Standard Deviations.

Z-DNA. A form of DNA in which the molecule spirals to the left, rather than to the right.

Zeatin or 2-methyl-4-[1H-purin-6-ylamino]-2-buten 1-ol [Z, $C_{10}H_{13}N_5O$ mw 215.21]: A natural cytokinin first isolated from corn. It is sometimes included in plant tissue culture media. It dissolves in acid [HCl ca. 1 M].

Zeatin riboside [mw 351.4]. A naturally occurring cytokinin sometimes used in plant tissue culture media.

Zeatin riboside. A naturally occurring cytokinin sometimes used in plant tissue culture media.

Zephiran. The brand name for a disinfecting agent for plant material containing benzalkonium chloride.

Zimmerman process. A sewage treatment process achieving oxidation of organic material in a closed vessel under high pressure and temperature.

Zoa. Pertaining to animal.

Zoidiophily. Pollination by means of animals.

Zonation. The distribution of organisms in zones; specifically, a stratification of certain kinds of algae at certain depths and locations in the ocean.

Zone electrophoresis. Type of electrophoresis in which the mobile ions or proteins move in a buffer through a support medium such as paper, cellulose acetate, starch gel, silica gel, polyurethane foam or acrylamide polymers. The sample is applied to the support in a narrow band approximately midway between the buffer reservoirs housing the electrodes. During electrophoresis, each component migrates towards the cathode or anode, depending on its charge, at a characteristic rate. Using this method charged species are separated into discrete zones. Subsequently, the support can be divided into sections and individual components may be separately eluted.

Zoochlorellae. Green endosymbiotic algae (e.g. *Chlorella*) present in certain protozoans e.g. species of *Paramecium, Stentor.*

Zooflagellate. An animallike form of flagellate. Compare Phytoflagellate.

Zoogamete. A motile gamete.

Zoogamy. Sexual procreation in animals.

Zoogloeal masses. Masses composed of microorganisms which are embedded in a common matrix of slime.

Zoogonous. Viviparous; giving birth direct to live young, as opposed to the production of eggs.

Zoology [Gr. *zoon*: animal + *logos*: discourse]. The scientific study of animals.

Zoonosis. An animal disease that is transmitted to humans by natural means.

Zoonosis. Zoonosis is infection by an organism that usually infects another species, for example, infection of humans by hanta viruses (which usually only infect rodents). It is an increasing concern for biotechnology for three reasons. New diseases. many of the `new' diseases which are claimed to be a major heath threat to Western cultures, such as Ebola, Marburg, and Lassa viruses, are probably zoonoses. They are `natural' diseases of tropical primates, but can also infect humans. With increasing travel between tropical and temperate zones, it is likely that more such cross-species infections will turn up. AIDS. The reason that this may be important is that AIDS, and possibly Lyme disease, may originally have been diseases of other species, which jumped the species barrier to infect humans as zoonoses, but which subsequently became *aturalized* human illnesses. AIDS in particular is similar to a range of illnesses that are endemic to old world monkeys. Understanding how that jump happened could both give insight into AIDS biology and into understanding whether diseases such as Ebola, which are now very rare, could become common amongst humans.

Xenografts. One way that AIDS was transmitted was by the most common transplant procedure, blood transfusion. There is concern that grafting animals organs (**xenograftng**) could transmit animal infections to humans through a route that would otherwise be impossible for the bacterium or virus to traverse. Viruses are the major concern here since, once introduced into a host, they might mutate very rapidly to give rise to a human version of a pig or goat disease. The possibility that a version of Creuztfeldt-Jacob disease has been passed on to humans from cows enhances people's concern about this type of zoonosis.

Zoophily. Pollination by animals.

Zooplankton [Gr. *zoon***:** animal + *planktos:* wandering]. The animal portion of the plankton. It is a collective term for nonphotosynthetic microorganisms found in flora and fauna of a body of water.

Zoospore [Gr. *zoon*: animal + *spora*: seed]. In algae and fungi, any swimming spore. May be diploid or haploid. It is a motile reproductive spore common to aquatic fungi.

Zovrax. (acyclovir) Nucleoside analogue (hydroxyethoxymethylguanine) with antiviral properties; active against both type 1 and 2 herpes virus. Inactive until phosphorylated by specific viral enzyme, thymidine kinase, and then blocks replication.

ZW. The symbol refers to sex chromosomes in which male is homogametic, *ZZ* and female heterogametic *ZW*, as opposed to *XY* type in which male and female are hetero-and homogametic respectively. *ZW* type in Drosophila, man, many dioecious plants etc.

Zwitterion. A dipolar ion, one with spatially separated positive and negative charges.

Zwitterionic buffer. From the German word meaning hybrid. They contain both positive and negative ionizable groups: secondary and tertiary amino groups provide the positive charges whilst sulphonic and carboxylic groups carry the negative charges.

Zygomycetes. The smallest class of fungi, including organisms such as the common bread mold.

Zygonema. The chromosome thread during the zygotene phase, when the chromosomes have come together in pairs.

Zygosis. Conjugation.

Zygosome. A newly fused chromosome pair produced by syndesis (zygotene); mixochromosome.

Zygospore. Sexual spore which results from the fusion of like gametes; characteristic of zygomycetes.

Zygotaxis. Attraction as between gametes of opposite sex.

Zygote [Gr. *zygotos;* yoked]. The cell created by the union of two gametes, in which the gamete nuclei are also fused. The earliest stage of the diploid generation.

Zygotene. The term proposed by Gregoire (1907) to describe the nuclei at a particular stage of prophase 1 of meiosis when the homologous chromosomes are associating side by side. (Gk. *zygon*=yoke; *taenia*= ribbon.).

Zygotic checkerboard. A Punnett square in which zygotes rather than gametes are used. Zygotes segregating for one gene (A/a) are placed on one axis, those of another (B/b) on the other axis, so that the proportions of all classes of a dihybrid segregation may be calculated directly.

Zygotic Incompatibility. A form of homomorphic incompatibility in which the incompatibility reaction depends directly on the genetic constitution of the zygote producing the male gamete, rather than that of the male gamete, rather than that of the male gamete itself, together with the constitution of the female zygote. cf. Gametic Incompatibility.

Zygotic induction. The lysis of a nonlysogenic (ly) recipient (F) bacterium following the transfer of a prophage from a lysogenic (ly^+) *Hfr* donor during conjugal mating. Lysis is inevitable under these conditions and depends on the reversion of the prophage to the vegetative state.

Zygotic lethal. A lethal gene whose effect is in the embryo, larva, or adult, in contrast to the gametic lethal effecting a gamete.

Zygotic meiosis. Meiosis which takes place in the first two divisions after fertilization, at the beginning of the life cycle of an organisms; initial meiosis; zygotic reduction.

Zygotic mutation. A mutation occurring in the zygote shortly after fertilization.

Zygotic number. The somatic ($2n$) chromosome number as opposed to the haploid or gametic (n) number.

Zygotic reduction. Meiosis when it takes place immediately after fertilization; zygote meiosis.

Zymogen granule. A secretory granule found in large numbers in enzyme secreting cells. The granules are vesicles containing an inactive precursor of the enzyme, the zymogen, e.g. trypsinogen in exocrine cells of the pancreas. It is activated after secretion via exocytosis at the plasma membrane. The vesicles are usually derived from the Golgi apparatus, where the enzyme is processed and concentrated after synthesis on the rough endoplasmic reticulum.

Zymogen. An inactive precursor of a digestive enzyme Secreted into the lumen of the gut, where a protease cleaves it, forming the active enzyme. This makes it unnecessary for some digestive enzymes to be formed inside cells, which they might then digest.

Zymogenic cells. Cells of the basal part of the gastric glands of the stomach. They contain extensive rough endoplasmic reticulum and zymogen granules and secrete pepsinogen, the inactive precursor of pepsin and rennin.

Zymogenous. Soil microorganisms, primarily rods and spore formers that readily decompose (ferment) organic matter.

Zymogenous. Soil microorganisms, primarily rods and spore formers that readily decompose (ferment) organic matter.

Zymogram. Electrophoretic gels used to reveal and compare protein polymorphism, as of isoenzymes such as lactate dehydrogenase subunits.

Zymosan. An extract of yeast cell walls.